*Probability and Mathematical Statistics (Cont*
ROHATGI • An Introduction to Proba
ematical Statistics
SCHEFFE • The Analysis of Variance
SEBER • Linear Regression Analysis
SERFLING • Approximation Theorems of Mathematical Statistics
TJUR • Probability Based on Radon Measures
WILLIAMS • Diffusions, Markov Processes, and Martingales, Volume I: Foundations
ZACKS • The Theory of Statistical Inference

*Applied Probability and Statistics*
ANDERSON, AUQUIER, HAUCK, OAKES, VANDAELE, and WEISBERG • Statistical Methods for Comparative Studies
BAILEY • The Elements of Stochastic Processes with Applications to the Natural Sciences
BAILEY • Mathematics, Statistics and Systems for Health
BARNETT and LEWIS • Outliers in Statistical Data
BARTHOLEMEW • Stochastic Models for Social Processes, *Second Edition*
BARTHOLOMEW and FORBES • Statistical Techniques for Manpower Planning
BECK and ARNOLD • Parameter Estimation in Engineering and Science
BELSLEY, KUH, and WELSCH • Regression Diagnostics: Identifying Influential Data and Sources of Collinearity
BENNETT and FRANKLIN • Statistical Analysis in Chemistry and the Chemical Industry
BHAT • Elements of Applied Stochastic Processes
BLOOMFIELD • Fourier Analysis of Time Series: An Introduction
BOX • R. A. Fisher, The Life of a Scientist
BOX and DRAPER • Evolutionary Operation: A Statistical Method for Process Improvement
BOX, HUNTER, and HUNTER • Statistics for Experimenters: An Introduction to Design, Data Analysis, and Model Building
BROWN and HOLLANDER • Statistics: A Biomedical Introduction
BROWNLEE • Statistical Theory and Methodology in Science and Engineering, *Second Edition*
BURY • Statistical Models in Applied Science
CHAMBERS • Computational Methods for Data Analysis
CHATTERJEE and PRICE • Regression Analysis by Example
CHERNOFF and MOSES • Elementary Decision Theory
CHOW • Analysis and Control of Dynamic Economic Systems
CLELLAND, BROWN, and deCANI • Basic Statistics with Business Applications, *Second Edition*
COCHRAN • Sampling Techniques, *Third Edition*
COCHRAN and COX • Experimental Designs, *Second Edition*
CONOVER • Practical Nonparametric Statistics, *Second Edition*
COX • Planning of Experiments
DANIEL • Biostatistics: A Foundation for Analysis in the Health Sciences, *Second Edition*
DANIEL • Applications of Statistics to Industrial Experimentation
DANIEL and WOOD • Fitting Equations to Data: Computer Analysis of Multifactor Data, *Second Edition*
DAVID • Order Statistics
DEMING • Sample Design in Business Research

*continued on back*

*Survival Models and Data Analysis*

# Survival Models and Data Analysis

**REGINA C. ELANDT-JOHNSON**

**Department of Biostatistics**
**School of Public Health**
**University of North Carolina at Chapel Hill**

**NORMAN L. JOHNSON**

**Department of Statistics**
**University of North Carolina at Chapel Hill**

**JOHN WILEY AND SONS,**
**New York · Chichester · Brisbane · Toronto**

**Library of Congress Cataloging in Publication Data**

Elandt-Johnson, Regina C. 1918-
Survival models and data analysis.

(Wiley series in probability and mathematical
statistics, applied section)
Includes index.
1. Failure time data analysis. 2. Mortality.
3. Medical statistics. 4. Competing risks.
I. Johnson, Norman Lloyd, joint author. II. Title.

QA276.E39 312'.01'51 79-22836
ISBN 0-471-03174-7

Printed in the United States of America

10 9 8 7 6 5 4 3 2 1

# Preface

This book contains material and techniques developed in several different disciplines: vital statistics, epidemiology, demography, actuarial science, reliability theory, statistical methods, among others. Despite this diversity of origin, these techniques are all relevant to aspects of the analysis of survival data.

Survival data can take so many different forms, spanning from results of small-scale laboratory tests to massive records from long-term clinical trials. Therefore it is impossible to lay down universal rules of procedure. Attempts to do so, even if apparently successful, are likely to lead to an uncritical, authoritarian approach, following whatever is currently regarded as the "correct approach." We have tried to set out general principles to be used in each particular case. A number of the exercises require considerable independent thought and cannot be said to have a unique "correct" answer. They call rather for sound appraisal of a situation.

We have also endeavored to avoid the use of hidden assumptions. Much of our analysis is, indeed, based on assumptions (of independence, stability, etc.), but we have always sought to make it clear what assumptions are being made, and we encourage the reader to consider what might be the effects of departures from these assumptions.

The remarkable increase of activity in the statistical analysis of survival data over the last two decades, largely stimulated by problems arising in the analysis of clinical trials, has resulted in a considerable volume of writing on the topic. A major purpose of this book is to act as a guide for using this literature, assist in the choice of appropriate methods, and warn against uncritical use.

The content of the book might be subclassified according to several different criteria—statistical approach, relevant scientific disciplines, types of applications, and so on. We have, in fact, divided the book into four broad parts.

Part 1 introduces the type of data to be analyzed and basic concepts used in their analysis.

Part 2 deals with problems related to univariate survival functions. These include construction of life tables from population (cross-sectional) data and from experimental-type follow-up data. Considerable space is devoted to fitting parametric distributions and comparisons of two or more mortality experiences.

Part 3 is concerned with multiple-failure data. Time as well as cause of death are identified. Parametric and nonparametric theories of competing causes and estimation of different kinds of failure distributions are presented in some detail.

Part 4 presents some more advanced topics, including speculative mathematical models of biological processes of disease progression and aging. These are not intended to be definitive. Rather, they give the reader some ideas of ways in which models may be constructed.

Some readers may find the mathematical level uneven. This is because mathematical techniques are used as they are needed, and never for their own sake.

We take this opportunity to acknowledge the help we have received while working on this book. The typing was done by Joyce Hill (in major part), June Maxwell, and Mary Riddick. We are especially grateful to Anna Colosi, who did all the calculations and obtained graphical presentations using an electronic computer.

REGINA C. ELANDT-JOHNSON
NORMAN L. JOHNSON

*Chapel Hill, North Carolina*
*October 1979*

# Contents

## PART 1. SURVIVAL MEASUREMENTS AND CONCEPTS


1. SURVIVAL DATA 3

- 1.1 Scope of the Book 3
- 1.2 Sources of Data 4
- 1.3 Types of Variables 5
- 1.4 Exposure to Risk 6
- 1.5 Use of Probability Theory 6
- 1.6 The Collection of Survival Data 7

2. MEASURES OF MORTALITY AND MORBIDITY. RATIOS, PROPORTIONS, AND MEANS 9

- 2.1 Introduction 9
- 2.2 Ratios and Proportions 10
  - 2.2.1 Ratios 10
  - 2.2.2 Proportion 11
- 2.3 Rates of Continuous Processes 12
  - 2.3.1 Absolute Rate 12
  - 2.3.2 Relative Rate 13
  - 2.3.3 Average (Central) Rate 14
- 2.4 Rates for Repetitive Events 16
- 2.5 Crude Birth Rate 17
- 2.6 Mortality Measures Used in Vital Statistics 18
  - 2.6.1 The Concept of Population Exposed to Risk 18
  - 2.6.2 Crude Death Rate 20
  - 2.6.3 Age Specific Death Rates 21
  - 2.6.4 Cause Specific Mortality Used in Vital Statistics 21
- 2.7 Relationships Between Crude and Age Specific Rates 22
- 2.8 Standardized Mortality Ratio (SMR): Indirect Standardization 22

2.9 Direct Standardization 25
2.10 Evaluation of Person-Years of Exposed to Risk in Long-Term Studies 25
2.10.1 'Exact' Dates for Each Individual Available 26
2.10.2 Only Years of Birth, Entry, and Departure Available 29
2.11 Prevalence and Incidence of a Disease 31
2.11.1 Prevalence 31
2.11.2 Incidence 32
2.12 Association Between Disease and Risk Factor. Relative Risk and Odds Ratio 35
2.12.1 Relative Risk 36
2.12.2 Odds Ratio 37

3. SURVIVAL DISTRIBUTIONS 50

3.1 Introduction 50
3.2 Survival Distribution Functions 50
3.3 Hazard Function (Force of Mortality) 51
3.4 Conditional Probabilities of Death (Failure) and Central Rate 52
3.5 Truncated Distributions 53
3.6 Expectation and Variance of Future Lifetime 55
3.7 Median of Future Lifetime 56
3.8 Transformations of Random Variables 57
3.9 Location-Scale Families of Distributions 58
3.10 Some Survival Distributions 59
3.11 Some Models of Failure 64
3.11.1 Series System 64
3.11.2 Parallel System 65
3.12 Probability Integral Transformation 66
3.13 Compound Distributions 67
3.14 Miscellanea 67
3.14.1 Interpolation 67
3.14.2 Method of Statistical Differentials 69
3.15 Maximum Likelihood Estimation and Likelihood Ratio Tests 72
3.15.1 Construction of Likelihood Functions 73
3.15.2 Maximum Likelihood Estimation 73
3.15.3 Expected Values, Variances and Covariances of the MLE's 74
3.15.4 Assessing Goodness of Fit 74

# PART 2. MORTALITY EXPERIENCES AND LIFE TABLES

4. LIFE TABLES: FUNDAMENTALS AND CONSTRUCTION 83

4.1 Introduction 83
4.2 Life Table: Basic Definition and Notation 83
4.3 Force of Mortality. Mathematical Relationships Among Basic Life Table Functions 93
4.4 Central Death Rate 95
4.5 Interpolation for Life Table Functions 96
4.6 Some Approximate Relationships Between ${}_nq_x$ and ${}_nm_x$ 99
4.6.1 Expected Fraction of the Last $n$ Years of Life 100
4.6.2 Special Cases 100
4.6.3 Exponential Approximation 101
4.7 Some Approximations to $\mu_x$ 101
4.8 Concepts of Stationary and Stable Populations 102
4.8.1 Stationary Population 103
4.8.2 Stable Population 103
4.9 Construction of an Abridged Life Table from Mortality Experience of a Current Population 104
4.9.1 Evaluation of ${}_nM_x$ 105
4.9.2 Estimation of ${}_nf_x$ 106
4.9.3 Estimation of ${}_nq_x$ 107
4.9.4 Evaluation of the Life Table Functions 108
4.10 Some Other Approximations Used in Construction of Abridged Life Tables 110
4.11 Construction of a Complete Life Table From an Abridged Life Table 111
4.12 Selection 114
4.13 Select Life Tables 115
4.14 Some Examples 117
4.15 Construction of Select Tables 119

5. COMPLETE MORTALITY DATA. ESTIMATION OF SURVIVAL FUNCTION 128

5.1 Introduction. Cohort Mortality Data 128
5.2 Empirical Survival Function 129
5.3 Estimation of Survival Function From Grouped Mortality Data 133

5.3.1 Grouping into Fixed Intervals 133
5.3.2 Grouping Based on Fixed Numbers of Deaths 135
5.4 Joint Distribution of the Numbers of Deaths 137
5.5 Distribution of $\hat{P}_i$ 138
5.6 Covariance of $\hat{P}_i$ and $\hat{P}_j$ $(i<j)$ 138
5.7 Conditional Distribution of $\hat{q}_i$ 138
5.8 Greenwood's Formula for the (Conditional) Variance of $\hat{P}_i$ 140
5.9 Estimation of Curve of Deaths 141
5.10 Estimation of Central Death Rate and Force of Mortality in $[t_i, t_{i+1})$ 142
5.11 Summary of Results 144

6. INCOMPLETE MORTALITY DATA: FOLLOW-UP STUDIES 150

6.1 Basic Concepts and Terminology 150
6.1.1 Incomplete Data 150
6.1.2 Truncation and Censoring 150
6.1.3 Follow-up Studies 151
6.1.4 Other Kinds of Follow-up 153
6.1.5 Topics of the Chapter 153
6.2 Actuarial Estimator of $q_i$ From Grouped Data 154
6.2.1 Amount of Person-Time Units. Central Death Rates 154
6.2.2 Effective Number of Initial Exposed to Risk. Estimation of $q_i$ 156
6.2.3 Estimation of Survival Function 157
6.3 Some Maximum Likelihood Estimators of $q_i$ 162
6.3.1 Failure Time Alone Regarded as a Random Variable 163
6.3.2 Failure Time and Censoring Time Regarded as Random Variables 166
6.4 Some Other Estimators of $q_i$ 169
6.4.1 Moment Estimator of $q_i$ 169
6.4.2 Estimator of $q_i$ Based on Reduced Sample 170
6.5 Comparison of Various Estimators of $q_i$ 170
6.6 Estimation of Curve of Deaths 172
6.7 Product-Limit Method of Estimating the Survival Function From Individual Times at Death 172
6.8 Estimation of Survival Function Using the Cumulative Hazard Function 174

7. FITTING PARAMETRIC SURVIVAL DISTRIBUTIONS 181

7.1 Introduction 181
7.2 Some Methods of Fitting Parametric Distribution Functions 182
7.3 Exploitation of Special Forms of Survival Function 182
7.4 Fitting Different Distribution Functions over Successive Periods of Time 185
7.5 Fitting a "Piece-Wise" Parametric Model to a Life Table: An Example 186
7.6 Mixture Distributions 191
7.7 Cumulative Hazard Function Plots—Nelson's Method for Ungrouped Data 196
7.7.1 Complete Data 196
7.7.2 Incomplete Data 200
7.8 Construction of the Likelihood Function for Survival Data: Some Examples 200
7.9 Minimum Chi-Square and Minimum Modified Chi-Square 209
7.10 Least Squares Fitting 209
7.11 Fitting a Gompertz Distribution to Grouped Data: An Example 211
7.12 Some Tests of Goodness of Fit 214
7.12.1 Graphical "Test" 214
7.12.2 Kolmogorov-Smirnov Statistics. Limiting Distribution 215
7.12.3 Anderson-Darling $A^2$-Statistic 218
7.12.4 Chi-Square Test for Grouped Data 219

8. COMPARISON OF MORTALITY EXPERIENCES 225

8.1 Introduction 225
8.2 Comparison of Two Life Tables 225
8.2.1 Graphical Displays 226
8.2.2 Conditional Probabilities $q_x$ 227
8.2.3 Conditional Expectations and Median of Future Lifetime 227
8.3 Comparison of Mortality Experience with a Population Life Table 228
8.3.1 Test Based on Median of Future Lifetime 228
8.3.2 Test Based on Expected Future Lifetime 229
8.4 Some Distribution-Free Methods for Ungrouped Data 231
8.4.1 Two Sample Kolmogorov-Smirnov Test 231

8.4.2 Two Sample Wilcoxon Test for Complete Data 234
8.4.3 Modified Wilcoxon Tests for Incomplete Data 236
8.5 Special Problems Arising in Clinical Trials and Progressive Life Testing 240
8.5.1 Early Decision 240
8.5.2 Multistage Testing 241
8.5.3 Testing for Trends in Mortality Patterns 242
8.5.4 Staggered Entries and Withdrawals 242
8.6 Censored Kolmogorov-Smirnov (or Tsao-Conover) Test 243
8.7 Truncated Data. Pearson's Conditional $X^2$ Test 247
8.7.1 Two Sample Problem 247
8.7.2 Extension to $k$ Treatments 249
8.8 Testing for Consistent Differences in Mortality. Mantel-Haenszel and Logrank Tests 251
8.8.1 Mantel-Haenszel Test 251
8.8.2 Logrank Test 258
8.8.3 Extension to $r$ Experiments (Classes) 260
8.9 Parametric Methods 261
8.10 Sequential Methods 261

PART 3. MULTIPLE TYPES OF FAILURE

9. THEORY OF COMPETING CAUSES: PROBABILISTIC APPROACH 269
9.1 Causes of Death: Basic Assumptions 269
9.2 Some Basic Problems 270
9.3 "Times Due to Die" 271
9.4 The Overall and 'Crude' Survival Functions 272
9.4.1 The Overall Survival Function 272
9.4.2 The Crude and Net Hazard Rates 272
9.4.3 The Crude Probability Distribution for Cause $C_\alpha$ 274
9.5 Case when $X_1,\ldots,X_k$ are Independent 276
9.6 Equivalence and Nonidentifiability Theorems in Competing Risks 277
9.6.1 Equivalent Models of Survival Distribution Functions 278

9.6.2 Nonidentifiability of the Member of a Parametric Family of Distributions 280
9.7 Proportional Hazard Rates 280
9.8 Examples 282
9.9 Heterogeneous Populations: Mixture of Survival Functions 288

10. MULTIPLE DECREMENT LIFE TABLES 294

10.1 Multiple Decrement Life Tables: Notation 294
10.2 Definitions of the MDLT Functions 295
10.3 Relationships Among Functions of Multiple Decrement Life Tables 295
10.4 Crude Forces of Mortality 297
10.5 Construction of Multiple Decrement Life Tables from Population (Cross-Sectional) Mortality Data 298
10.5.1 Mortality Data: Evaluation of ${}_na q_x$ and ${}_na q_{\alpha x}$ 298
10.5.2 Construction of MDLT 300
10.6 Some Major Causes of Death: An Example of Constructing the MDLT 301

11. SINGLE DECREMENT LIFE TABLES ASSOCIATED WITH MULTIPLE DECREMENT LIFE TABLES: THEIR INTERPRETATION AND MEANING 309

11.1 Elimination, Prevention, and Control of a Disease 309
11.2 Mortality Pattern from Cause $C_\alpha$ Alone: 'Private' Probabilities of Death 310
11.2.1 How Might the SDF from Cause $C_\alpha$ Alone be Interpreted 310
11.2.2 Estimable, Although not Observable, Waiting Time Distributions 311
11.3 Estimation of Waiting Time Distribution for Cause $C_\alpha$: Single Decrement Life Table 312

12. ESTIMATION AND TESTING HYPOTHESES IN COMPETING RISK ANALYSIS 323

12.1 Introduction. Experimental Data 323

12.2 Grouped Data. Nonparametric Estimation 324
12.2.1 Complete Data 324
12.2.2 Follow-up Data 326
12.2.3 Truncated Samples 326
12.3 Grouped Data. Fitting Parametric Models 329
12.3.1 Choice of a Joint Survival Function 329
12.3.2 Fitting Crude Parametric Distribution to Mortality Data from Each Cause Separately 331
12.4 Cohort Mortality Data with Recorded Times at Death or Censoring. Nonparametric Estimation 332
12.4.1 Complete Data 332
12.4.2 Incomplete Data 334
12.5 Cohort Mortality Data with Recorded Times at Death or Censoring. Parametric Estimation 335

PART 4. SOME MORE ADVANCED TOPICS

13. CONCOMITANT VARIABLES IN LIFETIME DISTRIBUTIONS MODELS 345

13.1 Concomitant Variables 345
13.2 The Role of Concomitant Variables in Planning Clinical Trials 346
13.3 General Parametric Model of Hazard Function with Observed Concomitant Variables 346
13.3.1 Types of Concomitant Variables 347
13.3.2 General Model 348
13.3.3 Some Other Expressions for the General Model 351
13.4 Additive Models of Hazard Rate Functions 353
13.5 Multiplicative Models 356
13.5.1 Exponential-Type Hazard Functions 356
13.5.2 Gompertz and Weibull Models with Covariates 358
13.6 Estimation in Multiplicative Models 359
13.6.1 Construction of the Likelihood Function 359
13.6.2 Multiple Failures 362
13.7 Assessment of the Adequacy of a Model: Tests of Goodness of Fit 365
13.7.1 Cumulative Hazard Plottings 366
13.7.2 Method of Half-Replicates 367

13.8 Selection of Concomitant Variables 367
13.8.1 Likelihood Ratio Tests of Composite Hypotheses 368
12.8.2 Step-Down Procedure 369
13.8.3 Step-Up Procedure 370
13.9 Treatment-Covariate Interaction 372
13.10 Logistic Linear Model 374
13.11 Time Dependent Concomitant Variables 377
13.12 Concomitant Variables Regarded as Random Variables 379
13.13 Posterior Distribution of Concomitant Variables 380
13.14 Concomitant Variables in Competing Risk Models 381

14. AGE OF ONSET DISTRIBUTIONS 392

14.1 Introduction 392
14.2 Models of Onset Distributions 392
14.2.1 Incidence Onset Distribution 393
14.2.2 Waiting Time Onset Distribution 394
14.2.3 Life Tables and Onset Distributions 394
14.3 Estimation of Incidence Onset Distribution from Cross-Sectional Incidence Data 395
14.4 Estimation of Incidence Onset Distribution from Prevalence Data 398
14.4.1 Estimation of Age Specific Incidence from Prevalence Data: No Differential Mortality 398
14.4.2 Affected Individuals Are Subject to Differential Mortality 399
14.5 Estimation of Waiting Time Onset Distribution from Population Data 403
14.6 Estimation of Waiting Time Onset Distribution from Retrospective Data 403
14.6.1 No Differential Mortality Between Affected and Unaffected 404
14.6.2 Effects of Differential Mortality 410

15. MODELS OF AGING AND CHRONIC DISEASES 414

15.1 Introduction 414
15.2 Aging and Chronic Diseases 415
15.2.1 Biological Aging 415

15.2.2 Hazard Rate: A Measure of Aging 415
15.2.3 Models of Chronic Diseases 416
15.3 Some Models of Carcinogenesis 416
15.3.1 A One-Hit Model of Carcinogenesis: Gompertz Distribution 416
15.3.2 Multi-Hit Models of Carcinogenesis: Parallel Systems and Weibull Distribution 417
15.4 Some "Mosaic" Models of a Chronic Disease 418
15.5 "Fatal Shock" Models of Failure 420
15.5.1 A Two Component Series System 420
15.5.2 Generalization of the "Fatal Shock" Model to $n$ Components 422
15.6 Irreversible Markov Processes in Illness-Death Modeling 423
15.6.1 Basic Concepts 423
15.6.2 A Two Component Parallel System: Two Distinct States Before Failure 424
15.6.3 Extension to $r$-Component Parallel System with $r$ States 426
15.6.4 A Model of Disease Progression 426
15.7 Reversible Models: The Fix-Neyman Model 431

Author index 433
Subject index 447

*Survival Models and Data Analysis*

# Part 1

# SURVIVAL MEASUREMENTS AND CONCEPTS

CHAPTER 1

# Survival Data

## 1.1 SCOPE OF THE BOOK

The title of this book indicates that we discuss the treatment of "mortality data." The direct meaning of this term is data that arise from recording times of death of individuals in a specified group. There will usually be additional data from observations of characters (other than survival or death) on the individuals in the group. These may be made at or near the moment of death (e.g., cause of death, length of illness, physical characteristics near the moment of death) or at earlier times (e.g., sex, age, family history, physical characteristics at earlier epochs). Certain of these variables—most commonly age (time elapsed since birth) and/or time elapsed since other important events (e.g., commencement of illness, date of operation)—are regarded as being of primary interest. It is often desired to assess the relationship between mortality and these *primary variables*, allowing, as far as possible for some of the other characteristics. The latter, in this context, are called *concomitant variables*. (Note that, for a given set of data, the distinction between primary and concomitant variables depends on the relationships to be studied.)

Individuals in the group may be humans, animals, fishes, insects, and so on. The group itself may be defined in various ways—by geographical location (e.g., population of a town or state, patients in a hospital or in a set of hospitals), by previous history (e.g., medical treatment, type of sickness, employment).

Occasionally we consider situations in which the replacement of "mortality" by the more general term "failure" is appropriate. In such contexts, the individuals are not necessarily (although they may be) living organisms. They may, for example, be mass-produced articles, such as electric lamps, with *failure* meaning inability to function in a specified role.

We are not primarily concerned with reversible changes of status, such as sickness causing temporary inability to work or repairable failure of electrical or mechanical systems. However there are occasional references to these matters, and Chapter 14 is devoted to discussing the distribution of age of onset of a disease.

Also, we are not concerned with statistics of birth, except as defining entry into a specific group of individuals and contributing to the assessment of mortality at juvenile ages. In particular, we do not study the measurement of fertility or the general province of demography.

Primarily, we are concerned with the study of failure data, and the relation of failure to a few important variables, such as age or time elapsed since some event (other than birth or manufacture). Other variables (concomitant variables) are introduced because of a possible relationship with failure but are not studied for their own sake.

## 1.2 SOURCES OF DATA

From the foregoing description, it can be seen that the methods discussed are applicable to a wide variety of situations. The sources of data are correspondingly varied. We first describe sources of mortality data, later turning to the topic of failure data in general.

A major subdivision of mortality data is between data relating to populations under more or less *uncontrolled* conditions (such as statistics of human deaths in a state or nation) and those observed under *controlled* conditions of a more or less experimental nature (as in a clinical trial).

Usually, the amount of data collected in the former situation is considerably greater than in the latter, though this need not be so. On the other hand, we almost always have more detailed information on each individual exposed to risk in the latter situation. In fact, in the uncontrolled situation we rarely have an exact enumeration of all the individuals who might be observed to fail (those exposed to risk). (A more precise discussion and definition of exposed to risk can be found in Chapter 2.)

When the date of death is recorded in a specific area over a specific period of time, estimates of the number exposed to risk are usually based on census data. For convenience, we use the term *census-type data* generally to describe data in which the numbers exposed to risk are estimated indirectly. When records are available from which the numbers exposed to risk can be ascertained directly we, again for convenience, use the term *experimental-type data*. Sometimes these terms may not appear to be very relevant to the data actually under consideration. Their function is to remind ourselves what type of data we are considering.

As we have already mentioned, experimental-type data are usually considerably smaller in volume than census-type data. An exception arises in the mortality experience of insurance companies. The records of such companies contain information on all persons insured with them, from which it is possible to determine exactly the exposed to risk, among whom the deaths (resulting in claims) are also recorded. For a given year of age, the numbers exposed to risk can quite easily be in the tens of thousands or more, and in practice some approximations to the exposed to risk may be used, corresponding to various groupings of ages (according to last birthday or nearest birthday), dates of entry, withdrawal, and death.

## 1.3 TYPES OF VARIABLES

We have already introduced the concept of concomitant variables in Section 1.1. Here we examine our classification of variables in somewhat greater detail.

The basic variable, representing survival or failure, is essentially a variable taking just two values (a binary variable) that can be chosen arbitrarily and are usually, and conveniently, taken to be 0 and 1. It can be measured by direct counting, as in experimental-type situations, or by indirect estimation, as in some census-type situations. We are most often concerned with studying the proportion of individuals surviving specified periods as a function of a few important variables. By far the most important variable is age, although in clinical trials, duration since an event such as initiation of treatment is sometimes taken as the "variable of interest."

Thus *life tables* (which will be discussed in Chapter 4) usually represent the pattern of mortality (or failure) as a function of age, for particular groups of individuals, but this is not always the case. For human or animal populations, the time since a specific event may be used as the variable of interest. Occasionally, as in select life tables (see Sections 4.12–4.14), both age and time since a specific event are variables of interest.

The remaining measured variables, beyond the basic (survival) variable and the variable(s) of interest are concomitant variables. In so far as they do affect the mortality (failure) pattern, it is desirable to allow for them. This may be done analytically, (1) by introducing some fairly simple model that (it is hoped) will represent adequately the influence of concomitant variables, or (2) by constructing separate life tables for different values of the concomitant variable(s). The latter is the safer method, but can be usefully applied only when the concomitant variable(s) can be defined in terms of very few categories. An important example, in living populations, is sex. It is quite common to have separate life tables for females and males.

Other important concomitant variables are geographical location, social class (often measured as an index based on income, education, etc.), and physical characteristics such as blood pressure, weight, and vital capacity. This last group is especially relevant in most clinical trial data.

In mechanical and electrical systems, the variable age (duration of effective service) is again of major importance. Other variables, mainly representing conditions of use, include temperature, pressure, operator training, chemical content of contact materials, and so on.

In particular, we study (in Chapter 13) the relationship between failure and one or more variables of interest. Age is very often the variable of interest, but in clinical trials time since some specified event, other than birth, is usually the variable of interest. There is a wide variety of other concomitant variables that may need attention.

A concomitant variable of some importance is chronological year (e.g., 1965, 1975). Although not always recognized as such, its importance is acknowledged, for example, in national life tables which always relate to a specific period of time (e.g., U. S. Life Tables for 1959–1961, 1969–1971, etc.).

## 1.4 EXPOSURE TO RISK

In most analyses of survival data we are interested in studying the *proportions* of failure among groups of individuals under specified conditions. Clearly, the longer the period for which an individual is under observation, the more likely it is that failure will be observed, sooner or later. Comparability of numbers of failures requires that they be referred to some unit period of observation. To do this, we would like to know, for each individual, the *period of exposure to risk*, that is, the period of time during which the failure (or death) of an individual will actually be recorded and contribute to the *observed failures*. For census-type data this period is usually not known and has to be estimated. For experimental-type data, information from which the period of exposure to risk can be determined for each individual is usually available, although when the volume of data is large, approximate evaluation may be used.

## 1.5 USE OF PROBABILITY THEORY

Since we are studying proportions, it is natural to represent them in terms of probabilities. We are then able to use the very well-developed techniques and concepts of the *theory of probability* to assist in understanding the data. It is assumed that readers of this book already have some knowledge of elementary statistics and probability theory. Chapter 3 provides a recapitulation of the necessary knowledge, together with some special definitions directed toward applications in survival analysis.

In all applications of statistical methods based on probabilities, the models and assumptions on which they are based are not usually fully satisfied. This is certainly so in applications to survival data. For this reason we wish to de-emphasize application of these methods—especially when they are complicated, for example, in many cases of maximum likelihood estimation. We do, indeed, give accounts of the more useful statistical techniques (see in particular Chapters 7 and 8), although in a somewhat condensed form, but we strongly urge the reader to regard direct comparison—for example, by use of graphs—between observed data and model(s) to be of first importance. We therefore give special prominence to the use of graphical methods.

Probability theory provides essential background and tools for survival analysis, but possibilities of departures from theoretical models are so great and varied that direct confrontation between data and model(s) in terms of adequate descriptive agreement, is essential.

## 1.6 THE COLLECTION OF SURVIVAL DATA

It is a feature of much data collection, and particularly of survival and failure data, that it depends heavily for accuracy on the meticulous compilation and preservation of records. This is especially important, since the relevant failures are, typically, spread out over considerable periods of time, and each failure requires a fresh entry in the records.

Experimental-type data require even more effort in recording and storing, because more or less detailed records have to be kept on individuals coming under observation during the period of study. Concomitant variables for each individual also need to be recorded, and up-dated when necessary. It is essential, of course, that the records state clearly, and as accurately as possible, the periods over which an individual was exposed to risk in the sense that failure would have been recorded if it had occurred at any moment during that period, but not if it occurred at any other time.

All observations should be recorded to the greatest apparent accuracy that is reasonable, practicable, and meaningful. This general principle, which is justifiable on the grounds of making use of as much of the available information as possible, is of great importance in regard to recording times of death or failure. Deaths are sometimes recorded only as occurring in rather broad time intervals (e.g., in a specified week as opposed to a specified day or hour). The broader the time intervals, the greater the chance that one time interval will contain more than one death (this is often referred to as "multiple deaths"). In such cases it is not possible to decide the order in which the deaths occur, and the apparently equal times of death are referred to as *tied* observations. Some statistical techniques (especially the nonparametric procedures described in Chapter 8) make use of the temporal order in which the failures occur, and they cannot be applied directly when the data include tied observations (multiple failures). Some ingenious methods of dealing with ties have been suggested, but they all involve extra trouble and are based on dubious assumptions. We feel that application of the following relatively straightforward principles is usually preferable to reliance on automatic rules for applying specific formulas:

1. For any sets of ties the actual ordering must be one of a finite number of possibilities.
2. If it is feasible, every possible ordering can be analyzed using methods appropriate when there are no ties.
3. It will often suffice to consider only a few, perhaps only two, extreme orderings. If these give concordant results they can be adopted with some confidence. Even if not, it may be possible to see that only a relatively minute proportion of possible orderings give results markedly different from those of the remaining orderings.

If there is no clear-cut preponderance of reasonable consistent results among the possible orderings, the conclusion must be faced that, in fact, the recording interval is too broad, and attempts should be made to narrow it.

Of course, it is possible to make various assumptions about resolving recorded ties. For example, if two persons are recorded as dying in a period of six months, and their ages last birthday at the beginning of the period were 28 and 58, it is reasonable (other things being equal) to give greater weight to the possibility of the second person actually being the first of the two to die.

CHAPTER 2

# Measures of Mortality and Morbidity. Ratios, Proportions, and Rates

## 2.1 INTRODUCTION

In Chapter 1, we discussed certain characteristics that describe an individual's health. Health, in turn, is closely related to chances of survival beyond a certain age. We discussed the collection of such data—commonly called *vital statistics*—over large populations. Phenomena such as deaths and diseases were our main concern. In this chapter, we discuss the construction of some summary measures from such data, and also the interpretation of the measures.

It is important to distinguish two aspects of description of a community of living organisms: *static* and *dynamic*. Measures appropriate to the description of the static state of a population at a point of time (or over a *specific*, short period of time) are usually *ratios* and/or *proportions*. Living processes are, of course, dynamic, proceeding from birth to death of individuals. We need to describe the *rapidity of change* in living communities, and this leads to the second class of measures. For example, we need to measure how fast a population is dying off and replenishing itself by new births and immigration; how rapidly epidemics spread, and how quickly they abate; or the number of new cases of specific diseases such as cancer, heart attack, or tuberculosis per year, or more precisely, per year and per 100 or 1000 individuals in the population. Appropriate measures of such phenomena are *rates*.

In almost every scientific discipline, there is some ambiguity and lack of precision in terminology. In the analysis of vital statistics data, misunderstanding and misuse of the terms "proportion" and "rate" are especially confusing. So much so that it is difficult to follow the text of some papers on epidemiology and survival analysis, in which these terms are used arbitrarily and often inconsistently. It is necessary, therefore, to start by clarifying this terminology before introducing any of the well-known measures. Sections 2.2, 2.3, and 2.4 are devoted to this clarification, and Sections 2.6 through 2.10 present the basic principles of calculating "exposed to risk" and their use in calculating death rates. Two important concepts—prevalence and incidence of a disease—are outlined in Section 2.11.

## 2.2 RATIOS AND PROPORTIONS

### 2.2.1 Ratio

In a very broad sense, a ratio results from dividing one quantity by another. In science, however, it is mostly used in a more specific sense, that is, when the numerator and the denominator are two distinct quantities. Two kinds of ratios are in common use.

1. A ratio is used in comparing the frequencies of two mutually exclusive classes. For example,

$$\text{sex ratio} = \frac{\text{number of males}}{\text{number of females}}$$

in a given population. Another example is

$$\text{fetal death ratio} = \frac{\text{number of fetal deaths}}{\text{number of live births}} \tag{2.1}$$

in a given year, for a given population. This expresses the number of fetal deaths as compared to live births in the same population. The composition of a heterogeneous population consisting of different ethnic groups can be described by ratios.

2. An *index* is a summary measure intended to reflect relationship among variables. It is often expressed as a ratio. For example, the well-known product correlation coefficient is, in fact, the ratio of covariance between two random variables to the product of standard deviations of

these variables. Another example is

$$\frac{\text{weight in kg}}{\text{height in cm} - 100},$$

which is called the weight-height index, and is used as a measure of obesity.

### 2.2.2 Proportion

*Proportion* is a special type of ratio in which the numerator is a part of the denominator, that is

$$p = \frac{a}{a+b}.$$

1. If the numerator and the denominator are integers and represent frequencies of certain events, then $p$ is a *relative frequency*. For example,

$$\frac{\text{number of males}}{\text{number of males} + \text{number of females}}$$

gives the proportion (relative frequency) of males, in a given community.

2. In a large population, proportion may determine the *probability* of a certain event; in a sample (experiment) proportion can be used as an *estimate* of probability of an event.

For example, the quantity

$$\frac{\text{number of fetal deaths}}{(\text{number of fetal deaths}) + (\text{number of live births})} \tag{2.2}$$

in a given population for a given year is clearly a proportion. It estimates the probability that a fetus might die before it is born. This quantity is sometimes called the *fetal death rate*, distinguishing it from fetal death *ratio* defined in (2.1). However, this quantity is not, in fact, a rate. The time unit "year" is used here only as a convenient period of counting.

3. Generally, the numerator and the denominator in $a/(a+b)$ do not need to be integers. They can be measurable quantities such as weight, length, volume, space. In such cases, proportions are also often called *fractions*. For example, in a chemical analysis, the mass of a given component can be expressed as a fraction of the total weight of the compound. *Percentage* is a proportion or a fraction per hundred units (in Latin, *per centum*). Thus a proportion $p$ corresponds to $100p$ percent ($100p\%$).

Ratios and proportions are useful *static* summary measures of phenomena that occurred under certain conditions. Particularly in studies of populations the conditions are commonly determined by demographic factors such as age, sex, or race, and are often averaged over a definite period of time (usually over a year).

## 2.3 RATES OF CONTINUOUS PROCESSES

The concept of rate is associated with the dynamics of phenomena such as chemical reactions (e.g., gain or loss of mass; changes in concentrations of certain constituents), growth, birth, death, spread of epidemics, etc. Generally, *rate* can be defined as a measure of change in one quantity ($y$) per unit of another quantity ($x$) on which $y$ depends. Usually the independent variable ($x$) is time, although it might represent some other physical quantities such as temperature or pressure. For convenience, we mostly confine ourselves to processes depending on time, and denote time by $t$ rather than by $x$.

### 2.3.1 Absolute Rate

Let

$$y = y(t) \tag{2.3}$$

be a continuous, strictly monotonic function of $t$ describing a time-dependent process. Let $\Delta t$ ($>0$) denote a (small) time interval, and

$$\Delta y = y(t+\Delta t) - y(t)$$

be the change of $y$ in time interval $\Delta t$. Then

$$a(t, t+\Delta t) = \frac{\Delta y}{\Delta t} = \left.\frac{dy}{dt}\right|_{t=t'}, \tag{2.4}$$

for some $t'$ in the interval $(t, t+\Delta t)$ will be called the *average absolute rate* this is the average change in $y$ per unit time over the interval $(t, t+\Delta t)$.

If

$$\lim_{\Delta t \to 0} \frac{y(t+\Delta t) - y(t)}{\Delta t} = \frac{dy}{dt} = \alpha(t) \tag{2.5}$$

exists, then $\alpha(t)$ is the *absolute instantaneous rate* at time $t$.

We notice that $a(t)$ is the average speed of the process over the interval $(t, t+\Delta t)$, whereas $\alpha(t)$ is the instantaneous speed at any time point $t$. The absolute rate function, $\alpha(t)=dy/dt$, describes the pattern of changes in $y(t)$ with time. It can be called a *speed function.*

**Example 2.1** *Linear motion*

If $s$ is the distance and $v$ is the speed (velocity), then $s=v \cdot t$

$$v=\frac{ds}{dt} \text{ (constant)} .$$

**Example 2.2** *Gravity law*

Distance:

$$s=\tfrac{1}{2}gt^2 .$$

Speed:

$$v(t)=\frac{ds}{dt}=gt$$

and

$$\frac{dv}{dt}=\frac{d^2s}{dt^2}=g \quad \text{represents } \textit{acceleration}.$$

### 2.3.2 Relative Rate

In most chemical and biological processes it is not the absolute change (i.e. the amount of loss or gain) in a substance per unit of time but the *relative* change per unit of time and per unit of mass available which is of interest.

Suppose that $y(t)$ represents a reaction law according to which a certain substance changes its mass (i.e., increases or decreases with time). The (relative) instantaneous rate per unit of mass and per unit of time at time point $t$ is

$$\beta(t)=\frac{1}{y(t)}\frac{dy}{dt}. \tag{2.6}$$

Since this kind of rate is widely used, we omit the word "relative" unless ambiguity arises. In biochemistry, $\beta(t)$ is often called the reaction velocity. In radioactive reactions, $\beta(t)$ has a negative sign and represents the instantaneous rate of radioactive decay.

We also notice that (2.6) is equivalent to

$$\beta(t)=\frac{d\log y(t)}{dt}. \tag{2.6a}$$

If we know the mathematical form of $\beta(t)$ we can evaluate $y(t)$ by integrating (2.6a). We obtain

$$y(t)=y(0)\exp\left[\int_0^t \beta(x)\,dx\right]. \tag{2.7}$$

**Example 2.3.** *Exponential growth model*

Let $N(t)$ denote the population size at time $t$. We assume that in the model population, $N(t)$ is a continuous function of $t$.

Suppose that the change in the population size, $dN(t)$, in a very small time interval, $dt$, is proportional to $N(t)$ and to $dt$, that is

$$dN(t)=\beta N(t)\,dt. \tag{2.8}$$

Then

$$\beta=\frac{1}{N(t)}\frac{dN(t)}{dt}$$

is a (constant) *growth rate*, and

$$N(t)=N(0)\exp\left[\int_0^t \beta\,dx\right]=N(0)e^{\beta t}, \tag{2.9}$$

where $N(0)$ is the population size at $t=0$.

Formula (2.9) represents the so-called *exponential growth model*.

**Example 2.4.** *Hazard rate function*

For any (continuous) distribution function $Pr\{X \leqslant x\}=F(x)$, the instantaneous (relative) rate, also called the *hazard rate function* (see Section 3.3), is

$$\mu(x)=\frac{1}{1-F(x)}\frac{dF(x)}{dx}=-\frac{d\log(1-F(x))}{dx}.$$

### 2.3.3 Average (Central) Rate

In practice, the mathematical form of the reaction process, $y(t)$, or distribution function $F(x)$, may not be known, but we may be able to calculate a sort of "average" rate, $b(t,t+\Delta t)$, say, from the formula

$$b(t,t+\Delta t) \doteqdot \frac{1}{y(t')}\frac{\Delta y}{\Delta t}=\frac{\Delta y}{y(t')\Delta t}, \tag{2.10}$$

for some $t'$ in the interval $t$ to $t+\Delta t$.

How do we estimate $y(t')$? We may use three kinds of approximation.

1. Note that $y(t')$ is the mass available at a certain point $t'$ between $t$ and $t+\Delta t$. This is not necessarily the midpoint of the interval $(t, t+\Delta t)$, but assuming that $y(t)$ is approximately linear over the interval $(t, t+\Delta t)$, we may take $t'$ to be the midpoint of this interval, that is, $t' = t + \frac{1}{2}\Delta t$, so that

$$y(t') \doteqdot y\left(t + \tfrac{1}{2}\Delta t\right). \tag{2.11}$$

2. The product $y(t')\Delta t$ represents the area of the shaded rectangle, $y(t') \times \Delta t$, expressed in (mass × time) units (Fig. 2.1).

Of course, if the form of $y(t)$ is known, $b(t, t+\Delta t)$, can be calculated more precisely; namely

$$b(t, t+\Delta t) = \frac{\Delta y}{\int_t^{t+\Delta t} y(x)\,dx}, \tag{2.12}$$

where $\int_t^{t+\Delta t} y(x)\,dx$ represents the area under the $AB$ part of the curve $y(x)$ and, of course, its measure also is in (mass × time) units.

Note that for small $\Delta t$, we may use the approximation

$$\begin{aligned}\int_t^{t+\Delta t} y(x)\,dx &\doteqdot \tfrac{1}{2}\left[\,y(t+\Delta t) + y(t)\right]\cdot\Delta t \\ &= \left\{y(t) + \tfrac{1}{2}\left[\,y(t+\Delta t) - y(t)\right]\right\}\Delta t \\ &= \left[\,y(t) + \tfrac{1}{2}\Delta y\right]\Delta t, \end{aligned} \tag{2.13}$$

and so

$$b(t, t+\Delta t) \doteqdot \frac{\Delta y}{\left[\,y(t) + \frac{1}{2}\Delta y\right]\cdot\Delta t} = \frac{1}{y(t) + \frac{1}{2}\Delta y}\cdot\frac{\Delta y}{\Delta t}. \tag{2.14}$$

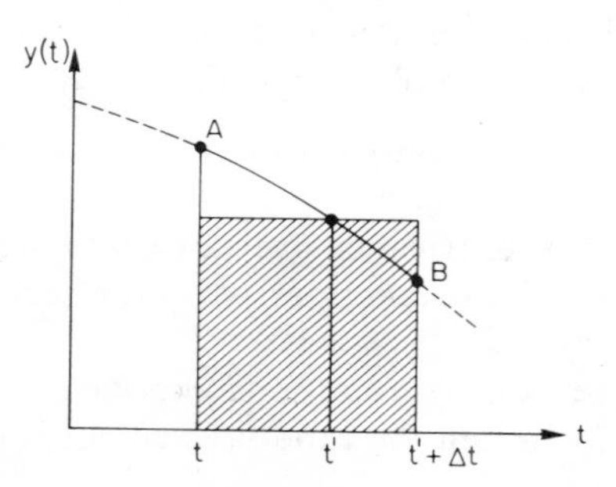


**Fig. 2.1**

Comparing (2.10) and (2.14), we have

$$y(t') \doteqdot y(t) + \tfrac{1}{2}\Delta y. \tag{2.15}$$

Note that for any decreasing function $y(t)$, as in Fig. 2.1, $\Delta y$ has a negative sign. Also note that (2.11) and (2.15) are not necessarily equal, but for sufficiently small $\Delta t$ they are nearly so.

The kind of rate defined in (2.12) and its approximations (2.10) or (2.14) are often called *central rates*. These rates play an essential role in construction of life tables (see Chapter 4). We note that the relative instantaneous rate expresses rapidity of change in mass per (mass × time) unit at a given point. The concept of amount (area) of mass × time is important in calculating average (central) rates.

It should be noted that for negative $\Delta y$, the ratio $|\Delta y|/y(t)$ represents the fraction of mass $y(t)$ that has been lost over the period $\Delta t$.

3. In certain situations $y(t')$ is conventionally approximated by the *initial value*, $y(t)$. For example, if $N(t)$ and $N(t+\Delta t)$ represent the sizes of the population at times $t$ and $(t+\Delta t)$, respectively, then the average growth rate per time unit per 100 individuals is usually defined as

$$r(t,t+\Delta t) = \frac{N(t+\Delta t) - N(t)}{N(t)\cdot \Delta t} \cdot 100. \tag{2.16}$$

This is often called a *percent growth rate*. Another example is an annual rate of interest expressed as a percentage of capital deposit at the beginning of an investment period.

## 2.4 RATES FOR REPETITIVE EVENTS

In the preceding section, we considered continuous processes such as motion, growth, chemical reactions, changes in concentrations of certain substances. We can also define rates for *events* that are *repetitive* in time or space. We consider a few examples for such events.

1. Average number of telephone calls per household per month in a given town.
2. Average number of hurricanes over a certain area per year.
3. Average number of customers arriving for service at a gas station per hour.

The above rates have a strong random component, but rates can also be calculated for events for which the random element is of little importance.

For example:

4. Pulse rate (average number of heart beats per minute) has only a relatively small random component.

5. Number of revolutions per hour of the seconds hand of a watch is a more extreme example.

We notice that rates in examples 2–5 are *absolute* rates, whereas 1 is an example of a *relative* rate, since the household is not specified. There, however, the number of telephone calls per household does not imply that the households themselves are changing in time; the number of households in a town represents a *reference* (or *target*) population of households to which the rate applies. We may also calculate absolute rate of telephone calls per month for a *specified* household. The specified gas station in example 3, and the restricted areas for which certain events are counted in examples 1 and 2 are only convenient reference areas to which these rates apply. We also notice that rates can be expressed per area unit as well as per time unit.

## 2.5 CRUDE BIRTH RATE

1. In the sense described above, birth is also a repetitive event. Thus a *birth rate* per (calendar) year can be defined as the number of live births per year and, perhaps, per 1000 persons of a reference (or target) population selected for convenience. A reference population can be, for example, a cohort of women born in a certain (calendar) time interval; or the population of married couples within a certain age group; or just a total population defined a certain time point (usually the midyear population). The birth rates can be *specific* (defined for a special subsets in a population), or *crude* (defined over the whole reference population).

In vital statistics, the total midyear population is used as a reference population, and we denote its size by $K$. Denoting by $B$ the number of live births during a (calendar) year, we define the crude annual birth rate per 1000 persons as

$$\text{Crude birth rate} = \frac{B}{K} \cdot 1000. \tag{2.17}$$

2. We may, however, look at this problem from a different point of view. We may consider a population as an *aggregate* (mass), which, on the average, increases by live births (cf. growth model in Example 2.3). Crude birth rate is still defined by (2.17), but the midyear population, $K$, in the denominator is an analogue of mass $y(t')$ in (2.11) (here $\Delta t = 1$).

Note that in both situations, (1) or (2), the average population in the denominator is not identical with the actual population directly involved in reproduction.

## 2.6 MORTALITY MEASURES USED IN VITAL STATISTICS

Phenomena observed in a population may be broadly classified into two categories: (1) those that change the size of the population (births, deaths), and (2) those that occur in the population, that is, the population represents a *target* for them, but they do not necessarily change its size (e.g., number of life insurances effected per year per 1000 people living on the average during this year in a community; number of marriages per year, etc.). In calculating rates for phenomena described in (1) we may use an approach of continuous changes in mass over time—the *mass-time* approach—discussed in Section 2.3; for phenomena of the kind described in (2), we can use the *repetitive event approach* discussed in Section 2.4. Clearly, deaths occur in a population almost every second and diminish population size; the *death process* may then be considered as a *continuous process* and the mass-time approach is appropriate in calculating death rates.

### 2.6.1 The Concept of Population Exposed to Risk

Central rates play an important role in describing population dynamics. We confine ourselves here to death rates, although concepts developed in this section are applicable to other time-dependent phenomena (e.g., onset of a disease). As a unit of time we conveniently select one year, so that we will be discussing *annual death rates*. As has already been mentioned, a living process is a continuous process, and, for practical purposes, we shall consider the population size $N(t)$, say, to be a continuous function of time $t$.

***Person-years.*** Let $N_T$ denote the number of individuals ever observed during a period of $T$ years (Fig. 2.2). Let $\tau_j$ denote the length of the period (in years) during which an individual ($j$) was under observation, that is, *exposed to risk* of being observed to die (horizontal lines in Fig. 2.2). Then the sum of lengths of such periods of exposure

$$A_T = \sum_{j=1}^{N_T} 1 \cdot \tau_j \tag{2.18}$$

gives the total amount of *person-years* (analogous to mass-time) exposed to risk.

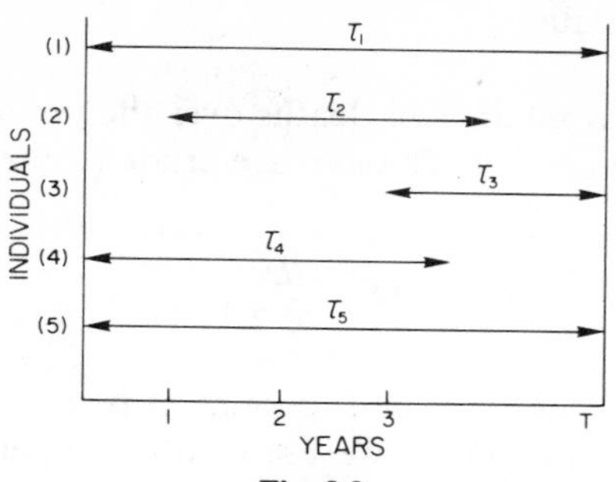


**Fig. 2.2**

What other interpretation can be assigned to (2.18)? Let $N(t)$ denote the population size at a given time $t$, and assume that $N(t)$ is a continuous function of $t$ (cf. Example 2.3). Then

$$A_T = \int_0^T N(t)\,dt \doteq T \cdot N(t') \tag{2.19}$$

for some $t'$ in $(0, T)$. Here

$$N(t') \doteq \frac{A_T}{T} \tag{2.19a}$$

is an analogue of $y(t')$ (the average mass) and is called the *average population of exposed to risk* (or *average population at risk*). Clearly, $N(t') \leqslant A_T$. Also note that $T$ is in time units (e.g., $T$ years).

Of course, the current population vital statistics data do not give the exact times $\tau_j$ for each member of the population. The midperiod population $[N(\frac{1}{2}t)]$ or census population $[N(t_c)$, say], which we now denote by $K$, is very often used as a conventional estimate of $N(t')$.

If the population is stratified by age groups, the average population of exposed to risk can be calculated for each age group in the same manner (see Exercise 2.4).

**Example 2.5** Suppose that the period of observation is $T = 3$ years. Suppose that $N_T = 10$ individuals were observed during this period for the following lengths of time $\tau_j$: 2.3, 1.5, 2.8, 2.5, 3.0, 1.8, 2.7, 2.5, 3.0, 3.0 years.

1. What is the number of person-years?

$$A_T = 2.3 + 1.5 + \cdots + 3.0 = 24.3.$$

2. What is the average size of the population exposed to risk?

$$N(t') \doteq \frac{A_T}{T} = \frac{24.3}{3} = 8.1 \approx 8.$$

### 2.6.2 Crude Death Rate

Let $D$ denote the total number of deaths over the period of $T$ years and $K$ be the midperiod population. The average crude (central) rate *per year per person* is

$$M = \frac{D}{K \cdot T}, \tag{2.20}$$

where $K \cdot T$ represents the estimated amount of person years.

Since $M$ is often rather small, we use death rate per 1000 or 10,000 (or any other convenient number $10^k$) persons. For example, $M' = M \cdot 1000 = (D/KT) \cdot 1000$ is the crude death rate per year per 1000 persons. A rate per 100 persons is called by some people the *percent rate*. It seems that this expression has no clear meaning and should be avoided.

In practice, we often have $T=1$. However, in the calculation of death rates from census data, $T=3$ years is used, with the census year being the middle. In the United States, the census population is evaluated as April 1, every 10 years with the last digit "0" (e.g., 1950, 1960, etc.). Three year death data are used to reduce the effects of year-to-year variation.

**Example 2.6** We evaluate the crude annual death rate over the period 1969–1971 with 1970 being a census year. (The data are obtained from U. S. Vital Statistics, 1969, 1970, 1971.) We have

$K^{(1970)} = 203{,}211{,}926$—census population

$D^{(1969)} = 1{,}921{,}990$—number of deaths, 1969

$D^{(1970)} = 1{,}921{,}031$—number of deaths, 1970

$D^{(1971)} = 1{,}927{,}524$—number of deaths, 1971

$D_{\text{Total}} = 5{,}770{,}563$

The average crude annual death rate per 1000 persons is

$$\frac{5{,}770{,}563}{203{,}211{,}926 \times 3} \times 1000 = 9.42 \text{ per year per 1000 population.}$$

If we calculate the crude death rate, using only census year data, we obtain

$$\frac{1{,}921{,}031}{203{,}211{,}926} \times 1000 = 9.45 \text{ per year per 1000 population.}$$

We may also evaluate the crude death rate using in the denominator the sum of midyear populations of these three consecutive years. From

U. S. Vital Statistics, we obtain the estimates $K^{(1969)}=201{,}921{,}000$, $K^{(1971)}=206{,}212{,}000$. Hence, $K^{(1969)}+K^{(1970)}+K^{(1971)}=611{,}344{,}926$, and the crude death rate is

$$\frac{5{,}770{,}563}{611{,}344{,}926}\times 1000=9.44 \text{ per year per 1000 population.}$$

The three methods of calculation give close results.

### 2.6.3 Age Specific Death Rates

Death rates associated with a specific age or specific age group are more informative about mortality pattern than the crude death rate.

Let $D_a$ denote the number of total deaths in age group $a$ (e.g., between 40–45) over a calendar year, and $K_a$ denote the midyear population in this age group. Very often age is recorded as age last birthday (lbd). For example, age group 40–45 corresponds to age 40–44 lbd.

The age specific annual death rate in age group $a$ is

$$M_a=\frac{D_a}{K_a} \tag{2.21}$$

per year per person (or per person-year), or

$$M_a'=\frac{D_a}{K_a}\cdot 1000 \tag{2.21a}$$

per year per 1000 persons (i.e., per 1000 person-years).

These rates play an important role in constructing life tables (Chapter 4).

### 2.6.4 Cause Specific Mortality Measures Used in Vital Statistics

As before, let $D$ denote the number of total deaths over a calendar year and $K$ the midyear population. Let $D_h$ denote number of total deaths from a specific cause $C_h$, say, over the same year. Then

$$M_h=\frac{D_h}{K} \tag{2.22}$$

is the *cause specific death rate* per person-year for $C_h$. Of course, one also can calculate a *cause and age specific death rate*

$$M_{ha}=\frac{D_{ha}}{K_a}, \tag{2.23}$$

where $D_{ha}$ is the number of deaths from cause $C_h$ in age group $a$.

In studying multiple causes of death (see Chapters 9–12) it is also useful to evaluate the proportions of deaths from different causes among all deaths. For example, for cause $C_h$

$$p_h = \frac{D_h}{D}. \tag{2.24}$$

This is often called *proportional mortality rate* or *cause specific death ratio*. Clearly, these terms are inappropriate and should be avoided.

## 2.7 RELATIONSHIPS BETWEEN CRUDE AND AGE SPECIFIC RATES

The results presented in this and the next two sections are not necessarily restricted to mortality rates, but since analysis of mortality data is the main topic of this book we discuss these results in terms of mortality rates. For a given calendar year we have

$$D = \sum_{a=1}^{r} D_a \qquad \text{and} \qquad K = \sum_{a=1}^{r} K_a, \tag{2.25}$$

where $r$ denotes the number of age groups.

If we know the age-specific death rates $M_a$, we have $D_a = M_a K_a$, so that the crude death rate $M$ is a weighted average of age-specific death rates

$$M = \frac{D}{K} = \frac{1}{K} \sum_{a=1}^{r} K_a M_a = \sum_{a=1}^{r} p_a M_a, \tag{2.26}$$

where $p_a = K_a / K$ is the proportion of midyear population in age group $a$.

## 2.8 STANDARDIZED MORTALITY RATIO (SMR): INDIRECT STANDARDIZATION

Suppose that we wish to compare the mortality experience of a particular population—a so called *study population*—with the mortality experience of another population chosen to be a *standard population*. Comparison of lifetime distributions of these two populations would be appropriate to use for this purpose. Such comparisons will be discussed in Chapter 8. In traditional epidemiological analysis, however, single-valued summary measures (indices) are commonly used. Among these the standardized mortality ratio (SMR) seems to be most popular.

Let $K_{Sa}$, $K_S$, $p_{Sa}$, $D_{Sa}$, $D_S$, $M_{Sa}$, and $M_S$ denote the analogues, for a standard population, of the corresponding quantities defined in Section 2.7 for a study population. (The symbol "$S$" is used for "standard.") We have,

of course,

$$M_S = \sum_a p_{Sa} M_{Sa}.$$

We notice from (2.26) that

$$\sum_a K_a M_a = K \sum_a p_a M_a = D \tag{2.27}$$

is the number of observed deaths in the study population.

If the deaths rates in the study population were the same as in the standard population, then the number of *expected* deaths in the study population would be

$$\sum_a K_a M_{Sa} = K \sum_a p_a M_{Sa} = E. \tag{2.28}$$

The ratio

$$\frac{\text{number of observed deaths}}{\text{number of expected deaths}} = \frac{D}{E} \tag{2.29}$$

$$= \frac{\sum_a K_a M_a}{\sum_a K_a M_{Sa}} = \frac{\sum_a p_a M_a}{\sum_a p_a M_{Sa}} = \text{SMR} \tag{2.29a}$$

is called the *standardized mortality ratio* (SMR). (It is often multiplied by 100.)

The value of SMR greater than 1 is regarded as indicating higher mortality in the study than in the standard population, and conversely, for SMR less than 1. In reality, however, the situation is not always straightforward. In common with other summary indices, SMR depends on age distribution as well as on mortality patterns in both populations. A critical discussion on the use of SMR is given by Gaffey (1975). A sounder approach to comparison of mortality in two populations, based on statistical methods, will be given in Chapter 8.

The quantity

$$M_{\text{indirect}} = \frac{D}{E} \cdot M_S = (\text{SMR}) \cdot M_S \tag{2.30}$$

is an *adjusted* rate, and the method used here is called *indirect standardization*. We notice that in indirect standardization, the age-specific rates and the crude rate of a standard population are used for adjustment.

*Remarks.* The following remarks on calculating the expected number of deaths might be useful:

1. The quantity $K_a M_a = D_a$ is the observed, and the quantity $K_a M_{Sa} = E_a$ is the expected (*conditional on* $K_a$) number of deaths in age group $a$. We have

$$\sum_a D_a = D \qquad \text{and} \qquad \sum_a E_a = E.$$

We also notice that for the observed deaths

$$D = \sum K_a M_a = K \cdot M,$$

whereas for the expected deaths

$$E = \sum K_a M_{Sa} \neq K \cdot M_S.$$

Clearly, $E$ is the sum of conditional (on $K_a$) expected numbers of deaths, $E_a$, and depends on the grouping intervals.

2. In calculating the observed and expected numbers of deaths, the rates $M_a$ and $M_{Sa}$ (per year per person), and not $M'_a$, $M'_{Sa}$ (per year per $10^k$ persons) are used. This, of course, is irrelevant in calculating the SMR (since it is a ratio), provided that $k$ is the same in obtaining the $M'_a$ and $M'_{Sa}$.

3. In our calculations we have tacitly assumed that the mortality data were collected over a single calendar year. We may, however, have data collected over a period of $T$ calendar years. If $D_a$ denotes now the observed and $E_a$ the expected numbers of deaths in age group $a$ over the period of $T$ calendar years while $K_a$ is the midperiod population in the age group $a$, then $E_a$ is calculated from the formula

$$E_a = K_a \cdot T \cdot M_{Sa} = A_{a(T)} \cdot M_{Sa}, \tag{2.31}$$

where $A_{a(T)}$ is the amount of person-years in age group $a$ over the period of $T$ years.

Moreover, the person-years of exposed to risk might be not exactly in the same calendar years; they should be, however, in the same calendar period for which the standard rates are applicable.

Therefore, not the midperiod population, but the amount of person-years of exposed to risk is essential in calculating the expected number of deaths. This is important in situations, when the study population is composed of individuals with different times at entry, and different times and modes of departure (cf. Section 2.10).

## 2.9 DIRECT STANDARDIZATION

In this method, the *age distribution* (age structure) of a standard population is used for adjustment—the overall rate so obtained is often called the *age adjusted rate*. It is obtained from the formula

$$M_{\text{direct}} = \frac{1}{K_S} \sum_a K_{Sa} M_a = \sum_a P_{Sa} M_a. \tag{2.32}$$

Of course, the new value, $M_{\text{direct}}$, depends on the age composition of the population selected as standard. Comparison of two age-adjusted rates has some meaning only in connection with the standard population. Using another population as standard may change our conclusions. We note that in direct standardization, the age distribution of the standard population is used as a base for adjustment.

As has been frequently indicated in the literature, standardizations are not often very helpful and are very often misleading. In particular, the age adjusted rate by direct standardization has no real interpretation. What meaning can be attached to the "standard population" if the actual population will never have such a structure? Comparison of two populations is better achieved by comparison of the two corresponding life tables (e.g., see Chapter 8). Comparison of two age adjusted (by direct standardization) rates has no real meaning and is not recommended. Meaningful information, however, can be obtained from comparison of two life tables.

## 2.10 EVALUATION OF PERSON-YEARS OF EXPOSED TO RISK IN LONG-TERM STUDIES

Ascertainment of exposed to risk, introduced in Section 2.6.1, is not always so simple. Depending on the information available on each individual in the experience, the evaluation of the amount of person-years might be quite complicated. Various aspects of this problem and exposure formulae for different kinds of data are extensively presented in the books by Gershenson (1961) and Batten (1978).

In this section, we restrict ourselves to long-term studies, in which it is usually desirable to stratify the mortality experience into shorter chronological periods (e.g., 5-year subperiods) as well as into age groups (e.g., 5-year age groups). This is especially useful when the study population consists of a fairly large number of workers who are exposed to occupational hazards (e.g., asbestos or rubber industry workers) and it is desired to compare their mortality with that in some relevant general population. If the duration of exposure plays an important role, we should also take it into account as a third variable for stratification. Here, however, we restrict ourselves to chronological time and age as stratification variables. (See Exercise 2.2 for including duration.)

The methods described below are applicable to any type of long-term experience (e.g., occupational studies, clinical trials, comparisons of mortality in different ethnic groups, holders of insurance policies, etc.), in which three dates are available—date of birth, date of entry to the experience, and date of departure, that is, date of death or date of leaving the experience for any other reason.

### 2.10.1 Exact Dates for Each Individual Are Available

Suppose that for each individual the three dates (birth, entry, and departure) are recorded exactly (up to 1 day, say). To show how the practical calculation of the amount of person-years can be carried out, it is convenient to use a numerical example.

**Example 2.7** An individual born March 4, 1931 enters an experience on June 10, 1963 and leaves December 2, 1974. We wish to determine this individual's contribution to the exposed to risk in quinquennial age groups (30–34, 35–39, etc., lbd). and quinquennial subperiods (1960–1964, 1965–1969, and 1970–1974).

We first express the three dates in decimal form

| | |
|---|---|
| Birth | 1931.18 |
| Entry | 1963.44 |
| Departure | 1974.92. |

Diagrammatic representation of this individual's experience, and contributions to each (age group)×(subperiod) combination is given in Fig. 2.3 [compare Hill (1972)].

$$\text{Age at entry is } 1963.44 - 1931.18 = 32.26$$
$$\text{Age at departure is } 1974.92 - 1931.18 = 43.74.$$

The line ED (Entry-Departure), with slope $-45°$, represents the period of the individual's inclusion in the experience. When this line crosses a

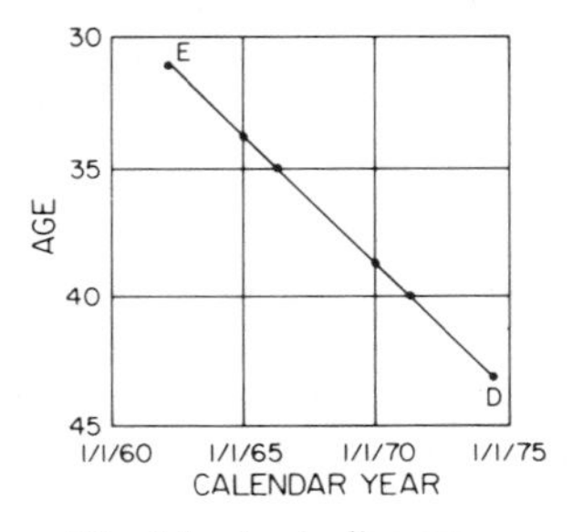


**Fig. 2.3** Lexis diagram.

**Table 2.1** Contribution of an individual to exposed to risk

| Age group (lbd) | Subperiod | Entry date | Departure date | Amount of person-years | |
|---|---|---|---|---|---|
| | | | | Exact | Approx. |
| 30–34 | 1960–1964 | 6/10/63 | 1/1/65 | 1.56 | 1.5 |
| 30–34 | 1965–1969 | 1/1/65 | 3/4/66 | 1.18 | 1.5 |
| 35–39 | 1965–1969 | 3/4/66 | 1/1/70 | 3.82 | 3.5 |
| 35–39 | 1970–1974 | 1/1/70 | 3/4/71 | 1.18 | 1.5 |
| 40–44 | 1970–1974 | 3/4/70 | 12/2/74 | 3.74 | 3.0 |

vertical line, the individual moves to a different 5-year subperiod; when it crosses a horizontal line, the individual moves to a different age group. The contributions to person-years are the lengths of projections of the appropriate portions of the line on either the vertical or horizontal sides of the cell. The type of diagram is sometimes called a *Lexis diagram*.

Initially, the individual is in age group 30–34 lbd and subperiod 1960–1964. On January 1, 1965 the subperiod changes to 1965–1969, but the age group remains unchanged. The contribution to 30–34 age group and 1960–1964 subperiod is $1965-1963.44=1.56$ years. On March 4, 1966 the 35th birthday is reached, so that the age group changes to 35–39 lbd but the subperiod 1965–1969 remains unchanged—the contribution to age group 30–34 and subperiod 1965–1969 is $1966.18-1965=1.18$ years. The next change is on January 1, 1970; the contribution to age group 35–39 and subperiod 1965–1969 is $1970-1966.18=3.82$ years. The following change is on March 4, 1971; the contribution to age group 35–39, and subperiod 1970–1974 is $1971.18-1970=1.18$ years. The final contribution to age group 40–44 and subperiod 1970–1974 is $1974.92-1971.18=3.74$. The results are summarized in Table 2.1. Note that the mode of departure (by death, withdrawal, transfer, etc.) does not affect the calculations.

An analysis similar to this can be done for each person in the experience. The individual contributions are summed to give the total exposed to risk for each (age-group)$\times$(subperiod) combination as illustrated in the following example.

**Example 2.8** Consider four individuals, $A$, $B$, $C$, and $D$, with dates shown below.

| Individual | Birth date | Entry date | Departure date |
|---|---|---|---|
| $A$ | 3/4/31 | 6/10/63 | 12/2/74 |
| $B$ | 11/22/38 | 9/30/59 | 1/4/64 |
| $C$ | 1/9/37 | 9/30/59 | 1/4/64 |
| $D$ | 1/1/30 | 7/1/57 | 7/1/72 |

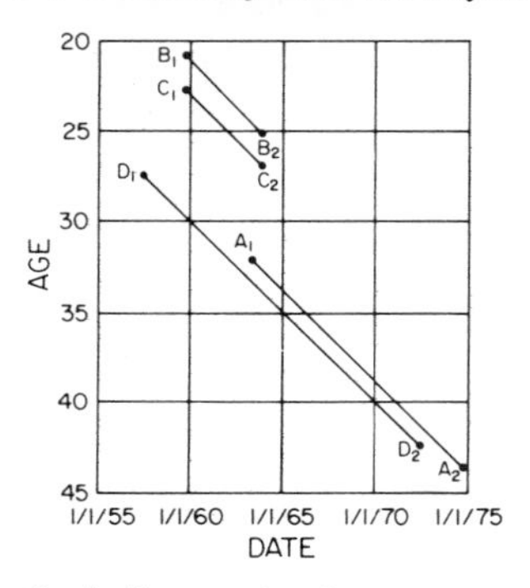


**Fig. 2.4** Lexis diagram for data from Table 2.2.

The data are represented diagrammatically in Fig. 2.4. The lines $A_1A_2$, $B_1B_2$, $C_1C_2$, and $D_1D_2$ represent the exposures to risk. The contributions to person-years of exposed to risk are given in Table 2.2a; the totals are given in Table 2.2b.

Dividing the number of deaths in each (age group)×(subperiod) stratum by the corresponding value of the person-years of exposed to risk, we obtain the observed death rate for this stratum. On the other hand,

**Table 2.2a** Contributions to exposed to risk

| Age | | 1/1/55–1/1/60 | 1/1/60–1/1/65 | 1/1/65–1/1/70 | 1/1/70–1/1/75 | Total |
|---|---|---|---|---|---|---|
| 20–25 | *A B C D* | 0.25, 0.25 } 0.50 | 3.89, 2.02 } 5.91 | | | 6.41 |
| 25–30 | *A B C D* | 2.50 | 0.12, 1.99 } 2.11 | | | 4.61 |
| 30–35 | *A B C D* | | 1.56, 5.00 } 6.56 | 1.18 | | 7.74 |
| 35–40 | *A B C D* | | | 3.82, 5.00 } 8.82 | 1.18 | 10.00 |
| 40–45 | *A B C D* | | | | 3.74, 2.50 } 6.24 | 6.24 |
| Total | | 3.00 | 14.58 | 10.00 | 7.42 | 35.00 |

**Table 2.2b** Summary of results

| Age group (lbd) | Exposed to risk (person-years) | Sub-period | Exposed to risk (person-years) |
|---|---|---|---|
| 20–24 | 6.41 | 1955–1959 | 3.00 |
| 25–29 | 4.61 | 1960–1964 | 14.58 |
| 30–34 | 7.74 | 1965–1969 | 10.00 |
| 35–39 | 10.00 | 1970–1974 | 7.42 |
| 40–44 | 6.24 | | |

multiplying the value of person-years in a given stratum by the corresponding death rate in the general population, we obtain the expected number of deaths for this stratum. These can be compared with observed deaths or used in standardization (see Section 2.8). We note that the person-years of exposed to risk in any stratum is calculated by summing the periods spent in that stratum by each individual.

### 2.10.2 Only Years of Birth, Entry, and Departure Are Available

The situation is not so simple if only the years of occurrence of birth, entry, and departure are given. The amount of person-years exposed to risk must then be evaluated approximately using certain assumptions. The usual assumptions are: (1) that the relevant event occurs at the middle (July 1) of the year specified, or (2) that the date of occurrence is uniformly distributed over the specified year. It should be emphasized that assumptions (1) and (2) are not equivalent.

Assumption (1) is often made because the computations are simple. However, when entry and departure occur in the same year, the calculated contribution to person-years of exposed to risk would be zero; it is then often arbitrarily assumed that this is approximately 0.25 person-years. For example, Monson (1974) uses this additional assumption in computing the amounts of person-years. We notice that in extreme cases the error in age at entry (or departure) might be + or −one year; sometimes a person may be misclassified even with respect to quinquennial age groups. For example, a man born on Nov. 5, 1930, who enters the study on June 10, 1960 is only 29 lbd and should be in age group 25–29 lbd, although according to the principle stated by assumption (1), he will be 30 lbd and classified in age group 30–34 lbd.

Assumption (2) seems to be somewhat more reasonable; we now examine its consequences on the calculation of person-years of exposed to risk.

Let the random variables $X'$, $Y'$, and $Z'$ represent dates (using decimals) and $\xi$, $\eta$, and $\zeta$ ($\xi<\eta<\zeta$) denote the recorded calendar years of birth, entry, and departure, respectively. (For example, for birth date March 3, $1931 \equiv 1931.18$, we have $\xi=1931$, $X'=0.18$.) We can write

$$X=\xi+X' \qquad Y=\eta+Y' \qquad Z=\zeta+Z',$$

and assume that they are mutually independent, where $X'$, $Y'$, $Z'$, each has standard uniform distribution as defined in Section 3.10. For the moment, we consider exposed to risk by 1 year age groups only.

$$\text{Age at entry is } Y-X=(\eta-\xi)+(Y'-X'),$$
$$\text{Age at departure is } Z-X=(\zeta-\xi)+(Z'-X').$$

Since $-1<Y'-X'<1$ and $-1<Z'-X'<1$, we see that an individual contributes one year to the exposed to risk for each age from $(\eta-\xi-1)$ to $(\zeta-\xi-2)$ lbd inclusive. The expected contribution to each of ages $(\eta-\xi-1)$ and $(\zeta-\xi)$ is $\frac{1}{6}$; and to each of ages $(\eta-\xi)$ and $(\zeta-\xi-1)$ is $\frac{5}{6}$ (compare Exercise 3.2).

Summarizing, we have for years of birth ($\xi$), entry ($\eta$), and departure ($\zeta$) the following expected contributions to exposed to risk in 1 year age groups

| Age lbd | $\eta-\xi-1$ | $\eta-\xi$ | $\eta-\xi+1$ | $\cdots$ | $\zeta-\xi-2$ | $\zeta-\xi-1$ | $\zeta-\xi$ |
|---|---|---|---|---|---|---|---|
| Average contribution | $\frac{1}{6}$ | $\frac{5}{6}$ | 1 | $\cdots$ | 1 | $\frac{5}{6}$ | $\frac{1}{6}$ |

The value for a 5-year age group is, of course, obtained by summing values for the five 1-year age groups it contains.

The analysis for calendar year exposed to risk is simpler. For each year $(\eta+1)$ through $(\zeta-1)$, the contribution is 1; for each of $\eta$ and $\zeta$ is $\frac{1}{2}$.

We now turn to the calculation of expected numbers of exposed to risk for (age group)$\times$(calendar year) combinations. It is convenient to consider 1-year age groups and single calendar years, and then to sum the appropriate values to obtain person-years of exposed to risk for wider groupings. Detailed calculations lead to the schedule shown in Table 2.3. The calculations are facilitated by noting that: (1) the marginal totals are known, and (2) only two entries in each column (and one or two in each row) can be nonzero.

Returning to our data in Example 2.7, we have $\xi=1931$, $\eta=1963$, $\zeta=1974$. In this case, under both assumptions (1) and (2), we obtain the same results given in the last column of Table 2.1. If, however, the year of birth were 1930, so that at the entry the individual were 30 lbd, then we

**Table 2.3** Expected contributions to person-years of exposed to risk

| Age lbd | Calendar year | | | | | | | | Total |
|---|---|---|---|---|---|---|---|---|---|
| | $\eta$ | $\eta+1$ | $\eta+2$ | $\eta+3$ | . . . | $\zeta-2$ | $\zeta-1$ | $\zeta$ | |
| $\eta-\xi-1$ | $\frac{1}{6}$ | | | | | | | | $\frac{1}{6}$ |
| $\eta-\xi$ | $\frac{1}{3}$ | $\frac{1}{2}$ | | | | | | | $\frac{5}{6}$ |
| $\eta-\xi-1$ | | $\frac{1}{2}$ | $\frac{1}{2}$ | | | | | | 1 |
| $\eta-\xi+2$ | | | $\frac{1}{2}$ | $\frac{1}{2}$ | | | | | 1 |
| ... | | | | | . | | | | . |
| ... | | | | | . | | | | . |
| $\zeta-\xi-2$ | | | | | . | $\frac{1}{2}$ | $\frac{1}{2}$ | | 1 |
| $\zeta-\xi-1$ | | | | | | | $\frac{1}{2}$ | $\frac{1}{3}$ | $\frac{5}{6}$ |
| $\zeta-\xi$ | | | | | | | | $\frac{1}{6}$ | $\frac{1}{6}$ |
| Total | $\frac{1}{2}$ | 1 | 1 | | . . . | 1 | 1 | $\frac{1}{2}$ | |

have also to consider age $\eta-\xi-1=1963-1933-1=29$ lbd which makes it necessary to consider age group 25–29 lbd. The calculations for such case are left to the reader.

## 2.11 PREVALENCE AND INCIDENCE OF A DISEASE

An individual who has contracted a disease is often referred to as a *case*. In studying morbidity, two indices are in common use: prevalence and incidence, both based on observed number of cases.

*Prevalence* is supposed to determine the status of a disease in a given community, whereas *incidence* is usually concerned with rapidity of development of epidemics in a relatively short time period, or with chronic disorders as functions of age (e.g., cancer incidence). Since these terms are not always precisely defined (especially "incidence"), it is useful to discuss their definitions and interpretation in some detail.

### 2.11.1 Prevalence

Consider a disease $A$ that occurs in a given population. Suppose that at a fixed date, the population is examined to find how many individuals actually have this disease. We define the *prevalence* of the disease $A$ as

$$\text{prevalence}=\frac{\text{number of existing cases of disease } A \text{ at a specific date}}{\text{number of persons in the community at this date}}. \tag{2.33}$$

Clearly, prevalence is the *proportion* of individuals who, at a given date, of examination, were affected by the disease $A$, although it is sometimes incorrectly called a "rate." If the disease is rare, we may multiply (2.33) by 1000 (obtaining prevalence per 1000 persons), but it still will not be a rate.

In practice, it might be difficult to examine the whole population on a given date. The counting of cases may then be done over a certain relatively short period over which the so-called period prevalence is defined

$$\begin{array}{l}\text{prevalence}\\ \text{(period)}\end{array} = \frac{\begin{array}{l}\text{number of existing cases of disease } A \text{ in}\\ \text{a given period of time}\end{array}}{\begin{array}{l}\text{average number of persons living in the}\\ \text{community during this period}\end{array}}. \tag{2.34}$$

This is not precisely a proportion, but it is nearly so, if the period of investigation is short. In our opinion, this is a misleading concept and should be avoided.

There are other kinds of prevalence indices. It is important to pay attention to what kind of prevalence is reported. *Lifetime prevalence* at age $x$ lbd is the proportion who have experienced an episode of the event (disease) in question at *any* time during their lives, no matter what may be their present status. It is to be distinguished from *treated prevalence*, the proportion now under treatment for the disease, and also from the *current prevalence*, the proportion now suffering from the disease, whether under treatment or not. There are other possibilities, for example, *hospital prevalence*, etc.

### 2.11.2 Incidence

Incidence of a disease, on the other hand, is associated with the frequency of inception of cases of this disease as time passes by. We would like to draw the reader's attention to different situations, for which incidence needs to be calculated from different formulae, depending on the type of disease we are studying. Many diseases are not necessarily lethal at their first occurrence. A person who contracts such a disease may be cured and later become ill again from the same disease. A number of infectious diseases belong to this category. We recall that a person, who contracts a given disease is called by epidemiologists a *case* of this disease. Thus the same individual may become a case more than once during a given period.

*1.* ***Repetitive Event Model.*** If a disease recurs rather frequently, as, for example, the common cold, a person may contract it, recover, become immune for a certain time, and contract it again during a relatively short period. If, for example, we count one year as the time unit, the same individual may become sick two, three, or even more times per year. Therefore, the number of cases occurring during a year may well exceed the total number of persons in the community.

This resembles the phenomenon of customers arriving for service (see Section 2.4) in which each customer is a "case," and over a certain period, the same person may become a customer several times.

We may define the annual incidence rate as

$$\begin{matrix}\text{annual}\\ \text{incidence}\\ \text{rate}\end{matrix} = \frac{\text{number of cases occurring during a period } T \text{ years}}{\begin{pmatrix}\text{average number of persons living}\\ \text{in the community during period } T\end{pmatrix} \times T}. \tag{2.35}$$

This can easily be greater than 1. Clearly, in this situation we are using the repetitive event approach discussed in Section 2.4. The denominator represents a target population for the attack of the disease; the midperiod population size of the community is a rough estimate of the "average population" in the denominator.

Even though a year as a unit of time is convenient, any other unit (e.g., month or week) can be used. We also may multiply (2.35) by $10^k$ (where $k$ is an integer), if the numerator is small, and obtain the incidence rate per $10^k$ persons, per time unit. We notice that the incidence rate defined in this way represents, in fact, the average number of disease attacks (cases) per person per year. Such a rate is useful for planning health care, predicting absenteeism from work, and so on. We also notice that the common cold model resembles the model of telephone calls per household per time unit.

Another example of a repetitive event type of incidence is cancer incidence. It is usually assumed that it is possible for a person with one type of cancer to get another type of cancer.

**Example 2.9** *Third National Cancer Survey.: Incidence Data National Cancer Institute* (1975).

The survey area consisted of seven metropolitan areas and two entire states. The information was collected on cancer diagnosed and/or treated during the 3-year period January 1, 1969–December 31, 1971. There were 58,586 cases of cancer diagnosed before January 1, 1969. If the size of the survey population at January 1, 1969 were known, we could estimate the prevalence of cancer.

The main objective of this survey was to provide information on the *incidence* of cancer in the United States, classified according to sex, race, and age. Individuals who developed more than one type of cancer were counted more than once, but there were no records as to whether a person who developed a cancer during the period 1969–1971, was free of, or had any cancer prior to 1969, that is, before the study started. In counting the "new cases" it was tacitly assumed that each person (whether free of or already having a cancer) is liable to get a new cancer. This implies that target population or population at risk was the entire population (cancer + noncancer persons). The number of all diagnosed cancers during the period 1969–1971 was 181,026, and the survey census population was 21,003.451. The crude annual incidence rate per 100,000 persons is

$$\frac{181{,}027}{21{,}003{,}451 \times 3} \times 100{,}000 = 287.3.$$

This means that, on the average, approximately 287 new cases per year of cancer will occur in 100,000 persons. Some persons develop more than one type of cancer.

**2.** ***Epidemics.*** A mass-time approach can be used when the disease under consideration is a contagious disease that spreads rapidly from person to person and has a relatively short duration period. Infected individuals become carriers, but are not themselves any longer exposed to risk of contracting the disease during the epidemic. An incidence rate (per week, or even per day) might be, perhaps, defined analogously to the death rate. If $f_A$ is the number of persons who contracted disease $A$ during a period of length $T$ time units, and $N_A$ is the initial population in the community exposed to disease $A$ (event), then the average incidence rate per person per time unit is calculated from

$$\begin{array}{c}\text{average}\\ \text{incidence}\\ \text{rate}\end{array} \doteq \frac{f_A}{\left(N_A - \frac{1}{2}f_A\right)\cdot T}. \tag{2.36}$$

Although, strictly speaking, $N_A$ should be evaluated at the date at which the epidemic started, in practice, it is only estimated. Also effects by migration are neglected by this formula.

We notice that if the epidemic spreads very rapidly, the average incidence rate defined in (2.36) will not be a good measure since, in fact, the incidence rate is a function of $t$. If we really wish to investigate the progress of the disease, rates for shorter time intervals should be calculated.

Clearly, the problems discussed in 1 and 2 are different. In the repetitive event model the population is a target population, whereas in the epidemic model it is a population for which the number of persons exposed to risk diminishes with time. Nevertheless, in both cases, we may express the incidence rate as

$$\begin{matrix}\text{average}\\ \text{incidence}\\ \text{rate}\end{matrix} = \frac{\text{no. of cases occurring during the period } T}{(\text{midperiod population}) \times T}, \qquad (2.37)$$

bearing in mind that the midperiod population should be appropriately defined for each particular problem.

*3.* ***Chronic Diseases. Age of First Occurrence.*** Chronic diseases require special attention. Some represent one-or-none phenomena: a person getting the disease becomes permanently affected (e.g., some genetic disorders). Others can be temporarily cured, and a new attack of the disease (e.g., some heart diseases) or a new form of the same disease (e.g. cancer) may occur. In such situations, it is very important to obtain data on the age of first occurrence (new cases). From such data, it is possible to construct the age of onset distributions that play significant roles in preventive medicine. This topic is discussed, in some detail, in Chapter 14.

## 2.12 Association Between Disease and Risk Factor. Relative Risk and Odds Ratio

In the present chapter, we have briefly discussed concepts and methods used in vital statistics and epidemiological studies, with emphasis on measures of mortality. We now mention two indices, widely used by epidemiologists in studies of diseases (mortality), and their association with risk factors.

Let $A$ denote a disease and $F$ a risk factor (attribute). Suppose that we have obtained a sample of $n$ individuals classified into four mutually exclusive classes. The numbers in the sample can be set out in a $2 \times 2$ table as in Table 2.4.

**Table 2.4**

| Risk factor | Disease $A$ | | Total |
|---|---|---|---|
| $F$ | $A$ | $\bar{A}$ | |
| $F$ | $n_{11}$ | $n_{12}$ | $n_{1\cdot}$ |
| $\bar{F}$ | $n_{21}$ | $n_{22}$ | $n_{2\cdot}$ |
| Total | $n_{\cdot 1}$ | $n_{\cdot 2}$ | $n$ |

***Notation.*** $F$=factor present; $\bar{F}$=factor absent; $A$=disease present; $\bar{A}$=disease absent; $n_{ij}$=observed frequency in an appropriate cell;

$$n_{i\cdot}=n_{i1}+n_{i2}, \qquad i=1,2,$$

$$n_{\cdot j}=n_{1j}+n_{2j}, \qquad j=1,2,$$

are marginal observed frequencies.

Such data can be obtained from three kinds of experimental design:

1. *Cross-sectional* study design, that is, when only $n$ is fixed; the distribution of $n_{ij}$'s is multinomial.
2. *Retrospective* study design, that is, the numbers of individuals with disease $A$ and controls (without disease $A$) are predetermined. In this design, the marginals $n_{\cdot 1}, n_{\cdot 2}$ are fixed, and $n_{11}, n_{12}$ are binomial variables from two independent samples.
3. *Prospective* study design, that is, the numbers of individuals with factor $F$ and controls (without factor $F$) are predetermined. They are followed to see whether disease $A$ occurs, or not. The marginals $n_{1\cdot}, n_{2\cdot}$ are fixed, and $n_{11}, n_{21}$ are binomial variables from two independent samples. For this design, we denote

$$\begin{aligned} &Pr\{A|F\}=P_1, \qquad Pr\{\bar{A}|F\}=Q_1,\\ &Pr\{A|\bar{F}\}=P_2, \qquad Pr\{\bar{A}|\bar{F}\}=Q_2, \end{aligned} \tag{2.38}$$

with $P_i+Q_i=1$, $i=1,2$.

### 2.12.1 Relative Risk

The probability $P_1$ is called the *risk* of getting disease $A$ when factor $F$ is present; $P_2$ is the risk of $A$ when factor $F$ is absent. The ratio

$$R=\frac{P_1}{P_2} \tag{2.39}$$

is called the *relative risk*, and its sample estimator is

$$\hat{R}=\frac{n_{11}/n_{1\cdot}}{n_{21}/n_{2\cdot}}. \tag{2.40}$$

It can be shown (see Exercise 3.9) that a reasonable estimator for the variance of $\hat{R}$ is

$$\text{est. var}(\hat{R}|n_{1\cdot},n_{2\cdot}) \doteq \hat{R}^2\frac{n_{12}n_{21}+n_{11}n_{22}}{n_{11}n_{21}}. \tag{2.41}$$

*Remark.* It is usually more reasonable to direct attention to the likely *increase* in probability of the disease when the risk factor is present rather than to the *ratio* of probabilities with and without risk factor. For example, suppose we have two samples with the following probabilities

| | Sample I | Sample II |
|---|---|---|
| With risk factor | 0.4 | 0.01 |
| Without risk factor | 0.2 | 0.002 |

The relative risk in situation I is 2, whereas in II it is 5; yet the effect of the risk factor, in terms of increased number of cases, is 25 times as much in I as it is in II.

### 2.12.2 Odds Ratio

The *odds* in the presence and absence of factor $F$ are defined as $P_1/Q_1$ and $P_2/Q_2$, respectively.

The *odds ratio* is

$$\omega = \frac{P_1/Q_1}{P_2/Q_2} = \frac{P_1 Q_2}{P_2 Q_1}, \tag{2.42}$$

and a natural estimator for $\omega$ is

$$\hat{\omega} = \frac{n_{11} \cdot n_{22}}{n_{12} \cdot n_{21}}. \tag{2.43}$$

The variance of $\hat{\omega}$ is estimated by

$$\text{est. var}(\hat{\omega}) \doteqdot \hat{\omega}^2 \left( \frac{1}{n_{11}} + \frac{1}{n_{12}} + \frac{1}{n_{21}} + \frac{1}{n_{22}} \right) \tag{2.44}$$

(see Exercise 3.10).

If the disease $A$ is independent of factor $F$, we should have $P_1 = P_2$, so that

$$R = \omega = 1. \tag{2.45}$$

Note that $\hat{\omega}$ can also be calculated from the results of retrospective studies whereas $\hat{R}$ cannot.

An excellent discussion on these (and some other) indices of association used in each of the three kinds of epidemiological study designs, with many references, is given by Walter (1976). Although relative risk and odds ratio are popular indices used in seeking for association between disease and risk factor, caution should be taken in their interpretation, since they easily can be misleading.

## REFERENCES


Batten, R. W. (1978). *Mortality Table Construction*. Prentice-Hall, Englewood Cliffs, NJ.

Dorn, M. F. (1955). Some applications in the collection and evaluation of medical data. *J. Chron. Dis.* **1**, 638–664.

Elandt-Johnson, R. C. (1975). Definition of rates: Some remarks on their use and misuse. *Amer. J. Epid.* **102**, 267–271.

Gaffey, W. R. (1975). A critique of the standardized mortality ratio. Paper presented at the Annual Meeting of the Occupational Medicine Association. San Francisco, April 17, 1975.

Gershenson, H. (1961). *Measurement of Mortality*. Society of Actuaries, Chicago.

Hill, A. B. (1956). *Principles of Medical Statistics*. Oxford University Press. Oxford.

Hill, I. D. (1972). Computing man years at risk. *Brit. J. Prev. Soc. Med.* **26**, 132–134.

Monson, R. R. (1974). Analysis of relative survival and proportional mortality. *Computers Biomed. Res.* **7**, 325–332.

National Cancer Institute (1975). *Third National Cancer Survey: Incidence Data*, Monograph No. 41, DHEW Publication No. (NIH) 75-787.

National Center for Health Statistics (1974). *Vital Statistics of the United States, 1970, Vol. II. Mortality. Part A*, Rockville, Md.

Shryock, H., Siegel, J. S., and Associates (1973). *The Methods and Materials in Demography*, Chapter 14. U. S. Department of Commerce, Washington, D.C.

Walter, S. D. (1976). The estimation and interpretation of attributable risk in health research. *Biometrics* **32**, 839–849.


**Table E2.1**

| Age group ($a$) | IBM employees $K_a$ | $D_a$ | $M'_{Sa}$ (per 100,000) |
|---|---|---|---|
| 15–20 | 179 | 0 | 149.2 |
| 20–25 | 8,198 | 5 | 189.9 |
| 25–30 | 28,998 | 19 | 164.9 |
| 30–35 | 26,361 | 16 | 181.5 |
| 35–40 | 21,780 | 33 | 256.1 |
| 40–45 | 15,945 | 21 | 408.2 |
| 45–50 | 10,635 | 37 | 658.7 |
| 50–55 | 6,345 | 53 | 1072.6 |
| 55–60 | 3,666 | 51 | 1718.4 |
| 60–65 | 1,904 | 39 | 2681.4 |
| Sum | 122,862 | 274 | |

## EXERCISES

**2.1.** The data displayed in Table E2.1, p. 38 [Miller and Pomper (1975) *J. Occup. Med.* **17**, 708] represent cross-sectional mortality data among IBM White Male employees in 1971. Using the death rates ($M'_{Sa}$), U. S. White Males, 1971 as standard rates, calculate the expected number of deaths for each group, and hence the SMR for these employees.

**2.2.** For data in Example 2.8 calculate the contributions to exposed to risk in 5-year age groups, 5-year calendar subperiods, and 1-year duration subperiods.

**2.3.** Consider three individuals $A$, $B$, and $C$ with recorded dates:

| Individual | Birth | Entry | Departure |
|---|---|---|---|
| $A$ | 2/18/25 | 1/1/50 | 9/1/78 |
| $B$ | 3/14/21 | 7/2/48 | 6/8/75 |
| $C$ | 12/ 1/30 | 3/15/45 | 4/6/72 |

(a) Draw a diagram and calculate the contributions of each of the individuals $A$, $B$, and $C$ to person-years exposed to risk in 5-year age groups and 5-year calendar periods. Find the amount of person-years in each (age group)×(calendar subperiod) combination.

(b) Repeat the calculations assuming that only integer years of birth, entry and departure are available.

**2.4.** Data in Table E2.2 [from Merrell and Schulman (1955). *J. Chron. Dis.* **1**, 12] represent observations on 99 patients affected with *Systemic Lupus Erythematosus*. There are 13 males and 86 females, but 1 female has unknown age. For analysis we use only the (complete) data on $N=85$ females (61 white and 24 black).

We note that: (*i*) age at diagnosis is given as age lbd, so that we use (age lbd$+0.5$) years as the recorded age; (*ii*) from the data, we can also obtain the age at onset, which is not the same as age at diagnosis, the latter representing age at entry. (Explain why we cannot use age at onset as age at entry.)

Divide the data for females into two groups: those patients for whom the period between onset and diagnosis was $\leqslant 5$ years, and those for whom this period was $>5$ years. For each group calculate the following:

(*a*) Age at diagnosis (i.e., age at entry).

(*b*) Age at onset.

**Table E2.2** Observations on the individual patients with SLE diagnosed at the Johns Hopkins Hospital, 1949–1953

| | | | | Onset to diagnosis | | Date of diagnosis | | Latest observation | | |
|---|---|---|---|---|---|---|---|---|---|---|
| | | | | | | | | Date | | |
| Case No. | Sex | Race | Age at diagnosis (year) | Years | Months | Month | Year | Month | Year | Status (living or dead) |
| 1 | F | N | 26 | 3 | 11 | 2 | 49 | 11 | 49 | L |
| 2 | F | N | 28 | 2 | 9 | 3 | 49 | 4 | 49 | D |
| 3 | F | N | 35 | 5 | 5 | 3 | 49 | 12 | 53 | L |
| 4 | F | W | 21 | 2 | 5 | 3 | 49 | 6 | 52 | D |
| 5 | F | W | 35 | 25 | | 3 | 49 | 9 | 53 | D |
| 6 | F | N | 31 | 2 | 7 | 4 | 49 | 5 | 49 | D |
| 7 | M | W | 47 | | 6 | 5 | 49 | 1 | 54 | L |
| 8 | F | N | 30 | 17 | | 6 | 49 | 6 | 49 | D |
| 9 | M | W | 33 | | 11 | 7 | 49 | 5 | 53 | L |
| 10 | F | W | 62 | | 8 | 8 | 49 | 8 | 50 | D |
| 11 | F | W | 43 | 1 | 4 | 8 | 49 | 1 | 54 | L |
| 12 | M | W | 37 | 1 | 6 | 9 | 49 | 8 | 50 | D |
| 13 | F | W | 60 | 2 | 7 | 10 | 49 | 1 | 50 | D |
| 14 | F | W | 14 | 3 | | 11 | 49 | 12 | 49 | D |
| 15 | F | W | 21 | 2 | 11 | 11 | 49 | 6 | 50 | D |
| 16 | F | W | 47 | 1 | 6 | 11 | 49 | 1 | 50 | D |
| 17 | M | W | 36 | 8 | | 1 | 50 | 5 | 53 | L |
| 18 | F | W | 30 | 3 | | 1 | 50 | 10 | 53 | L |
| 19 | F | W | 18 | 5 | 1 | 2 | 50 | 12 | 53 | L |
| 20 | F | W | 40 | 2 | 7 | 2 | 50 | 2 | 53 | L |

| | | | | | | | | | | |
|---|---|---|---|---|---|---|---|---|---|---|
| 21 | F | N | 19 | | 1 | 3 | 50 | 11 | 50 | D |
| 22 | F | W | 37 | | 1 | 3 | 50 | 6 | 53 | L |
| 23 | M | N | 50 | 1 | 10 | 5 | 50 | 9 | 53 | L |
| 24 | F | W | 62 | 8 | 11 | 5 | 50 | 5 | 50 | D |
| 25 | F | W | 30 | 6 | | 5 | 50 | 6 | 51 | L |
| 26 | F | W | 32 | 6 | | 5 | 50 | 1 | 54 | L |
| 27 | F | W | 32 | | 6 | 7 | 50 | 8 | 52 | D |
| 28 | M | W | 22 | | 10 | 7 | 50 | 7 | 50 | D |
| 29 | F | W | 29 | 1 | 2 | 7 | 50 | 3 | 53 | L |
| 30 | F | W | 21 | 1 | 3 | 7 | 50 | 11 | 53 | L |
| 31 | F | W | 29 | 7 | 7 | 7 | 50 | 11 | 53 | L |
| 32 | F | W | 71 | 4 | | 8 | 50 | 10 | 53 | D |
| 33 | F | N | 27 | | 11 | 10 | 50 | 7 | 52 | D |
| 34 | F | W | 61 | 35 | | 10 | 50 | 2 | 54 | L |
| 35 | M | W | 39 | 10 | | 10 | 50 | 8 | 52 | D |
| 36 | F | W | 19 | 2 | | 10 | 50 | 11 | 50 | L |
| 37 | F | W | 24 | 13 | 1 | 11 | 50 | 7 | 53 | D |
| 38 | F | W | 38 | 4 | 7 | 11 | 50 | 12 | 53 | L |
| 39 | F | W | 33 | 6 | 7 | 11 | 50 | 11 | 51 | D |
| 40 | F | N | 39 | 1 | 6 | 12 | 50 | 1 | 51 | D |
| 41 | M | W | 64 | | 9 | 2 | 51 | 3 | 51 | D |
| 42 | F | W | 18 | | 7 | 3 | 51 | 8 | 52 | D |
| 43 | F | N | 44 | 5 | | 4 | 51 | 1 | 54 | L |
| 44 | F | W | 40 | 7 | 10 | 4 | 51 | 7 | 51 | D |
| 45 | F | N | 37 | 7 | | 5 | 51 | 1 | 54 | L |
| 46 | M | W | 23 | 5 | 6 | 5 | 51 | 6 | 53 | D |
| 47 | M | W | 57 | | 5 | 6 | 51 | 5 | 53 | L |
| 48 | M | W | 51 | 1 | 5 | 6 | 51 | 1 | 54 | L |
| 49 | F | W | Unknown | 2 | | 6 | 51 | 2 | 54 | L |
| 50 | F | W | 45 | 6 | 11 | 6 | 51 | 7 | 53 | L |

**Table E2.2** (continued)

| Case | | | Age at | Onset to Diagnosis | | Date of Diagnosis | | Latest observation | | |
|---|---|---|---|---|---|---|---|---|---|---|
| | | | diagnosis | | | | | Date | | Status |
| No. | Sex | Race | (year) | Years | Months | Month | Year | Month | Year | (living or dead) |
| 51 | F | W | 50 | 20 | | 7 | 51 | 6 | 53 | L |
| 52 | M | W | 35 | 4 | 1 | 7 | 51 | 2 | 54 | L |
| 53 | F | N | 47 | | 2 | 8 | 51 | 10 | 51 | D |
| 54 | F | W | 37 | 2 | 1 | 8 | 51 | 3 | 53 | L |
| 55 | F | W | 49 | | 9 | 8 | 51 | 10 | 53 | L |
| 56 | F | N | 24 | 1 | 7 | 10 | 51 | 12 | 51 | D |
| 57 | F | W | 31 | 4 | 9 | 10 | 51 | 1 | 54 | L |
| 58 | F | W | 44 | 16 | | 10 | 51 | 7 | 53 | L |
| 59 | F | W | 44 | 3 | 5 | 11 | 51 | 8 | 53 | L |
| 60 | F | W | 13 | 1 | 6 | 12 | 51 | 4 | 52 | L |
| 61 | F | N | 34 | 4 | 5 | 3 | 52 | 12 | 53 | L |
| 62 | F | W | 39 | 12 | | 3 | 52 | 12 | 53 | L |
| 63 | F | W | 47 | 17 | | 4 | 52 | 2 | 54 | L |
| 64 | F | N | 43 | 1 | 11 | 5 | 52 | 4 | 53 | D |
| 65 | F | W | 24 | 4 | 11 | 5 | 52 | 9 | 52 | L |
| 66 | F | N | 48 | 16 | | 6 | 52 | 2 | 54 | L |
| 67 | M | W | 44 | | 8 | 6 | 52 | 2 | 53 | L |
| 68 | F | W | 21 | 3 | | 6 | 52 | 10 | 53 | L |
| 69 | F | W | 27 | 5 | | 6 | 52 | 10 | 53 | L |
| 70 | F | W | 34 | 2 | | 6 | 52 | 4 | 53 | L |

| 71 | F | W | 39 | 16 |  | 6 | 52 | 1 | 54 | L |
|---|---|---|---|---|---|---|---|---|---|---|
| 72 | F | N | 30 |  | 4 | 8 | 52 | 8 | 53 | D |
| 73 | F | W | 27 |  | 7 | 8 | 52 | 7 | 53 | L |
| 74 | F | W | 40 | 18 |  | 9 | 52 | 12 | 53 | L |
| 75 | F | W | 13 | 1 | 2 | 10 | 52 | 7 | 53 | L |
| 76 | F | W | 23 | 1 |  | 10 | 52 | 7 | 53 | L |
| 77 | F | W | 43 | 2 | 3 | 10 | 52 | 1 | 54 | L |
| 78 | F | W | 47 | 2 | 8 | 10 | 52 | 1 | 54 | L |
| 79 | F | N | 15 |  | 1 | 11 | 52 | 2 | 54 | L |
| 80 | F | W | 24 | 3 | 2 | 12 | 52 | 2 | 54 | L |
| 81 | F | W | 30 | 1 | 11 | 1 | 53 | 12 | 53 | D |
| 82 | F | N | 46 | 1 |  | 4 | 53 | 10 | 53 | L |
| 83 | F | N | 51 |  | 8 | 4 | 53 | 1 | 54 | L |
| 84 | F | W | 18 |  | 1 | 4 | 53 | 1 | 54 | L |
| 85 | F | W | 25 |  | 5 | 4 | 53 | 11 | 53 | L |
| 86 | F | W | 34 |  | 5 | 4 | 53 | 6 | 53 | L |
| 87 | F | N | 34 | 1 | 10 | 5 | 53 | 2 | 54 | L |
| 88 | F | W | 41 | 7 |  | 5 | 53 | 1 | 54 | L |
| 89 | F | W | 40 | 10 |  | 5 | 53 | 1 | 54 | L |
| 90 | F | N | 42 | 1 | 11 | 6 | 53 | 1 | 54 | L |
| 91 | F | W | 45 | 1 | 4 | 7 | 53 | 7 | 53 | L |
| 92 | F | N | 31 | 1 | 6 | 9 | 53 | 2 | 54 | L |
| 93 | F | N | 37 | 2 | 2 | 9 | 53 | 1 | 54 | L |
| 94 | F | W | 56 | 4 |  | 9 | 53 | 1 | 54 | L |
| 95 | F | W | 12 |  | 3 | 9 | 53 | 2 | 54 | L |
| 96 | F | W | 41 | 5 | 3 | 9 | 53 | 1 | 54 | L |
| 97 | F | N | 28 | 2 | 2 | 10 | 53 | 1 | 54 | L |
| 98 | F | N | 41 | 10 |  | 10 | 53 | 2 | 54 | L |
| 99 | F | W | 44 | 15 |  | 12 | 53 | 2 | 54 | D |

(*c*) Age of the last observation (i.e., age at departure).
(*d*) The appropriate contribution to exposed to risk in combination (5-year age group)×(1-year duration subperiod).
(*e*) Using the central death rates ($_nm_x$) from the U. S. Life Table (1950) as the "standard" death rate for Females (Table E2.3.), obtain the expected numbers of deaths in each (age×duration)−cell, and calculate the standardized mortality ratio (SMR).
(*f*) Compare the observed and expected numbers of deaths in each group, regardless of duration.

**Table E2.3** U. S. Life Table, 1950, Females

| Age ($x$) | $_nq_x$ | $l_x$ | $_nd_x$ | $_nL_x$ | $_nm_x$ | $\mathring{e}_x$ |
|---|---|---|---|---|---|---|
| 0–1 | 0.02766 | 100000 | 2766 | 97561 | 0.02835 | 71.0 |
| 1–5 | 0.00502 | 97234 | 488 | 387716 | 0.00126 | 72.0 |
| 5–10 | 0.00258 | 96746 | 250 | 483105 | 0.00052 | 68.4 |
| 10–15 | 0.00223 | 96496 | 215 | 481966 | 0.00045 | 63.6 |
| 15–20 | 0.00379 | 96281 | 365 | 480547 | 0.00076 | 58.7 |
| 20–25 | 0.00498 | 95916 | 478 | 478430 | 0.00100 | 53.9 |
| 25–30 | 0.00610 | 95438 | 582 | 475795 | 0.00122 | 49.2 |
| 30–35 | 0.00809 | 94856 | 767 | 472469 | 0.00162 | 44.5 |
| 35–40 | 0.01160 | 94089 | 1091 | 467895 | 0.00233 | 39.8 |
| 40–45 | 0.01743 | 92998 | 1621 | 461201 | 0.00352 | 35.2 |
| 45–50 | 0.02577 | 91377 | 2355 | 451359 | 0.00522 | 30.8 |
| 50–55 | 0.03768 | 89022 | 3354 | 437217 | 0.00767 | 26.6 |
| 55–60 | 0.05505 | 85668 | 4716 | 417262 | 0.01130 | 22.5 |
| 60–65 | 0.08367 | 80952 | 6773 | 388726 | 0.01742 | 18.7 |
| 65–70 | 0.12175 | 74179 | 9031 | 349566 | 0.02584 | 15.1 |
| 70–75 | 0.19594 | 65148 | 12765 | 255190 | 0.04324 | 11.9 |
| 75–80 | 0.29723 | 52383 | 15570 | 223652 | 0.06962 | 9.1 |
| 80–85 | 0.43311 | 36813 | 15944 | 143768 | 0.11090 | 6.9 |
| 85+ | 1.00000 | 20869 | 20869 | 109437 | 0.19069 | 5.2 |

From Preston et al. (1972).

**2.5.** (*i*) The following table gives the ratios, for a recent year, obtained by dividing the mortality rates for females of the stated marital status by the mortality rates for all females in the same age group irrespective of marital status. The ratios for divorced women have not been shown.

| Age group | Single | Married | Widowed |
|---|---|---|---|
| 15–24 | 1.10 | 0.75 | 10.00 |
| 25–34 | 1.90 | 0.85 | 4.00 |
| 35–44 | 1.70 | 0.90 | 1.75 |
| 45–54 | 1.35 | 0.93 | 1.50 |
| 55–64 | 1.15 | 0.93 | 1.30 |
| 65–74 | 1.00 | 0.93 | 1.10 |
| 75–84 | 0.98 | 0.93 | 1.05 |

Discuss possible reasons for the patterns shown by these figures.

(*ii*) A student has been asked how he would project the proportion married in each of the next 40 years for females now aged 15, assuming that he has access to any statistics he likes relating to past experience by single ages. He replies as follows:

> 'I would obtain marriage rates for the latest available year by single ages for each status. I would similarly obtain rates of widowhood and of divorce for married women. I would then simply apply all these rates year by year, starting off with a proportion single of 1.00 at age 15, and calculating the proportion in each status at each successive age.'

Criticize this reply. *(Institute of Actuaries, 1975)*

**2.6.** The personnel manager of a company wishes to compare the losses of staff from its subsidiary with those from the parent company. Three single figure indices have been suggested as providing a suitable method of making the comparison:

(*a*) The rate of turnover of staff, defined as the total number of staff leaving during the year divided by the mean number employed.

(*b*) Two indices, calculated in the same manner as the standardized mortality ratio, one based on the data subdivided by age and the other based on the data subdivided by duration of service.

(*i*) Calculate these three indices from the data given below.

(*ii*) Discuss the results obtained in (*i*).

(*iii*) What further enquiries and calculations, if any, would you make before coming to any conclusions?

| Age | Mean numbers employed | Numbers leaving | Duration of service | Mean numbers employed | Numbers leaving |
|---|---|---|---|---|---|
| | | *Data for parent company* | | | |
| 25–34 | 25,000 | 5000 | 0–4 | 19,000 | 3800 |
| 35–44 | 10,000 | 700 | 5–9 | 11,000 | 1650 |
| 45–54 | 8,000 | 120 | 10–14 | 6,000 | 300 |
| 55–64 | 7,000 | 70 | 15 & over | 14,000 | 140 |
| | 50,000 | 5890 | | 50,000 | 5890 |
| | | *Data for subsidiary company* | | | |
| 25–34 | 1500 | 270 | 0–4 | 4700 | 850 |
| 35–44 | 3200 | 480 | 5–9 | 1900 | 250 |
| 45–54 | 3000 | 320 | 10–14 | 1100 | 50 |
| 55–64 | 2300 | 110 | 15 & over | 2300 | 30 |
| | 10,000 | 1180 | | 10,000 | 1180 |

[*Institute of Actuaries (1973)*]

**2.7.** Tables E2.4 and E2.5 are parts of tables in the Report from Select Committee on Life Annuities (J. Finlaison, London House of Commons, 1829). The data in Table E2.4 relate to an English tontine started in 1693. Ages lbd at entrance and death are given for every member. The last member died in 1783. The data in Table E2.5 relate to three Irish tontines started in 1773, 1775, and 1778, respectively. Ages lbd at entrance are given for every member and also:

(*i*) at death of those members who died before January 1, 1826, and
(*ii*) on January 1, 1826 of those members then still alive.

(*a*) In what sense can the data recorded in Tables E2.4 and E2.5 be regarded as cohort data?
(*b*) Do the data in these tables constitute complete mortality data?

**2.8.** Obtain recurrence formulae relating the (central) exposed to risk in successive yearly age groups for the data in Tables E2.4 and E2.5 on the four possible combinations of assumptions set out below:

Enrollment:

(*i*) "Age $x$" means exact age $x$, that is, enrollment takes place only on birthdays.
(*ii*) "Age $x$ lbd" means "exact" age $(x+\frac{1}{2})$.

**Table E2.4** English tontine

| | Males | | Females | | | Males | | Females | | | Males | Females |
|---|---|---|---|---|---|---|---|---|---|---|---|---|
| Age | (a) | (b) | (a) | (b) | Age | (a) | (b) | (a) | (b) | Age | (b) | (b) |
| 0 | 20 | 1 | 7 | — | 30 | 6 | 11 | 5 | 7 | 60 | 7 | 5 |
| 1 | 20 | 2 | 10 | — | 31 | 2 | 11 | — | 8 | 61 | 9 | 6 |
| 2 | 27 | — | 23 | 2 | 32 | 2 | 5 | — | 5 | 62 | 10 | 3 |
| 3 | 34 | — | 24 | 1 | 33 | 1 | 9 | — | 3 | 63 | 11 | 4 |
| 4 | 34 | 1 | 25 | — | 34 | 2 | 5 | 1 | 4 | 64 | 3 | 5 |
| 5 | 43 | 1 | 24 | — | 35 | 2 | 9 | 1 | 6 | 65 | 10 | 5 |
| 6 | 33 | 3 | 23 | 2 | 36 | 3 | 4 | 2 | 3 | 66 | 13 | 3 |
| 7 | 29 | 1 | 20 | 3 | 37 | — | 7 | 1 | 7 | 67 | 6 | 7 |
| 8 | 45 | 3 | 19 | 3 | 38 | 1 | 10 | — | 6 | 68 | 9 | 9 |
| 9 | 33 | 4 | 26 | 2 | 39 | 1 | 8 | — | 5 | 69 | 6 | 8 |
| 10 | 38 | 2 | 30 | 4 | 40 | 2 | 13 | 1 | 7 | 70 | 11 | 9 |
| 11 | 28 | 2 | 23 | 2 | 41 | 1 | 10 | 1 | — | 71 | 6 | 6 |
| 12 | 27 | 2 | 24 | 1 | 42 | — | 7 | — | 4 | 72 | 11 | 12 |
| 13 | 28 | 1 | 12 | 3 | 43 | — | 8 | — | 3 | 73 | 5 | 4 |
| 14 | 20 | 4 | 18 | — | 44 | — | 2 | — | 4 | 74 | 5 | 5 |
| 15 | 16 | 4 | 19 | 2 | 45 | 1 | 6 | — | 7 | 75 | 6 | 6 |
| 16 | 15 | 10 | 9 | 4 | 46 | 1 | 8 | — | 6 | 76 | 11 | 8 |
| 17 | 10 | 4 | 3 | 1 | 47 | * | 6 | — | 10 | 77 | 7 | 6 |
| 18 | 10 | 8 | 12 | 4 | 48 | | 11 | — | 7 | 78 | 8 | 3 |
| 19 | 6 | 7 | 8 | 5 | 49 | | 10 | 1 | 5 | 79 | 6 | 3 |
| 20 | 8 | 11 | 7 | 1 | 50 | | 8 | — | 3 | 80 | 6 | 10 |
| 21 | 10 | 19 | 5 | 4 | 51 | | 12 | 1 | 3 | 81 | 5 | 7 |
| 22 | 8 | 6 | 5 | 4 | 52 | | 5 | * | 9 | 82 | 1 | 3 |
| 23 | 6 | 15 | 5 | 6 | 53 | | 9 | | 7 | 83 | 2 | 8 |
| 24 | 9 | 10 | 2 | 6 | 54 | | 2 | | 11 | 84 | 4 | 3 |
| 25 | 2 | 11 | 2 | 8 | 55 | | 8 | | 5 | 85 | 1 | 2 |
| 26 | 3 | 12 | 3 | 3 | 56 | | 9 | | 4 | 86 | 4 | 2 |
| 27 | 5 | 11 | 3 | 5 | 57 | | 7 | | 9 | 87 | 1 | 1 |
| 28 | 1 | 10 | 3 | 2 | 58 | | 2 | | 4 | 88 | — | 1 |
| 29 | 1 | 8 | — | 6 | 59 | | 12 | | 6 | 89 | — | 1 |
| | | | | | | | | | | 90 | 2† | 1† |

(a) Number enrolled at each age. (b) Number who died at each age.
*There are no (a) entries for males after age 46, or for females after age 51.
†There are additional (b) entries, '1' each, for males, age 93 and females, age 100.

**Table E2.5** Irish tontine

| | Males | | Females | | | Males | | | Females | | | | Males | | Females | |
|---|---|---|---|---|---|---|---|---|---|---|---|---|---|---|---|---|
| Age | (a) | (b) | (a) | (b) | Age | (a) | (b) | (c) | (a) | (b) | (c) | Age | (b) | (c) | (b) | (c) |
| 0 | 36 | 3 | 29 | — | 30 | 15 | 16 | | 29 | 16 | | 60 | 15 | 23 | 19 | 40 |
| 1 | 55 | 2 | 67 | — | 31 | 14 | 12 | | 16 | 22 | | 61 | 12 | 23 | 15 | 34 |
| 2 | 72 | 1 | 101 | 4 | 32 | 11 | 10 | | 17 | 21 | | 62 | 15 | 22 | 17 | 26 |
| 3 | 70 | 1 | 104 | 5 | 33 | 11 | 11 | | 16 | 14 | | 63 | 15 | 20 | 13 | 27 |
| 4 | 87 | 2 | 106 | 1 | 34 | 14 | 7 | | 14 | 11 | | 64 | 8 | 12 | 20 | 28 |
| 5 | 73 | 1 | 115 | 6 | 35 | 10 | 15 | | 9 | 14 | | 65 | 15 | 6 | 19 | 14 |
| 6 | 58 | 3 | 112 | 4 | 36 | 9 | 17 | | 13 | 21 | | 66 | 8 | 13 | 21 | 29 |
| 7 | 61 | 2 | 89 | 6 | 37 | 6 | 15 | | 10 | 21 | | 67 | 17 | 13 | 9 | 18 |
| 8 | 70 | 1 | 91 | 1 | 38 | 1 | 11 | | 9 | 18 | | 68 | 11 | 5 | 17 | 19 |
| 9 | 58 | 7 | 85 | 2 | 39 | 3 | 12 | | 7 | 11 | | 69 | 5 | 9 | 13 | 11 |
| 10 | 44 | 5 | 82 | 4 | 40 | 42 | 14 | | 22 | 24 | | 70 | 15 | 11 | 10 | 18 |
| 11 | 57 | 2 | 63 | 4 | 41 | 33 | 9 | | 14 | 13 | | 71 | 15 | 7 | 20 | 11 |
| 12 | 56 | 7 | 80 | 5 | 42 | 19 | 13 | | 19 | 21 | | 72 | 13 | 6 | 11 | 13 |
| 13 | 42 | 7 | 62 | 6 | 53 | 25 | 16 | | 18 | 15 | | 73 | 16 | 4 | 10 | 10 |
| 14 | 34 | 4 | 59 | 9 | 44 | 16 | 19 | | 18 | 19 | | 74 | 9 | 6 | 11 | 11 |
| 15 | 35 | — | 58 | 9 | 45 | 11 | 11 | * | 22 | 19 | * | 75 | 6 | 5 | 22 | 7 |
| 16 | 30 | 7 | 45 | 7 | 46 | 9 | 13 | 6 | 9 | 10 | 15 | 76 | 4 | 4 | 9 | 5 |
| 17 | 25 | 10 | 40 | 9 | 47 | 8 | 10 | 10 | 4 | 23 | 23 | 77 | 3 | 1 | 13 | 7 |
| 18 | 21 | 7 | 31 | 13 | 48 | 3 | 15 | 21 | 4 | 20 | 35 | 78 | 4 | 4 | 8 | 4 |
| 19 | 18 | 11 | 27 | 11 | 49 | 6 | 16 | 25 | 13 | 24 | 43 | 79 | 9 | 1 | 10 | 7 |
| 20 | 32 | 18 | 42 | 10 | 50 | 4 | 14 | 29 | 5 | 11 | 35 | 80 | 9 | 3 | 12 | 3 |
| 21 | 35 | 13 | 46 | 17 | 51 | 1 | 15 | 31 | 4 | 15 | 53 | 81 | 9 | 3 | 5 | 5 |
| 22 | 21 | 19 | 45 | 12 | 52 | 3 | 20 | 20 | 4 | 18 | 59 | 82 | 3 | 1 | 6 | 2 |
| 23 | 24 | 14 | 29 | 14 | 53 | 2 | 23 | 36 | — | 15 | 54 | 83 | 2 | 2 | 3 | — |
| 24 | 18 | 11 | 26 | 18 | 54 | 2 | 17 | 38 | 2 | 15 | 57 | 84 | 4 | — | 3 | — |
| 25 | 18 | 14 | 23 | 17 | 55 | — | 16 | 29 | — | 12 | 60 | 85 | 3 | 1 | 2 | 1 |
| 26 | 18 | 13 | 34 | 13 | 56 | — | 13 | 27 | 1 | 14 | 62 | 86 | 6 | — | 2 | 1 |
| 27 | 10 | 14 | 25 | 10 | 57 | 1 | 16 | 36 | — | 13 | 49 | 87 | 2 | 1 | 2 | 1 |
| 28 | 12 | 10 | 26 | 15 | 58 | — | 11 | 29 | — | 13 | 49 | 88 | 1 | — | 3 | 1 |
| 29 | 16 | 10 | 17 | 13 | 59 | — | 16 | 21 | 1 | 15 | 41 | 89 | 1 | — | 2 | — |
| | | | | | | † | | | † | | | 90 | 2 | 1 | 2† | 1 |

(a) Number enrolled at each age. (b) Number who died at each age. (c) Number alive after passing each age, in January 1826.

*There are no (c) entries below age 46.

†Additional entries under (a)—'1' for males, age 61 and females, ages 63 and 68
under (b)—'1' for females, age 93.

Date of enrollment:

(*i*) July 1.

(*ii*) Uniformly spaced over calendar year.

**2.9.** Calculate central exposed to risk for the four sets of data (two for males and two for females) in Tables E2.4 and E2.5 on the four bases described in Exercise 2.8.

CHAPTER 3

# Survival Distributions

## 3.1 INTRODUCTION

We assume that readers of this book are already familiar with basic concepts of probability theory and statistical methods. In this chapter, we briefly review some concepts and terminology, with special relevance to survival analysis.

## 3.2 SURVIVAL DISTRIBUTION FUNCTION

Suppose the random variable $X$ denotes the *lifetime* of a living organism or an inanimate device. It is also called *age at death* or *age at failure* or, briefly, *age*. Of course, other symbols can be used. For example, we use the symbol $T$ to denote time elapsed between some specified event (e.g., inclusion into a study) and the time of death or failure.

The cumulative distrubution function (CDF)

$$F_X(x) = \Pr\{X \leqslant x\} \tag{3.1}$$

is called the *lifetime distribution* or *failure distribution*. If $X$ represents age of first occurrence of a certain event (e.g., chronic disease), then $F_X(x)$ represents *age of onset distribution* of the event (disease) (see Chapter 14).

In survival analysis, however, the complementary function,

$$S_X(x) = 1 - F_X(x) = \Pr\{X > x\} \tag{3.2}$$

is more commonly used. It is called the *survival distribution function* (SDF) or briefly, the *survival function* (SF).

In survival analysis applications, usually $X$ cannot be negative, and so $S_X(0)=1$ [i.e., $F_X(0)=0$]. The probability density function (PDF)

$$f_X(x)=\frac{dF_X(x)}{dx}=-\frac{dS_X(x)}{dx} \tag{3.3}$$

is sometimes called the *curve of deaths* in survival analysis. By definition, $f_X(x)$ is, in fact, an absolute instantaneous rate of dying (compare Section 2.3.1).

## 3.3 HAZARD FUNCTION (FORCE OF MORTALITY)

The instantaneous (relative) failure rate at time point $x$ is

$$\lambda_X(x)=\frac{f_X(x)}{S_X(x)}=-\frac{d\log S_X(x)}{dx} \tag{3.4}$$

(cf. Section 2.3.2).

In demography and actuarial science, it is called the *force of mortality* (and is denoted by $\mu_x$); in reliability theory, the terms *hazard rate* (HR) or *intensity rate* are well established. Although we are mostly dealing with mortality data, "hazard rate" will be the term most often used in this book.

The hazard rate defined in (3.4) considered as a function of $x$, is the *hazard rate function* (HRF).

On the other hand,

$$\Lambda_X(x)=\int_0^x \lambda_X(u)\,du=-\log S_X(x) \tag{3.5}$$

is called the *cumulative hazard function* (note: no term "rate" is used here).

The conditional probability of failure in $(x,x+dx)$, given surviving till age $x$, is approximately $\lambda_X(x)\,dx$, whereas the unconditional probability of failure in $(x,x+dx)$ is approximately

$$f_X(x)\,dx=S_X(x)\lambda_X(x)\,dx. \tag{3.6}$$

Hence, the PDF is

$$f_X(x)=\lambda_X(x)S_X(x). \tag{3.6a}$$

The SDF can be expressed in two forms:

1. From (3.5), we have

$$S_X(x)=\exp\left[-\int_0^x \lambda_X(u)\,du\right]=\exp[-\Lambda_X(x)]. \tag{3.7}$$

Noticing that lifetime cannot be negative, CDF's (or SDF's) are called *proper*, if

$$\left.\begin{aligned}&\lim_{x\to 0} F_X(x)=0\\ \text{and}\quad &\\ &\lim_{x\to\infty} F_X(x)=1\end{aligned}\right\}\quad \text{or equivalently}\quad \left.\begin{aligned}&\lim_{x\to 0} S_X(x)=1\\ \text{and}\quad &\\ &\lim_{x\to\infty} S_X(x)=0\end{aligned}\right\}. \tag{3.8}$$

Using (3.7), we note that for $S_X(x)$ to be a proper survival function, we must have

$$\lim_{x\to\infty} \Lambda_X(x)=\lim_{x\to\infty}\int_0^x \lambda_X(u)\,du=\infty,$$

that is, the CHF must diverge.

2. On the other hand, from (3.6), the SDF can also be defined as

$$S_X(x)=\int_x^\infty f_X(u)\,du=\int_x^\infty \lambda_X(u)S_X(u)\,du. \tag{3.9}$$

Note that the SDF in (3.7) is expressed in terms of past lifetime, whereas the SDF in (3.9) is expressed in terms of future lifetime.

## 3.4 CONDITIONAL PROBABILITIES OF DEATH (FAILURE) AND CENTRAL RATE

In the analysis of mortality data, the conditional probability of death (failure) in the age interval $(x, x+t)$, given alive at age $x$, plays an essential role. In actuarial literature it is denoted by ${}_tq_x$. The prefix $t$ denotes the length of the period over which this probability is calculated. The notation $q(x, x+t)$ seems to be more appropriate (compare notation for average rates in Sections 2.3.1 and 2.3.2). However, we retain the traditional notation, ${}_tq_x$. For $t=1$, it is customary to use $q_x$ rather than ${}_1q_x$.

We have

$$
{}_tq_x = \Pr\{x < X \leqslant x+t \mid X > x\} = \frac{\int_x^{x+t} f_X(u)\,du}{\int_x^{\infty} f_X(u)\,du}
$$

$$
= \frac{\int_x^{x+t} \lambda_X(u) S_X(u)\,du}{\int_x^{\infty} \lambda_X(u) S_X(u)\,du} = \frac{S_X(x) - S_X(x+t)}{S_X(x)}. \tag{3.10}
$$

Note that $S_X(x) - S_X(x+t)$ is the proportion of deaths between age $x$ and $x+t$. On the other hand, the central rate—denoted usually by ${}_tm_x$—is

$$
{}_tm_x = \frac{\int_x^{x+t} \lambda_X(u) S_X(u)\,du}{\int_x^{x+t} S_X(u)\,du} = \frac{S_X(x) - S_X(x+t)}{\int_x^{x+t} S_X(u)\,du} \tag{3.11}
$$

(cf. Section 2.3.2). The conditional probability of surviving the age interval $x$ to $x+t$, given alive at age $x$, is

$$
{}_tp_x = 1 - {}_tq_x = \frac{S_x(x+t)}{S_x(x)}. \tag{3.10a}
$$

For a set of points $x_0 = 0 < x_1 < x_2 < \cdots < x_k$ with $x_{i+1} - x_i = h_i$, we have

$$
S_X(x_k) = \exp\left[ -\int_0^{x_k} \lambda_X(u)\,du \right] = \exp\left[ -\sum_{i=0}^{k-1} \int_{x_i}^{x_{i+1}} \lambda_X(u)\,du \right],
$$

$$
= \prod_{i=0}^{k-1} \exp\left[ -\int_{x_i}^{x_{i+1}} \lambda_X(u)\,du \right] = \prod_{i=0}^{k-1} {}_{h_i}p_{x_i} = \prod_{i=0}^{k-1} (1 - {}_{h_i}q_{x_i}). \tag{3.12}
$$

Formula (3.12) is very important in the construction of life tables (Chapter 4).

## 3.5 TRUNCATED DISTRIBUTIONS

Mortality data are often collected over a restricted period of time. This is so, for example, in clinical trials, when there is a fixed termination date for the trial, and in studies of insurance data or failure data of mechanical devices. Appropriate models for such situations often require the use of *truncated* distributions.

Suppose that we can observe the value of a random variable $X$ only over a range $a<x\leqslant b$. If $F_X(x)$ is the original (untruncated) CDF of $X$, then the truncated CDF is

$$F_X(x|a<X\leqslant b)=\begin{cases} 0 & \text{for } x\leqslant a, \\ \dfrac{F_X(x)-F_X(a)}{F_X(b)-F_X(a)} & \text{for } a<x\leqslant b, \\ 1 & \text{for } x>b. \end{cases} \tag{3.13}$$

In terms of SDF, we have

$$S_X(x|a<X\leqslant b)=\begin{cases} 1 & \text{for } x\leqslant a, \\ \dfrac{S_X(x)-S_X(b)}{S_X(a)-S_X(b)} & \text{for } a<x\leqslant b, \\ 0 & \text{for } x>b. \end{cases} \tag{3.13a}$$

Truncation by exclusion of values less than $a$ is called *truncation from below*, or *left-hand truncation*; truncation by exclusion values greater than $b$ is called *truncation from above* or *right-hand truncation*. In (3.13) and (3.13a) we have a double truncation.

If the original distribution has PDF $f_X(x)$, then the truncated distribution has PDF

$$f_X(x|a<X\leqslant b)=\begin{cases} \dfrac{f_X(x)}{F_X(b)-F_X(a)}=\dfrac{f_X(x)}{S_X(a)-S_X(b)} & \text{for } a<x\leqslant b, \\ 0 & \text{elsewhere.} \end{cases} \tag{3.14}$$

The hazard function is

$$\lambda_X(x|a<X\leqslant b)=\frac{f_X(x|a<X\leqslant b)}{S_X(x|a<X\leqslant b)}=\frac{f_X(x)}{S_X(x)-S(b)}. \tag{3.15}$$

This can also be written as

$$\lambda_X(x|a<X\leqslant b)=\frac{f_X(x)}{S_X(x)}\cdot\frac{S_X(x)}{S_X(x)-S_X(b)}=\lambda_X(x)\frac{S_X(x)}{S_X(x)-S_X(b)}. \tag{3.15a}$$

Note that the ratio of hazard functions for the truncated and untruncated distributions depends only on $b$, the upper truncation point, and not on $a$, the lower truncation point.

If the distribution is truncated only from below (left-hand truncation), we take $b=\infty$ [and so $S_X(b)=0$], and obtain

$$\lambda_X(x|X>a)=\lambda_X(x). \tag{3.16}$$

Note that the left-hand truncation does not change the hazard function, whereas right-hand truncation increases the hazard function.

## 3.6 EXPECTATION AND VARIANCE OF FUTURE LIFETIME

The (unconditional) *expected value* of a continuous random variable $X$ is

$$\mathcal{E}(X)=\int_{-\infty}^{\infty} xf_X(x)\,dx=\int_{-\infty}^{\infty} x\lambda_X(x)S_X(x)\,dx, \tag{3.17}$$

and the (unconditional) variance is

$$\begin{aligned}\mathrm{var}(X)&=\mathcal{E}\{[X-\mathcal{E}(X)]^2\}=\mathcal{E}(X^2)-[\mathcal{E}(X)]^2,\\ &=\int_{-\infty}^{\infty} x^2f_X(x)\,dx-[\mathcal{E}(X)]^2.\end{aligned} \tag{3.18}$$

If, as is often the case in survival analysis, $X$ cannot be negative, so that for $x<0$, $f_X(x)=0$, then the lower limit of the integrals in (3.17) and (3.18) can be taken as zero. When $X$ is discrete, the integration sign is replaced by summation sign. These two unconditional moments are not much used in survival analysis; more meaningful are certain conditional means and variances.

In addition to the random variable $X$ $(>0)$ representing age, we introduce a new random variable $T$ $(>0)$, representing time elapsed beyond attainment of a given age $x$. The conditional distribution of $T$ given $x$ is the *distribution of future lifetime*. Its PDF is

$$f_T(t)=\frac{f_X(x+t)}{S_X(x)}, \qquad t>0, \tag{3.19}$$

and its SDF is

$$S_T(t)=\frac{S_X(x+t)}{S_X(x)}, \qquad t>0, \tag{3.19a}$$

Hence, the corresponding hazard rate is

$$\lambda_T(t)=\lambda_X(x+t), \qquad t>0. \tag{3.19b}$$

The expected value of $T$ is called the (conditional) expectation of future lifetime (at age $x$) or *life expectancy* of an individual of age $x$. This is denoted by $\mathring{e}_x$.

We then have

$$\mathring{e}_x=\mathcal{E}(T)=\mathcal{E}(X-x|X>x)=\int_0^{\infty} tf_T(t)\,dt=\int_0^{\infty} t\lambda_T(t)S_T(t)\,dt. \tag{3.20}$$

In view of (3.19) and (3.19b), this can be written as

$$\mathring{e}_x=\frac{1}{S_X(x)}\int_0^{\infty} tf_X(x+t)\,dt=\frac{1}{S_X(x)}\int_0^{\infty} t\lambda_X(x+t)S_X(x+t)\,dt. \tag{3.20a}$$

The average age at death is, of course,

$$x+\mathring{e}_x. \tag{3.21}$$

The variance of $T$ is

$$\begin{aligned}\operatorname{var}(T)=\operatorname{var}(X|X>x)&=\int_0^{\infty} t^2f_T(t)\,dt-\mathring{e}_x^2\\ &=\frac{1}{S_X(x)}\int_0^{\infty} t^2f_X(x+t)\,dt-\mathring{e}_x^2.\end{aligned} \tag{3.22}$$

If we take $x=0$, we obtain the unconditional expected lifetime and variance of the lifetime for a newborn individual.

## 3.7 MEDIAN OF FUTURE LIFETIME

We may ask the question: for a given age $x$, what would be age $y$, say, up to which there is 50% chance of surviving, assuming that $S_X(x)$ is the survival distribution? The difference $y-x=Me_x$, say, is the *median future lifetime*. Clearly, $y$ can be obtained by solving the equation

$$\frac{S_X(y)}{S_X(x)}=0.50. \tag{3.23}$$

In a similar way, for a given age $x$, we can obtain the age $y_p$, say, up to which there is $100p\%$ chance of surviving, by solving the equation

$$\frac{S_X(y_p)}{S_X(x)}=p. \tag{3.24}$$

## 3.8 TRANSFORMATIONS OF RANDOM VARIABLES

It is sometimes convenient to relate the distribution of a random variable $X$ to some other distribution. An important consideration is that the latter may be better known and, perhaps, values of its CDF may be tabulated. The most common way of establishing a useful relationship is to seek a *transformation* of $x$ to some other variable $y$

$$y=g(x), \tag{3.25}$$

where $g(x)$ is a monotonic (increasing or decreasing) function of $x$, so that (3.25) is a one-to-one transformation, and the inverse function

$$x=g^{-1}(y)=h(y) \tag{3.26}$$

exists. If the transformation is increasing (so that $dx/dy>0$), then we have

$$F_Y(y)=F_X(h(y)), \qquad S_Y(y)=S_X(h(y))$$

and

$$f_Y(y)=f_X(h(y))\frac{dx}{dy}. \tag{3.27}$$

The hazard rate function of the transformed variable $Y$ is

$$\lambda_Y(y)=\frac{f_Y(y)}{S_Y(y)}=\frac{f_X(h(y))}{S_X(h(y))}\frac{dx}{dy}=\lambda_X(h(y))\frac{dx}{dy}. \tag{3.28}$$

Transformations belonging to the special class for which

$$y=\frac{x-\alpha}{\beta} \qquad \text{or} \qquad x=\alpha+\beta y, \tag{3.29}$$

with $\beta\neq 0$, are called *linear transformations*. If $\beta>0$, the hazard rate in (3.28) becomes

$$\lambda_Y(y)=\beta\lambda_X(\alpha+\beta y), \tag{3.30}$$

or equivalently

$$\lambda_X(x)=\frac{1}{\beta}\lambda_Y\left(\frac{x-\alpha}{\beta}\right). \tag{3.30a}$$

This means that a linear transformation multiplies the original hazard rate by a constant, and transforms the argument linearly.

## 3.9 LOCATION-SCALE FAMILIES OF DISTRIBUTION

If $S_X(x)$ [and so $F_X(x)$] depends on the values of parameters $\xi$, $\theta$ ($\theta>0$) through the values of $(X-\xi)/\theta$, then $\xi$ is called a *location parameter*, and $\theta$ is called a *scale parameter*. The SDF (CDF) may depend on other parameters $\gamma_1,\gamma_2,\ldots$, say. These are *shape parameters*.

Formally, we can write

$$S_X(x)=g\left(\frac{x-\xi}{\theta};\gamma_1,\gamma_2,\ldots\right), \tag{3.31}$$

where $g((x-\xi)/\theta;\ \gamma_1,\gamma_2,\ldots)$ is a mathematical function.

The class of distributions obtained by assigning to $\xi$ and $\theta$ all possible values is called a *location-scale family* of distributions.

We now apply the linear transformation

$$x=\alpha+\beta y, \qquad \beta>0. \tag{3.32}$$

Then since

$$\Pr\{Y>y\}=\Pr\{X>\alpha+\beta y\},$$

we have

$$S_Y(y)=S_X(\alpha+\beta y)=g\left(\frac{\alpha+\beta y-\xi}{\theta};\gamma_1,\gamma_2,\ldots\right)$$

$$=g\left(\frac{y-\xi_1}{\theta_1};\gamma_1,\gamma_2,\ldots\right), \tag{3.33}$$

with

$$\xi_1=\frac{\xi-\alpha}{\beta} \quad \text{and} \quad \theta_1=\frac{\theta}{\beta}. \tag{3.34}$$

Note that a linear transform of a random variable with distribution belonging to a location-scale family, has a distribution belonging to the same family. (The shape parameters are unaltered.)

## 3.10 SOME SURVIVAL DISTRIBUTIONS

Theoretically, any proper distribution for which $S_X(0)=1$ might be used as a lifetime distribution. In real situations, however, certain families of distributions are especially useful for fitting survival data. In this section, we give analytical forms of some better known distributions of this kind.

***Uniform and Exponential Distributions.*** Of special interest are the well-known uniform and exponential distributions, which are often used to approximate lifetime distributions, especially when truncated over fairly short intervals (see Sections 4.6 and 5.10).

***Uniform Distribution.*** The PDF of the uniform distribution of a continuous random variable $X$ over the interval $(a,b]$ is

$$f_X(x)=\begin{cases} \dfrac{1}{b-a} & \text{for } a<x\leqslant b, \\ 0 & \text{elsewhere.} \end{cases} \tag{3.35}$$

The corresponding SDF is

$$S_X(x)=\begin{cases} 1 & \text{for } x\leqslant a, \\ \dfrac{b-x}{b-a} & \text{for } a<x\leqslant b, \\ 0 & \text{for } x>b. \end{cases} \tag{3.36}$$

The expected value and variance are, respectively,

$$\mathcal{E}(X)=\tfrac{1}{2}(a+b), \qquad \operatorname{Var}(X)=\tfrac{1}{12}(a+b)^2. \tag{3.37}$$

The hazard rate is

$$\lambda_X(x)=\frac{1}{b-x} \qquad \text{for } a<x\leqslant b, \tag{3.38}$$

is an increasing function of $x$. Similarly, the conditional probability of death within a period $h$ [defined in (3.10)]

$$_hq_x = 1 - \frac{S_X(x+h)}{S_X(x)} = \frac{h}{b-x} \tag{3.39}$$

is an increasing function of $x$.

When $a=0$ and $b=1$, the distribution is called the *standard uniform distribution*. Of course, the truncated uniform distribution is still uniform.

***Exponential Distribution.*** The PDF of an exponential distribution is

$$f_X(x) = \lambda e^{-\lambda x}, \qquad \lambda > 0, \qquad x > 0, \tag{3.40}$$

and the SDF is

$$S_X(x) = e^{-\lambda x}. \tag{3.41}$$

The expected value and variance are, respectively,

$$\mathcal{E}(X) = \frac{1}{\lambda} \quad \text{and} \quad \text{var}(X) = \frac{1}{\lambda^2}. \tag{3.42}$$

The hazard rate is equal to $\lambda$; it is *constant* for all $x>0$. Similarly, the conditional probability of death within a period $h$,

$$_hq_x = 1 - e^{-\lambda h}, \tag{3.43}$$

is also constant, whatever the value of $x$.

If the exponential distribution is *truncated from above* at $x=c$, then the SDF of the truncated distribution is

$$S_X(x|X \leqslant c) = \frac{e^{-\lambda x} - e^{-\lambda c}}{1 - e^{-\lambda c}}, \qquad 0 < x \leqslant c. \tag{3.44}$$

This is not an exponential distribution. On the other hand, if truncation at $x=c$ is *from below*, then

$$S_X(x|X > c) = e^{-\lambda(x-c)} \qquad \text{for } x > c, \tag{3.45}$$

which is itself an exponential distribution.

***Extreme Value Distributions.*** This name applies to three types of *limiting distributions*, which approximate the shape of distributions of extreme values (the least or the greatest) in large random samples. Detailed discussion of geneses of these distributions can be found in a number of books, for example, see Gumbel (1958) and Johnson and Kotz (1970).

Here we only give the distribution forms of Type 1 and Type 3 of *least value*. These two distributions are widely used in reliability theory and survival analysis.

***Type 1 Extreme (Least) Value Distribution.*** Its SDF is

$$S_X(x)=\exp[-e^{(x+\xi)/\theta}] \qquad \text{for } -\infty<x<\infty, \qquad \theta>0. \tag{3.46}$$

Letting

$$\frac{1}{\theta}=a \qquad \text{and} \qquad \frac{1}{\theta}e^{\xi/\theta}=R, \tag{3.47}$$

we can write (3.46) in the form

$$S_X(x)=\exp\left[-\frac{R}{a}e^{ax}\right], \qquad -\infty<x<\infty, \qquad R>0, \qquad a>0. \tag{3.48}$$

The hazard rate function (HRF) of this distribution is

$$\lambda_X(x)=Re^{ax}. \tag{3.49}$$

***Gompertz and Makeham-Gompertz Distribution.*** Note that Type 1 extreme value distribution is defined for $-\infty<x<\infty$. Although for appropriately selected parameters, $R$ and $a$, $\Pr\{-\infty<x<0\}$ might be very small, it is more convenient to consider the corresponding, *truncated from below at* $x=0$, distribution. Its SDF is

$$S_X(x)=\exp\left[\frac{R}{a}(1-e^{ax})\right], \qquad x>0, \qquad R>0, \qquad a>0. \tag{3.50}$$

This is known as the *Gompertz distribution*.

Its hazard rate is

$$\lambda_X(x)=Re^{ax}, \tag{3.51}$$

which, in view of (3.16), is, of course, the same as that given by (3.51).

The Gompertz distribution is widely used in actuarial work, but usually expressed in the following notation. Put $R=b$ and $e^a=c$. Then

$$\lambda_X(x)=bc^x. \tag{3.52}$$

Adding a constant $A$ to the hazard rate (to allow for accidental deaths in addition to deaths from natural causes), that is, taking

$$\lambda_X(x) = A + Re^{ax} = A + bc^x, \tag{3.53}$$

we obtain a Makeham-Gompertz distribution.

Both Gompertz and Makeham-Gompertz, are location-scale families.

***Weibull Distribution (Type 3 Extreme Value).*** Its SDF is

$$S_X(x) = \exp\left[-\left(\frac{x-\xi}{\theta}\right)^c\right], \qquad x > \xi, \qquad \theta > 0, \qquad c > 0 \tag{3.54}$$

and the hazard rate is

$$\lambda_X(x|X>\xi) = \frac{c}{\theta}\left(\frac{x-\xi}{\theta}\right)^{c-1} = \frac{c}{\theta^c}(x-\xi)^{c-1}. \tag{3.55}$$

Here $\xi$ and $\theta$ are location and scale parameters, respectively, whereas $c$ is a shape parameter.

It can be shown (see Exercise 3.6) that if $X$ has a Weibull distribution then $Y = \log(X-\xi)$ has a Type 1 distribution.

***Some Other Distributions.*** Three other distributions are often used in fitting survival data. These are: gamma, lognormal, and logistic distributions.

***Gamma Distribution.*** The general form for the PDF of this distribution is

$$f_X(x) = \frac{1}{\beta^\alpha \Gamma(\alpha)}(x-\xi)^{\alpha-1}\exp\left(-\frac{x-\xi}{\beta}\right), \qquad x > \xi, \qquad \alpha > 0, \qquad \beta > 0, \tag{3.56}$$

where $\xi$ and $\beta$ are location and scale parameters, respectively, and $\alpha$ determines the shape of the PDF.

In most applications, $\xi = 0$, and we have the two parameter gamma PDF

$$f_X(x) = \frac{1}{\beta^\alpha \Gamma(\alpha)} x^{\alpha-1}\exp\left(-\frac{x}{\beta}\right). \tag{3.57}$$

The mean and variance are respectively

$$\mathcal{E}(X) = \beta\alpha, \qquad \text{var}(X) = \beta^2\alpha. \tag{3.58}$$

For general values of $\alpha$, special tables are needed to evaluate the SDF, which is of the form

$$S_X(x)=\frac{1}{\beta^\alpha\Gamma(\alpha)}\int_x^\infty t^{\alpha-1}\exp\left(-\frac{t}{\beta}\right)dt=\frac{1}{\Gamma(\alpha)}\int_{x/\beta}^\infty t^{\alpha-1}e^{-t}\,dt. \quad (3.59)$$

Tables of the integrals in the right-hand side of (3.59) are available [for references, see Johnson and Kotz (1970)].

When $\alpha=\frac{1}{2}\nu$ and $\beta=2$, we have the *Chi-square distribution* with $\nu$ degrees of freedom. Its PDF is

$$f_X(x)=\frac{1}{2^{\nu/2}\Gamma(\nu/2)}x^{(\nu/2)-1}e^{-x/2}, \quad (3.60)$$

and

$$\mathcal{E}(X)=\nu, \qquad \operatorname{var}(X)=2\nu. \quad (3.61)$$

In particular, if $\alpha=\frac{1}{2}\nu=1$ and $\beta=2$, we obtain

$$f_X(x)=\tfrac{1}{2}\exp\left(-\tfrac{1}{2}x\right). \quad (3.62)$$

This is the exponential distribution with parameter $\lambda=\frac{1}{2}$; but it is also a Chi-square distribution with $\nu=2$ degrees of freedom.

***Lognormal Distribution.*** The random variable $X$ is said to have a lognormal distribution if $Y=\log X$ [more generally, $Y=\log(X-\alpha)$] has a normal distribution. If $\xi$ and $\sigma^2$ are, respectively, the mean and variance of $Y$, then the PDF of $X$ is

$$f_X(x)=(\sqrt{2\pi}\,\sigma x)^{-1}\exp\left[-\frac{1}{2}\left(\frac{\log x-\xi}{\sigma}\right)^2\right], \qquad x>0, \qquad \sigma>0. \quad (3.63)$$

The CDF is

$$F_X(x)=\Phi\left(\frac{\log x-\xi}{\sigma}\right).$$

***Logistic Distribution.*** The CDF of a logistic distribution can be written as

$$F_X(x)=\frac{1}{1+\exp[-(x-\xi)/\theta]}, \qquad -\infty<x<\infty, \qquad \theta>0, \quad (3.64)$$

where $\xi$ is a location parameter and $\theta$ is a scale parameter.

Writing $1/\theta = b$ and $\xi/\theta = a$, we can present (3.64) in the form

$$F_X(x) = \frac{1}{1+\exp(a-bx)}, \qquad -\infty < x < \infty, \qquad b > 0. \qquad (3.64a)$$

This distribution has the convenient property that the SDF, PDF, and HRF can be expressed as simple functions of CDF.

$$S_X(x) = 1 - F_X(x) = \frac{1}{1+\exp(bx-a)} = e^{a-bx}F_X(x), \qquad (3.65)$$

$$\lambda_X(x) = -\frac{d\log S_X(x)}{dx} = \frac{b}{1+\exp(a-bx)} = bF_X(x), \qquad (3.66)$$

and

$$f_X(x) = \lambda_X(x)S_X(x) = be^{a-bx}\left[1+\exp(a-bx)\right]^{-2}$$
$$= bF_X(x)\left[1-F_X(x)\right]. \qquad (3.67)$$

Note that

$$\log\frac{F_X(x)}{1-F_X(x)} = bx - a. \qquad (3.68)$$

## 3.11 SOME MODELS OF FAILURE

Consider an organism or mechanical device that is composed of $n$ parts. Without being specific, we will call such an entity a *system*. Suppose that each part has its own lifetime distribution. For the $i$th part we introduce lifetime $X_i$, and denote its CDF by $F_i(x_i)$ and its SDF by $S_i(x_i)$. Furthermore, we assume that the lifetimes of parts are *mutually independent*.

### 3.11.1 Series System

Suppose that the system is so constructed that it only operates when all parts operate. *In other words, the system fails if at least one part fails*. This is called a *series system*.

We introduce a random variable

$$Y = \min(X_1, X_2, \ldots, X_n). \qquad (3.69)$$

The survival function of the system is

$$S_Y(t) = \Pr\{\text{system fails after time } t\}$$

$$= \Pr\{Y > t\} = \Pr\left\{\bigcap_{i=1}^{n} (X_i > t)\right\} = \prod_{i=1}^{n} S_i(t). \quad (3.70)$$

In particular, if $S_i(t) = S(t)$ for all $i$, then (3.70) takes the form

$$S_Y(t) = [S(t)]^n. \quad (3.70a)$$

**Example 3.1** If $S_i(t) = \exp(-\lambda_i t)$, $\lambda_i > 0$, $t > 0$ is exponential, then

$$S_Y(t) = \exp\left(-\sum_{i=1}^{n} \lambda_i t\right) = \exp(-\lambda t),$$

where

$$\lambda = \sum_{i=1}^{n} \lambda_i.$$

This means that the SDF of the system is also exponential.

### 3.11.2 Parallel System

Suppose that a system is so constructed that it operates so long as at least one part operates. In other words, the *system fails if and only if all parts fail*. This is called a *parallel system*.

Let

$$Z = \max(X_1, X_2, \ldots, X_n). \quad (3.71)$$

Then the failure distribution of the system is

$$F_Z(t) = \Pr\{\text{system fails before time } t\}$$

$$= \Pr\{Z \leqslant t\} = \Pr\left\{\bigcap_{i=1}^{n} (X_i \leqslant t)\right\} = \prod_{i=1}^{n} F_i(t). \quad (3.72)$$

In particular, if $F_i(t) = F(t)$ for all $i$, we have

$$F_Z(t) = [F(t)]^n. \quad (3.72a)$$

Using Example 3.1 with exponential CDF's, we obtain

$$F_Z(t) = \prod_{i=1}^{n} [1 - \exp - (\lambda_i t)],$$

which is *not exponential*.

We notice that the failure distribution of a series (parallel) system is the distribution of minimum (maximum) of lifetimes of component parts.

## 3.12 PROBABILITY INTEGRAL TRANSFORMATION

1. The random variable

$$Y = S_X(X) \tag{3.73}$$

is called the probability integral transform of $X$. If $y = S_X(x)$, then

$$\frac{dy}{dx} = -f_X(x),$$

and so

$$f_Y(y) = \frac{f_X(x)}{|dy/dx|} = 1 \qquad \text{for } 0 \leqslant y \leqslant 1, \qquad f_Y(y) = 0 \quad \text{elsewhere.}$$

This is the PDF of the *standard uniform distribution*. We have

$$\Pr\{Y \leqslant y\} = y \qquad \text{for } 0 \leqslant y \leqslant 1.$$

The same result holds for $Y_1 = F_X(X)$.

2. Let

$$Z = -\log S_X(X) = -\log Y, \tag{3.74}$$

where $Y$ has the standard uniform distribution for $0 < y \leqslant 1$. From (3.74),

$$Y = e^{-Z} \qquad \text{and} \qquad \left|\frac{dy}{dz}\right| = e^{-z}.$$

Hence

$$f_Z(z) = f_Y(y)\left|\frac{dy}{dz}\right| = e^{-z}.$$

This is the density of a *standard exponential* distribution with the parameter $\lambda = 1$.

We note from (3.5) that, in fact we have $Z = \Lambda_X(X)$, where $\Lambda_X(X)$ is the cumulative hazard function.

3. Using a similar argument as in (2), it can be shown that

$$U = -2\log S_X(X) \tag{3.75}$$

has an *exponential* distribution with the parameter $\lambda = \frac{1}{2}$.

## 3.13 COMPOUND DISTRIBUTIONS

Sometimes we assume that a parameter of a distribution varies according to a certain probability law. Suppose that $S_X(x;\theta)$ is the so called *parental* SDF, and assume that $\theta$ has the PDF $g(\theta)$. The SDF of $X$ alone, $S_X(x)$, say, is, in fact, a *mixture* of distributions. Clearly

$$S_X(x) = \int_{-\infty}^{\infty} S_X(x;\theta) g(\theta)\, d\theta. \tag{3.76}$$

The survival function, $S_X(x)$, is also called the *compound* survival distribution, while $g(\theta)$ is the PDF of the *compounding* distribution. Extension to multivariate cases is straightforward.

## 3.14 MISCELLANEA

Although the next two subsections are not specifically associated with survival distributions, their contents are very useful in survival analysis. *Interpolation* is used in construction of life tables (Chapter 4). *The method of statistical differentials* is useful in calculation of approximate variances (and covariances) of various life table functions. Since these topics are not discussed in many statistical textbooks, we give a brief account of them here.

### 3.14.1 Interpolation

Tables of a function give values of the function for some specified values of the argument. From time to time, it becomes necessary to evaluate the function at some nontabulated values. If the function is specified mathematically, this might be done by a special calculation. Generally, in life tables, the function is not specified mathematically and a process of *interpolation*, that is, estimating a nontabulated value from tabulated values, must be applied. Interpolation is also sometimes used when the function is specified mathematically but is very different to evaluate.

Without some additional assumption, tabulated values tell us nothing about nontabulated values. Most simple interpolation formulae are obtained by fitting a *polynomial* through the tabulated values of the argument. The basic formula is *Lagrange's interpolation formula*, which expresses the function $u(x)$ (denoted here for simplicity by $u_x$) as a polynomial of degree $k$ in $x$, which takes tabulated values $u_{x_1}, u_{x_2}, \ldots, u_{x_{k+1}}$, at $x = x_1, x_2, \ldots, x_{k+1}$, respectively.

This formula is

$$u_x = \sum_{i=1}^{k+1} \frac{\prod_{j\neq i} (x - x_j)}{\prod_{j \neq i} (x_i - x_j)} u_{x_i}. \tag{3.77}$$

For example, taking

$$x_1 = 1 \qquad x_2 = 2 \qquad x_3 = 5 \qquad x_4 = 10,$$

we get for the polynomial of the third order

$$u_x = \frac{(x-2)(x-5)(x-10)}{(-1)(-4)(-9)} u_1 + \frac{(x-1)(x-5)(x-10)}{1\cdot(-3)(-8)} u_2$$
$$+ \frac{(x-1)(x-2)(x-10)}{4\cdot 3\cdot(-5)} u_5 + \frac{(x-1)(x-2)(x-5)}{9\cdot 8\cdot 5} u_{10}.$$

The interpolated value for $x = 7$ would be

$$u_7 = \left(\frac{-30}{-36}\right)u_1 + \left(\frac{-36}{24}\right)u_2 + \left(\frac{-90}{-60}\right)u_5 + \left(\frac{60}{360}\right)u_{10}$$
$$= 0.833333u_1 - 1.500000u_2 + 1.50000u_5 + 0.166667u_{10}.$$

Of course, this is not, in general, the exact value of $u_7$; it is an estimated value.

Note that the coefficients in the expression for $u_7$ add up to 1. It can be shown that this is true generally for the coefficients in (3.77). It corresponds to the fact that if all the tabulated values are equal (to $u$, say) then every interpolated value is equal to $u$. When the $x_i$'s are equally spaced, formula (3.77) can be expressed in simpler forms.

When only two tabulated values, $x_1$, $x_2$, are used the fitted polynomial is linear, and the resultant interpolation is called *linear interpolation*. The linear interpolation formula can be written

$$u_x = \frac{x_2 - x}{x_2 - x_1} u_{x_1} + \frac{x - x_1}{x_2 - x_1} u_{x_2}. \tag{3.78}$$

It is convenient to write $x - x_1 = \theta(x_2 - x_1)$; then (3.78) becomes

$$u_x = (1-\theta) u_{x_1} + \theta u_{x_2}, \qquad 0 < \theta < 1. \tag{3.78a}$$

The assumption of uniform distribution of failures between ages $x_1$ and $x_2$ (see Section 3.10) corresponds to linear interpolation for $S_X(x)$ between $S_X(x_1)$ and $S_X(x_2)$.

Extensive discussion, with derivations and examples on polynomial interpolation formulae, can be found in Abramowitz and Stegun (1965).

If the curve $u(x) = u_x$ is concave from above rather than linear over the interval $(x_1, x_2)$, we may assume $z = \log u_x$ to be linear, and use linear interpolation on $\log u_x$, giving

$$\log u_x \doteq (1-\theta)\log u_{x_1} + \theta \log u_{x_2}, \qquad 0 < \theta < 1. \tag{3.79}$$

This is equivalent to

$$u_x \doteq u_{x_1}^{1-\theta} u_{x_2}^{\theta},$$

or

$$u_x \doteq u_{x_1}\left(\frac{u_{x_2}}{u_{x_1}}\right)^{\theta} = u_{x_1} e^{\alpha\theta}, \tag{3.79a}$$

where

$$\alpha = \log\left(\frac{u_{x_2}}{u_{x_1}}\right).$$

This is called *exponential interpolation* and is often used in construction of life tables.

### 3.14.2 The Method of Statistical Differentials

***Univariate Case.*** Let $g(X)$ be a function of a random variable $X$. If $X$ is continuous and we know its PDF, $f_X(x)$, then the expected value of $g(X)$

is

$$\mathcal{E}[g(X)] = \int_{-\infty}^{\infty} g(x) f_X(x)\,dx.$$

In the discrete case, integration is replaced by summation. By definition, the variance of $g(X)$ is

$$\text{var}[g(X)] = \mathcal{E}\{[g(X) - \mathcal{E}(g(X))]^2\}.$$

Sometimes the exact calculations of these moments are difficult or even impossible. Provided that the first and the second derivatives of $g(X)$ exist, and also $\mathcal{E}(X) = \xi$, and $\mathcal{E}[(X-\xi)^2] = \text{var}(X)$ exist and are *known* we can use the following method of approximate calculations.

We write $X = \xi + (X - \xi)$ in powers of $(X - \xi)$, using Taylor series. This gives

$$g(X) = g(\xi) + (X-\xi)g'(\xi) + \tfrac{1}{2}(X-\xi)^2 g''(\xi) + \cdots. \tag{3.80}$$

Taking expectations of each side of (3.80) and noticing that $\mathcal{E}(X-\xi) = 0$, we obtain (using only the first two terms of Taylor expansion) the approximate formula

$$\mathcal{E}[g(X)] \doteqdot g(\xi) + \tfrac{1}{2} g''(\xi)\,\text{var}(X). \tag{3.81}$$

In deriving (3.81) we have used formal manipulations without checking their validity. For example, the Taylor series in (3.80) may not converge for all values of $(X-\xi)$, yet in taking expected values, we have integrated (or summed) over all possible values of $X$. However, this process often does give useful approximations. The less variable $X$ is, better one might expect the approximation to be.

Quite often we confine ourselves to the linear term only in Taylor expansion. In this case

$$\mathcal{E}[g(X)] \doteqdot g(\xi).$$

The quantities $(X-\xi)$, which are deviations of $X$ from its expected value $\xi$, are called *statistical differentials*, and this method of approximation is called the *method of statistical differentials*. [Another name is the *delta method*, since some authors represent $X-\xi$ by $(\delta X)$.]

Subtracting (3.81) from (3.80), taking squares, and then expected values of squares on both sides, and confining ourselves to the terms of second order, we obtain

$$\text{var}[g(X)] \doteq [g'(\xi)]^2 \text{var}(X). \tag{3.82}$$

Equivalently, the approximate formula

$$\text{SD}[g(X)] = |g'(\xi)| \text{SD}(X) \tag{3.83}$$

for the standard deviation is in common use.

**Example 3.2** Let $g(X)=1/X$. Find $\mathcal{E}(1/X)$ and $\text{var}(1/X)$. We have

$$\left.\frac{dg}{dx}\right|_{x=\xi} = -\frac{1}{\xi^2} \qquad \left.\frac{d^2g}{dx^2}\right|_{x=\xi} = \frac{2}{\xi^3}.$$

Thus from (3.81)

$$\mathcal{E}\left(\frac{1}{X}\right) \doteq \frac{1}{\xi} + \frac{1}{2}\frac{2}{\xi^3}\text{var}(X) = \frac{1}{\xi}\left[1 + \frac{\text{var}(X)}{\xi^2}\right].$$

If the coefficient of variation $[\text{SD}(X)/\xi]$ is not small, the second term in the right-hand expression cannot be neglected. From (3.82)

$$\text{var}\left(\frac{1}{X}\right) \doteq \frac{\text{var}(X)}{\xi^2}.$$

***Multivariate Case.*** We now extend the method to the multivariate case. Let $g(X_1,\ldots,X_n)$ be a continuous function of $n$ random variables, $X_1,\ldots,X_n$. Let

$$\mathcal{E}(X_i) = \xi_i, \qquad \mathcal{E}\left[(X_i - \xi_i)^2\right] = \text{var}(X_i),$$
$$\mathcal{E}\left[(X_i - \xi_i)(X_j - \xi_j)\right] = \text{cov}(X_i, X_j),$$

and suppose that their values are *known*.

We introduce vector notation $\mathbf{X}' = (X_1,\ldots,X_n)$ and $\boldsymbol{\xi}' = (\xi_1,\ldots,\xi_n)$, and denote also for simplicity

$$\left.\frac{\partial g(\mathbf{x})}{\partial x_i}\right|_{\mathbf{x}=\boldsymbol{\xi}} = \frac{\partial g}{\partial \xi_i}, \qquad \left.\frac{\partial^2 g(\mathbf{x})}{\partial x_i \partial x_j}\right|_{\mathbf{x}=\boldsymbol{\xi}} = \frac{\partial^2 g}{\partial \xi_i \partial \xi_j}, \qquad \text{etc.}$$

(This notation is not strictly legitimate, although it is customary to use it.)

We expand $g(\mathbf{X})$ in a multivariate Taylor series

$$g(\mathbf{X})=g(\boldsymbol{\xi})+\sum_{i=1}^{n}\frac{\partial g}{\partial \xi_i}(X_i-\xi_i)+\tfrac{1}{2}\sum_{i=1}^{n}\sum_{j=1}^{n}\frac{\partial g}{\partial \xi_i}\cdot\frac{\partial g}{\partial \xi_j}(X_i-\xi_i)(X_j-\xi_j)+\dots \tag{3.87}$$

Taking expectations of each side of (3.87) and confining ourselves only to the terms of second order, we obtain

$$\mathcal{E}[g(\mathbf{X})]\doteq g(\boldsymbol{\xi})+\frac{1}{2}\left[\sum_{i=1}^{n}\left(\frac{\partial g}{\partial \xi_i}\right)^2\operatorname{var}(X_i)+\sum_{i\neq j}\sum\frac{\partial g}{\partial \xi_i}\cdot\frac{\partial g}{\partial \xi_j}\operatorname{cov}(X_i,X_j)\right]. \tag{3.88}$$

Often the approximation

$$\mathcal{E}[g(\mathbf{X})]\doteq g(\boldsymbol{\xi}) \tag{3.89}$$

is used.

By similar arguments as in univariate case, we obtain

$$\operatorname{var}[g(\mathbf{X})]=\mathcal{E}\{[g(\mathbf{X})-g(\boldsymbol{\xi})]^2\}\doteq\sum_{i=1}^{n}\left(\frac{\partial g}{\partial \xi_i}\right)^2\operatorname{var}(X_i)$$

$$+2\sum_{i<j}\sum\frac{\partial g}{\partial \xi_i}\cdot\frac{\partial g}{\partial \xi_j}\operatorname{cov}(X_i,X_j). \tag{3.90}$$

***Covariance.*** Consider two functions, $g_1(X_1,\dots,X_n)=g_1(\mathbf{X})$ and $g_2(X_1,\dots,X_n)=g_2(\mathbf{X})$. Using the method of statistical differentials, it can be shown (the proof is left to the reader) that

$$\operatorname{cov}[g_1(\mathbf{X}),g_2(\mathbf{X})]\doteq\sum_{i=1}^{n}\frac{\partial g_1}{\partial \xi_2}\cdot\frac{\partial g_2}{\partial \xi_i}\operatorname{var}(X_i)+\sum_{i\neq j}\sum\frac{\partial g_1}{\partial \xi_i}\cdot\frac{\partial g_2}{\partial \xi_j}\operatorname{cov}(X_i,X_j). \tag{3.91}$$

## 3.15 MAXIMUM LIKELIHOOD ESTIMATION AND LIKELIHOOD RATIO TESTS

As remarked in Section 3.1, the reader is assumed to be familiar with the ideas of statistical inference. We wish, however, to draw attention to a few techniques, frequently used in this book, which are especially relevant to analysis of survival data. These include construction of likelihood functions which is a necessary preliminary to derivation of maximum likelihood estimators (MLE), and likelihood ratio tests used as tests of "goodness of fit."

### 3.15.1 Construction of Likelihood Functions

A *statistical model* is intended to represent the generation of random variables, realization of which corresponds to observed data. Construction of such a model is equivalent to ascribing joint distributions to the random variables concerned. The models often involve *unknown* parameters, which we wish to estimate from the data.

Given a statistical model and a set of observed data, the *likelihood function* is proportional (or equal) to the joint probability (or density) for that set of observed data regarded as a function of the parameters involved.

**Example 3.3** Suppose, starting from time zero, $n$ individuals are observed until all fail, and suppose that the exact times of failure, $x_1, x_2, \dots, x_n$ are recorded. Assuming that the failure model is given by the PDF $f_X(x;\boldsymbol{\theta})$, where $\boldsymbol{\theta}' = (\theta_1, \dots, \theta_s)$ represents $s$ unknown parameters, and the failure times are mutually independent, the likelihood function takes the form

$$L(\boldsymbol{\theta}; x_1, \dots, x_n) = L(\boldsymbol{\theta}) = \prod_{i=1}^{n} f_X(x_i; \boldsymbol{\theta}). \tag{3.92}$$

**Example 3.4** Suppose, however, that we only know that out of $n$ individuals starting at time zero, $r$ died before time $x'$, and $n-r$ survived beyond $x'$. In this case, we use the facts that the probability of dying before fixed time point $x'$ is $F_X(x';\boldsymbol{\theta})$, and the probability of surviving beyond this point is $S_X(x';\boldsymbol{\theta})$, where $F_X(x';\boldsymbol{\theta})$ and $S_X(x';\boldsymbol{\theta})$ are the CDF and SDF, respectively.

The statistical model for this set of data is binomial, so that the likelihood function is

$$L(\boldsymbol{\theta}; r) = L(\boldsymbol{\theta}) = \binom{n}{r} [F_X(x';\boldsymbol{\theta})]^r [S_X(x';\boldsymbol{\theta})]^{n-r}. \tag{3.93}$$

### 3.15.2 Maximum Likelihood Estimation

If there is a set of parameter values, $\hat{\boldsymbol{\theta}}$, say, which maximize the likelihood, they are the *maximum likelihood estimates* of the true parameter values based on the data and the model. Regarding $\hat{\theta}$'s as functions of random sample, they are called the *maximum likelihood estimators* (MLE's). It is usually more convenient to maximize $\log L(\boldsymbol{\theta})$ rather than $L(\boldsymbol{\theta})$, since $\log L(\boldsymbol{\theta})$ is a monotonic (increasing) function of $L(\boldsymbol{\theta})$, and so the values of the $\hat{\theta}$'s are unchanged.

Maximum likelihood estimators do have certain desirable properties, but it has to be remembered that: (1) these properties depend on the model being a sufficiently accurate presentation of reality, and (2) even if the model is valid, the desirable properties are *asymptotic*—they apply when the volume of data is sufficiently large, and it is not always clear what is "sufficiently large."

Nevertheless, MLE's are usually good estimators. They are sometimes difficult to calculate but many computer programs are available to aid this work.

### 3.15.3 Expected Values, Variances, and Covariances of the MLE's

It can be proved that

$$\lim_{n\to\infty} \mathcal{E}(\hat{\theta}_i) = \theta_i, \qquad i = 1, 2, \ldots, s,$$

that is, the MLE's are asymptotically *unbiased*.

To find the asymptotic variances and covariances of the estimators, $\hat{\theta}_i, \ldots, \hat{\theta}_s$, we first construct the so called *Fisher information matrix* (or Hessian matrix), $\mathbf{I}$, regarding the likelihood as a function of random variables observed in a given sample. For example, the likelihood in (3.92) is regarded as $L(\boldsymbol{\theta}; X_1, \ldots, X_n)$; and in (3.93) as $L(\boldsymbol{\theta}; R)$. Generally, we may write $L(\boldsymbol{\theta}; \mathbf{X})$.

The $(ij)$th element of the information matrix $\mathbf{I}$ is

$$I_{ij} = \mathcal{E}\left[ -\frac{\partial^2 \log L(\boldsymbol{\theta}; \mathbf{X})}{\partial\theta_i \partial\theta_j} \right]. \tag{3.94}$$

The inverse matrix, $\mathbf{I}^{-1}$, with the $(ij)$th element denoted by $I^{ij}$, is the variance-covariance matrix of the $\hat{\theta}$'s, so that

$$\operatorname{var}(\hat{\theta}_i) \doteqdot I^{ii} \qquad \text{and} \qquad \operatorname{cov}(\hat{\theta}_i, \hat{\theta}_j) \doteqdot I^{ij}. \tag{3.95}$$

### 3.15.4 Assessing Goodness of Fit

When fitting parametric distributions to survival data, we would like to know how closely the models fit the data. Although formal tests are not our main concern, nevertheless several tests of goodness of fit will be mentioned in this book. Some of these are described in many other statistical books. The *approximate likelihood ratio tests* have been, perhaps, somewhat less generally described.

Let $\boldsymbol{\alpha}' = (\alpha_1, \ldots, \alpha_k)$ be a vector of $k$ parameters, and $\boldsymbol{\beta}' = (\beta_1, \ldots, \beta_r)$ be a vector of $r$ additional parameters. Let the null hypothesis $(H_0)$ specify values of the $\alpha$'s.

Let $L(\hat{\boldsymbol{\alpha}}; \hat{\boldsymbol{\beta}})$ be the maximized likelihood when all $(k+r)$ parameters are estimated, and let $L(\boldsymbol{\alpha}; \tilde{\boldsymbol{\beta}})$ be the maximized likelihood when only the $r$ parameters $\beta_1, \ldots, \beta_r$ are estimated, the remaining ($\alpha$'s) being fixed at the values specified by $H_0$. *When* $\mathrm{H}_0$ *is true*, the statistic

$$\chi^2(k) = -2\log \frac{L(\boldsymbol{\alpha}; \tilde{\boldsymbol{\beta}})}{L(\hat{\boldsymbol{\alpha}}; \hat{\boldsymbol{\beta}})} \tag{3.96}$$

is approximately distributed as Chi-square with $k$ degrees of freedom [see, for example, Wilks (1962), Section 13.4]. Some applications and modifications of these tests are given in Chapters 7, 8, and 13.

## REFERENCES

Abramowitz, M., and Stegun, I. A. (1965). *Handbook of Mathematical Functions with Formulas, Graphs and Mathematical Tables*. National Bureau of Standards, Applied Mathematics Series 55, U. S. Government Printing Office, Washington, D. C.

Gumbel, E. J. (1958). *Statistics of Extremes*, Columbia University Press, New York.

Johnson, N. L., and Kotz, S. (1970). *Distributions in Statistics: Continuous Univariate Distributions*, – 1, 2, Wiley, New York.

Wilks, S. S. (1962). *Mathematical Statistics*. Wiley, New York.

## EXERCISES

**3.1.** Let $X$ and $Y$ be two independent random variables, each having a uniform distribution over the interval $(0, 1)$.

(*a*) Find the probability density function (PDF) of $T = Y - X$ and represent it graphically.

(*b*) Find the cumulative distribution function (CDF), $F_T(t)$, and represent it graphically.

(*c*) Suppose that $F_T(t)$ represents a law according to which a mass undergoes changes. Find the two values of mass $\times$ time for the intervals $-1 < t < 0$, and $0 < t < 1$, respectively (see Section 2.3.3, formula 2.12).

**3.2.** Let $\xi$ and $\eta$ $(\eta > \xi)$ denote the recorded calendar years of birth and entry to an experience, and let $X$ and $Y$ represent dates (in decimals of year) of birth and entry, respectively.

Let $X = \xi + X'$ and $Y = \eta + Y'$, and assume that $X'$ and $Y'$ are independent random variables, each uniformly distributed $(0, 1)$.

(*a*) Find PDF of *age at entry* $T = Y - X$.

(*b*) Find the probability that an individual is exposed to risk at age $\eta - \xi - 1 + u$, where $0 < u < 1$, and calculate its expected contribution to person-years of exposed to risk in the year $\eta - \xi - 1$ to $\eta - \xi$.

(*c*) Find the probability that an individual is exposed to risk at age $\eta - \xi + u$, $0 < u < 1$, and calculate its contribution to person-years of exposed to risk in the year from $\eta - \xi$ to $\eta - \xi + 1$.

**3.3.** Let a random variable $Y$ be distributed as Chi-square with $\nu$ degrees of freedom.

(*a*) Show that the expected value of $Y^{-r}$ $(r > 0)$ is

$$\mathcal{E}(Y^{-r}) = \frac{\Gamma\left[\frac{1}{2}(\nu - 2r)\right]}{2^r \Gamma(\nu/2)}, \qquad r < \tfrac{1}{2}\nu.$$

(*b*) Suppose that $\nu$ and $r$ are positive integers. Obtain $\mathcal{E}(Y^{-r})$ in this case.

**3.4.**

(*a*) Let $u(x)$ be a monotonic increasing function of $x$ such that $u(x) \to 0$ as $x \to x_0$, and $u(x) \to \infty$ and $x \to \infty$. Let

$$F_X(x) = 1 - \exp[-u(x)], \qquad x > x_0.$$

be the CDF of a random variable $X$. Prove that $Y = 2u(X)$ has a Chi-square distribution with 2 degrees of freedom.

(*b*) Let $X_1, X_2, \ldots, X_k$ be $k$ $(>2)$ independent random variables $X_j$ having the distribution

$$F_{X_j}(x_j) = 1 - \exp[-\lambda u_j(x_j)], \qquad \lambda > 0, \qquad j = 1, 2, \ldots, k,$$

where $u_j(x_j)$ satisfy the conditions of $u(x)$ in (*a*) and $\lambda$ $(>0)$ is a constant.

Using the result obtained in (*a*), show that

$$Y = 2\lambda \sum_{j=1}^{k} u_j(X_j)$$

has a Chi-square distribution with $2k$ degrees of freedom.

(*c*) Suppose that $\lambda$ is an unknown parameter and is estimated from the formula

$$\hat{\lambda}=(k-1)\left[\sum_{j=1}^{k} u_j(x_j)\right]^{-1}.$$

Show that $\hat{\lambda}$ is an unbiased estimator of $\lambda$. *Hint:* Notice that $\hat{\lambda}=2\lambda(k-1)/\chi^2_{2k}$.

(*d*) Find the variance of $\hat{\lambda}$.

**3.5.** Suppose that the lifetime distribution of a certain group of experimental mice can be approximated by a normal distribution with mean $\xi$ and variance $\sigma^2$ truncated from below at the point $A$.

(*a*) Find the survival function $S_X(x|X>A)$, and the hazard rate $\lambda_X(x|x>A)$.

(*b*) Derive formulae for the mean lifetime and variance of this (truncated) distribution.

(*c*) Evaluate the proportion $p_1$ of mice that survive beyond time $x=x_1>A$.

(*d*) What is the average age of the $100p\%$ longest-lived mice?

(*e*) Suppose that $x$ is recorded in months (1 month = 30 days). Suppose that $\xi=10.84$ months, $\sigma=5.78$ months, $A=0.0$ months; $x_1=20$ months; $p=0.25$. Calculate the appropriate quantities in (*b*)–(*d*).

**3.6.** Let

$$S_X(x)=\exp\left[-\left(\frac{x-\xi}{\theta}\right)^c\right], \qquad x>\xi, \qquad \theta>0, \qquad c>0,$$

be the SDF of a Weibull distribution. Show that the variable

$$Y=\log(X-\xi)$$

has an extreme value Type 1 distribution.

**3.7.** The hazard rate of the so called linear-exponential distribution is

$$\lambda_X(x)=\alpha+\beta x, \qquad x>0, \qquad \alpha>0, \qquad \beta>0.$$

(*a*) Obtain the SDF and PDF of this distribution.

(*b*) Find mode $t_{max}$ of the PDF (Kodlin, *Biometrics* **23** (1967), 227).

**3.8.** Let a random variable $Y=\log X$ ($X>0$) be normally distributed with expected value $\xi$ and standard deviation $\sigma$.

(*a*) Obtain the expected value of $X$, $E(X)=\mu$, say, in terms of $\xi$ and $\sigma$.
(*b*) Let $X_1, X_2, \ldots, X_n$ be a random sample from this distribution. Find the maximum likelihood estimator of $\mu$ ($\hat{\mu}$, say).
(*c*) Find the expected value and variance of $\hat{\mu}$.

**3.9.** The estimator of the relative risk, $\hat{R}$, defined in (2.40) is (conditional on $n_{1\cdot}, n_{2\cdot}$) a ratio of two binomial proportions. Find its approximate variance, and show that it is reasonable to estimate it by (2.41).

**3.10.** Consider a set of data on association between a disease ($A$) and risk factor ($F$) arranged in a form of a $2\times2$ table as shown in Table 2.4.

(*a*) Assume first that data are obtained from a *cross-sectional* study, so that only $n$ is fixed and $n_{ij}$'s represent *multinomial* variables. Let $p_{ij}$ be the multinomial parameter for the ($ij$)th cell ($i=1,2$, $j=1,2$). The *odds ratio* (or *cross-product ratio*) can be defined now

$$\omega = \frac{p_{11} \cdot p_{22}}{p_{12} \cdot p_{21}},$$

and its sample estimator $\hat{\omega}$ is given by (2.43), that is

$$\hat{\omega} = \frac{n_{11} \cdot n_{22}}{n_{12} \cdot n_{21}}.$$

Show that the variance of $\log\hat{\omega}$ is approximately

$$\operatorname{var}(\log\hat{\omega}) \doteq \frac{1}{n}\left(\frac{1}{p_{11}} + \frac{1}{p_{12}} + \frac{1}{p_{21}} + \frac{1}{p_{22}}\right),$$

and its estimator, which may be obtained by substituting $n_{ij}/n$ for $p_{ij}$, is

$$\text{est. var}(\log\hat{\omega}) \doteq \frac{1}{n_{11}} + \frac{1}{n_{12}} + \frac{1}{n_{21}} + \frac{1}{n_{22}}.$$

Also show that (approximate) estimated variance of $\hat{\omega}$ is given by (2.44). *Hint:* Use the method of statistical differentials.

(*b*) Suppose that the data in Table 2.4 are from a *prospective* study (i.e., the marginals $n_{1\cdot}$ and $n_{2\cdot}$ are fixed). The odds ratio is given by (2.42), where $P_1$ and $P_2$ are binomial parameters in two independent samples [($F$) and ($\bar{F}$), respectively]. Estimate the approximate variance of $\log\hat{\omega}$. Show that its estimated variance is the same as in (*a*).

(*c*) Using normal approximation construct a test of the hypothesis $H_0: \log\omega = 0$ against alternatives: (*i*) $\log\omega > 0$, (*ii*) $\log\omega \neq 0$.

**3.11.** Lomax (*J. Amer. Statist. Assoc.* **49** (1954), 847) used distributions with hazard rate functions

(*i*) $ae^{-bt}$, $t>0$, $a>0$, $b>0$;

(*ii*) $b(t+a)^{-1}$, $t>0$, $a>0$, $b>0$,

to fit survival distributions of businesses. Find the corresponding SDF's and show that (*i*) gives an improper SDF.

**3.12.** Let $g_1(X_1,\ldots,X_k)$ and $g_2(X_1,\ldots,X_k)$ be two continuous functions of $k$ random variables, $X_1,\ldots, X_k$, with $\mathcal{E}(X_i) = \xi_i$, $\text{var}(X_i) = \sigma_i^2$, and $\text{cov}(X_i, X_j) = \sigma_{ij}$, for $i,j = 1,2,\ldots,k$, $i \neq j$. Derive the approximate formula for $\text{cov}[g_1(X_1,\ldots,X_k), g_2(X_1,\ldots,X_k)]$ given in (3.91).

**3.13.** A compound SDF is generated from the Makeham-Gompertz SDF with $\mu_x = A + Re^{ax}$ by ascribing to $A$ an exponential distribution with hazard rate $\theta$. Find the compound SDF.

**3.14.** Consider two SDF's $S_1(x)$, $S_2(x)$ which are Gompertz with parameters $(R_1, a)$, $(R_2, a)$ respectively. If $S_1(x_1) = S_2(x_2)$, express $x_1$ in terms of $x_2$. Generalize this result to the case when the parameters are $(R_1, a_1)$, $(R_2, a_2)$.

# Part 2

# MORTALITY EXPERIENCES AND LIFE TABLES

CHAPTER 4

# Life Tables: Fundamentals and Construction

## 4.1 INTRODUCTION

Imagine a *cohort* of $l_0$ newborn individuals. No new births, and no immigration or emigration, are observed in this ideal cohort. As time goes by, the cohort gradually decreases in size, as its members die. The mortality experience of such an ideal cohort can be represented by a *cohort life table*. In practice, however, we use the mortality data of a current population and construct a *hypothetical* cohort, experiencing mortality estimated from that of the actually observed population.

The form of a life table is usually nonparametric. It is a model of survival, not expressed in terms of proportions, but by *expected numbers of survivors* out of $l_0$ starters at age zero. The value of $l_0$ is called the *radix* of the life table, and is often conveniently taken as 100,000 or 1,000,000. Of course, the life table can start from any age $\alpha$; the radix is then denoted by $l_\alpha$.

Although life tables were (and still are) constructed without using probability theory, we wish to emphasize analogies between life tables and the survival distribution functions discussed in Chapter 3.

In this chapter, we first define the basic functions of life tables and discuss their interpretation. Construction of life tables is discussed toward the end of the chapter. Once the life table model is understood, it is possible to organize mortality data so as to facilitate construction of life tables. Applications of life table techniques in survival analysis will appear throughout this book.

## 4.2 LIFE TABLE: BASIC DEFINITIONS AND NOTATION

We describe a life table emphasizing its analogy with the probability functions introduced in Chapter 3. It is useful to distinguish between: $x$ the *exact* age of an individual, and $x'$ lbd – $x'$ *last birthday*; $x'$ is the integer part of $x$.

For example, if a person is aged 50 lbd, this means that his exact age $x$ must be between 50 and 51. It also means that he is in his 51st year of life.

$l_x$ is the expected number of survivors out of $l_0$ at *exact* age $x$. This is the fundamental function of the life table. It has meaning only in conjunction with $l_0$. For theoretical purposes, it is convenient to assume that $l_x$ is a continuous function of $x$. For practical purposes, however, it is convenient to tabulate $l_x$ for integer values of $x$, although in many tables $l_x$ is given for fractional values of $x$ at very young ($x<1$) ages.

We denote by $P_x$ the *proportion* of survivors at exact age $x$, that is

$$P_x = \frac{l_x}{l_0}. \tag{4.1}$$

This is, in fact, the survival function, $S_X(x)$, but when the mathematical form of $S_X(x)$ is not specified, it is customary to use the notation $P_x$. Some further life table functions are defined over the age *interval* determined by two adjacent values $x$ and $x+t$, for example, ${}_tq_x$ defined in (3.10). Except for the very first year of life, in which the mortality changes very rapidly (see Table 4.1), it is customary to take $t$ as an integer.

There are two kinds of life tables in common use: *complete* life tables in which the tabulation is for every year (i.e., $t=1$), and *abridged* life tables in which the tabulation is for every $n$ (i.e., $t=n$) years. It is a common practice to construct quinquennial abridged life tables, that is, with $n=5$ (except for the first 5 years, which is usually split up into two intervals: 0–1 and 1–5).

We recall (from Section 3.4) that

$q_x$ is the *conditional* probability of death in $(x, x+1)$ given an individual is alive at exact age $x$.

$p_x$ is the conditional probability that an individual alive at exact age $x$ *will not die* in $(x, x+1)$, that is, will survive beyond age $x+1$. Of course, $p_x = 1 - q_x$.

$d_x$ is the expected number of *deaths* in $(x, x+1)$.

We notice that

$$d_x = l_x - l_{x+1}, \tag{4.2}$$

$$q_x = \frac{d_x}{l_x} = \frac{l_x - l_{x+1}}{l_x}, \tag{4.3}$$

$$p_x = 1 - q_x = \frac{l_{x+1}}{l_x}. \tag{4.4}$$

From (4.4) (changing $x$ to $x-1$)

$$l_x = l_{x-1}(1-q_{x-1}) = l_{x-1}p_{x-1} \tag{4.5}$$

$$= l_0(1-q_0)(1-q_1)\dots(1-q_{x-1})$$

$$= l_0 p_0 p_1 \cdots p_{x-1}. \tag{4.5a}$$

The expected proportion of survivors at exact age $x$ is

$$P_x = \frac{l_x}{l_0} = \prod_{y=0}^{x-1}(1-q_y) = \prod_{y=0}^{x-1} p_y, \tag{4.6}$$

with $P_0 = 1$.

Also the expected proportion surviving for $n$ years, given alive at exact age $x$ is

$${}_np_x = (1-q_x)(1-q_{x+1})\dots(1-q_{x+n-1})$$

$$= p_x \cdot p_{x+1} \cdots p_{x+n-1} \tag{4.7}$$

$$= \frac{l_{x+1}}{l_x}\cdot\frac{l_{x+2}}{l_{x+1}}\cdots\frac{l_{x+n}}{l_{x+n-1}} = \frac{l_{x+n}}{l_x}. \tag{4.7a}$$

It can be seen from (4.5), that once the $q_x$'s ($x = 0, 1, 2, \dots$) are obtained (e.g., by estimation from population mortality data), the evaluation of the $l_x$ and $d_x$ columns in the life table is straightforward.

There are some further basic functions of a life table.

$L_x$ is the expected total number of years lived between age $x$ and $x+1$ or, more precisely, the number of person-years that $l_x$ persons, aged $x$ exactly, are expected to live through $(x, x+1)$.

Each member in the cohort who survives the full year $x$ to $x+1$ contributes one year to $L_x$; on the other hand, each member who dies in $[x, x+1)$ contributes only a fraction of a year to $L_x$ (see Fig. 4.1).

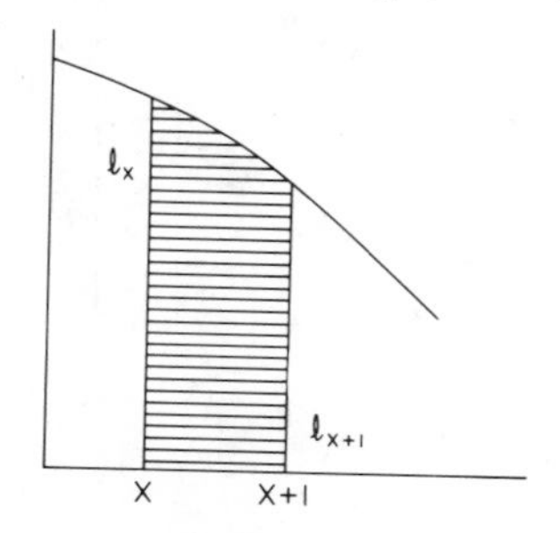


**Fig. 4.1**

**Table 4.1** Life Table for White Males: United States, 1969–1971

| Age interval | Proportion dying | Of 100,000 born alive | | Stationary population | | Average remaining lifetime |
|---|---|---|---|---|---|---|
| Period of life between exact ages | Proportion of persons alive at beginning of age interval dying during interval | Number living at beginning of age interval | Number dying during age interval | In the age interval | In this and all subsequent age intervals | Average number of years of life remaining at beginning of age interval |
| (1) | (2) | (3) | (4) | (5) | (6) | (7) |
| $x$ to $x+t$ | ${}_tq_x$ | $l_x$ | ${}_td_x$ | ${}_tL_x$ | $T_x$ | $\mathring{e}_x$ |
| Days | | | | | | |
| 0–1 | 0.00892 | 100,000 | 892 | 273 | 6,793,828 | 67.94 |
| 1–7 | 0.00527 | 99,108 | 522 | 1,625 | 6,793,555 | 68.55 |
| 7–28 | 0.00139 | 98,586 | 138 | 5,668 | 6,791,930 | 68.89 |
| 28–365 | 0.00462 | 98,448 | 454 | 90,686 | 6,786,262 | 68.93 |
| Years | | | | | | |
| 0–1 | 0.02006 | 100,000 | 2,006 | 98,252 | 6,793,828 | 67.94 |
| 1–2 | 0.00116 | 97,994 | 114 | 97,937 | 6,695,576 | 68.33 |
| 2–3 | 0.00083 | 97,880 | 81 | 97,840 | 6,597,639 | 67.41 |
| 3–4 | 0.00072 | 97,799 | 71 | 97,763 | 6,499,799 | 66.46 |
| 4–5 | 0.00059 | 97,728 | 57 | 97,700 | 6,402,036 | 65.51 |

|  |  |  |  |  |  |  |
|---|---|---|---|---|---|---|
| 5–6 | 0.00054 | 97,671 | 52 | 97,645 | 6,304,336 | 64.55 |
| 6–7 | 0.00051 | 97,619 | 50 | 97,594 | 6,206,691 | 63.58 |
| 7–8 | 0.00048 | 97,569 | 47 | 97,546 | 6,109,097 | 62.61 |
| 8–9 | 0.00044 | 97,522 | 43 | 97,500 | 6,011,551 | 61.64 |
| 9–10 | 0.00039 | 97,479 | 38 | 97,460 | 5,914,051 | 60.67 |
| 10–11 | 0.00034 | 97,441 | 32 | 97,425 | 5,816,591 | 59.69 |
| 11–12 | 0.00032 | 97,409 | 32 | 97,393 | 5,719,166 | 58.71 |
| 12–13 | 0.00039 | 97,377 | 38 | 97,358 | 5,621,773 | 57.73 |
| 13–14 | 0.00055 | 97,339 | 54 | 97,313 | 5,524,415 | 56.75 |
| 14–15 | 0.00080 | 97,285 | 77 | 97,246 | 5,427,102 | 55.79 |
| 15–16 | 0.00107 | 97,208 | 104 | 97,156 | 5,329,856 | 54.83 |
| 16–17 | 0.00134 | 97,104 | 130 | 97,039 | 5,232,700 | 53.89 |
| 17–18 | 0.00156 | 96,974 | 152 | 96,897 | 5,135,661 | 52.96 |
| 18–19 | 0.00172 | 96,822 | 167 | 96,739 | 5,038,764 | 52.04 |
| 19–20 | 0.00181 | 96,655 | 175 | 96,568 | 4,942,025 | 51,13 |
| 20–21 | 0.00190 | 96,480 | 183 | 96,388 | 4,845,457 | 50.22 |
| 21–22 | 0.00201 | 96,297 | 193 | 96,200 | 4,749,069 | 49.32 |
| 22–23 | 0.00205 | 96,104 | 198 | 96,005 | 4,652,869 | 48.42 |
| 23–24 | 0.00203 | 95,906 | 195 | 95,809 | 4,556;864 | 47.51 |
| 24–25 | 0.00195 | 95,711 | 187 | 95,618 | 4,461,055 | 46.61 |
| 25–26 | 0.00184 | 95,524 | 175 | 95,436 | 4,365,437 | 45.70 |
| 26–27 | 0.00173 | 95,349 | 165 | 95,267 | 4,270,001 | 44.78 |
| 27–28 | 0.00165 | 95,184 | 157 | 95,105 | 4,174,734 | 43.86 |
| 28–29 | 0.00162 | 95,027 | 154 | 94,950 | 4,079,629 | 42.93 |
| 29–30 | 0.00165 | 94,873 | 157 | 94,795 | 3,984,679 | 42.00 |

**Table 4.1** (continued)

| Age interval | Proportion dying | Of 100,000 born alive | | Stationary population | | Average remaining lifetime |
|---|---|---|---|---|---|---|
| Period of life between exact ages | Proportion of persons alive at beginning of age interval dying during interval | Number living at beginning of age interval | Number dying during age interval | In the age interval | In this and all subsequent age intervals | Average number of years of life remaining at beginning of age interval |
| (1) | (2) | (3) | (4) | (5) | (6) | (7) |
| Years | | | | | | |
| 30–31 | 0.00170 | 94,716 | 161 | 94,635 | 3,889,884 | 41.07 |
| 31–32 | 0.00176 | 94,555 | 167 | 94,471 | 3,795,249 | 40.14 |
| 32–33 | 0.00183 | 94,388 | 173 | 94,302 | 3,700,778 | 39.21 |
| 33–34 | 0.00192 | 94,215 | 181 | 94,125 | 3,606,476 | 38.28 |
| 34–35 | 0.00203 | 94,034 | 191 | 93,939 | 3,512,351 | 37.35 |
| 35–36 | 0.00217 | 93,843 | 204 | 93,741 | 3,418,412 | 36.43 |
| 36–37 | 0.00235 | 93,639 | 219 | 93,529 | 3,324,671 | 35.51 |
| 37–39 | 0.00256 | 93,420 | 239 | 93,300 | 3,231,142 | 34.59 |
| 38–39 | 0.00281 | 93,181 | 262 | 93,050 | 3,137,842 | 33.67 |
| 39–40 | 0.00310 | 92,919 | 288 | 92,775 | 3,044,792 | 32.77 |
| 40–41 | 0.00340 | 92,631 | 315 | 92,474 | 2,952,017 | 31.87 |
| 41–42 | 0.00372 | 92,316 | 343 | 92,144 | 2,859,543 | 30.98 |
| 42–43 | 0.00409 | 91,973 | 377 | 91,785 | 2,767,399 | 30.09 |
| 43–44 | 0.00452 | 91,596 | 414 | 91,389 | 2,675,614 | 29.21 |
| 44–45 | 0.00501 | 91,182 | 457 | 90,953 | 2,584,225 | 28.34 |

| | | | | | | |
|---|---|---|---|---|---|---|
| 45–46 | 0.00555 | 90,725 | 504 | 90,473 | 2,493,272 | 27.48 |
| 46–47 | 0.00612 | 90,221 | 552 | 89,945 | 2,402,799 | 26.63 |
| 47–48 | 0.00673 | 89,669 | 603 | 89,368 | 2,312,854 | 25.79 |
| 48–49 | 0.00739 | 89,066 | 658 | 88,737 | 2,223,486 | 24.96 |
| 49–50 | 0.00812 | 88,408 | 718 | 88,049 | 2,134,749 | 24.15 |
| 50–51 | 0.00892 | 87,690 | 782 | 87,300 | 2,046,700 | 23.34 |
| 51–52 | 0.00980 | 86,908 | 852 | 86,482 | 1,959,400 | 22.55 |
| 52–53 | 0.01081 | 86,056 | 930 | 85,591 | 1,872,918 | 21.76 |
| 53–54 | 0.01194 | 85,126 | 1,016 | 84,618 | 1,787,327 | 21.00 |
| 54–55 | 0.01318 | 84,110 | 1,109 | 83,556 | 1,702,709 | 20.24 |
| 55–56 | 0.01452 | 83,001 | 1,205 | 82,399 | 1,619,153 | 19.51 |
| 56–57 | 0.01594 | 81,796 | 1,304 | 81,144 | 1,536,754 | 18.79 |
| 57–58 | 0.01745 | 80,492 | 1,404 | 79,790 | 1,455,610 | 18.08 |
| 58–59 | 0.01906 | 79,088 | 1,508 | 78,334 | 1,375,820 | 17.40 |
| 59–60 | 0.02077 | 77,580 | 1,611 | 76,775 | 1,297,486 | 16.72 |
| 60–61 | 0.02258 | 75,969 | 1,716 | 75,111 | 1,220,711 | 16.07 |
| 61–62 | 0.02451 | 74,253 | 1,820 | 73,344 | 1,145,600 | 15.43 |
| 62–63 | 0.02657 | 72,433 | 1,924 | 71,471 | 1,072,256 | 14.80 |
| 63–64 | 0.02879 | 70,509 | 2,030 | 69,494 | 1,000,765 | 14.19 |
| 64–65 | 0.03120 | 68,479 | 2,136 | 67,411 | 931,291 | 13.60 |
| 65–66 | 0.03386 | 66,343 | 2,246 | 65,220 | 863,880 | 13.02 |
| 66–67 | 0.03674 | 64,097 | 2,355 | 62,919 | 798,660 | 12.46 |
| 67–68 | 0.03977 | 61,742 | 2,456 | 60,514 | 735,741 | 11.92 |
| 68–69 | 0.04284 | 59,286 | 2,540 | 58,016 | 675,227 | 11.39 |
| 69–70 | 0.04597 | 56,746 | 2,608 | 55,442 | 617,211 | 10.88 |

**Table 4.1** (continued)

| Age interval | Proportion dying | Of 100,000 born alive | | Stationary population | | Average remaining lifetime |
|---|---|---|---|---|---|---|
| Period of life between exact ages | Proportion of persons alive at beginning of age interval dying during interval | Number living at beginning of age interval | Number dying during age interval | In the age interval | In this and all subsequent age intervals | Average number of years of life remaining at beginning of age interval |
| (1) | (2) | (3) | (4) | (5) | (6) | (7) |
| Years | | | | | | |
| 70–71 | 0.04916 | 54,138 | 2,662 | 52,807 | 561,769 | 10.38 |
| 71–72 | 0.05262 | 51,476 | 2,708 | 50,122 | 508,962 | 9.89 |
| 72–73 | 0.05655 | 48,768 | 2,758 | 47,389 | 458,840 | 9.41 |
| 73–74 | 0.06118 | 46,010 | 2,815 | 44,603 | 411,451 | 8.94 |
| 74–75 | 0.06647 | 43,195 | 2,871 | 41,759 | 366,848 | 8.49 |
| 75–76 | 0.07231 | 40,324 | 2,916 | 38,866 | 325,089 | 8.06 |
| 76–77 | 0.07843 | 37,408 | 2,934 | 35,941 | 286,223 | 7.65 |
| 77–78 | 0.08472 | 34,474 | 2,921 | 33,014 | 250,282 | 7.26 |
| 78–79 | 0.09103 | 31,553 | 2,872 | 30,117 | 217,268 | 6.89 |
| 79–80 | 0.09749 | 28,681 | 2,796 | 27,283 | 187,151 | 6.53 |
| 80–81 | 0.10466 | 25,885 | 2,709 | 24,530 | 159,868 | 6.18 |
| 81–82 | 0.11273 | 23,176 | 2,613 | 21,870 | 135,338 | 5.84 |
| 82–83 | 0.12127 | 20,563 | 2,494 | 19,316 | 113,468 | 5.52 |
| 83–84 | 0.13012 | 18,069 | 2,351 | 16,894 | 94,152 | 5.21 |
| 84–85 | 0.13942 | 15,718 | 2,191 | 14,623 | 77,258 | 4.92 |

| | | | | | | |
|---|---|---|---|---|---|---|
| 85–86 | 0.15033 | 13,527 | 2,034 | 12,510 | 62,635 | 4.63 |
| 86–87 | 0.16321 | 11,493 | 1,875 | 10,555 | 50,125 | 4.36 |
| 87–88 | 0.17666 | 9,618 | 1,699 | 8,768 | 39,570 | 4.11 |
| 88–89 | 0.18947 | 7,919 | 1,501 | 7,169 | 30,802 | 3.89 |
| 89–90 | 0.20145 | 6,418 | 1,293 | 5,771 | 23,633 | 3.68 |
| 90–91 | 0.21344 | 5,125 | 1,094 | 4,579 | 17,862 | 3.49 |
| 91–92 | 0.22684 | 4,031 | 914 | 3,574 | 13,283 | 3.30 |
| 92–93 | 0.24152 | 3,117 | 753 | 2,740 | 9,709 | 3.12 |
| 93–94 | 0.25767 | 2,364 | 609 | 2,060 | 6,969 | 2.95 |
| 94–95 | 0.27426 | 1,755 | 481 | 1,514 | 4,909 | 2.80 |
| 95–96 | 0.29014 | 1,274 | 370 | 1,089 | 3,395 | 2.67 |
| 96–97 | 0.30431 | 904 | 275 | 766 | 2,306 | 2.55 |
| 97–98 | 0.31784 | 629 | 200 | 529 | 1,540 | 2.45 |
| 98–99 | 0.33085 | 429 | 142 | 358 | 1,011 | 2.36 |
| 99–100 | 0.34324 | 287 | 98 | 238 | 653 | 2.27 |
| 100–101 | 0.35479 | 189 | 67 | 155 | 415 | 2.20 |
| 101–102 | 0.36553 | 122 | 45 | 100 | 260 | 2.13 |
| 102–103 | 0.37550 | 77 | 29 | 62 | 160 | 2.08 |
| 103–104 | 0.38471 | 48 | 18 | 39 | 98 | 2.02 |
| 104–105 | 0.39320 | 30 | 12 | 24 | 59 | 1.98 |
| 105–106 | 0.40101 | 18 | 7 | 15 | 35 | 1.94 |
| 106–107 | 0.40818 | 11 | 5 | 8 | 20 | 1.90 |
| 107–108 | 0.41475 | 6 | 2 | 5 | 12 | 1.86 |
| 108–109 | 0.42075 | 4 | 2 | 3 | 7 | 1.82 |
| 109–110 | 0.42624 | 2 | 1 | 2 | 4 | 1.79 |

Formally,

$$L_x = \int_x^{x+1} l_y \, dy = \int_0^1 l_{x+t} \, dt. \tag{4.8}$$

(cf. Section 2.3.3).

Let $\omega$ denote the last year of life ever observed in a population (so that $l_{\omega-1} > l_\omega = 0$. Thus the interval $(0, \omega)$ represents the lifespan for this population.

$T_x$ is the expected total number of years lived beyond exact age $x$ by $l_x$ persons alive at that age.

We have

$$T_x = L_x + L_{x+1} + \cdots + L_{\omega-1} = \sum_{i=0}^{\omega-x-1} L_{x+i} = \sum_{i=0}^{\infty} L_{x+i}$$

$$= \int_0^{\omega-x} l_{x+t} \, dt = \int_0^{\infty} l_{x+t} \, dt, \tag{4.9}$$

since

$$\sum_{i=\omega-x}^{\infty} L_{x+i} = \int_{\omega-x}^{\infty} l_{x+t} \, dt = 0. \tag{4.10}$$

Of course,

$$T_x = L_x + T_{x+1}. \tag{4.11}$$

Another important basic function of the life table is

$\mathring{e}_x$ - the (conditional) expectation of future lifetime of an individual alive at exact age $x$ [cf. (3.20)].

This is the (average) number of years expected to be lived by a person of present age $x$, and so

$$\mathring{e}_x = \frac{T_x}{l_x} = \frac{1}{l_x} \sum_{i=0}^{\infty} L_{x+i}. \tag{4.12}$$

The expected age at death of a person aged $x$ is

$$x + \mathring{e}_x. \tag{4.13}$$

[cf. (3.21)].

The expected number of *full* (complete) years lived after age $x$ of an individual alive at exact age $x$ is known as the *curtate* expectation of life and is denoted by $e_x$. We have

$$e_x = \frac{1}{l_x} \sum_{i=1}^{\infty} l_{x+i}, \tag{4.14}$$

where $i$ is an integer. (As we will see in Example 4.4, $e_x \doteq \mathring{e}_x - \frac{1}{2}$.)

The basic functions $q_x$, $l_x$, $d_x$, $L_x$, $T_x$, and $\mathring{e}_x$ are usually tabulated in a standard format as in Table 4.1, which is the Life Table 1969–1971 for U. S. White Males. It was constructed using data from the 1970 census and death statistics for the 3 years 1969–1971. Note that $l_x$ refers to exact age $x$; $q_x$, $d_x$, and $L_x$ refer to the interval $(x, x+1)$, whereas $T_x$ and $\mathring{e}_x$ refer to the interval $[x, \infty)$, that is, beyond $x$.

In an abridged life table, the functions defined over the interval $[x, x+n)$ are denoted by ${}_nq_x$, ${}_nd_x$, and ${}_nL_x$, (usually with $n=5$), and generally for the interval $[x, x+t)$, by ${}_tq_x$, ${}_td_x$, and ${}_tL_x$, respectively. Note that in Table 4.1 there is a short life table for age 0–1 with intervals that are fractions of the 0–1 interval.

**Example 4.1** Suppose that a person aged 30 exactly is subject to mortality experience according to U. S. White Males Life Table 1969–1971,

(a) What is his expected age at death?
From Life Table WM, 1969–1971 we obtain $\mathring{e}_{30} = 40.07$, so that his expected age at death is $x + \mathring{e}_x = 30 + 41.07 = 71.07$.

(b) What is the probability that he dies in his 45th year?

$$\Pr\{44 < X \leqslant 45 | X > 30\} = \frac{l_{44} - l_{45}}{l_{30}} = \frac{d_{44}}{l_{30}} = \frac{457}{94716} = 0.0048250.$$

## 4.3 FORCE OF MORTALITY. MATHEMATICAL RELATIONSHIPS AMONG BASIC LIFE TABLE FUNCTIONS

As has been mentioned before, for theoretical purposes $l_x / l_0$ can be considered as a *continuous* function and is an analogue of the survival function $S_X(x)$. We may then set up some mathematical relationships among the life table functions analogous to those discussed in Section 3.3.

The *hazard rate* of the survival function, $\lambda_X(x)$, defined in (3.4) is called in the actuarial literature the *force of mortality* and is customarily denoted by $\mu_x$. We have

$$\mu_x = -\frac{1}{l_x} \frac{dl_x}{dx} = -\frac{d \log l_x}{dx}. \tag{4.15}$$

The function

$$-\frac{dl_x}{dx}=\mu_x l_x \tag{4.16}$$

is the density function [cf. (3.6a)]. It is sometimes called the *curve of deaths* in this context.

Integrating (4.16), we obtain

$$l_x=\int_x^{\infty}\mu_y l_y\,dy=\int_0^{\infty}\mu_{x+t}l_{x+t}\,dt, \tag{4.17}$$

and integrating (4.15), we obtain

$$l_x=l_0\exp\left[-\int_0^x \mu_y\,dy\right]. \tag{4.18}$$

[cf. (3.7)]. From (4.17) we obtain the expected number of deaths, $d_x$, between ages $x$ and $x+1$,

$$d_x=\int_x^{x+1}\mu_y l_y\,dy=\int_0^1 \mu_{x+t}l_{x+t}\,dt. \tag{4.19}$$

Using a probabilistic approach we notice that the conditional expectation of future life time $\mathring{e}_x$ can be expressed as

$$\begin{aligned}\mathring{e}_x&=\frac{1}{l_x}\int_0^{\infty}t\mu_{x+t}l_{x+t}\,dt=\frac{1}{l_x}\int_0^{\infty}t\left[-\frac{1}{l_{x+t}}\frac{dl_{x+t}}{dt}\right]l_{x+t}\,dt\\&=-\frac{1}{l_x}\int_0^{\infty}t\frac{dl_{x+t}}{dt}\,dt.\end{aligned} \tag{4.20}$$

[cf. (3.20)]. Integrating (4.20) by parts, we obtain

$$\mathring{e}_x=\frac{1}{l_x}\left[-tl_{x+t}\right]_{t=0}^{t=\infty}+\frac{1}{l_x}\int_0^{\infty}l_{x+t}\,dt=\frac{T_x}{l_x}, \tag{4.21}$$

which is identical with (4.12).

***Conditional Variance of Future Lifetime.*** Some life tables include an additional column of the quantity

$$Y_x=\int_0^{\infty}T_{x+t}\,dt, \tag{4.22}$$

which is useful in calculating (conditional) variance of future lifetime. We have [cf. (3.22)]

$$\operatorname{var}(X|X>x)=\frac{1}{l_x}\int_0^{\infty} t^2\mu_{x+t}l_{x+t}\,dt-\mathring{e}_x^2$$

$$=-\frac{1}{l_x}\int_0^{\infty} t^2\frac{dl_{x+t}}{dt}\,dt-\left(\frac{T_x}{l_x}\right)^2. \qquad (4.23)$$

Integrating by parts, we obtain

$$-\int_0^{\infty} t^2\frac{dl_{x+t}}{dt}\,dt=2\int_0^{\infty} T_{x+t}\,dt=2Y_x. \qquad (4.24)$$

Hence

$$\operatorname{var}(X|X>x)=\frac{2Y_x}{l_x}-\left(\frac{T_x}{l_x}\right)^2. \qquad (4.25)$$

In evaluating $Y_x$ we may use the approximation

$$\int_0^1 T_{x+t}\,dt \doteqdot \tfrac{1}{2}(T_x+T_{x+1}),$$

so that

$$Y_x=\int_0^{\infty} T_{x+t}\,dt \doteqdot \tfrac{1}{2}T_x+\sum_{i=1}^{\infty} T_{x+i}. \qquad (4.26)$$

## 4.4 CENTRAL DEATH RATE

Besides the force of mortality, $\mu_x$, which is an instantaneous death rate, we also define an "average" death rate over the age interval $[x,x+1)$, called the *central death rate* and denoted by $m_x$. We have

$$m_x=d_x/L_x. \qquad (4.27)$$

This is the "number of deaths per person-year lived in $[x,x+1)$." As will be shown in Section 4.6, there is an approximate relationship between $m_x$ and $q_x$.

Of course, $m_x$ is often a very small (fractional) number. If we multiply it by 1000, say, then we obtain the "average" number of deaths per 1000 persons per year.

## 4.5 INTERPOLATION FOR LIFE TABLE FUNCTIONS

We recall that $l_x/l_0$ is an analogue of the survival function. Any fitting, or approximation by means of parametric distributions should be judged primarily by the accuracy of approximation to $l_x/l_0$ or $l_x$.

We now consider calculation of approximate values of conditional probabilities of death over intervals $(x_1, x_2)$ when both $x_1$ and $x_2$ might not be integers. We ask a general question:

*What is the (approximate) probability that a person of exact age $x+\theta$, experiencing mortality according to a certain life table, will die before age $x+n+\phi$, where $x$ and $n$ are integers, and $0<\theta<1$, $0<\phi<1$?*

Using standard notation this probability can be denoted by ${}_{n+(\phi-\theta)}q_{x+\theta}$ [the interval is of length $(x+n+\phi)-(x+\theta)=n+(\phi-\theta)$]. We then have

$$ {}_{n+(\phi-\theta)}q_{x+\theta}=\Pr\{x+\theta \leqslant X \leqslant x+n+\phi | X > x+\theta\}=1-\frac{l_{x+n+\phi}}{l_{x+\theta}}. \tag{4.28}$$

Both $l_{x+n+\theta}$ and $l_{x+\theta}$ have to be obtained from interpolation formulae.

***Linear Interpolation.*** In this case [from (3.78a)], we have

$$ l_{z+\zeta} \doteqdot l_z(1-\zeta q_z), \qquad 0 \leqslant \zeta \leqslant 1. \tag{4.29}$$

Hence

$$ {}_{n+(\phi-\theta)}q_{x+\theta} \doteqdot \frac{l_{x+n}(1-\phi q_{x+n})}{l_x(1-\theta q_x)} = 1 - {}_np_x \frac{1-\phi q_{x+n}}{1-\theta q_x}, \tag{4.30}$$

where ${}_np_x = l_{x+n}/l_x$.

We note that although the interpolation is on $l_x$, the right-hand side of (4.30) is in terms of tabulated values of $q_x$. In particular, for $n=1$ and $\theta=\phi$, we obtain

$$ q_{x+\theta} \doteqdot 1 - p_x \frac{1-\theta q_{x+1}}{1-\theta q_x}. \tag{4.31}$$

For $n=0$ and $\theta=0$, we obtain

$$ {}_\phi q_x \doteqdot \phi q_x. \tag{4.32}$$

In particular,

$$_{1/2}q_x \doteq \tfrac{1}{2}q_x,$$

whereas

$$q_{x+1/2} \doteq 1 - p_x \frac{1-\frac{1}{2}q_{x+1}}{1-\frac{1}{2}q_x}. \tag{4.33}$$

When $n=0$ and $0<\theta<\phi<1$, we have $_np_x = 1$, so that

$$_{\phi-\theta}q_{x+\theta} \doteq 1 - \frac{1-\phi q_x}{1-\theta q_x}. \tag{4.34}$$

For $\phi=1$, (4.34) takes the form

$$_{1-\phi}q_{x+\theta} \doteq \frac{(1-\theta)q_x}{1-\theta q_x}. \tag{4.35}$$

In particular, for $\theta=\frac{1}{2}$,

$$_{1/2}q_{x+1/2} \doteq \frac{\frac{1}{2}q_x}{1-\frac{1}{2}q_x} = \frac{q_x}{2-q_x}. \tag{4.36}$$

Note that $_{1/2}q_x$ is the conditional probability of dying in the first half of the interval $x$ to $x+1$, whereas $_{1/2}q_{x+1/2}$ is the (conditional) probability of dying in the second half of the interval given surviving the first half. In general, these two probabilities are not the same.

***Exponential Interpolation.*** In this case

$$l_{z+\zeta} \doteq l_z^{1-\zeta} l_{z+1}^{\zeta} = l_z(1-q_z)^{\zeta}, \qquad 0 \leqslant \zeta \leqslant 1. \tag{4.37}$$

The approximation to the probability defined in (4.28) is now

$$_{n+(\phi-\theta)}q_{x+\theta} \doteq 1 - \frac{l_{x+n}(1-q_{x+n})^{\phi}}{l_x(1-q_x)^{\theta}} = 1 - {}_np_x \frac{(1-q_{x+n})^{\phi}}{(1-q_x)^{\theta}}. \tag{4.38}$$

We notice that we now have

$$_{1/2}q_x \doteq {}_{1/2}q_{x+1/2} \doteq 1-(1-q_x)^{1/2} = 1-p_x^{1/2}. \tag{4.39}$$

It should be emphasized that interpolation on $q_x$ (instead of $l_x$) gives inconsistent results and thus should be avoided.

**Example 4.2** A man aged 56 years and 22 weeks (a week is the smallest recorded time unit) enters a study 2 years and 14 weeks after the starting date of the study. The duration of the study is supposed to be 8 years.

(a) What is the probability that he will still be alive at the end of the study, assuming that his mortality experience is expected to be that described by U. S. Life Table 1969–1971, White Males? We have $x=56$, $\theta=22/52=0.42$. Since he enters 2 years and 14 weeks after the study started, then the period for which he can be under observation (the so called follow-up period) is 5 years and 38 weeks (i.e., 5.73 years). At the end of the follow-up period his age would be $56.42+5.73=62.15$. We then have $n=6$, $\phi=0.15$.

$$
\begin{aligned}
{}_{5.73}p_{56.42} &= \Pr\{X>62.15|X>56.42\}\\
&= \frac{l_{x+n+\phi}}{l_{x+\theta}} = \frac{l_{62.15}}{l_{56.42}} = \frac{l_{62}(1-0.15q_{62})}{l_{56}(1-0.42q_{56})}.
\end{aligned}
$$

From U. S. Life Table 1969–1971, White Males, we have

$$
l_{56}=81{,}796, \qquad q_{56}=0.01594, \qquad l_{62}=72{,}433, \qquad q_{62}=0.02657,
$$

and

$$
l_{62}/l_{56}={}_6p_{56}=\frac{72433}{81796}=0.88553.
$$

Hence

$$
{}_{5.73}p_{56.42} \doteq 0.88553\cdot\frac{1-0.15\cdot 0.02657}{1-0.42\cdot 0.01594}=0.88795.
$$

(b) What is the probability that he dies before the end of the study? We calculate

$$
\begin{aligned}
{}_{5.73}q_{56.42} &= \Pr\{56.42<X\leqslant 62.15|X>56.42\}\\
&= 1-{}_{5.73}p_{56.42} \doteq 0.11205.
\end{aligned}
$$

If we use exponential interpolation, we obtain ${}_{5.73}q_{56.42}=0.11228$, which is fairly close to that when a uniform distribution is assumed.

**Example 4.3** *Median of future lifetime*

Let $y$ denote the age up to which a person present age $x$ has 50% chance of surviving, that is $y-x$ is the *median* of future lifetime. The value of $y$ can be calculated using formula (3.23), that is, by solving the equation

$$\frac{l_y}{l_x}=0.50 \qquad \text{or} \qquad l_y=0.50l_x.$$

From the appropriate life table we can find only an integer $a$, say, such that

$$l_a<l_y<l_{a+1}.$$

An approximate value of $l_y$ can be found using linear interpolation on $l_x$.
Put $y=a+\theta$. Since

$$l_y=l_{a+\theta}=(1-\theta)l_a+\theta l_{a+1},$$

then

$$\theta=\frac{l_a-l_y}{l_a-l_{a+1}}. \tag{4.40}$$

For example, suppose we wish to find $y$ for a White Male, present age $x'=60$ lbd. We assume $x=60.5$, so that

$$l_x=l_{60.5}\doteqdot\tfrac{1}{2}(l_{60}+l_{61})=\tfrac{1}{2}(75{,}969+74{,}253)=75{,}111.$$

Hence $l_y\doteqdot 0.50l_{60.5}=37{,}556$, so that

$$l_{75}<l_y<l_{76} \qquad \text{or} \qquad 75<y<76.$$

Therefore, $y'=75$ lbd, so that there is a 50% chance of surviving another 15 to 16 years.

More precisely, we calculate [from (4.40)]

$$\theta\doteqdot\frac{40{,}324-37{,}984}{40{,}324-37{,}408}=\frac{2340}{2916}=0.80,$$

whence $y=75+0.80=75.80$.

It is left to the reader to calculate the median using exponential interpolation. In a similar way we can evaluate various percentiles of the distribution of future lifetime.

## 4.6 SOME APPROXIMATE RELATIONSHIPS BETWEEN ${}_nq_x$ AND ${}_nm_x$

We now give some approximate relationships between the conditional probability of death, ${}_nq_x$, and the central death rate, ${}_nm_x$, introduced for abridged life tables. The formulae are useful in construction of a life table from cross-sectional (census) mortality data (see Section 4.9).

### 4.6.1 Expected Fraction of the Last *n* Years of Life

${}_na_x$ denotes the expected (average) number of years lived by an individual age $x$ who dies in age interval $[x,x+n)$. We define

$$ {}_nf_x = \frac{1}{n}\,{}_na_x, \tag{4.41}$$

which is the expected *fraction* of this last $n$-years-lived interval. The *expected* total number of years lived by $l_x$ individuals, now aged $x$, over the years $x$ to $x+n$, will be ${}_nL_x$, and it is made up of: (1) full $n$ years for each $n \cdot l_{x+n}$ survivors out of $l_x$, and (2) ${}_na_x \cdot {}_nd_x$ (average) years for individuals who die in $[x,x+n)$.

We then have

$$\begin{aligned} {}_nL_x &\doteq n \cdot l_{x+n} + {}_na_x \cdot {}_nd_x = n(l_x - {}_nd_x) + n\,{}_nf_x \cdot {}_nd_x \\ &= n\left[\,l_x - (1 - {}_nf_x)\,{}_nd_x\right], \end{aligned} \tag{4.42}$$

or

$$ l_x \doteq \frac{1}{n}\left[\,{}_nL_x + n(1 - {}_nf_x)\,{}_nd_x\right]. \tag{4.43}$$

Whence (noticing that ${}_nd_x = l_x \cdot {}_nq_x$)

$$ {}_nm_x \doteq \frac{{}_nd_x}{{}_nL_x} = \frac{{}_nd_x}{n\left[\,l_x - (1 - {}_nf_x)\,{}_nd_x\right]} = \frac{{}_nq_x}{n\left[1 - (1 - {}_nf_x)\,{}_nq_x\right]}. \tag{4.44}$$

Similarly (noticing that ${}_nd_x = {}_nL_x \cdot {}_nm_x$),

$$ {}_nq_x \doteq \frac{{}_nd_x}{l_x} = \frac{{}_nd_x}{\frac{1}{n}\left[\,{}_nL_x + n(1 - {}_nf_x)\,{}_nd_x\right]} = \frac{{}_nm_x}{\frac{1}{n}\left[1 + n(1 - {}_nf_x)\,{}_nm_x\right]} \tag{4.45}$$

[see also Chiang (1968, Chapter 9)].

### 4.6.2 Special Cases

Assuming an approximate *uniform* distribution of time at death over the interval $[x,x+n)$, we have ${}_na_x \doteq n/2$ (or ${}_nf_x \doteq \frac{1}{2}$). With this assumption we obtain

$$ {}_n m_x \doteq \frac{{}_n q_x}{n\left(1-\frac{1}{2}{}_n q_x\right)} \quad \text{and} \quad {}_n q_x \doteq \frac{{}_n m_x}{\frac{1}{n}\left(1+\frac{n}{2}{}_n m_x\right)}. \tag{4.46} $$

Taking $n=1$, we obtain the approximate formulae for complete life tables. In particular, when $n=1$ and also ${}_n f_x = f_x = \frac{1}{2}$, we have the well-known formulae

$$ m_x \doteq \frac{q_x}{1-\frac{1}{2}q_x} = \frac{2q_x}{2-q_x} \quad \text{and} \quad q_x \doteq \frac{m_x}{1+\frac{1}{2}m_x} = \frac{2m_x}{2+m_x}. \tag{4.47} $$

**Example 4.4** We now show that the curtate expectation of life, $e_x$, defined in (4.14) is approximately $\mathring{e}_x - \frac{1}{2}$. Taking $n=1$ and assuming $f_x = \frac{1}{2}$, we obtain [from (4.42)] $L_x \doteq \frac{1}{2}(l_x + l_{x+1})$. We have

$$ \mathring{e}_x = \frac{1}{l_x}\sum_{i=0}^{\infty} L_{x+i} \doteq \frac{1}{l_x}\sum_{i=0}^{\infty} \tfrac{1}{2}(l_{x+i} + l_{x+i+1}) $$

$$ = \tfrac{1}{2} + \frac{1}{l_x}\sum_{i=1}^{\infty} l_{x+i} = \tfrac{1}{2} + e_x, \tag{4.48} $$

or

$$ e_x \doteq \mathring{e}_x - \tfrac{1}{2}. \tag{4.48a} $$

### 4.6.3 Exponential Approximation

Slightly different results are obtained if we assume an approximately *exponential* distribution of age at death over $[x, x+n)$, with constant force of mortality $\mu_x$. In this case

$$ {}_n m_x \doteq \mu_x \quad \text{and} \quad {}_n q_x \doteq 1 - e^{-n\,{}_n m_x}. \tag{4.49} $$

## 4.7 SOME APPROXIMATIONS TO $\mu_x$

Assuming that the force of mortality is constant over age interval $[x, x+n)$, we could approximate $\mu_x$ by ${}_n m_x$ [see (4.49)]. Here are some further approximations to $\mu_x$, *not* subject to this assumption.

1. From the relation

$$ \int_x^{x+n} \mu_y \, dy \doteq n\mu_{x+\frac{1}{2}n}, $$

and from (3.10a), we obtain

$$\mu_{x+\frac{1}{2}n} \doteqdot -\frac{1}{n}\log {}_n p_x. \tag{4.50}$$

In particular, for $n=1$, we have

$$\mu_{x+\frac{1}{2}} \doteqdot -\log p_x. \tag{4.50a}$$

These are the most common approximations to $\mu_x$ over short intervals.

2. Using

$$2\mu_x \doteqdot \int_{x-1}^{x+1} \mu_y \, dy = -\log p_{x-1} - \log p_x,$$

we obtain

$$\mu_x \doteqdot -\tfrac{1}{2}(\log p_{x-1} + \log p_x). \tag{4.51}$$

3. If $\mu_x$ (and so $q_x$) is very small

$$\log p_x = \log(1-q_x) = -q_x + \tfrac{1}{2}q_x^2 - \dots.$$

Taking only first terms, (4.50a) and (4.51) give

$$\mu_{x+\frac{1}{2}} \doteqdot q_x, \qquad \text{and} \qquad \mu_x \doteqdot \tfrac{1}{2}(q_{x-1}+q_x), \tag{4.52}$$

respectively.

## 4.8 CONCEPTS OF STATIONARY AND STABLE POPULATIONS

As we have said before, many life tables have to be constructed from mortality data of current populations. How can this be done?

First, we realize that the available population data usually are: age distribution, number of births, and number of deaths in each age group of the *current* population over a certain period.

Second, we should be aware that, for example, the death rate in age group 60–65, say, for the current population cannot be expected to be the same when the presently newborn individuals reach that age. Therefore, we make some assumptions which, although almost invariably untrue, help us to appreciate the nature of the information in the life table. These assumptions are the basis for two concepts in common use: concepts of *stationary* and *stable* populations.

### 4.8.1 Stationary Population

Associated with a given life table, we suppose that we have a population in which the following are true:

1. There are exactly $l_0$ births per (calendar) year, and they are uniformly distributed over the year, so that the number of births in a period of length $t$ is $tl_0$.
2. The mortality experienced is exactly that of the life table and deaths ($d_x$) are uniformly distributed in the same way as births.

This is clearly a *hypothetical* population (for example, one cannot in reality have a fractional number of births), but it is a helpful concept.

In a population of this kind, the age composition never changes and the total size remains constant, since $d_0 + d_1 + \ldots = l_0$ deaths each year are exactly matched by $l_0$ births. Such a population is called a *stationary population*. The number of individuals aged $x$ lbd in a stationary population is $L_x$. This can be seen by noticing that at any moment, individuals aged $x$ lbd are survivors of births between $x$ and $(x+1)$ years ago, and summing these gives

$$\int_x^{x+1} l_y \, dy = L_x$$

(since there are $l_y\,dy$ survivors out of $l_0 dy$ births $y$ years ago). Similarly, $T_x$ is the number of individuals in the population aged $x$ *or more*, and, in particular $T_0$ is the total size of the population.

Of course, real populations are not stationary because: (1) the number of births changes from year to year, (2) the mortality experience changes from year to year, and (3) births and deaths are not uniformly distributed over each year.

### 4.8.2 Stable Population

A restricted amount of variation from a stationary population can be introduced by supposing that the number of births is increasing at a *constant rate*, $r$, say, per year ($r$ may be negative).

This implies that the total number of births in one year is $e^r$ times the total number in the preceding year. It is still supposed that the age-specific death rates remain unaltered. As a consequence: (1) the total *size* of population will be increasing at proportional rate $e^r$ per year, and (2) the *proportionate composition by age* will be stable, with

$$L_x \propto \int_x^{x+1} e^{-ry} l_y \, dy. \tag{4.53}$$

Such a population is called a *stable* population.

Models of stationary and stable populations can be useful in providing insight into the implications of the corresponding life tables. They exhibit the consequences of the mortality pattern under hypothetical, but easily comprehended, circumstances.

## 4.9 CONSTRUCTION OF AN ABRIDGED LIFE TABLE FROM MORTALITY EXPERIENCE OF A CURRENT POPULATION

We are now ready to discuss methods of constructing a life table using mortality data from a *current* population. We first discuss construction of an *abridged* life table giving $l_x$ at quinquennial intervals (with $l_1$ in addition) until $l_{85}$ as is done in abridged U. S. tables. The last age class *and over* (denoted by 85+) appears to be rather wide; it seems that at least class 95+, or even 100+ should be used. We will first discuss the methods of constructing abridged life tables, all having a common feature: estimating ${}_nq_x$ from the observed age specific (central) death rate ${}_nM_x$. Each method represents essentially an approximate formula relating ${}_nq_x$ and ${}_nM_x$. (Note that for $x \geqslant 5$, $n=5$, but for $x=0$, $n=1$ and for $x=1$, $n=4$.)

The following books and papers might be useful for learning about these techniques: Benjamin and Haycocks, (1970), Chiang (1968, 1972), Hooker and Longley-Cook (1953), and Jordan (1967). Some specific techniques can be found in Greville (1943, 1967), Keyfitz (1966, 1968), King (1914), Reed and Merrell (1939), and Sirken (1964). We found especially useful Chapter 15 of Shryock et al. (1973), in which a number of these techniques is presented, with examples.

In this book we restrict detailed discussion to the simplest methods, and only briefly outline a few others. We are concerned mainly to provide some background and basic concepts rather than full technical details.

Let ${}_nM_x$ denote the *observed* (central) rate in age group $x$ to $x+n$ *per person per year*. The basic assumption on which the construction of a life table is based is

$${}_nm_x \doteqdot {}_nM_x, \tag{4.54}$$

that is, that the observed age-specific (central) death rate ${}_nM_x$ is a good approximation to the central death rate of the life table ${}_nm_x$.

### 4.9.1 Evaluation of ${}_nM_x$

If $T$ is the length of the observation period, then ${}_nM_x$ is calculated from

$$ {}_nM_x = \frac{\begin{array}{l}\text{no. of deaths in age group } x \text{ to } x+n\\ \text{over the period of length } T\end{array}}{\left[\begin{array}{l}\text{population size at time } t'\\ \text{for age group } x \text{ to } x+n\end{array},\right]\times T} $$

where $0<t'<T$.

Of special interest is the situation for $T=1$ (one calendar year). We now introduce special notation.

${}_nK_x$ denotes the "average population exposed to risk" for the age group $x$ to $x+n$ over a single *calendar* year. Generally, this population is evaluated at time point $t'$, $0<t'<1$. In practice, however, it is usually assumed that $t'=\frac{1}{2}$, so that ${}_nK_x$ is usually the *midyear* population as of July 1 for the age group $x$ to $x+n$. For census year, ${}_nK_x$ is often taken as the *census* population in age group $x$ to $x+n$ (in the United States as of April 1), without correcting it to the midyear population. It appears that there are other effects that tend to counterbalance those arising from omitting this correction (Greville (1967)).

${}_nD_x$ is the observed number of deaths in age group $x$ to $x+n$ over the specified calendar year.

We then have

$$ {}_nM_x = \frac{{}_nD_x}{{}_nK_x}. \tag{4.55} $$

In construction of national life tables, the specified period of observation is usually 3 calendar years ($T=3$), the census year being the middle one.

Let ${}_nK_x^{(j)}$ and ${}_nD_x^{(j)}$ denote the "population exposed to risk," and the number of deaths in age group $x$ to $x+n$, respectively, for the $j$th calendar year. Then ${}_nM_x$ is often calculated from the formula

$$ {}_nM_x \doteq \frac{{}_nD_x^{(1)}+{}_nD_x^{(2)}+{}_nD_x^{(3)}}{{}_nK_x^{(1)}+{}_nK_x^{(2)}+{}_nK_x^{(3)}}, \tag{4.56} $$

or

$$ {}_nM_x \doteq = \frac{{}_nD_x^{(1)}+{}_nD_x^{(2)}+{}_nD_x^{(3)}}{3\,{}_nK_x^{(2)}}, \tag{4.57} $$

where ${}_nK_x$ is the census population and is often taken as the *midperiod* (for the period of three years) population (cf. Example 2.6).

### 4.9.2 Estimation of ${}_nf_x$

We discuss here three methods of estimating ${}_nf_x$—the expected fraction of the $[x,x+n)$ interval for those aged $x$ who die in $[x,x+n)$. We need this quantity for calculating ${}_nq_x$ from mortality data of current populations.

1. Assuming *uniform* distribution of time at death in $[x,x+n)$, we obtain ${}_nf_x=\frac{1}{2}$ (see Section 4.6.2).

2. Taking ${}_nf_x=\frac{1}{2}$ for all $x$, is likely to be an oversimplification. Chiang (1972) evaluated ${}_nf_x$ *empirically* from death certificates in two large studies (California, 1960, and 10% sample of U. S. deaths, 1963). In each study, the exact number of days lived beyond age $x$ by a person, who died between age $x$ and $x+n$, was recorded to obtain the estimate ${}_na_x$, and then ${}_nf_x={}_na_x/n$ [cf. (4.41)]. Both studies gave similar results. Chiang's values

**Table 4.2** Fractions of the last $n$-year interval of life, ${}_nf_x$, from Chiang's study and evaluated for different U. S. Life Tables

| Age group $x$ to $x+n$ | Empirical (Chiang's study) | Total population 1959–1961 | Total population 1969–1971 | White males 1969–1971 | White females 1969–1971 |
|---|---|---|---|---|---|
| 0–1 | 0.10 | 0.157 | 0.142 | 0.129 | 0.134 |
| 1–5 | 0.39 | 0.402 | 0.419 | 0.430 | 0.420 |
| 5–10 | 0.46 | 0.460 | 0.462 | 0.470 | 0.460 |
| 10–15 | 0.54 | 0.549 | 0.578 | 0.597 | 0.546 |
| 15–20 | 0.57 | 0.542 | 0.546 | 0.549 | 0.532 |
| 20–25 | 0.49 | 0.509 | 0.506 | 0.502 | 0.505 |
| 25–30 | 0.50 | 0.507 | 0.500 | 0.488 | 0.515 |
| 30–35 | 0.52 | 0.523 | 0.522 | 0.517 | 0.530 |
| 35–40 | 0.54 | 0.533 | 0.532 | 0.535 | 0.537 |
| 40–45 | 0.54 | 0.536 | 0.533 | 0.537 | 0.536 |
| 45–50 | 0.54 | 0.536 | 0.531 | 0.535 | 0.532 |
| 50–55 | 0.53 | 0.530 | 0.531 | 0.535 | 0.530 |
| 55–60 | 0.52 | 0.529 | 0.527 | 0.529 | 0.528 |
| 60–65 | 0.52 | 0.524 | 0.523 | 0.522 | 0.528 |
| 65–70 | 0.52 | 0.519 | 0.519 | 0.515 | 0.531 |
| 70–75 | 0.51 | 0.514 | 0.516 | 0.508 | 0.530 |
| 75–80 | 0.51 | 0.508 | 0.507 | 0.496 | 0.521 |
| 80–85 | 0.48 | 0.490 | 0.492 | 0.479 | 0.503 |
| 85+ | — | — | — | — | — |

taken from his book (1968) are exhibited in Table 4.2 (column 2).

3. Values of ${}_nf_x$ are sometimes obtained from a so-called *standard* life table, which is believed to be comparable with the life table to be constructed. For example, U. S. Life Table 1959–1961 might be used as a standard life table for calculating ${}_nf_x$, which can be used in constructing the U. S. abridged life table 1969–1971.

In many complete life tables $L_x$ (except for $L_0$) is calculated from the approximate formula $L_x \doteq \frac{1}{2}(l_x + l_{x+1})$, which implies that $f_x = \frac{1}{2}$. However, using values of ${}_nL_x = L_x + L_{x+1} + \dots + L_{x+n-1}$, ${}_nf_x$ is not necessarily equal to $\frac{1}{2}$. It is possible to estimate it by using (4.42) in the form

$$ {}_n\hat{f}_x = \frac{{}_nL_x - nl_{x+n}}{n\cdot{}_nd_x} = \frac{{}_nL_x - n(l_x - {}_nd_x)}{n\cdot{}_nd_x}. \tag{4.58} $$

Table 4.2 gives ${}_n\hat{f}_x$ values calculated from three different life tables.

### 4.9.3 Estimation of ${}_nq_x$

The next step is to evaluate ${}_nq_x$. Assuming that the relation (4.54) is approximately correct, ${}_nq_x$ can be estimated from (4.45), substituting ${}_nM_x$ for ${}_nm_x$, that is

$$ {}_n\hat{q}_x \doteq \frac{{}_nM_x}{\frac{1}{n}\left[1 + n(1 - {}_n\hat{f}_x){}_nM_x\right]}, \tag{4.59} $$

or substituting for ${}_nM_x$ from (4.55), we obtain

$$ {}_n\hat{q}_x = \frac{{}_nD_x}{\frac{1}{n}\left[{}_nK_x + n(1 - {}_n\hat{f}_x){}_nD_x\right]} = \frac{{}_nD_x}{N'_x}, \tag{4.60} $$

where

$$ N'_x = \frac{1}{n}\left[{}_nK_x + n(1 - {}_n\hat{f}_x){}_nD_x\right] \tag{4.61} $$

is the estimated number of individuals in the actual population who passed through exact age $x$ during a given calendar year (or effective number of lives at age $x$).

Note that although (4.61) resembles (4.43), $N'_x$ is not equivalent to $l_x$ (why? cf. Section 2.6.1 and Exercises 4.2 and 4.3).

### 4.9.4 Evaluation of the Life Table Functions

Once the basic life table function, ${}_nq_x$, has been estimated the remaining functions are calculated in a straightforward manner. For convenience, we now summarize the steps in calculation of abridged life tables.

***Data.*** ${}_nK_x$, ${}_nD_x$, ${}_n\hat{f}_x$ (or equivalently, ${}_nM_x$, ${}_n\hat{f}_x$) and $\omega'$ $(=85)$ is the last *recorded* age.

***Calculations.***

$$ {}_nM_x = \frac{{}_nD_x}{{}_nK_x}, \qquad x=0,1,5,10,\ldots,85, $$

and

$$ {}_nq_x \doteq \frac{{}_nM_x}{\frac{1}{n}\left[1+n(1-{}_n\hat{f}_x){}_nM_x\right]} \qquad \text{for } x=0,1,5,10,\ldots,80, \quad \text{and} \quad {}_\infty q_{85}=1 $$

or equivalently,

$$ N_x' = \frac{1}{n}\left[{}_nK_x + n(1-{}_n\hat{f}_x){}_nD_x\right], $$

$$ {}_nq_x \doteq \frac{{}_nD_x}{N_x'}, \qquad x=0,1,5,10,\ldots,80, \qquad \text{and} \qquad {}_\infty q_{85}=1. $$

Take $l_0 = 100{,}000$, (for example), and calculate

$$ l_{x+n} = l_x(1-{}_nq_x), \qquad x=0,1,5,10,\ldots,80, $$

$$ {}_nd_x = l_x - l_{x+n} = l_x \cdot {}_nq_x, \qquad x=0,1,5,10,\ldots,80, \qquad \text{and} \qquad {}_\infty d_{85} = l_{85}; $$

$$ {}_nL_x = n\left[l_x - (1-{}_n\hat{f}_x){}_nd_x\right], \qquad x=0,1,5,10,\ldots,80, $$

$$ \text{and} \qquad {}_\infty L_{85} = \frac{{}_\infty d_{85}}{{}_\infty M_{85}} = \frac{l_{85}}{{}_\infty M_{85}}; $$

$$ T_x = {}_nL_x + {}_nL_{x+n} + \cdots + {}_\infty L_{85}, \qquad x=0,1,5,10,\ldots,85, $$

$$ \mathring{e}_x = \frac{T_x}{l_x}, \qquad x=0,1,5,10,\ldots,85. $$

Additionally,

$$_na_x \doteq n \cdot {}_n\hat{f}_x, \qquad x=0,1,5,10,\ldots,80, \quad \text{and} \quad {}_\infty a_{85} = \mathring{e}_{85}.$$

Note that if $\omega'$ is much less than $\omega$ (where $\omega$ is the last actually observed age in a population), then ${}_\infty L_x$, and so $T_x$, are not very reliable values. Thus the whole column of $\mathring{e}_x$ could be misleading (see Exercise 4.9).

**Example 4.5** The data in Table 4.3 represent the estimated midyear population (${}_nK_x$) and deaths (${}_nD_x$) for U. S. White Males, 1972 (from U. S. Vital Statistics, 1972, Vol. II. Part A).

**Table 4.3** U. S. White Males, 1972. Population data. (U. S. Vital Statistics 1972. Vol. II. Part A)

| Age group $x$ to $x+n$ | ${}_nK_x$ | ${}_nD_x$ | ${}_nM_x \cdot 10^3$ | ${}_nf_x$ | ${}_n\hat{q}_x$ | ${}_n\tilde{q}_x$* | ${}_n\tilde{\tilde{q}}_x$† |
|---|---|---|---|---|---|---|---|
| 0–1 | 1,412,000 | 25,422 | 18.004 | 0.129 | 0.01773 | 0.01784 | 0.0182 |
| 1–5 | 5,889,000 | 4,858 | 0.825 | 0.430 | 0.00329 | 0.00329 | 0.0033 |
| 5–10 | 8,058,000 | 3,632 | 0.451 | 0.470 | 0.00225 | 0.00225 | 0.0023 |
| 10–15 | 9,103,000 | 4,442 | 0.488 | 0.597 | 0.00244 | 0.00244 | 0.0024 |
| 15–20 | 8,727,000 | 13,210 | 1.514 | 0.549 | 0.00754 | 0.00754 | 0.0075 |
| 20–25 | 7,650,000 | 14,744 | 1.927 | 0.502 | 0.00959 | 0.00959 | 0.0096 |
| 25–30 | 6,580,000 | 11,040 | 1.676 | 0.488 | 0.00834 | 0.00834 | 0.0083 |
| 30–35 | 5,295,000 | 9,462 | 1.787 | 0.517 | 0.00890 | 0.00890 | 0.0089 |
| 35–40 | 4,777,000 | 11,782 | 2.466 | 0.535 | 0.01226 | 0.01226 | 0.0123 |
| 40–45 | 5,071,000 | 20,264 | 3.966 | 0.537 | 0.01980 | 0.01978 | 0.0198 |
| 45–50 | 5,205,000 | 34,950 | 6.715 | 0.535 | 0.03306 | 0.03302 | 0.0331 |
| 50–55 | 5,061,000 | 53,478 | 10.567 | 0.535 | 0.05157 | 0.05147 | 0.0516 |
| 55–60 | 4,393,000 | 75,640 | 17.218 | 0.529 | 0.08274 | 0.08254 | 0.0827 |
| 60–65 | 3,788,000 | 101,132 | 26.698 | 0.522 | 0.12548 | 0.12514 | 0.1255 |
| 65–70 | 2,971,000 | 116,682 | 39.274 | 0.515 | 0.17929 | 0.17881 | 0.1793 |
| 70–75 | 2,117,000 | 124,830 | 58.966 | 0.508 | 0.25748 | 0.25695 | 0.2575 |
| 75–80 | 1,453,000 | 128,614 | 88.516 | 0.496 | 0.36186 | 0.36239 | 0.3624 |
| 80–85 | 845,000 | 106,410 | 125.929 | 0.479 | 0.47411 | 0.47888 | 0.4745 |
| 85+ | 477,000 | 96,670 | 202.662 | — | 1.00000 | 1.00000 | 1.0000 |
| All | 88,880,000 | 957,262 | 13.773 | | | | |

*Estimator of ${}_nq_x$ when ${}_nf_x = \frac{1}{2}$ for each interval.

†Values of ${}_nq_x$ from the abridged life table published in U. S. Vital Statistics, 1972, Vol. II. Part A, p. 5-5.

The life table (using values of ${}_nf_x$ estimated from U. S. Life Table 1969–1971, White Males (see Table 4.2 column 5)) is presented in Table 4.4, which shows only small discrepancies from the corresponding life table for White Males published in U. S. Vital Statistics, 1972, Vol. II, Part A p. 55, which was probably calculated using some further interpolation.

**Table 4.4** Abridged life table. U. S. White Males, 1972.

| Age group $x$ to $x+n$ | ${}_nq_x$ | $l_x$ | ${}_nd_x$ | ${}_nL_x$ | $T_x$ | $\mathring{e}_x$ | ${}_na_x$ |
|---|---|---|---|---|---|---|---|
| 0–1 | 0.01773 | 100,000 | 1,773 | 98,456 | 6,832,777 | 68.33 | 0.13 |
| 1–5 | 0.00329 | 98,227 | 324 | 392,172 | 6,734,321 | 68.56 | 1.72 |
| 5–10 | 0.00225 | 97,904 | 220 | 488,935 | 6,342,149 | 64.78 | 2.35 |
| 10–15 | 0.00224 | 97,683 | 238 | 487,938 | 5,853,213 | 59.92 | 2.99 |
| 15–20 | 0.00754 | 97,445 | 735 | 485,569 | 5,365,276 | 55.06 | 2.75 |
| 20–25 | 0.00959 | 96,710 | 928 | 481,242 | 4,879,706 | 50.46 | 2.51 |
| 25–30 | 0.00834 | 95,783 | 799 | 476,869 | 4,398,464 | 45,92 | 2.44 |
| 30–35 | 0.00890 | 94,984 | 845 | 472,878 | 3,921,595 | 41.29 | 2.59 |
| 35–40 | 0.01226 | 94,139 | 1,154 | 468,010 | 3,448,717 | 36.63 | 2.68 |
| 40–45 | 0.01980 | 92,984 | 1,841 | 460,661 | 2,980,707 | 32.06 | 2.69 |
| 45–50 | 0.03306 | 91,144 | 3,013 | 448,713 | 2,520,047 | 27.65 | 2.68 |
| 50–55 | 0.05157 | 88,131 | 4,545 | 430,087 | 2,071,334 | 23.50 | 2.68 |
| 55–60 | 0.08274 | 83,586 | 6,916 | 401,644 | 1,641,247 | 19.64 | 2.65 |
| 60–65 | 0.12548 | 76,670 | 9,621 | 360,358 | 1,239,603 | 16.17 | 2.61 |
| 65–70 | 0.17929 | 67,050 | 12,021 | 306,096 | 879,245 | 13.11 | 2.58 |
| 670–75 | 0.25748 | 55,028 | 14,169 | 240,286 | 573,149 | 10.42 | 2.54 |
| 75–80 | 0.36186 | 40,860 | 14,786 | 167.038 | 332,863 | 8.15 | 2.48 |
| 80–85 | 0.47411 | 26,074 | 12,362 | 98,167 | 165,826 | 6.36 | 2.40 |
| 85+ | 1.00000 | 13,712 | 13,712 | 67,569 | 67,659 | 4.93 | 4.93 |

## 4.10 SOME OTHER APPROXIMATIONS USED IN CONSTRUCTION OF ABRIDGED LIFE TABLES

***Greville's Cubic Approximation.*** In this method, a first draft of an abridged life table is constructed using ${}_nf_x=\frac{1}{2}$. The correction suggested by Greville (1943) is based on a third-order polynomial approximation. It turns out that the ${}_nL_x$ can be estimated from the formula

$$ {}_nL_x \doteq n\left[\tfrac{1}{2}(l_x + l_{x+n}) + \frac{1}{24}({}_nd_{x+n} - {}_nd_{x-n})\right] \tag{4.62}$$

for $x>5$. When $x<5$ special formulae are used. Using now ${}_nL_x$ calculated from (4.42), a new ${}_nf_x$ can be obtained. The procedure is repeated until all life table functions converge with a desired precision.

***Keyfitz' method.*** The method has been described in detail by Keyfitz in various publications (e.g., Keyfitz (1966), Preston et al. (1972)). Its essential feature is the allowance for a supposed constant growth rate. In this method, the growth rate specific for the age group $x$ to $x+n$ is represented by ${}_nr_x$. The value of ${}_nr_x$ is estimated directly from midyear population age structure, and mortality data in two adjacent age groups $x-5$ to $x$, and $x$ to $x+5$, applying Greville's cubic approximation just described, and using iterative procedures. The resulting life table usually differs only little from a life table constructed by the simpler methods described in Section 4.9, but the Keyfitz' method seems to be interesting on its own merits.

## 4.11 CONSTRUCTION OF A COMPLETE LIFE TABLE FROM AN ABRIDGED LIFE TABLE

A complete life table should really be constructed from the original mortality data grouped in single year (or even narrower) intervals. When such data are not available in published form, a complete life table can often be constructed from an abridged life table using some 'smoothing' formulae. Essentially, we need three kinds of interpolation schemes: (1) for very young (0–1), and young ages (1–9); (2) for the broad class of ages (10–74); and (3) for very old ages (75 and above). Different methods have been used by different authors. From our experience, it appears that the use of six-point Lagrangian interpolation formulae over the age range 1–74 and smoothing by fitting a Gompertz curve for ages above 75 represents a satisfactory method. Special methods are needed for the first year of life. Table 4.5 gives the coefficients for the six-point Lagrangian formula for interpolating on any tabulated function $u_x$. For details of calculation of these coefficients see, for example, Abramowitz and Stegun (1965), and also Section 3.14.1. In this interpolation, $u_x$ always represents $l_x$. The use of this table is as follows.

*1. Ages 1–9.* If the abridged life table includes only quinquennial values of $l_x$, we use the coefficients in the upper part of Table 4.5 (for $x<10$), that is, we use the values of $l_0$, $l_5$, $l_{10}$, $l_{15}$, $l_{20}$, and $l_{25}$. For example, we calculate $l_2$

$$l_2=0.344448l_0+1.148160l_5-0.861120l_{10}$$
$$+0.529920l_{15}-0.191360l_{20}+0.029952l_{25}.$$

**Table 4.5** Six point Lagrangian coefficients for interpolation on values $u_x$

| Coefficients of $u_x$ to obtain: | Coefficients to be used for $x<10$ | | | | | |
|---|---|---|---|---|---|---|
| | $x=0$ | $x=5$ | $x=10$ | $x=15$ | $x=20$ | $x=25$ |
| $u_1$ | 0.612864 | 0.766080 | −0.680960 | 0.437760 | −0.161280 | 0.025536 |
| $u_2$ | 0.344448 | 1.148160 | −0.861120 | 0.529920 | −0.191360 | 0.029952 |
| $u_3$ | 0.167552 | 1.256640 | −0.718080 | 0.418880 | −0.147840 | 0.022848 |
| $u_4$ | 0.059136 | 1.182720 | −0.394240 | 0.215040 | −0.073920 | 0.011264 |
| $u_6$ | −0.025536 | 0.766080 | 0.383040 | −0.170240 | 0.054720 | −0.008064 |
| $u_7$ | −0.029952 | 0.524160 | 0.698880 | −0.262080 | 0.080640 | −0.011648 |
| $u_8$ | −0.022848 | 0.304640 | 0.913920 | −0.261120 | 0.076160 | −0.010752 |
| $u_9$ | −0.011264 | 0.126720 | 1.013760 | −0.168960 | 0.046080 | −0.006336 |
| when $u_1$ is used | $x=1$ | $x=5$ | $x=10$ | $x=15$ | $x=20$ | $x=25$ |
| $u_2$ | 0.562030 | 0.717600 | −0.478400 | 0.283886 | −0.100716 | 0.015600 |
| $u_3$ | 0.273392 | 1.047199 | −0.531911 | 0.299200 | −0.103747 | 0.015867 |
| $u_4$ | 0.096491 | 1.108800 | −0.328533 | 0.172800 | −0.058358 | 0.008800 |
| $u_6$ | −0.041667 | 0.798000 | 0.354667 | −0.152000 | 0.048000 | −0.007000 |
| $u_7$ | −0.048872 | 0.561600 | 0.665600 | −0.240686 | 0.072758 | −0.010400 |
| $u_8$ | −0.037281 | 0.333200 | 0.888533 | −0.244800 | 0.070147 | −0.009800 |
| $u_9$ | −0.018379 | 0.140800 | 1.001244 | −0.160914 | 0.043116 | −0.005867 |

| Coefficients of $u_x$ to obtain: | Coefficients to be used for $x>10$ | | | | | |
|---|---|---|---|---|---|---|
| | $x=5m-10$ | $x=5m-5$ | $x=5m$ | $x=5m+5$ | $x=5m+10$ | $x=5m+15$ |
| $u_{5m+1}$ | 0.008064 | −0.073920 | 0.887040 | 0.221760 | −0.049280 | 0.006336 |
| $u_{5m+2}$ | 0.011648 | −0.099840 | 0.698880 | 0.465920 | −0.087360 | 0.010752 |
| $u_{5m+3}$ | 0.010752 | −0.087360 | 0.465920 | 0.698880 | −0.099840 | 0.011648 |
| $u_{5m+4}$ | 0.006336 | −0.049280 | 0.221760 | 0.887040 | −0.073920 | 0.008064 |

Substituting the corresponding values for $l_0$ through $l_{25}$ from Table 4.4, we obtain $l_2 = 98{,}738$.

However, in most abridged life tables we also have $l_1$. In this case, we use the second set of coefficients in Table 4.5 and obtain $l_2 = 98{,}148$.

The mortality in the first year of life is very different from that in the second and sebsequent years, and if $l_1$ is available, the second set of coefficients should always be used.

**2.** ***Ages 10–74.*** The last set (for $x > 10$) of coefficients in Table 4.5 is used in this case. For example, to calculate $l_{11}$, we put $m = 2$, so that we have

$$l_{11} = 0.008064 l_0 - 0.073920 l_5 + 0.887040 l_{10} + 0.221760 l_{15} - 0.049280 l_{20} + 0.006336 l_{25}.$$

**3.** ***Ages 75 and above.—Fitting Gompertz distribution.*** The method described below is applicable in the special circumstances prevailing. A more general discussion of fitting Gompertz distributions is in Section 8.4. We use the following argument.

The Gompertz survival function can be written as

$$S(x|a,R) = \exp\left[\frac{R}{a}(1 - e^{ax})\right] = b^{1-c^x}, \tag{4.63}$$

where $b = e^{R/a}$ and $c = e^a$. We take values of $l_x$ at two adjacent tabulated points, $x_1$ and $x_1 + 5$, say, and calculate

$$y_1 = \log\frac{l_{x_1}}{l_{x_1+5}} = c^{x_1}(c^5 - 1)\log b. \tag{4.64}$$

Similarly, for another pair, $x_2$ and $x_2 + 5$, we have

$$y_2 = \log\frac{l_{x_2}}{l_{x_2+5}} = c^{x_2}(c^5 - 1)\log b. \tag{4.65}$$

Solving these two simultaneous equations, we obtain estimates of $b$ and $c$ (and so of $a$ and $R$). (Note that $y_1/y_2 = c^{x_1 - x_2}$.) Of course, we may have $x_2 = x_1 + 5$, so that, in fact, we use only three points; in this case $x_1 - x_2 = -5$.

We may repeat this procedure for several pairs of available values, $x_1$, $x_2$, and estimate $c$ and $b$ as *averages*. We may also calculate the corresponding $y_1$, $y_2$ [from (4.64) and (4.65), respectively], and estimate $b$ and $c$

by fitting a regression, or by some other appropriate methods (see Chapter 7). If, indeed, a Gompertz survival function is a good fit over the range of ages under consideration, the values of $b$ and $c$ obtained from different pairs, $x_1$, $x_2$, should be reasonably consistent with each other.

In the present case, there are only three tabulated values at age 75 and above, so we cannot make a comparison of this kind. We just use the Gompertz fitted to the values tabulated at 75, 80, and 85 (i.e., with $x_1=75$, $x_2=80$). (Although we do not directly test whether Gompertz is a good fit for ages 75 and above, similar calculations using the U. S. Life Table 1969–1971 did, in fact, produce good agreement with tabulated values of $l_x$ for ages $85<x\leqslant 105$.)

If the Gompertz survival function applies for all ages exceeding $x$, then the fitted value of $l_{x+t}$ will be

$$l'_{x+t}=l_x\cdot{}_tp_x=l_x\frac{S(x+t|\hat{a},\hat{R})}{S(x|\hat{a},\hat{R})},\tag{4.66}$$

where $l_x$ is tabulated value at age $x$, and $\hat{a}$ and $\hat{R}$ are the estimated values of $a$ and $R$.

Formula (4.66) can be used to *extrapolate* to very old ages, although the uncertainties inherent in extrapolation should always be born in mind. For our example we obtain $\hat{b}=1.004857$, $\hat{c}=1.074254$, and so $\hat{a}=0.071626$, $\hat{R}=0.00034707$. It is left to the reader to check that, for instance, $l_{90}=5468$.

Several other methods of interpolation and extrapolation have been used in the construction of complete from abridged life tables. Essentially, most are based on fitting polynomials for ages 0–74, and fitting some functions or using experience of a specified subpopulation, for very old ages. There are also many mathematical formulae designed so as to reproduce especially important features of the data.

## 4.12 SELECTION

Except when the census method is used, data from which life tables are constructed are collected from some special source, which restricts the nature of the lives constituting the experience. From the discussion in the preceding chapters, for example, we have seen that data can be obtained from participants in a clinical trial, from individuals insured by an insurance company, from persons employed by a certain company or industry, and so on. In all these cases the individuals, whose survival or death is measured, belong to a restricted class. In many cases, this restriction itself can affect the mortality experience. The fact of being included in the group which is measured, in itself, may imply expectation of higher or lower mortality. This effect is known as *selection*.

We first consider, as an example, selection operating among persons holding life insurance policies with a certain company. Before the company accepts these persons for insurance, it tries to ensure (perhaps, by medical examination) that they are not in such poor health that they are a "bad risk." Consequently, one would expect that, at least in the period immediately following acceptance, such persons will experience lighter mortality than that of persons of the same age in the general population.

This is, indeed, found to be the case. However, as time goes by, the disparity diminishes, unless there are further checks on the state of health of the insured persons (and this is hardly ever the case). Eventually, the difference becomes so small that it is not worthwhile, for practical purposes, to distinguish the "select" group (of insured persons) from nonselected persons. The period, subsequent to acceptance, for which the difference is not, for practical purposes, negligible is called the *select period*. Clearly the select period depends on the purposes for which the life table is to be used, and there is scope for personal judgment in deciding upon it.

If selection is to be allowed for, it is generally necessary to calculate death probabilities as a function both of age and of time elapsed since entry into select status. "Entry into select status" has various meanings specific to the situations to which the life table applies. It can mean acceptance for life insurance, as we have just discussed, it can also mean, for example, (1) being accepted for membership of a police force, (2) retiring from an industrial organization before normal retirement age, or (3) being given some treatment (e.g., an operation) in a clinical trial. The effect of selection can be so great that often the effect of age is neglected and only the time elapsed since entry (here, "time in the trial") is considered. (Sometimes age is introduced as a concomitant variable, see Chapter 13.)

The effect of selection is not always a reduction in mortality. If it is, it is often described as *positive* selection. If selection results in increased mortality [as is usually the case in (2), and in (3) if only persons suffering from a specific disease are included in the trial], it is called *negative* selection.

## 4.13 SELECT LIFE TABLES

When selection is of sufficient importance, it may be allowed for by the use of special life tables called *select* life tables. There is a special notation that allows for effects of selection. The essential feature of this is that age at selection, $x$ say, is enclosed in square brackets. Time since selection, $t$

say, is indicated by addition, so that $[x]+t$ means that $t$ years have elapsed since entry to select status at age $x$.

An individual aged $x$, included in the experience, can be represented by

$[x]$ if just entering select status,

$[x-1]+1$ if select status was entered 1 year ago, and generally,

$[x-t]+t$ if select status was entered $t$ years ago.

This notation is used in conjunction with the standard life table functions. For example, $q_{[x]}$ is the probability that an individual aged $x$, who has just entered select status, will not survive till age $(x+1)$. Similarly $q_{[x]+t}$ is the probability that an individual who entered select status at age $x$, and has survived $t$ years, will not survive one further year.

Note that

$$q_x, q_{[x]}, q_{[x-1]+1}, \ldots, q_{[x-t]+t}$$

all represent probabilities that an individual aged $x$ will not survive 1 year. The first $(q_x)$ applies to an individual who has never entered select status or who entered so long ago that the select period has ended.

With positive selection, and a select period of more than $(t+1)$ $(>2)$ years

$$q_{[x]} < q_{[x-1]+1} < \cdots < q_{[x-t]+t} < \cdots < q_x. \tag{4.67}$$

If the select period, $\delta$ say, is less than or equal to $t$ then

$$q_{[x-t]+t} = q_x.$$

This is called the *ultimate* value of the probability of death within a year for a person aged $x$. It is, indeed, the value to which all persons who enter into select status "ultimately" revert on expiry of the select period.

Since the values of $q_{[x-t]+t}$ depend on both the attained age $x$, and the time elapsed since selection, it appears to be necessary to have separate life tables, one for each integer age at entry, if the same information is to be provided as in a standard life table. This might mean as many as 40 or more separate life tables.

Fortunately, if the select period $(\delta)$ is not too long, it is possible to compress this information so that the life tables do not require an excessive amount of space. We now describe how this can be done.

We first consider the construction of a table of the $q$ functions. For $t \geqslant \delta$,

$$q_{[x-t]+t} = q_x. \tag{4.68}$$

The set of values of the $q_x$'s define a life table called the *ultimate* table.

Note that the ultimate table does not reflect mortality in the population generally, because individuals are excluded for the select period after entry.

One may use the term *aggregate* for the population mortality, to distinguish it from ultimate mortality, corresponding to durations over $\delta$. These are set out in the penultimate column of Table 4.6 while the $q$'s for the preceding $\delta$ years of select period occupy the preceding $\delta$ columns. The death probabilities experienced in successive years by an individual entering select status at age $x$ (which are $q_{[x]}, q_{[x]+1}, \ldots$) are found by entering the *left-hand* Age column at $x$ and moving to the right until reaching the Ultimate column, thereafter until moving down that column (with the appropriate ages given by the *right-hand* Age column).

**Table 4.6** Section of a select life table (select period $\delta$ years) values of $q$'s

| | Select duration | | | | | |
|---|---|---|---|---|---|---|
| Age | 0 | 1 | ... | $(\delta-1)$ | Ultimate | Age |
| $\vdots$ | $\vdots$ | $\vdots$ | | $\vdots$ | $\vdots$ | $\vdots$ |
| $x$ | $q_{[x]}$ | $q_{[x]+1}$ | ... | $q_{[x]+\delta-1}$ | $q_{x+\delta}$ | $x+\delta$ |
| $x+1$ | $q_{[x+1]}$ | $q_{[x+1]+1}$ | ... | $q_{[x+1]+\delta-1}$ | $q_{x+\delta+1}$ | $x+\delta+1$ |
| $\vdots$ | $\vdots$ | $\vdots$ | | $\vdots$ | $\vdots$ | $\vdots$ |

Although Table 4.6 provides all the needed information, it is very convenient to have a corresponding table of $l$ functions (i.e., with $l_{[x]+t}$, $l_x$, etc. appearing where $q_{[x]+t}$, $q_x$, etc. appear in Table 4.6).

### 4.14 SOME EXAMPLES

In this section we describe a few situations in which select tables are relevant. It will be noted that in some of these cases there is no finite select period after which the effect of selection is negligible. In such cases, we really need a considerable number of completely different life tables, one for each age at selection. For this reason, when selection cannot be neglected there is often a considerable amount of effort expended in attempts to obtain mathematical formulae that express the life table parameters as functions of age and time since selection.

1. In the A1967–70 (1976) life tables (based on experience of British life insurance offices) there is a select period of only 2 years. The life table

parameters are given by the formula

$$\frac{q_x}{p_x} = A + Bc^x - Hx, \tag{4.69}$$

with three different sets of values for $A$, $B$, $c$, and $H$—one for "duration 0" (i.e., the first year of insurance), one for "duration 1," and one for "durations 2 and over" (the "ultimate" values).

It is, of course, possible, and indeed likely, that the effective select period will be different at different ages. One would expect that positive selection at younger ages, where the great majority of lives would be reasonably healthy (and so the selected group could not be all that much healthier), would be less, and would not last so long as, at older ages. In fact, it is very unusual to have select tables with different select periods for different entry ages (although this is not impossible from a technical point of view). The common select period is then of the nature of a maximum select period, so that the ultimate and select mortality probabilities at younger ages agree quite closely in the later years of the select period. In practice, of course, there may be an element of compromise, with the select period at advanced ages being somewhat inadequate.

2. Tallis et al. (1973) have produced a set of life tables based on data for 9096 women diagnosed as having breast cancer, from the Central Cancer Registry, Melbourne, Australia, during the period 1946–1970. In this case the selection is negative and there is no limit to the select period. A mathematical model is fitted from which a life table for any date of first observation of breast cancer can be built up, and also gives separate life tables for different ages of first observation.

3. Cutler and Axtell (1963) consider three sets of data for female cancer patients, aged 45–54 at time of diagnosis:

(a) 198 with localized cancer of the uterine cervix, treated by radiation.
(b) 254 with localized cancer of the breast, treated by surgery.
(c) 287 with cancer of the breast with regional spread, treated by surgery.

(Data from the Connecticut Tumor Registry for patients seen in 1935–1944 and followed until 1960.) In their analysis they are essentially concerned with assessing the way in which negative selection wears off. From the data they estimated survival probabilities over each of 20 years from first observation, and then calculated the ratios (the survival ratios) of these to the corresponding values for a (hypothetical) group of women of the same ages from the general population ("estimated from general population life tables, taking into account the calendar period of observation.")

As one might expect these survival ratios are considerably less than 1 at short durations, they increase with duration, and tend to stabilize as time passes. However, only when the stable value is 1 can it be said that selection has worn off. Otherwise, if the stable value is approximately 0.9, say, then we have to say that there is a residual 10% increased mortality in the affected group.

In fact, Cutler and Axtell found that in one case (group (a), with cancer of the cervix) there was evidence of effective wearing-off of selection, while in others (groups (b) and (c)), there was evidence of residual excess mortality that gave no indication of any further decline. (Very roughly, the ratio to normal survival probabilities, over a period of a year, seemed to stabilize at approximately 97% for group (b), 94% for group (c).)

While no explicit life tables were calculated from these data, it would be possible to construct such tables using (possibly smoothed and extrapolated) values of the survival ratios.

## 4.15 CONSTRUCTION OF SELECT TABLES

As we have already emphasized, a select table is really a collection of separate life tables. Each of these can, in principle, be constructed in the same way as we have described in Section 4.12. This would require preliminary sorting of the data according to age at selection and then fitting separate life tables for each of the subgroups of data so obtained. In fact, however, there are two major factors that have to be taken into account:

1. Assessment of the select period (or perhaps concluding there is no effective end to the effects of selection).
2. Then combining data to estimate the ultimate values of the life table function.

The statistical analogue of (1) is testing whether a number of population proportions are equal, but in practice we are testing rather whether the population proportions do not differ sufficiently seriously to affect calculations based on them.

Once a select period has been decided upon we can construct the ultimate life table in the way described in Section 4.12, using only data for durations greater than the select period. We could, of course, also use the data for durations within the select period for each age at selection, to estimate the select period life table values separately for each age at selection. However, the amount of data for some (and maybe most) ages of selection will be relatively sparse, and it is desirable to investigate whether some form of grouping neighboring ages of selection is feasible. Sometimes, indeed, it may even be possible to fit a mathematical formula to

$$\frac{q_{[x-t]+t}}{q_x}$$

as a function of $t$, using *all* the select period data.

To sum up, select tables are constructed on the same principles as nonselect tables, using data segregated according to age at entry during the select period. Graduation may use data in neighboring select periods for smoothing purposes.

## REFERENCES

*A1967–70 Tables for Assured Lives* (1975). Institute of Actuaries, London; Faculty of Actuaries, Edinburgh.

Abramowitz, M. and Stegun, I. A. (1965). *Handbook of Mathematical Functions with Formulas, Graphs, and Mathematical Tables*, National Bureau of Standards, Applied Mathematics Series 55, Fourth Printing, U. S. Government Printing Office, Washington, D.C.

Barclay, G. W. (1958). *Techniques of Population Analysis*, Chapter 4 and Appendix. Wiley, New York.

Beers, H. S. (1944). Six-term formulas for routine actuarial interpolation. *Record Amer. Inst. Actuaries* **33**, Part II, 245–257.

Benjamin, B. and Haycocks, H. W. (1970). *The Anaylsis of Mortality and Other Actuarial Statistics*. Cambridge University Press, Cambridge.

Chiang, C. L. (1968). *Introduction to Stochastic Processes in Biostatistics*. Wiley, New York.

Chiang, C. L. (1972). On constructing current life tables. *J. Amer. Statist. Assoc.* **67**, 538–541.

Cutler, S. J. and Axtell, L. M. (1963). Partitioning of a patient population with respect to differential mortality risks. *J. Amer. Statist. Assoc.* **58**, 701–712.

Dawson, M. M. (1904). *Practical Lessons in Actuarial Science*, Vol. 2. The Spectator Company, New York.

Greville, T. N. E. (1943). Short method of constructing abridged life tables. *Record Amer. Inst. Actuaries*, **32**, 29–43.

Greville, T. N. E. (1967). Methodology of the national, regional, and state life tables for the United States: 1959–61. Publication No. 1252, Vol. 1, No. 4, 1–14. U. S. Dept. of Health, Education and Welfare. Public Health Service.

Hooker, P. F. and Longley-Cook, L. H. (1953). *Life and Other Contingencies*, Vol. 1. Cambridge University Press, Cambridge.

Joint Mortality Investigation Committee (J.M.I.C.) (1974). Considerations affecting the preparation of standard tables of mortality. *J. Inst. Actuaries London*, **101**, 135–201.

Jordan, C. W. (1967). *Life Contingencies*. Society of Actuaries, Chicago.

Keyfitz, N. (1966). A life table that agrees with the data. *J. Amer. Statist. Assoc.* **61**, 305–311; (1968). II. *J. Amer. Statist. Assoc.* **63**, 1253–1268.

King, G. (1914). On a short method of constructing an abridged mortality table. *J. Inst. Actuaries*, **48**, 294.

Preston, S. H., Keyfitz, N., and Schoen, R. (1972). *Causes of Death. Life Tables for National Populations*. Seminar Press. New York and London.

Reed, J. L. and Merrell, M. (1939). A short method of constructing an abridged life table. *Amer. J. Hyg*. **30**, 30–62.

Shryock, H., Siegel, J. S., and Associates (1973). *The Methods and Materials in Demography*, Chapter 15. U. S. Department of Commerce, Washington, D.C.

Sirken, M. (1964). Comparison of two methods of constructing abridged life tables by reference to a "standard" life table. Publication No. 1000-Ser. 2, No. 4, 1–14. U. S. Dept. of Health Education and Welfare, Public Health Service.

Tallis, G. M., Sarfaty, G., and Leppard, P. (1973). The use of a probability model for the construction of age specific life tables for women with breast cancer. Endocrine Research Unit, Cancer Institute, Melbourne, Australia.

United States Life Tables: 1969–71, Vol. I, No. 1, (1975). U. S. Dept. of Health, Education, and Welfare, Publication No. (HRA) 75-1150, National Center for Health Statistics, Rockville, MD.

Vital Statistics of the United States, 1972. Vol. II-Mortality. Part A. (1976). U. S. Dept. of Health, Education, and Welfare, National Center for Health Statistics, Rockville, MD.

## EXERCISES

**4.1.** In a certain family the grandfather is 60 years old, the father is 38 years old, and the daughter is 16 years old, Suppose that the daughter intends to get married when she is 20 years old.

Give solutions to the following questions assuming that:

(i) ages are "exact";

(ii) ages are last birthday;

and using the U. S. Life Table 1969–1971, for White Males and White Females.

(a) What is the probability that there will be a wedding (i.e., daughter will survive till age 20), and the father and grandfather will be alive for the wedding?

(b) What is the probability that the grandfather will die before the wedding (given that the wedding will take place)?

(c) What is the probability that the grandfather will be alive for the wedding (given the wedding takes place) and will die within the next three years?

**4.2.** In formula (4.60) explain the meaning of the following:

(*a*) The terms ${}_nK_x$, ${}_nf_x$, ${}_nD_x$.

(*b*) Expression ${}_nK_x + n(1 - {}_nf_x){}_nD_x$.

(*c*) What does $N_x'$ represent in the terminology of Section 2.6.1?
(*d*) What does $\sum_{y=0}^{\infty} {}_nD_{ny}$ represent?
(*e*) What does $\sum_{y=0}^{\infty} {}_nd_{ny}$ represent in life tables?
(*f*) Could you explain why $N_x' \neq l_x$?

**4.3.** The following data are available from a certain occupational study over a calendar year:

| Age $x$ | No. of deaths ${}_nD_x$ | Mortality rate ${}_nM_x$ |
|---|---|---|
| 25–30 | 2 | 0.00192 |
| 30–35 | 1 | 0.00177 |
| 35–40 | 4 | 0.00215 |
| 40–45 | 5 | 0.00238 |
| 45–50 | 6 | 0.00442 |
| 50–55 | 4 | 0.00823 |
| 55–60 | 3 | 0.01221 |
| 60–65 | 3 | 0.02253 |

(*a*) Find the amount of person-years in each group.
(*b*) Find the population at risk ${}_nK_x$, say, and the number of initial exposed to risk, $N_x'$ (use ${}_nf_x = \frac{1}{2}$). Interpret these quantities.
(*c*) Construct a life table (with radix $l_{25} = 100{,}000$) for this group.

**4.4.** In the calendar year 1900, the mortality experience of the population is well described by the Gompertz formula $\mu_x = Re^{ax}$. In subsequent years there has been a constant proportional rate of decrease of $100r\%$ per year in $\mu_x$.

(*a*) Under these conditions what form of distribution represents cohort mortality for individuals born in the year $(1900 + y)$?
(*b*) Give a diagram showing estimated values of $\mu_x$ for cohorts born in 1900, 1910, and 1920.

**4.5.** As in Exercise 4.4, but with $\mu_x = A + Re^{ax}$.

**4.6.** Construct a complete life table for U. S. White Males, 1972, from the abridged life table given in Table 4.4. (Use tables of Lagrangian coefficients for ages 0–74, and fit a Gompertz distribution for ages above 75.)

**4.7.** Using the complete life table from Exercise 4.6, calculate the following:

(*a*) The column $Y_x$ and the variance of future lifetime.

(*b*) Median future lifetime for each age.
(*c*) Twenty-fifth percentile of future lifetime for each age.
(*d*) Represent (*b*) and (*c*) graphically.

**4.8.** Use the vital statistics data given in Table E4.1 to construct an abridged life table for each of the four sets of data. Compare graphically the following sets of mortality patterns:

(*i*) Australia, 1911 and Australia 1960;
(*ii*) U. S. 1910 and 1960;
(*iii*) Australia 1911 and U. S. 1910;
(*iv*) Australia 1960 and U. S. 1960.

In each comparison use: (a) $q_x$; (b) $l_x$. Comment on your results.

**Table E4.1** Population and Deaths, Males.
(1) Australia, 1911, 1960. (2) U. S. 1910 (Registration States), 1960 (census).

| Age | Australia, Males | | | | United States, Males | | | |
|---|---|---|---|---|---|---|---|---|
| | 1911 | | 1960 | | 1910 | | 1960 | |
| $x$ to $x+n$ | $_nK_x$ | $_nD_x$ | $_nK_x$ | $_nD_x$ | $_nK_x$ | $_nD_x$ | $_xK_n$ | $_nD_x$ |
| 0–1 | 60,553 | 4,750 | 115,347 | 2,653 | 521,945 | 76,290 | 2,096,966 | 63,957 |
| 1–5 | 213,422 | 1,386 | 444,609 | 522 | 1,964,987 | 28,946 | 8,267,645 | 9,851 |
| 5–10 | 235,222 | 515 | 522,115 | 291 | 2,260,512 | 8,163 | 9,536,463 | 5,354 |
| 10–15 | 221,100 | 357 | 502,772 | 252 | 2,157,475 | 5,239 | 8,553,074 | 4,693 |
| 15–20 | 232,399 | 556 | 394,470 | 512 | 2,222,736 | 8,844 | 6,656,062 | 8,636 |
| 20–25 | 233,779 | 852 | 347,068 | 549 | 2,397,351 | 13,731 | 5,290,144 | 9,484 |
| 25–30 | 204,829 | 867 | 342,331 | 513 | 2,324,456 | 14,966 | 5,351,084 | 9,286 |
| 30–35 | 175,319 | 900 | 392,283 | 667 | 2,045,875 | 15,509 | 5,865,966 | 11,739 |
| 35–40 | 155,599 | 986 | 388,023 | 847 | 1,887,488 | 17,328 | 6,100,042 | 17,605 |
| 40–45 | 148,423 | 1,262 | 344,779 | 1,209 | 1,616,494 | 17,617 | 5,695,048 | 26,250 |
| 45–50 | 136,345 | 1,471 | 333,034 | 1,923 | 1,372,795 | 18,589 | 5,376,018 | 40,847 |
| 50–55 | 110,786 | 1,677 | 286,647 | 2,847 | 1,171,527 | 20,272 | 4,750,818 | 59,344 |
| 55–60 | 73,827 | 1,575 | 234,220 | 3,853 | 831,385 | 20,182 | 4,141,182 | 76,541 |
| 60–65 | 52,678 | 1,627 | 184,478 | 4,905 | 658,306 | 22,641 | 3,420,832 | 97,590 |
| 65–70 | 41,214 | 1,958 | 148,703 | 6,309 | 491,796 | 24,268 | 2,940,986 | 121,385 |
| 70–75 | 25,777 | 2,116 | 115,952 | 7,236 | 327,264 | 23,838 | 2,192,595 | 130,157 |
| 75–80 | 19,293 | 2,158 | 64,067 | 6,276 | 193,905 | 21,151 | 1,364,015 | 117,185 |
| 80–85 | 9,024 | 1,570 | 32,116 | 4,552 | 89,263 | 14,924 | 667,339 | 88,960 |
| 85+ | 3,513 | 1,008 | 13,551 | 3,713 | 42,411 | 10,851 | 363,499 | 76,784 |
| All | 2,357,102 | 27,591 | 5,196,565 | 49,629 | 24,577,971 | 383,349 | 88,629,778 | 975,648 |

*From Preston et al. (1972), *Causes of Death. Life Tables for National Populations*. Seminar Press, New York and London.

**4.9.** (*i*) Show that

$$0 \leqslant {}_{\infty}L_{\omega'} \leqslant (\omega-\omega')l_{\omega'},$$

and hence, for $x<\omega'$, $\mathring{e}_x$ lies between

$$\left(\sum_{t=0}^{\omega'-x-1} L_{x+1}\right)l_x^{-1} \quad \text{and} \quad \left(\sum_{t=0}^{\omega'-x-1} L_{x+1}\right)l_x^{-1}+(\omega-\omega')_{\omega'-x}p_x.$$

(*ii*) If $\mu_y$ is an increasing function of $y$ for $y>\omega'$, would you expect $(\Sigma_{t=0}^{\omega'-x-1}L_{x+1})l_x^{-1}+\frac{1}{2}(\omega-\omega')_{\omega'-x}p_x$ to be greater or less than $\mathring{e}_x$?

(*iii*) Show that the proportional error in taking $(\Sigma_{t=0}^{\omega'-x-1}L_{x+1})l_x^{-1}$ as $\mathring{e}_x$ increases with $x$.

**4.10.** Part of the life tables for two populations, A and B, are shown below.

| | Population A | | Population B | |
|---|---|---|---|---|
| $x$ | $l_x$ | $_5d_x$ | $l_x$ | $_5d_x$ |
| 20 | 1000 | 200 | 1000 | 100 |
| 25 | 800 | 80 | 900 | 90 |
| 30 | 720 | 80 | 810 | 90 |
| 35 | 640 | 80 | 720 | 128 |
| 40 | 560 | | 576 | |

Each population is maintained by recruiting 1000 individuals at exact age 20 uniformly through each year. Show that if direct standardization is used for comparison over the age range 20–40, then population A appears to have the higher mortality if the age distribution in A is used as standard, but the lower mortality if B is used as standard. Comment on this result.

**4.11.** In the populations A and B of Exercise 4.10, it is desired to adjust the annual intake into B, so that the total population aged 20–40 exactly, is the same for B as for A. What should be the new annual intake for B?

**4.12.** The *Balducci hypothesis* assumes that for integer $x$, $l_{x+t}^{-1}$ is a linear function of $t$ for $0 \leqslant t \leqslant 1$. It is a third alternative to the assumptions leading to linear and exponential interpolations (see Section 4.5).

Obtain formulae parallel to (4.29), (4.30), and (4.35) on the assumption of the Balducci hypothesis. [See Batten (1978), Chapter 2.]

**4.13.** Show that on the Balducci hypothesis

$$\mathcal{E}\{\text{Number of deaths in } [x, x+1)\}$$
$$= \{\text{Initial exposed to risk in } [x, x+1)\} \cdot q_x.$$

[See Batten (1978), Chapter 2.]

**4.14.** Using select life table symbols, what is the probability that an individual aged exactly 35 years and just selected will:

(*i*) survive at least 10 years?
(*ii*) die between ages 45 and 50 exactly?

**4.15.** Point out the inaccuracy in the following argument. "The difference in curtate expectation of life over the select period, between a select life and one chosen at random, each aged exactly, is

$$\frac{1}{l_{[x]}l_x} \sum_{t=1}^{k} (l_{[x]+t} - l_{x+t})$$

(select period is $k$ years). Once the select period is over, there is no difference in (curtate) expectation of life, and so the above formula gives the difference in life expectation at the moment of selection."

**4.16.** An organization recruits 300 persons per year, 100 each aged 21, 22, and 23 lbd, nearly uniformly over the year. Assuming there are no losses by withdrawal and that mortality follows that of a select table with select period 2 years, obtain an expression for the central death rate at age 24 lbd among members of the organization. (The stated conditions imply that we may regard the members of the organization as forming a stationary population.)

**4.17.** It is claimed that "ultimate mortality is the mortality in the (unselected) general population." It is replied that this not so because "all the individuals contributing to the ultimate mortality experience were selected at some time in their life." Explain why this is a valid objection, but does not exhaust the reasons why the first statement is not correct.

**4.18.** Table E4.2 contains extracts from certain select life tables. Comment on these figures with special reference to the reasonableness of the select period, having regard to possible uses of the tables. (Extracted from

**Table E4.2**

| Canada Life | | Gotha | |
|---|---|---|---|
| $q_{[40]}$ | 0.00392 | $q_{[40]}$ | 0.00483 |
| $q_{[40]+1}$ | 0.00532 | $q_{[39]+1}$ | 0.00619 |
| $q_{[40]+2}$ | 0.00666 | $q_{[38]+2}$ | 0.00688 |
| $q_{[40]+3}$ | 0.00778 | $q_{[37]+3}$ | 0.00736 |
| $q_{[40]+4}$ | 0.00862 | $q_{[36]+4}$ | 0.00770 |
| $q_{45}$ | 0.00919 | $q_{[35]+5}$ | 0.00796 |
| | | $q_{[34]+6}$ | 0.00831 |
| | | $q_{40}$ | 0.00860 |
| $q_{[39]+1}$ | 0.00518 | | |
| $q_{[38]+2}$ | 0.00630 | | |
| $q_{[37]+3}$ | 0.00710 | | |
| $q_{[36]+4}$ | 0.00754 | $q_{41}$ | 0.00910 |
| $q_{40}$ | 0.00770 | $q_{42}$ | 0.00957 |
| $q_{41}$ | 0.00794 | $q_{43}$ | 0.01003 |
| $q_{42}$ | 0.00820 | $q_{44}$ | 0.01048 |
| $q_{43}$ | 0.00849 | $q_{45}$ | 0.01096 |
| $q_{44}$ | 0.00882 | $q_{46}$ | 0.01152 |

Dawson, M. M. (1905), *Practical Lessons in Actuarial Science*, Vol. 2. Spectator, New York.)

**4.19.** Using the information in Exercise 4.18, estimate the probability that a person aged 40, who just entered select status, will survive a person now aged 40, who entered select status 10 years ago, assuming their mortality to be represented by the Canada Life table. State explicitly any further assumptions you find you need to make.

**4.20.** Given that $\mathring{e}_{45}=26.13$ according to the Canada Life table, estimate complete expectation of life for each of the two individuals described in Exercise 4.19.

**4.21.** Show that if $l_t$ is a quadratic function of $t$ for $x-1 \leqslant t \leqslant x+1$, then

$$q_x = \frac{m_x}{1-q_{x-1}} \cdot \frac{12-13q_{x-1}}{12+5m_x}$$

(J. J. McCutcheon, *Transactions of the Faculty of Actuaries in Scotland*, **35**, 297–301 (1977))

**4.22.** The formula given in Exercise 4.21 could be used to obtain values of $q_x$ (and so form a life table) from a set of values of $m_x$ (for integer $x$) and an initial value $q_\alpha$ for a low value of $\alpha$. It can be objected that this means that the whole table depends on the value chosen for $q_\alpha$, and that there might be an accumulation of errors. Answer these criticisms. [This method was used in calculation of English Life Table No. 12 (1960–2). J. J. McCutcheon, *Transactions of the Faculty of Actuaries in Scotland*, **35**, 281–296 (1977).]

CHAPTER 5

# Complete Mortality Data. Estimation of Survival Function

## 5.1 INTRODUCTION. COHORT MORTALITY DATA

In Chapter 4 we discussed methods of construction of life tables from cross-sectional type mortality data observed in the current population. A life table constructed in this way serves as a useful model of survivorship of an actual population.

In the present chapter, we discuss problems associated with *estimation* of the survival function from a *sample* that represents the mortality experience of a *cohort*. The term cohort is used here in a rather special way. We define it as a sufficiently homogeneous group of individuals for which a certain initial event has already occurred. Furthermore, this group is under observation until the last member of the group fails—so we have *complete* records of each life subsequent to the initial event. We refer to such data as *complete mortality data* (though complete survival data might be used sometimes). The methods of analysis described in this chapter make no allowance for loss or withdrawal from observation. (Methods developed in Chapter 6 do take account of these features.)

Data of this kind can arise from studies of mortality in a group of persons (classified, for example, by sex and race) each of whom has contracted a disease at about the same specified age. For each person, the age at diagnosis (the initial event) and age at death are recorded. The difference between these two ages represents the survival time (see Example 5.1). More often, cohort data are obtained from experimental-type investigations using laboratory animals (e.g., mice, rats, flies). The lifespan of such individuals is usually short, and it is possible to follow such a cohort until the last individual dies. We notice that in the first of these two cases, our cohort was not necessarily restricted to a common calendar period, whereas in the second case, a homogeneous group of laboratory animals is (approximately) a cohort by birth.

A random variable of special interest is the *survival time* (the time elapsed from the initial event to failure). We will denote this by $T$ rather than $X$, even though survival time may sometimes be just chronological age. The methods of estimating the survival function presented in this chapter (and in Chapter 6) are *distribution free*. Techniques of fitting parametric SDFs are presented in Chapter 7.

Essentially, our task is to present methods appropriate for: (1) *small sample* data, with each individual time of death recorded effectively exactly; (2) *large sample* data, with times of death usually grouped into fixed intervals.

It is assumed that the reader is familiar with the common statistical estimation techniques—in particular with the method of maximum likelihood (see Section 3.15.1). In this chapter we only need maximum likelihood estimation of binomial and multinomial parameters.

We first consider the situation in which individual times at death are recorded.

## 5.2 EMPIRICAL SURVIVAL FUNCTION

Let $t'_1 < t'_2 < \ldots < t'_N$ represent the $N$ (distinct) ordered times at death. Let

$$\Pr\{T \leqslant t\} = F(t) \tag{5.1}$$

be the theoretical (unknown) failure distribution (CDF), and

$$\Pr\{T > t\} = 1 - F(t) = S(t) \tag{5.1a}$$

be the corresponding survival distribution function (SDF).

The *empirical failure distribution* is

$$F_N^o(t) = \begin{cases} 0 & \text{for } t < t'_1 \\ \dfrac{i}{N} & \text{for } t'_i \leqslant t < t'_{i+1} \\ 1 & \text{for } t \geqslant t'_N, \end{cases} \tag{5.2}$$

where $i$ is the *rank* of the $i$th (ordered) observation (Fig. 5.1a). Note that $F_N^o(t)$ is a *right continuous* function of $t$, and estimates $F(t)$, that is, the probability $\Pr\{T \leqslant t\}$.

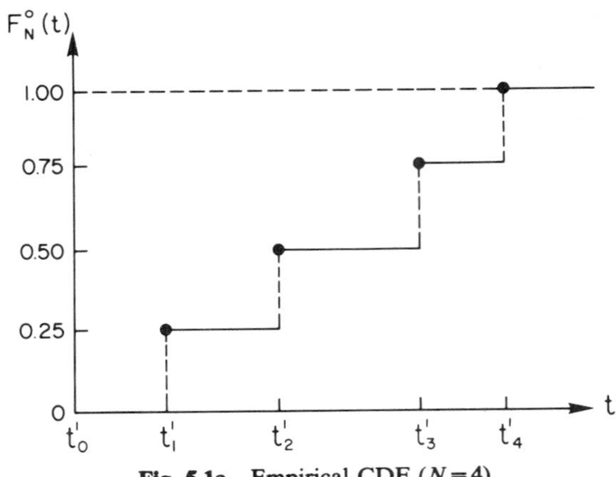


**Fig. 5.1a** Empirical CDF ($N=4$).

The *empirical survival function*, $S_N^\circ(t)=1-F_N^\circ(t)$, is

$$S_N^\circ(t)=\begin{cases} 1 & \text{for } t<t'_1 \\ \dfrac{N-i}{N} & \text{for } t'_i \leqslant t<t'_{i+1} \\ 0 & \text{for } t\geqslant t'_N\,. \end{cases} \tag{5.2a}$$

The function $S_N^\circ(t)$ is also *right continuous* and estimates $\Pr\{T>t\}$.

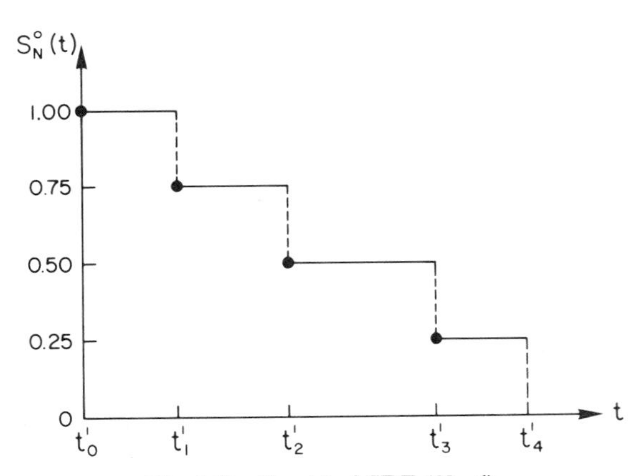


**Fig. 5.1b** Empirical SDF ($N=4$).

Since $t'_1, \ldots, t'_N$ can be considered as a set of random variables, so $F_N^{\circ}(t)$ [and $S_N^{\circ}(t)$] is also a random variable. In fact, *conditional on t* being fixed, $F_N^{\circ}(t)$ is a *binomial proportion* with parameters $N$ and $F(t)$. We then have

$$\mathcal{E}\left[F_N^{\circ}(t)\right] = F(t), \qquad \mathcal{E}\left[S_N^{\circ}(t)\right] = S(t), \tag{5.3}$$

$$\operatorname{var}\left[F_N^{\circ}(t)\right] = \operatorname{var}\left[S_N^{\circ}(t)\right] = \frac{F(t)S(t)}{N}. \tag{5.4}$$

Empirical SDF should be used when only a single death is recorded at any one time point. When units, in which time is recorded, are not sufficiently small (e.g., week rather than day), we may sometimes have "multiple deaths" at a given recorded time point; this is briefly referred to as "multiplicity."

In practice, this means that all we know of the exact times of death is that they occur during the *time-recording unit* concluding at the recorded time. If $\delta_i$ ($\geqslant 1$) is the number of deaths recorded at $t'_i$ then $S_N^{\circ}(t)$ cannot be evaluated for $t'_{i-1} < t < t'_i$, but

$$S_N^{\circ}(t'_i) = 1 - \frac{1}{N}\sum_{j=1}^{i} \delta_j, \qquad i = 1, 2, \ldots \tag{5.5}$$

If there are no multiple deaths at $t'_j$, then $\delta_j = 1$. If there are some multiple deaths, so that $\delta_j > 1$ for some $j$, then the number, $K$ say, of recorded (distinct) times of death must be less then $N$, and

$$\sum_{j=1}^{K} \delta_j = N.$$

As an approximation one may spread out the deaths evenly over the time unit. For example, if $\delta$ is the number of deaths in a given unit, we may spread them out randomly as single deaths at $(1/2\delta)$, $(3/2\delta), \ldots, [1 - 1/(2\delta)]$ fractions of this unit. This, of course, is quite arbitrary.

Generally, if there are too many time points with multiple deaths, the use of an empirical survival function is not advised; smaller time units should be used if possible.

**Example 5.1** The data in Table 5.1 represent survival times (in years) after diagnosis of dementia (a mental disorder) in a group of 97 Swiss females with age at diagnosis 70–74 inclusive. This is part of a larger set of data collected at the University Psychiatric Clinic of Geneva over the period January 1, 1960 to December 31, 1961 described by Todorov et al.

**Table 5.1** Survival time (in years) after diagnosis of dementia in 97 Swiss females with age at diagnosis 70–74

| Age at diagnosis ($x$) | Number of cases | Length of survival time in years ($t$) | | | | | | | | | |
|---|---|---|---|---|---|---|---|---|---|---|---|
| 70 | 15 | 0.50 | 0.83 | 1.17 | 1.25 | 1.66 | 1.67 | 1.83 | 3.17 | 4.17 | 4.92 |
| | | 5.58 | 6.58 | 6.92 | 8.16 | 12.50 | | | | | |
| 71 | 18 | 0.50 | 0.83 | 1.25 | 1.33 | 1.67 | 1.83 | 3.42 | 3.42 | 3.92 | 4.92 |
| | | 5.75 | 6.92 | 7.83 | 7.84 | 9.00 | 9.17 | 13.83 | 18.08 | | |
| 72 | 22 | 0.50 | 1.00 | 1.08 | 1.42 | 1.58 | 1.59 | 2.00 | 2.33 | 2.67 | 3.58 |
| | | 4.08 | 4.17 | 5.75 | 6.25 | 7.00 | 7.08 | 9.92 | 10.17 | 11.25 | 12.67 |
| | | 21.00 | 21.83 | | | | | | | | |
| 73 | 17 | 0.58 | 1.08 | 1.25 | 1.67 | 2.00 | 2.08 | 2.17 | 3.75 | 4.25 | 4.58 |
| | | 4.92 | 5.25 | 6.75 | 7.83 | 7.84 | 9.17 | 11.50 | | | |
| 74 | 25 | 0.50 | 1.00 | 1.25 | 1.41 | 1.42 | 1.58 | 1.66 | 1.67 | 2.25 | 2.33 |
| | | 2.92 | 3.08 | 3.75 | 3.92 | 4.17 | 4.67 | 5.00 | 5.25 | 5.83 | 7.25 |
| | | 8.50 | 9.33 | 10.33 | 11.25 | 12.50 | | | | | |

**Table 5.2** Empirical survival function $S_N^{\circ}(t)$ for data given in Table 5.1 with age at diagnosis 74

| $i$ | $t_i'$ | $N-i$ | $S_N^{\circ}(t_i')$ | $i$ | $t_i'$ | $N-i$ | $S_N^{\circ}(t_i')$ |
|---|---|---|---|---|---|---|---|
| 0 | 0 | 25 | 1.00 | 13 | 3.75 | 12 | 0.48 |
| 1 | 0.50 | 24 | 0.96 | 14 | 3.92 | 11 | 0.44 |
| 2 | 1.00 | 23 | 0.92 | 15 | 4.17 | 10 | 0.40 |
| 3 | 1.25 | 22 | 0.88 | 16 | 4.67 | 9 | 0.36 |
| 4 | 1.41 | 21 | 0.84 | 17 | 5.00 | 8 | 0.32 |
| 5 | 1.42 | 20 | 0.80 | 18 | 5.25 | 7 | 0.28 |
| 6 | 1.58 | 19 | 0.76 | 19 | 5.83 | 6 | 0.24 |
| 7 | 1.66 | 18 | 0.72 | 20 | 7.25 | 5 | 0.20 |
| 8 | 1.67 | 17 | 0.68 | 21 | 8.50 | 4 | 0.16 |
| 9 | 2.25 | 16 | 0.64 | 22 | 9.33 | 3 | 0.12 |
| 10 | 2.33 | 15 | 0.60 | 23 | 10.33 | 2 | 0.08 |
| 11 | 2.92 | 14 | 0.56 | 24 | 11.25 | 1 | 0.04 |
| 12 | 3.08 | 13 | 0.52 | 25 | 12.50 | 0 | 0.00 |

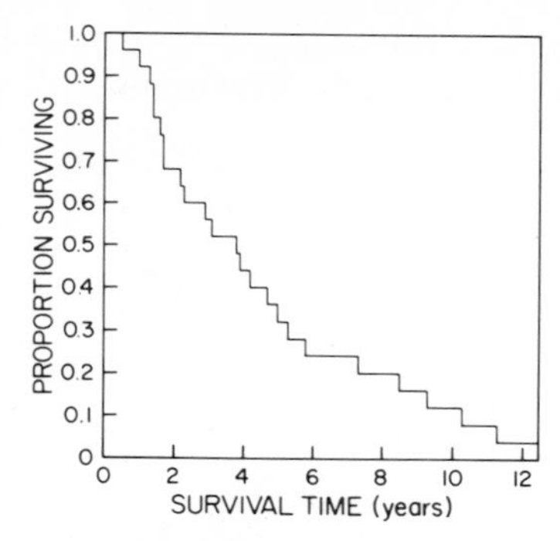


**Fig. 5.2** Empirical distribution of survival time in dementia.

(1975), *J. Neurol. Sci.* **26**, 81. We obtained the original data by courtesy of Dr. Todorov.

The 'age at diagnosis' ($x'$lbd) is used here as exact age at onset. For practical purposes, it seems to be sufficient to use this approximation. The age at death obtained from the death certificate is effectively exact. The survival time ($t$), which is the difference between age at diagnosis and age at death, is clearly subject to a certain error, but it was used as a fair approximation to the survival time.

In the present example, we use only the data for age at diagnosis $x=74$ with $N=25$. The remaining data will be used later. Values of the empirical survival function, $S_N^{\circ}(t)$, are given in Table 5.2, and the graphical representation of the step-function $S_N^{\circ}(t)$ in Fig. 5.2.

## 5.3 ESTIMATION OF SURVIVAL FUNCTION FROM GROUPED MORTALITY DATA

### 5.3.1 Grouping into Fixed Intervals

When the sample is large enough, the data can be grouped into *M fixed* intervals, $[t_i, t_{i+1})$, $i=0,1,\ldots,M-1$. (The notation, [ ), for the right-hand side open interval is equivalent to the notation $t_i \leqslant t < t_{i+1}$.) The *length* of the interval is $h_i = t_{i+1} - t_i$. It is convenient to have intervals of the same length, but this is not essential.

***Notation.*** We use here the following notation:

$d_i$ is the observed number of deaths in the interval $[t_i, t_{i+1})$. It should be emphasized that $d_i$ is here the *observed* number of deaths in the interval $[t_i, t_{i+1})$ and should not be confused with $d_x$ used in life table model. (A more precise notation might be ${}_{h_i}d_i$, if $h_i$ is not a time unit.)

$N_i$ is the number of survivors at the beginning of the interval $[t_i, t_{i+1})$. In the present case we have

$$N_i = \sum_{j=i}^{M-1} d_j = \sum_{j=i}^{\infty} d_j, \tag{5.6}$$

since $d_j = 0$ for $j$ greater than $M-1$. In particular,

$$N_0 = \sum_{j=0}^{M-1} d_j = N \tag{5.7}$$

is the total sample size, that is, the size of the group that starts at time $t_0$ and is followed over the period $t_M - t_0$, during which all members of the group die.

The observed proportion of survivors at time $t_i$ out of $N_0$ starters is

$$\hat{P}_i = \frac{N_i}{N_0}; \tag{5.8}$$

this estimates the survival function $S(t_i)$, that is $\hat{P}_i = \hat{S}(t_i)$.

We may also estimate the conditional probability of death in the interval $[t_i, t_{i+1})$, given alive at $t_i$.

$$\hat{q}_i = \frac{d_i}{N_i}, \tag{5.9}$$

and also

$$\hat{p}_i = 1 - \hat{q}_i = \frac{N_i - d_i}{N_i} = \frac{N_{i+1}}{N_i}. \tag{5.10}$$

(A more precise notation might be ${}_{h_i}\hat{q}_{t_i}$, ${}_{h_i}\hat{p}_{t_i}$— by analogy to ${}_tq_x$ and ${}_tp_x$, as used in life tables.) Furthermore, we notice that

$$\hat{S}(t_i) = \hat{P}_i = \hat{p}_0 \cdot \hat{p}_1 \cdots \hat{p}_{i-1} = \frac{N_1}{N_0} \cdot \frac{N_2}{N_1} \cdots \frac{N_i}{N_{i-1}} = \frac{N_i}{N_0}, \tag{5.11}$$

which is, of course, the same as (5.8).

The results can be exhibited in a form of a sample life table as shown in Table 5.3. In fact, Table 5.3 represents the mortality experience of a cohort

**Table 5.3** Estimation of survival function from (grouped) mortality data

| $i$ | $t_{i+1}-t_i$ | $h_i$ | $d_i$ | $N_i$ | $\hat{q}_i$ | $\hat{p}_i$ | $\hat{P}_i$ |
|---|---|---|---|---|---|---|---|
| 0 | $t_1-t_0$ | $h_0$ | $d_0$ | $N_0=N$ | $\hat{q}_0$ | $\hat{p}_0$ | $\hat{P}_0=1$ |
| 1 | $t_2-t_1$ | $h_1$ | $d_1$ | $N_1$ | $\hat{q}_1$ | $\hat{p}_1$ | $\hat{P}_1$ |
| 2 | $t_3-t_2$ | $h_2$ | $d_2$ | $N_2$ | $\hat{q}_2$ | $\hat{p}_2$ | $\hat{P}_2$ |
| ⋮ | ⋮ | ⋮ | ⋮ | ⋮ | ⋮ | ⋮ | ⋮ |
| $M-1$ | $t_M-t_{M-1}$ | $h_{M-1}$ | $d_{M-1}$ | $d_{M-1}$ | 1 | 0 | $P_{M-1}$ |
| $M$ | — | — | 0 | 0 | — | — | 0 |

of $N_0=N$ individuals. The column $d_i$ is here a sample analogue of the life table column $d_x$; the column $N_i$ is the sample analogue of the column $l_x$, etc. (cf. Section 4.2). There is, however, an essential difference between the life tables discussed in Section 4.2 and a mortality experience, represented in the form of a sample life table in Table 5.3. The former is a model of a hypothetical cohort, the latter represent actual cohort data. Quantities such as $d_i$, $N_i$, $\hat{q}_i$ can be regarded as random variables.

**Example 5.2** For illustration, we use data from Fürth et al. (1959) (given by Kimball, (1960). *Biometrics* **16**, 514). These are mortality data for 208 male mice between 6–12 weeks of age, which were exposed to 240 rad of gamma radiation and died between the 40th and 177th weeks after radiation (Table 5.4). We grouped these data into intervals of fixed length $h_i=15$ weeks, starting from $t=0$. Since, however, there are no deaths in the two first intervals, Table 5.5 starts from $t_0=30$. The estimated survival function is displayed in Fig. 5.3 ("cross").

### 5.3.2 Grouping Based on Fixed Numbers of Deaths

It is sometimes convenient to use a grouping based on *fixed numbers of deaths*. In contrast to the previous scheme based on intervals of fixed length, $d_i$ is now no longer a random variable, but the length of the interval needed to produce the $d_i$ deaths is a random variable. We still can estimate the survival function *as if* the lengths of the intervals were fixed, but the distribution theory for estimators of various life table functions differs from that when $d_i$'s are random variables.

We have used the same mice data for grouping based on fixed numbers of deaths (in most cases 21 deaths; in fact, the resultant grouping is indicated in Table 5.4). The calculations are straightforward and are not presented here, but estimates of the survival function are marked with

**Table 5.4** Frequency distribution of survival time for male mice exposed to 240 rad of gamma radiation [Furth et. al. (1959)] (Dotted lines indicate groupings based on fixed number of deaths)

| Survival time (wk) | Frequency | Survival time (wk) | Frequency | Survival time (wk) | Frequency | Survival time (wk) | Frequency | Survival time (wk) | Frequency | Survival time (wk) | Frequency |
|---|---|---|---|---|---|---|---|---|---|---|---|
| 40 | 1 | 83 | 1 | 103 | 5 | 120 | 3 | 138 | 1 | 157 | 1 |
| 48 | 1 | 84 | 1 | 104 | 3 | 121 | 2 | 139 | 2 | · · · | · · · |
| 50 | 1 | 86 | 2 | 105 | 2 | 123 | 2 | · · · | · · · | 158 | 2 |
| 54 | 1 | 87 | 1 | · · · | · · · | 124 | 3 | 140 | 2 | 160 | 1 |
| 56 | 1 | 88 | 5 | 106 | 3 | 125 | 2 | 141 | 5 | 161 | 1 |
| 59 | 1 | 89 | 1 | 107 | 1 | 126 | 5 | 142 | 1 | 162 | 2 |
| 62 | 1 | 90 | 2 | 108 | 1 | · · · | · · · | 144 | 5 | 163 | 2 |
| 63 | 1 | 91 | 1 | 109 | 2 | 127 | 4 | 145 | 2 | 164 | 1 |
| 67 | 2 | 93 | 1 | 110 | 3 | 128 | 4 | 146 | 4 | 165 | 2 |
| 69 | 1 | 94 | 1 | 111 | 3 | 129 | 6 | · · · | · · · | 166 | 1 |
| 70 | 1 | 95 | 1 | 112 | 1 | 130 | 4 | 147 | 4 | 168 | 1 |
| 71 | 1 | 96 | 1 | 113 | 2 | 131 | 2 | 148 | 4 | 169 | 1 |
| 73 | 2 | 97 | 2 | 114 | 2 | · · · | · · · | 149 | 1 | 171 | 2 |
| 76 | 1 | 98 | 1 | 115 | 1 | 132 | 1 | 150 | 1 | 172 | 2 |
| 77 | 1 | · · · | · · · | 116 | 2 | 133 | 3 | 151 | 4 | 174 | 1 |
| 80 | 1 | 99 | 2 | · · · | · · · | 134 | 4 | 152 | 2 | 177 | 2 |
| 81 | 2 | 100 | 4 | 117 | 1 | 135 | 3 | 153 | 1 | | |
| 82 | 1 | 101 | 3 | 118 | 3 | 136 | 4 | 155 | 1 | | |
| · · · | · · · | 102 | 2 | 119 | 2 | 137 | 3 | 156 | 1 | | |

**Table 5.5** Estimation of survival function. Data grouped in fixed intervals, $h_i = 15$ weeks (mice data in Example 5.2)

| $i$ | $[t_i, t_{i+1})$ | $d_i$ | $N_i$ | $\hat{q}_i$ | $\hat{p}_i$ | $\hat{P}_i$ | SE($\hat{P}_i$) |
|---|---|---|---|---|---|---|---|
| 0 | 30–45 | 1 | 208 | 0.00481 | 0.99519 | 1.0000 | — |
| 1 | 45–60 | 5 | 207 | 0.02415 | 0.97585 | 0.9952 | 0.0048 |
| 2 | 60–75 | 9 | 202 | 0.04455 | 0.95545 | 0.9712 | 0.0116 |
| 3 | 75–90 | 17 | 193 | 0.08808 | 0.91192 | 0.9279 | 0.0179 |
| 4 | 90–105 | 29 | 176 | 0.16477 | 0.83523 | 0.8462 | 0.0250 |
| 5 | 105–120 | 29 | 147 | 0.19728 | 0.80272 | 0.7067 | 0.0316 |
| 6 | 120–135 | 45 | 118 | 0.38136 | 0.61864 | 0.5673 | 0.0344 |
| 7 | 135–150 | 41 | 73 | 0.56164 | 0.43836 | 0.3510 | 0.0331 |
| 8 | 150–165 | 20 | 32 | 0.62500 | 0.37500 | 0.1538 | 0.0250 |
| 9 | 165–180 | 12 | 12 | 1.00000 | 0.00000 | 0.0577 | 0.0162 |
| Total | | 208 | | | | 0.0000 | |

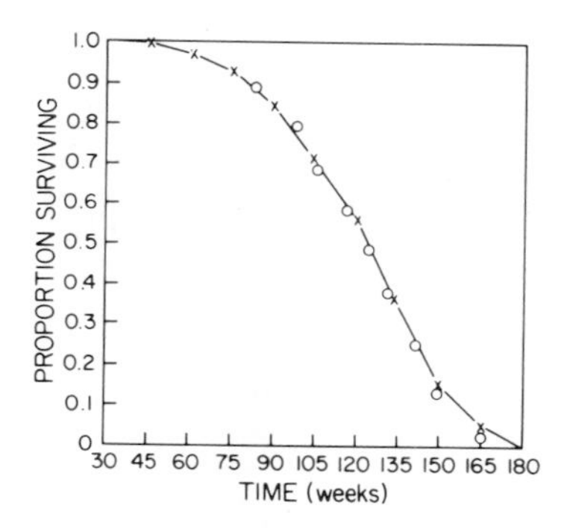


**Fig. 5.3** Estimated survival function (mice data from Example 5.2).

circles in Fig. 5.3. We see that they are quite consistent with those obtained from grouping based on fixed intervals. In further analysis we will use grouping based on fixed intervals.

## 5.4 JOINT DISTRIBUTION OF THE NUMBERS OF DEATHS

We recall that $d_i$ denotes the number of deaths in $[t_i, t_{i+1})$ among the $N = N_0$ starters, and can be regarded as a random variable. Since there are $M$ nonoverlapping intervals, we have

$$d_0 + d_1 + d_2 + \dots + d_{M-1} = N_0. \tag{5.12}$$

The unconditional probability of death in $[t_i, t_{i+1})$ is

$$P_i - P_{i+1} = P_i(1-p_i) = P_i q_i$$

with $\Sigma_{i=0}^{M-1} P_i q_i = 1$. We can consider $d_0, d_1, \ldots, d_{M-1}$ as multinomial variables, with joint distribution

$$\Pr\{d_0, d_1, \ldots, d_{M-1}\} = N_0! \prod_{i=0}^{M-1} \frac{(P_i q_i)^{d_i}}{d_i!} \tag{5.13}$$

It follows that

$$\mathcal{E}(d_i) = N_0 P_i q_i, \qquad \operatorname{var}(d_i) = N_0 P_i q_i (1 - P_i q_i),$$
$$\operatorname{cov}(d_i, d_j) = -N_0 P_i P_j q_i q_j. \tag{5.14}$$

## 5.5 DISTRIBUTION OF $\hat{P}_i$

For *fixed* value $t_i$, the estimator of the survival function $S(t_i)$, $\hat{P}_i = N_i / N_0$, is distributed as a *binomial proportion* with parameters $N_0$, $P_i$. We then have

$$\mathcal{E}(\hat{P}_i) = P_i \tag{5.15}$$

and

$$\operatorname{var}(\hat{P}_i) = \frac{P_i(1-P_i)}{N_0} \tag{5.16}$$

for $i = 0, 1, \ldots, M-1$ [see also (5.3) and (5.4)].

## 5.6 COVARIANCE OF $\hat{P}_i$ AND $\hat{P}_j$ $(i<j)$

For $t_i < t_j$ we must have $\hat{P}_i \leqslant \hat{P}_j$, and $\hat{P}_i$ so $\hat{P}_j$ are correlated. It can be shown (see Exercise 5.9) that

$$\operatorname{cov}(\hat{P}_i, \hat{P}_j) = \frac{(1-P_i)P_j}{N_0}. \tag{5.17}$$

## 5.7 CONDITIONAL DISTRIBUTION OF $\hat{q}_i$

The preceding discussion relates to the very special case of complete mortality data in which a cohort of $N_0$ individuals is followed from starting point until failure, the numbers of failures in each of a set of time intervals being observed. In particular the relation $N_{i+1}=N_i-d_i$ holds for all $i$. Later we will need a method of analysis which is of more general application. We may, for example, have data from several similar experiments or a sequence of experiments and wish to calculate variances of different life table functions. This can be done by considering distributions *conditional on the actually observed numbers* $N_0, N_1, \ldots$ $(\equiv\{N_j\})$ at the beginning of each interval. This leads to formulae for variances that reflect the accuracy provided by the actual amount of data in the experiment, rather than average accuracy over all experiments. This approach is appropriate in *all* cases, including that of complete mortality data, at present under study. Consideration of unconditional variances (allowing for distribution of the $N_j$'s), when possible, is relevant in planning stages, when assessing the likely accuracy of estimates to be calculated from the (as yet unavailable) data.

*Conditional on* $N_i$, the proportion of deaths in $[t_i, t_{i+1})$, $\hat{q}_i = d_i/N_i$ is distributed as a *binomial proportion* with parameters $N_i$, $q_i$.

We have

$$\mathcal{E}(\hat{q}_i|N_i)=q_i, \qquad \mathcal{E}(\hat{p}_i|N_i)=p_i, \tag{5.18}$$

and

$$\operatorname{var}(\hat{q}_i|N_i)=\operatorname{var}(\hat{p}_i|N_i)=\frac{p_iq_i}{N_i}. \tag{5.19}$$

Note that conditional on $N_0, N_1, \ldots, N_i$ the $(i+1)$ random variables, $\hat{q}_0, \hat{q}_1, \ldots, \hat{q}_i$ are mutually independent.

Similar remarks apply to the joint distribution of deaths $d_0, d_1, \ldots, d_i$, conditional on the numbers of survivors at the beginning of each interval, $N_0, N_1, \ldots, N_i$, respectively. In the last case we have

$$\Pr\{d_0, d_1, \ldots, d_i|N_0, N_1, \ldots, N_i\} = \prod_{j=0}^{i}\binom{N_j}{d_j}q_j^{d_j}p_j^{N_j-d_j} \tag{5.20}$$

for $i=0, 1, \ldots, M-1$.

Note that this distribution also corresponds to a sequence of experiments or cross-sectional data in which the $N_j$'s are fixed at the values actually obtained, but the $d_j$'s are random, so that we do not necessarily have $N_{i+1}=N_i-d_i$.

To illustrate the remarks made at the beginning of this section, suppose that $N_i$ has an expected value of 10, whereas the actual $N_i$ is 40. It would be inappropriate to claim that formula (5.19)—in this case,

$p_iq_i/40$—underestimates the variance applicable to the $\hat{q}_i$ calculated from these data, just because in *other* experiments we would expect to get values of $N_i$ generally considerably less than 40.

## 5.8 GREENWOOD'S FORMULA FOR THE (CONDITIONAL) VARIANCE OF $\hat{P}_i$

We now derive a formula for the *conditional* (on set $\{N_k\}$) variance of $\hat{P}_i$. From (5.11), we have

$$\hat{P}_i = \hat{p}_0 \cdot \hat{p}_1 \cdots \hat{p}_{i-1}.$$

Conditionally on the $N_j$'s, the $\hat{p}_j$'s are mutually independent. It follows that $\text{cov}(\hat{p}_j, \hat{p}_{j'} | \{N_k\}) = 0$ (and, in fact, since $\mathcal{E}(\hat{p}_j | \{N_k\}) = p_j$, the *unconditional* covariance is also zero).

Using the method of statistical differentials [see (3.90)], we then have

$$\text{var}(\hat{P}_i | N_1, N_2, \ldots, N_{i-1}) \doteq \sum_{j=0}^{i-1} \left( \frac{\partial P_i}{\partial p_j} \right)^2 \text{var}(\hat{p}_j | N_j), \tag{5.21}$$

where $\partial P_i / \partial p_j$ is an abbreviated expression for the partial derivative $\partial \hat{P}_i / \partial \hat{p}_j$ in which the set of values $\hat{p}_0, \hat{p}_1, \ldots, \hat{p}_{i-1}$ is replaced by their expected values $p_0, p_1, \ldots, p_{i-1}$, respectively.

We have $\partial P_i / \partial p_j = P_i / p_j$ and $\text{var}(\hat{p}_j | N_j) = p_j q_j / N_j$. Substituting these into (5.21) we obtain

$$\text{var}(\hat{P}_i | N_1, \ldots, N_{i-1}) \doteq P_i^2 \sum_{j=0}^{i-1} p_j^{-2} \frac{p_j q_j}{N_j} = P_i^2 \sum_{j=0}^{i-1} \frac{q_j}{N_j p_j}. \tag{5.22}$$

This is known as *Greenwood's formula*. Of course, in practice, the true values of $P_i$, $P_j$, and $q_j$ are unknown and we substitute their estimates.

In this case, we obtain the approximate *estimated* variance

$$\begin{aligned} \text{est. var}(\hat{P}_i | N_0, \ldots, N_{i-1}) &\doteq \hat{P}_i^2 \sum_{j=0}^{i-1} \frac{1}{N_j} \cdot \frac{d_j}{N_j} \cdot \frac{N_j}{N_{j+1}} \\ &= \hat{P}_i^2 \sum_{j=0}^{i-1} \left( \frac{1}{N_{j+1}} - \frac{1}{N_j} \right) \end{aligned} \tag{5.23}$$

$$= \hat{P}_i^2 \left( \frac{1}{N_i} - \frac{1}{N_0} \right) = \frac{\hat{P}_i^2 (N_0 - N_i)}{N_0 N_i} = \frac{\hat{P}_i (1 - \hat{P}_i)}{N_0}, \tag{5.23a}$$

since $N_i / N_0 = \hat{P}_i$.

Incidentally, this also estimates the unconditional (exact) variance of $\hat{P}_i$ given in (5.16), substituting $\hat{P}_i$ for $P_i$.

## 5.9 ESTIMATION OF CURVE OF DEATHS

Let $t_i^* = t_i + \frac{1}{2}h_i$ denote the midpoint of the interval $h_i = t_{i\pm 1} - t_i$. The estimated probability density function (PDF) represents the estimated *curve of deaths*. First, we use the (well-known) approximation to the derivative of a function in a small interval. Provided $h_i$ is sufficiently small, we have

$$\widehat{f(t_i^*)} = -\left.\frac{dS(t)}{dt}\right|_{t=t_i^*} \doteq \frac{S(t_i) - S(t_{i+1})}{h_i}$$

$$= \frac{P_i - P_{i+1}}{h_i} = \frac{P_i - P_i p_i}{h_i} = \frac{P_i q_i}{h_i}. \tag{5.24}$$

Next, substituting the estimators $\hat{q}_i = d_i / N_i$ and $\hat{P}_i = N_i / N_0$, we obtain an estimator of the density function at the midpoint of the interval, that is

$$\widehat{f(t_i^*)} = \frac{\hat{P}_i \hat{q}_i}{h_i} = \frac{1}{h_i}\frac{N_i}{N_0}\cdot\frac{d_i}{N_i} = \frac{d_i}{h_i N_0}. \tag{5.25}$$

Clearly, from (5.14) we have $\mathcal{E}(d_i) = N_0 P_i q_i$ and $\text{var}(d_i) = N_0 P_i q_i (1 - P_i q_i)$, so that the *unconditional* variance of $\widehat{f(t_i^*)}$ is

$$\text{var}[\widehat{f(t_i^*)}] \doteq \left(\frac{1}{h_i N_0}\right)^2 \text{var}(d_i) = \frac{P_i q_i (1 - P_i q_i)}{h_i^2 N_0}. \tag{5.26}$$

We can also obtain the *conditional* (on set $\{N_k\}$) variance of $\widehat{f(t_i^*)}$. We recall (from Section 5.7) that in this case it is not necessarily true that $N_{i+1} = N_i - d_i$, so that the right-hand side of (5.25) might not be valid. We then represent (5.25) in the general form

$$\widehat{f(t_i^*)} = \frac{\hat{P}_i \hat{q}_i}{h_i} = \frac{\hat{p}_0 \cdot \hat{p}_1 \cdots \hat{p}_{i-1} \cdot \hat{q}_i}{h_i} = g(\hat{p}_0, \hat{p}_1, \ldots, \hat{p}_{i-1}, \hat{q}_i). \tag{5.25a}$$

(Note $\hat{q}_i = 1 - \hat{p}_i$.) Its (conditional) variance can be obtained by the method of statistical differentials; this is similar to the derivation of Greenwood's formula.

We have

$$\frac{\partial g}{\partial p_j}=\frac{P_i q_i}{h_j p_j},\qquad j=0,1,\ldots,i-1;\qquad \frac{\partial g}{\partial q_i}=\frac{P_i}{h_i};$$

$$\operatorname{var}(\hat{p}_j|N_j)=\operatorname{var}(\hat{q}_j|N_j)=\frac{p_j q_j}{N_j},\qquad j=0,1,\ldots,i;$$

and

$$\operatorname{cov}(\hat{p}_j,\hat{p}_k)=\operatorname{cov}(\hat{p}_j,\hat{q}_k)=0,\qquad j\neq k;\qquad j,k=0,1,\ldots,i-1.$$

Hence [from (3.90)] and following the techniques of Section 5.8, we obtain

$$\begin{aligned}\operatorname{var}[\widehat{f(t_i^*)}|\{N_k\}]&\doteqdot\frac{1}{h_i^2}\left[P_i^2q_i^2\sum_{j=0}^{i-1}\frac{q_j}{N_jp_j}+P_i^2\frac{q_ip_i}{N_i}\right]\\&=\left(\frac{P_iq_i}{h_i}\right)^2\left[\sum_{j=0}^{i-1}\frac{q_j}{N_jp_j}+\frac{p_i}{N_iq_i}\right].\end{aligned}\tag{5.27}$$

[See Gehan (1969).]

## 5.10 ESTIMATION OF CENTRAL DEATH RATE AND FORCE OF MORTALITY IN $[t_i,t_{i+1})$

The central death rate in $[t_i,t_{i+1})$ can be calculated from the approximate formula

$$\hat{m}_i\doteqdot\frac{d_i}{h_i(N_i-\frac{1}{2}d_i)}=\frac{\hat{q}_i}{h_i(1-\frac{1}{2}\hat{q}_i)}=\frac{2\hat{q}_i}{h_i(2-\hat{q}_i)}.\tag{5.28}$$

[cf. (4.46) with $n=h_i$] since $\hat{q}_i=d_i/N_i$. Now $\hat{m}_i$ is a random variable; (conditional on $N_i$) its variance can be obtained using the method of statistical differentials. We have

$$\operatorname{var}(\hat{m}_i|N_i)\doteqdot\left[\frac{4}{h_i(2-q_i)^2}\right]^2\frac{q_i(1-q_i)}{N_i}=\frac{16q_i(1-q_i)}{h_i^2N_i(2-q_i)^4}=\frac{16p_i(1-p_i)}{h_i^2N_i(1+p_i)^4}.\tag{5.29}$$

The estimator of the hazard rate at the midpoint $t_i^*=t_i+\frac{1}{2}h_i$, and denoted here by $\widehat{\mu(t_i^*)}$, can be evaluated from a formula analogous to that in (4.50), that is

$$\widehat{\mu(t_i^*)} = -\frac{1}{h_i}\log \hat{p}_i = -\frac{1}{h_i}\log\frac{N_{i+1}}{N_i} = -\frac{1}{h_i}\log\frac{N_i - d_i}{N_i}. \tag{5.30}$$

Its variance, conditional on $N_i$, is

$$\text{var}[\widehat{\mu(t_i^*)}|N_i] \doteq \left(\frac{1}{h_i p_i}\right)^2 \frac{p_i q_i}{N_i} = \frac{1}{h_i^2}\frac{q_i}{N_i p_i}. \tag{5.31}$$

**Table 5.6** Estimation of curve of deaths, central mortality rate and force of mortality (mice data in Example 5.2)

| | | Curve of deaths | | Central rate | | Force of mortality | |
|---|---|---|---|---|---|---|---|
| $i$ | $[t_i, t_{i+1})$ | $\widehat{f(t_i^*)}$ | $\text{SE}[\widehat{f(t_i^*)}]$ | $\hat{m}_i$ | $\text{SE}(\hat{m}_i)$ | $\widehat{\mu(t_i^*)}$ | $\text{SE}[\widehat{\mu(t_i^*)}]$ |
| 0 | 30–45 | 0.00032 | 0.00032 | 0.00032 | 0.00032 | 0.00032 | 0.00032 |
| 1 | 45–60 | 0.00160 | 0.00071 | 0.00163 | 0.00073 | 0.00163 | 0.00073 |
| 2 | 60–75 | 0.00288 | 0.00094 | 0.00304 | 0.00101 | 0.00304 | 0.00101 |
| 3 | 75–90 | 0.00545 | 0.00127 | 0.00614 | 0.00149 | 0.00615 | 0.00149 |
| 4 | 90–105 | 0.00929 | 0.00160 | 0.01197 | 0.00221 | 0.01200 | 0.00223 |
| 5 | 105–120 | 0.00929 | 0.00160 | 0.01459 | 0.00269 | 0.01465 | 0.00273 |
| 6 | 120–135 | 0.01442 | 0.00190 | 0.03141 | 0.00455 | 0.03202 | 0.00482 |
| 7 | 135–150 | 0.01314 | 0.00184 | 0.05206 | 0.00749 | 0.05498 | 0.00883 |
| 8 | 150–165 | 0.00641 | 0.00136 | 0.06061 | 0.01207 | 0.06539 | 0.01521 |
| 9 | 165–180 | 0.00385 | 0.00108 | 0.13333 | — | — | — |

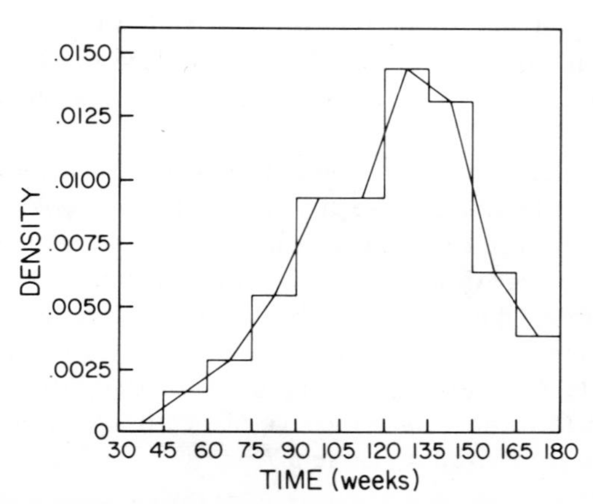


**Fig. 5.4a** Estimated probability density function (mice data from Example 5.2).

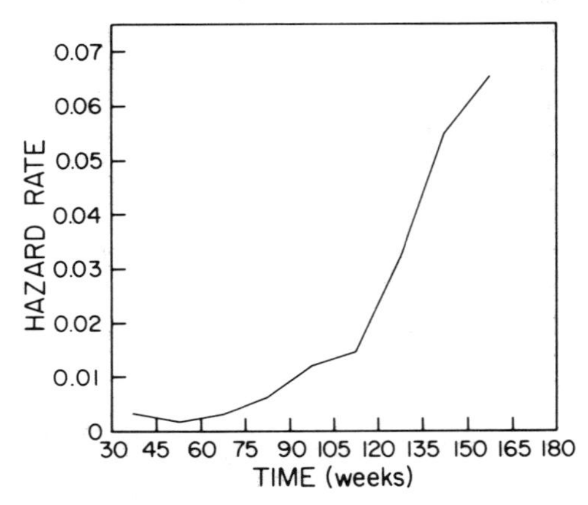


**Fig. 5.4b** Estimated hazard rate function (mice data from Example 5.2).

Table 5.6 exhibits values of the estimators of $f(t_i^*)$ and $\mu(t_i^*)$ and their standard errors for the mice data from Example 5.2. Graphical representations of $\widehat{f(t_i^*)}$ and $\widehat{\mu(t_i^*)}$ are displayed in Figures 5.4a and 5.4b, respectively.

## 5.11 SUMMARY OF RESULTS

In this chapter, we have essentially considered complete (cohort) mortality data and derived estimators of various life table functions and their variances. For grouped data, the survival function is estimated as the proportion of survivors to time $t_i$, $\hat{P}_i = N_i / N_0$ [see (5.8)], and its variance (conditional on $t_i$)—referred to as *unconditional* variance—is given by (5.16).

We have also derived the variance of $\hat{P}_i$ *conditional on the set* $\{N_k\}$ of the number of individuals alive at time $t_i$ (*Greenwood's formula*). This variance should be used generally, whether or not we have cohort data, because it gives approximate variances appropriate to the data actually collected. Generally, the data might be obtained from several experiments. In these cases, the simple formula (5.8) for $\hat{P}_i$ is no longer valid and the standard formula $\hat{P}_i = \hat{p}_0 \cdot \hat{p}_i \ldots \hat{p}_{i-1}$ was used for derivation of its variance. Similar remarks to estimating the curve of deaths, $\widehat{f(t_i^*)}$, and its variances. The results are summarized in Table 5.7.

**Table 5.7** Summary results on estimation of survival probabilities from complete mortality data

| Description | True or approx. value | Time | Estimator | Variance of the estimator |
|---|---|---|---|---|
| 1. Conditional probability of dying | $q_i$ | $[t_i, t_{i+1})$ | $\frac{d_i}{N_i}$ (unbiased) | $\frac{p_i q_i}{N_i}$ (cond. on $N_i$) |
| 2. Conditional probability of surviving | $p_i$ | $[t_i, t_{i+1})$ | $\frac{N_i - d_i}{N_i}$ (unbiased) | same as above |
| 3. Survival function (SDF) | $P_i$ | $t_i$ | (a) $\hat{p}_0 \cdot \hat{p}_1 \cdots \hat{p}_{i-1}$ (cond. on $N_1, N_2, \ldots, N_{i-1}$)<br>(b) $\frac{N_i}{N_0}$ (uncond.) | (a) $P_i^2 \sum_{j=0}^{i-1} (q_j / p_j N_j)$ (cond. on $N_1, N_2, \ldots, N_{i-1}$)<br>(b) $\frac{P_i(1-P_i)}{N_0}$ (uncond.) |
| 4. Curve of deaths (PDF) | $f(t_i^*)$ (approx.) | $t_i^* = t_i + \frac{1}{2}h_i$ | (a) $\frac{\hat{p}_0 \cdot \hat{p}_1 \cdots \hat{p}_{i-1} \cdot \hat{q}_i}{h_i}$ (cond. on $N_1, N_2, \ldots, N_i$)<br>(b) $\frac{d_i}{h_i N_0}$ (uncond.) | (a) $(P_i q_i / h_i)^2 \left[ \left( \sum_{j=0}^{i-1} q_j / p_j N_j \right) + p_i / q_i N_i \right]$ (cond. on $N_1, N_2, \ldots, N_i$)<br>(b) $P_i q_i (1 - P_i q_i) / h_i^2 N_0$ (uncond.) |
| 5. Central rate | $m_i$ | $[t_i, t_{i+1})$ | $\frac{d_i}{h_i\left(N_i - \frac{1}{2}d_i\right)}$ | $\frac{16 q_i (1-q_i)}{h_i^2 N_i (2-q_i)^4}$ (approx.) (cond. on $N_i$) |
| 6. Force of mortality (hazard rate) | $\mu(t_i^*)$ | $t_i^* = t_i + \frac{1}{2}h_i$ | $-\frac{1}{h_i} \log \frac{N_i - d_i}{N_i}$ | $\frac{1}{h_i^2} \frac{q_i}{N_i p_i}$ (approx.) (cond. on $N_i$) |

## REFERENCES

Chiang, C. L. (1960). A stochastic study of the life table and its applications. I. Probability distributions of the biometric functions. *Biometrics* **16**, 618–635.

Chiang, C. L. (1960). A stochastic study of the life table functions and its applications. II. Sample variance of the observed expectation of life and other biometric functions. *Hum. Biol.* **32**, 221–238.

Chiang, C. L. (1968). *Introduction to Stochastic Processes*, Wiley, New York, Chapter 10.

Fürth, J., Upton, A. C., and Kimball, A. W. (1959). Late pathological effects of atomic detonation and their pathogenesis. *Radiation Res. Suppl* **1**, 243–264.

Gehan, E. A. (1969). Estimating survival functions from the life table. *J. Chron. Dis.* **21**, 629–644.

Kimball, A. W. (1960) Estimation of mortality intensities in animal experiments, *Biometrics*, **16**, 505–521.

Seal, H. L. (1949). Mortality data and the binomial probability law, *Skand. Actuar.-tidskr.* **32**, 188–216.

Todorov, A. B., Go, R. C. P., Constantidinis, J., and Elston, R. C. (1975). Specificity of the clinical diagnosis of dementia. *J. Neurol. Sci.* **98**, 26–81.

## EXERCISES

**5.1.** Consider the data in Table 5.1 representing the survival time of 97 Swiss females diagnosed with dementia. Suppose that the Geneva life table 1968–1972 for females can be used as a model of general population mortality for comparison. Some values of $l_x$ from this life table are given in Table 8.1.

(*a*) Estimate the "true" median future survival time for each age of diagnosis and compare with the observed median survival time for the corresponding age. (No formal test is required.)

(*b*) Do the same for the 25th percentile. *Hint:* See Example 4.3. Comment on the results.

**5.2.** It appears from Table 5.1 that the length of survival time after diagnosis of dementia is not heavily affected by age of diagnosis in each group. Consider the 97 cases in Table 5.1 as a random sample from a fairly large and homogeneous population.

(*a*) Group the mortality data into fixed intervals of length $h_i = 1$ year. Estimate $q_i$, $p_i$, and $P_i$ from the grouped data. Calculate standard errors for these estimates.

(*b*) Estimate the curve of deaths and force of mortality.

(*c*) Represent estimates of survival function, curve of deaths (and frequency histogram) and force of mortality on separate graphs. Describe the results verbally. Draw some conclusions.

**5.3.** We recall formula (5.11) for the estimator of survival function

$$\hat{P}_i = \hat{p}_0 \hat{p}_1 \ldots \hat{p}_{i-1}.$$

(*a*) Since the $\hat{p}_j$'s are binomial proportions and, *conditional* on set $\{N_j\}$, are independent, show that the *exact* variance of $\hat{P}_i$ conditional on $N_0, N_1, \ldots, N_i$ is

$$\operatorname{var}(\hat{P}_i | N_0, N_1, \ldots, N_i) = P_i^2 \left[ \prod_{j=0}^{i-1} \left( 1 + \frac{q_j}{N_j p_j} \right) - 1 \right].$$

(*b*) Obtain the approximate variance of $\hat{P}_i$ given by Greenwood's formula (5.22) from the exact formula given in (a).

**5.4.** Consider complete cohort mortality data grouped into fixed intervals as in Table 5.1.

(*i*) The random variables, $d_0, d_1, \ldots, d_{M-1}$, have a joint multinomial distribution with parameters $P_0 q_0, P_1 q_1, \ldots, P_{M-1} q_{M-1}$, respectively (formula (5.12)).Thus the overall (*unconditional*) likelihood

$$L \propto \prod_{i=0}^{M-1} (P_i q_i)^{d_i} \tag{1}$$

(*ii*) *Conditional on* $N_i$, $d_i$ has a binomial distribution with parameters $N_i$, $q_i$. The conditional likelihood

$$L_i \propto q_i^{d_i} p_i^{N_i - d_i},$$

and the overall likelihood, conditional on the set $\{N_i\}$

$$L_{\text{cond}} \propto \prod_{i=0}^{M-1} L_i = \prod_{i=0}^{M-1} q_i^{d_i} \cdot p_i^{N_i - d_i}. \tag{2}$$

Show that the right-hand side of (i) is identical with the right-hand side of (ii). *Hint:* Notice that $P_i = p_0 p_1 \cdots p_{i-1}$.

**5.5.** The data below are from Bryson, M. C. and Siddiqui, M. M., (1969), (*JASA* **64**, 1472–1483) and represent the ordered times at death $t_j'$ (in days) of 43 patients suffering from granulocytic leukemia with $t_0'=0$ taken as date of diagnosis.

| | | | | | | | | | |
|---|---|---|---|---|---|---|---|---|---|
| 7 | 47 | 58 | 74 | 177 | 232 | 273 | 285 | 317 | 429 |
| 440 | 445 | 455 | 468 | 495 | 497 | 532 | 571 | 579 | 581 |
| 650 | 702 | 715 | 779 | 881 | 900 | 930 | 968 | 1077 | 1109 |
| 1314 | 1334 | 1367 | 1534 | 1712 | 1784 | 1877 | 1886 | 2045 | 2056 |
| 2260 | 2429 | 2509 | | | | | | | |

(*a*) Calculate the empirical survival function and represent it graphically.

(*b*) Group the data into intervals of length 200 days (i.e., 0–200, 200–400, etc). Evaluate the SDF for the grouped data and plot it on the same graph as in (a).

(*c*) Estimate PDF from grouped data. Draw a histogram and the estimated PDF on the same graph.

(*d*) Estimate the hazard rate (force of mortality) function from the grouped data. Represent it graphically. Comment on the results.

**5.6.** $N_g$ individuals are observed from time $gh$ until each has failed $(g=0,1,\ldots)$, but the data are incomplete because it is only known in which of the intervals $I_j((j-1)h, jh)$ failure occurs. Show that

$$\prod_{i=1}^{j}\left(\frac{d}{N_0-\Sigma_{g=1}^{i-1}(d_g-N_g)}\right)$$

is an unbiased estimator of $S(jh)$. ($d_j$ is the number of deaths occurring in $I_j$.)

Find the standard deviation of this estimator, on the assumption that $\Sigma_{g=1}^{i-1}(d_g-N_g)$ remains constant (i.e., that $N_g$ is adjusted to make this so—even if negative values of $N_g$, corresponding to removal of individuals from the experience, are needed).

**5.7.** The assumption, in Exercise 5.6, that $\Sigma_{g=1}^{i-1}(d_g-N_g)$ is kept constant has been criticized on the grounds that these conditions are unlikely to be satisfied in actual investigations. Explain why, nevertheless, it may still provide results that are relevant in assessing accuracy of estimates of $S(t_j)$. In particular, give examples of situations in which the more natural assumption that $N_0$, $N_1$, $N_2,\ldots$ remain fixed may give misleading results.

**5.8.** For a certain population, there are data available on numbers of deaths (by age lbd) in each of 50 successive years. Assuming the population is *stable* (see Section 4.8.2) give a method of estimating:

(*a*) The age-specific mortality probabilities.
(*b*) The rate of growth for this population.

**5.9.** Let $\hat{P}_i = N_i/N_0$ and $\hat{P}_j = N_j/N_0$ $(\hat{P}_j < \hat{P}_i)$ be the observed proportions of survivors at times $t_i$ and $t_j$ $(t_i < t_j)$ in a cohort of $N_0$ individuals. Show that

$$\operatorname{cov}(\hat{P}_i, \hat{P}_j) = \frac{(1-P_i)P_j}{N_0}. \qquad \text{[formula(5.17)]}$$

*Hint:* Consider a trinomial distribution of deaths $d_{0i}$ in $(t_0, t_i)$, $d_{ij}$ in $(t_i, t_j)$ and survivors $s_j = N_0 - (d_{0i} + d_{ij})$ beyond $t_j$.

CHAPTER 6

# Incomplete Mortality Data: Follow-up Studies

## 6.1 BASIC CONCEPTS AND TERMINOLOGY

### 6.1.1 Incomplete Data

When experimental units (individuals) are still in operation (alive) at the closing date of an investigation, and their subsequent times of failure are not known, the mortality data are *incomplete*. A similar situation arises if units are lost to observation voluntarily or involuntarily, even before the planned date of termination. Although methods described in this chapter are applicable to various kinds of incomplete failure data, we will, for convenience, mostly use the terminology appropriate for mortality data.

### 6.1.2 Truncation and Censoring

Termination of observation may be controlled in many ways. The two most common are effected by the following schemes:

*Scheme 1.* Terminate observation at a *preassigned time point* (date); the resultant data are termed *truncated*.

*Scheme 2.* Terminate observation when a *preassigned number of deaths* have occurred. The data are called *censored*.

Some authors call scheme 1—censoring type 1, and scheme 2—censoring type 2. It is important to realize that in scheme 1, the *numbers of deaths* are random variables, whereas in scheme 2 the *time of termination* is a random variable. Individual times of death are random variables in both cases.

In typical clinical trials, patients enter the study at different time points. In the general case of multiple times of entry combined with scheme 1 or scheme 2, it is customary to speak about *progressive time censoring* or *progressive failure censoring*, respectively [Nelson (1972)]. (Perhaps, progressive truncation or progressive censoring would be more consistent with the previously stated definitions, but these terms are not used in the statistical literature.) For the remainder of this chapter, we confine our discussion to scheme 1.

A distinction is sometimes made between *planned* and *unplanned* termination of observation of an individual. In scheme 1, the time due for withdrawal (planned) is known *in advance* for each individual. If, however, the observation is not continued for the determined period, because the person has withdrawn his participation for reasons other than termination of the study, this person is said to be *lost*—his termination is unplanned. In this book, we will use the common term *withdrawals* for those who left the study *alive*, irrespective of whether this was planned or unplanned termination. A great deal of discussion of how to treat cases with unplanned and planned censoring has appeared in the statistical literature, and some of these ideas will be presented in this chapter. In our opinion, if subsequent mortality does not depend on mode of withdrawal—a very common assumption—distinguishing between planned and unplanned withdrawals is irrelevant, once the data have been collected. This also will be apparent from statistical analyses presented in this chapter.

In summary, truncation and censoring are the terms used to indicate whether a fixed calendar date or a fixed number of deaths has been used for termination of the study. Withdrawals are those individuals who are no longer under observation and were still alive when last observed, irrespective of reasons for which observation was terminated.

### 6.1.3 Follow-up Studies

The simplest cases are those in which all individuals came under observation at the same time. Then the fixed closing date is a *single truncation* of the data. In clinical investigations, however, times of entry for individual patients are usually variable. We shall distinguish between: (a) chronological time, and (b) time since an individual entered into observation—the so called follow-up time.

***Chronological Time of Study ($y$).*** This is usually calendar time. For example, a study starting on January 1, 1970 and ending December 31, 1979 has duration 10 years, and a fixed date of termination. On this scale, each participant has his own date of entry. The patients present on

January 1, 1970, are called *beginners*, those entering after this date are called *new entrants*. During the study, observations on some individuals are discontinued—these are *withdrawals*. Persons still alive on December 31, 1979 can be distinguished as *enders*.

***Time of Follow-up (t).*** With this arrangement, time $t=0$, say, is assigned to each individual *at entry*. The individual is followed up until one of three possible events occur: (i) death; (ii) leaving the study for some reasons not known at entry; (iii) termination of the study. In this situation we do not distinguish between beginners and new entrants. Effectively, all participants are beginners; there are no new entrants.

Figures 6.1a and 6.1b represent the same experience in terms of chronological time and follow-up time, respectively.

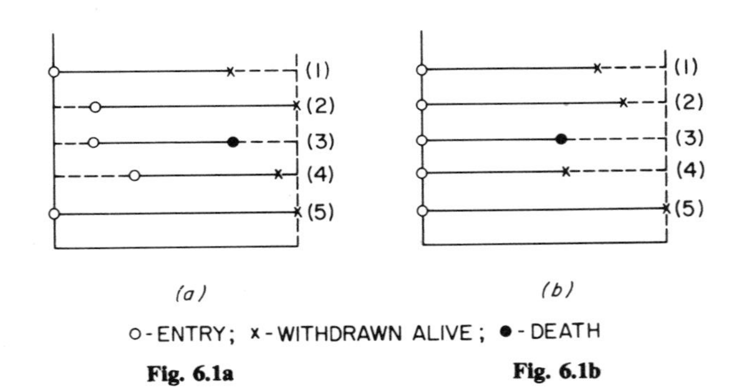


**Fig. 6.1a** **Fig. 6.1b**

Note that although individuals (3) and (4) have the same follow-up time, (3) died while (4) is lost.

There are also some modifications of the idea of a fixed termination time.

(i) Each individual is followed for the *same length of time* since his entry (e.g., 5 or 7 years). This is the so called *anniversary method* of collecting the data and is quite commonly used in long-term studies. The closing time for the whole study is determined by the last entry.

(ii) Combination of fixed termination date for the study with anniversary date, whichever comes last, the so called *method of the last record*, is also frequently used. In both cases each individual entering the study still has his own planned withdrawal time.

### 6.1.4 Other Kinds of Follow-up

Mortality data can also be obtained from records kept by organizations such as commercial enterprises, police forces, or life insurance companies. A noticeable feature distinguishing data from such records is the large number of *beginners*, that is, persons already on the records. The lost individuals will usually include persons leaving employment or policy holders surrendering their policies during the period of the study. In some cases, efforts are made to find these lost individuals. These efforts are often described as follow-up as, indeed, they are, although the nature of this following differs somewhat from that in clinical trials.

The primary difference between the kinds of follow-up in clinical trials and those just described lies in their prospective and retrospective nature. In clinical trials, the ways of collecting information are planned in advance, whereas when data are collected from existing records a special effort is needed to recover information on individuals who are already lost before the study started. However, after the data are collected, they are usually similar in structure to those in clinical trials, and the principles involved in analyzing such data are similar to those employed for data from clinical trials.

### 6.1.5 Topics of the Chapter

The purpose of this chapter is to review methods that have been proposed for estimating the survival function from incomplete data—that is, data arising from experiments in which censoring (withdrawal) occurs.

Time at death (or survival time), $T$, is, of course, a random variable. On the other hand, withdrawal time for an individual is the time point (or interval), where the individual was actually under observation and alive for the last time seen. Some authors also regard censoring time as a random variable, $U$, say, independent of $T$, and having its own distribution (cf. Sections 6.3.1 and 6.3.2).

We recall that the basic function in construction of a survival function (life table) from grouped data is the conditional death probability, $q_i$. Most of this chapter is devoted to a review of methods used in estimating $q_i$, discussing the assumptions and techniques involved (Sections 3 and 4). A handy summary of results is given in Table 6.3. Estimation of survival function from observations of individual times at death is presented in Section 6.5.

At this point we reiterate our attitude that distributions conditional on actually observed numbers of exposed to risk (including survivors and withdrawals) are appropriate for statistical inference from observed data.

## 6.2 ACTUARIAL ESTIMATOR OF $q_i$ FROM GROUPED DATA

$N_i$ denotes the number of survivors at the beginning of the interval $[t_i, t_{i+1})$;

$d_i$ denotes the observed number of deaths in $[t_i, t_{i+1})$ (do not confuse with $d_x$ in life tables);

$w_i$ denotes the number of withdrawals (losses or planned) in $[t_i, t_{i+1})$;

$t_{i+1} - t_i = h_i$ is the length of the interval.

Clearly,

$$N_{i+1} = N_i - d_i - w_i. \tag{6.1}$$

Our first problem is the estimation of (central) death rate in $[t_i, t_{i+1})$.

### 6.2.1 Amount of Person-Time Units. Central Death Rates

There are many systems for recording mortality data. Consideration of each system separately is necessary to evolve an appropriate formula for exposed to risk. Excellent discussions of exposed to risk are given by Batten (1978) and Gershenson (1961). Here we confine ourselves to simple techniques similar to those presented in Section 2.6.1.

Even if the data are grouped, we may still have the records of the exact times of death or withdrawal. Let $t_{ij}^{\circ}$ denote the recorded (observed) time of death or withdrawal for the $j$th individual in the interval $[t_i, t_{i+1})$, and let

$$t_{ij}^{\circ} = \begin{cases} t_{ij} & \text{if } j\text{th individual is observed to die;} \\ t_{ij}^{*} & \text{if } j\text{th individual is withdrawn;} \\ t_{i+1} & \text{if } j\text{th individual survives till } t_{i+1}. \end{cases} \tag{6.2}$$

The total amount of person-time units is clearly

$$\tilde{A}_i = \sum_{j=1}^{N_i} (t_{ij}^{\circ} - t_i) = \sum_j (t_{ij} - t_i) + \sum_j (t_{ij}^{*} - t_i) + N_{i+1} h_i. \tag{6.3}$$

Therefore, the estimated central death rate in $[t_i, t_{i+1})$ is

$$\tilde{m}_i = \frac{d_i}{\tilde{A}_i} = \frac{\text{number of deaths}}{\text{amount of exposed to risk}}. \tag{6.4}$$

Note that $\tilde{m}_i$ is expressed as the *number of deaths per person per unit time.*

We may present (6.3) in a different form. Let

$$\theta_{ij}^{\circ} = \frac{t_{ij}^{\circ} - t_i}{h_i} \tag{6.5}$$

denote the *fraction* of the interval $h_i$, in which the individual was under observation and was alive, and let

$$\theta_{ij}^{\circ} = \begin{cases} \theta_{ij} & \text{if } j\text{th individual is observed to die;} \\ \theta_{ij}^{*} & \text{if } j\text{th individual is withdrawn;} \\ 1 & \text{if } j\text{th individual survives till } t_{i+1}. \end{cases} \tag{6.6}$$

Then (6.3) can be written in the form

$$\tilde{A}_i = h_i \sum_{j=1}^{N_i} \theta_{ij}^{\circ} = h_i \left[ \sum_j \theta_{ij} + \sum_j \theta_{ij}^{*} + N_{i+1} \right]. \tag{6.7}$$

Also, $\theta_{ij}^{\circ}$ may be regarded as a *fraction* of the $j$th individual exposed to risk of dying over the full period $h_i$.

The quantity

$$N_i^c = \frac{\tilde{A}_i}{h_i} \tag{6.8}$$

is equivalent to the average population of exposed to risk [cf. (2.19a)]. This is sometimes called the number of *central* exposed to risk [Benjamin and Haycocks (1970), p. 40]. Although the average (or midperiod) population is a useful concept, the idea of "central exposed to risk" seems to have no special meaning, and we shall not use this phrase in the book.

Assuming additionally, that over $[t_i, t_{i+1})$: (1) time at death had expected value $\frac{1}{2}(t_i + t_{i+1})$, and (2) time at withdrawal is $\frac{1}{2}(t_i + t_{i+1})$, we may replace (6.7) by the *average* amount of person-years time units

$$A_i = h_i\left(N_{i+1} + \tfrac{1}{2}d_i + \tfrac{1}{2}w_i\right) = h_i\left(N_i - \tfrac{1}{2}d_i - \tfrac{1}{2}w_i\right), \tag{6.9}$$

so that the central (observed) death rate (6.4) takes the form

$$\hat{m}_i = \frac{d_i}{A_i} = \frac{d_i}{h_i\left(N_i - \frac{1}{2}d_i - \frac{1}{2}w_i\right)}. \tag{6.10}$$

### 6.2.2 Effective Number of Initial Exposed to Risk. Estimation of $q_i$

For practical purposes, we assume that the individuals who died in the interval $[t_i, t_{i+1})$ would have been exposed to risk for the full interval if they had not died. For those, who withdrew alive in $[t_i, t_{i+1})$, we may regard $\theta_{ij}^*$ as a "fraction" of individual ($j$) exposed to risk for the *full* interval. Therefore, the estimated number of individuals who passed through time point $t_i$ alive is

$$\tilde{N}_i' = (N_i - w_i) + \sum_j \theta_{ij}^* = N_i - \sum_j (1 - \theta_{ij}^*). \tag{6.11}$$

This is sometimes called the *effective number of initial exposed to risk*.

The conditional probability of death in $[t_i, t_{i+1})$, given alive at $t_i$, is estimated from

$$\tilde{q}_i = \frac{d_i}{\tilde{N}_i'}. \tag{6.12}$$

Formula (6.12) is especially useful when the time of withdrawal is known. We note the following:

1. If an individual who died has a (potential) planned time due to withdraw between $t_i$ and $t_{i+1}$, one may count this individual as a "fraction" equal to the fraction of the interval for which he would have been exposed to risk if he had not died.
2. One may also use the approximation

$$\tilde{\tilde{q}}_i \doteq \frac{h_i \cdot \tilde{m}_i}{1 + \frac{1}{2} h_i \cdot \tilde{m}_i} = \frac{d_i}{\left(\tilde{A}_i + \frac{1}{2} h_i \cdot d_i\right)/h_i}, \tag{6.13}$$

[cf. (4.59), (4.60)] with $n = h_i$ and ${}_nf_x = \frac{1}{2}$, so that the estimated number of initial exposed to risk defined in (6.11) will now be expressed by the formula

$$\tilde{\tilde{N}}_i' = \frac{1}{h_i}\left(\tilde{A}_i + \tfrac{1}{2} h_i \cdot d_i\right)$$

[cf. (4.61) with $f_i = \frac{1}{2}$].

It is easy to show (see Exercise 6.9) that $\tilde{\tilde{N}}_i' \neq \tilde{N}_i'$, but since $\mathcal{E}(\theta_{ij}) = \frac{1}{2}$, (6.13) is often used to estimate $q_i$.

If the exact time at withdrawal is not recorded or not utilized, and we assume $\mathcal{E}(\theta_{ij}) = \frac{1}{2}$, then we replace (6.11) by the average number of initial exposed to risk

$$N_i' = N_i - \tfrac{1}{2}w_i, \tag{6.14}$$

and the estimator of $q_i$ defined in (6.12) takes the form

$$\hat{q}_i = \frac{d_i}{N_i - \frac{1}{2}w_i}. \tag{6.15}$$

This is often called the *actuarial* estimator of $q_i$. Note that no probability theory is used in its derivation. The only assumption made is that, *on the average*, the withdrawal time is in the middle of the interval; no distinction between planned and unplanned withdrawals has been made. Clearly, (6.15) is a special case of (6.12).

Breslow and Crowley (1974) have investigated the properties of the actuarial estimator (6.15) under random censorship. They have shown that, in general, $\hat{q}_i$ is an *inconsistent* estimator of $q_i$ in the sense that in large samples the probability that $\hat{q}_i$ differs from $q_i$ by less than a specified amount $\varepsilon$, does not tend to 1. These authors have shown that also, in general, $\mathcal{E}(\hat{q}_i) \neq q_i$, that is, $q_i$ is a *biased* estimator of $q_i$; the bias is *negative*. However, conditional on $N_i$ and $w_i$, and for $N_i$ sufficiently large, it is easy to see that $\mathcal{E}(\hat{q}_i | N_i, w_i) \doteq q_i$, that is, $\hat{q}_i$ is approximately unbiased.

We also notice that from (6.10) and (6.15) we obtain the well-known relationships

$$\hat{q}_i \doteq \frac{h_i \hat{m}_i}{1 + \frac{1}{2}h_i \hat{m}_i} \qquad \text{and} \qquad \hat{m}_i \doteq \frac{\hat{q}_i}{h_i\left(1 + \frac{1}{2}\hat{q}_i\right)}. \tag{6.16}$$

[cf. (4.46)].

In summary, the actuarial estimator of $q_i$ is based on the assumption that the withdrawals are exposed to risk, on the average, for half of the interval.

### 6.2.3 Estimation of Survival Function

Once $q_i$ is estimated, the SDF [denoted here by $P_i = S(t_i)$] is estimated from the well known formula

$$\hat{P}_i = \hat{p}_0 \cdot \hat{p}_1 \cdots \hat{p}_{i-1} = \hat{P}_{i-1}\hat{p}_{i-1}, \tag{6.17}$$

where $\hat{p}_j = 1 - \hat{q}_j$ and $P_0 = 1$.

*Conditional on* $N_j$, $w_j$, and assuming that $\hat{q}_i$ is approximately a binomial proportion,

$$\operatorname{var}(\hat{q}_j|N_j,w_j)=\operatorname{var}(\hat{p}_j|N_j,w_j)$$

$$\doteq\frac{p_jq_j}{N_j'}=\frac{p_jq_j}{N_j-\frac{1}{2}w_j}. \tag{6.18}$$

Thus *conditional on set* $\{N_j\}=(N_0,N_1,\ldots,N_{i-1})$ *and set* $\{w_j\}=(w_0,w_1,\ldots,w_{i-1})$, Greenwood's formula for the conditional variance of $\hat{P}_i$, derived for complete data in Section 5.8, is also approximately valid here if we replace $N_j$ by $N_j'=N_j-\frac{1}{2}w_j$, that is

$$\operatorname{var}(\hat{P}_i|\{N_j\},\{w_j\})\doteq P_i^2\sum_{j=0}^{i-1}\frac{q_j}{p_j\left(N_j-\frac{1}{2}w_j\right)}. \tag{6.19}$$

Of course, in practice, we can only obtain an estimate of this variance by substituting in (6.19), appropriately $\hat{P}_i$, $\hat{p}_j$, and $\hat{q}_j$.

**Example 6.1** The data in the first part of Table 6.1 are adapted from Crowley and Hu (1974) and represent mortality experience of 68 patients enrolled in the Stanford Heart Transplantation Program. In our example, the date at entry is taken to be the *date of operation* (heart transplant).

The investigation covers the period: January 6, 1968 to April 1, 1974, the fixed closing date. The recorded time unit is *one day*. The total length of the study was 2276 days.

Column 1 in Table 6.1 gives (in parentheses) the identification number (ID) assigned to the patients at entry. In column 2, the time of death or withdrawal (the latter marked with the asterisk), counted from the date at entry, is given. Note that patient No. 1 (38) died during the operation, and his time at death is marked as 0+. In column 3, the number of deaths at $t_i$, $\delta_i$, is given. These individual records are used later, in Example 6.2. Perhaps, the data are too few to be grouped, but we have done so for illustrative purposes. We have grouped the data into nine intervals of different lengths as shown in Table 6.2a.

(a) For each interval $[t_i,t_{i+1})$, we first utilize the exact time of withdrawal and calculate the effective number of initial exposed to risk, $\tilde{N}_i'$, from (6.11).

**Table 6.1** Heart transplant mortality data

| No. | (ID) | Follow-up time (in days) $t_i$ | No. of deaths $\delta_i$ | Risk set $R_i$ | $\hat{p}_i = \frac{R_i - \delta_i}{R_i}$ | $\hat{S}(t_i)$ | Time due for withdrawal |
|---|---|---|---|---|---|---|---|
| 1 | (38) | 0+ | 1 | 68 | 0.98529 | 0.98529 | 1422 |
| 2 | (28) | 1 | 1 | 67 | 0.98507 | 0.97059 | 1688 |
| 3 | (100) | 1* | 0 | (66) | 1.0 | ↓ | 1 |
| 4 | (4) | 3 | 1 | 65 | 0.98462 | 0.95566 | 2159 |
| 5 | (20) | 10 | 1 | 64 | 0.98438 | 0.94072 | 1870 |
| 6 | (74) | 12 | 1 | 63 | 0.98413 | 0.92579 | 555 |
| 7 | (98) | 13* | 0 | (62) | 1.0 | ↓ | 13 |
| 8 | (95) | 14 | 1 | 61 | 0.98361 | 0.91061 | 189 |
| 9 | (3) | 15 | 1 | 60 | 0.98333 | 0.89544 | 2276 |
| 10 | (18) | 23 | 1 | 59 | 0.98305 | 0.88026 | 1957 |
| 11 | (70) | 25 | 1 | 58 | 0.98276 | 0.86508 | 648 |
| 12 | (90) | 26 | 1 | 57 | 0.98246 | 0.84991 | 200 |
| 13 | (79) | 29 | 1 | 56 | 0.98214 | 0.83473 | 420 |
| 14 | (92) | 30* | 0 | (55) | 1.0 | ↓ | 30 |
| 15 | (22) | 39 | 1 | 54 | 0.98148 | 0.81927 | 1828 |
| 16 | (45) | 44 | 1 | 53 | 0.98113 | 0.80381 | 1181 |
| 17 | (10) | 46 | 1 | 52 | 0.98077 | 0.78836 | 2047 |
| 18 | (37) | 47 | 1 | 51 | 0.98039 | 0.77290 | 1418 |
| 19 | (83) | 48 | 1 | 50 | 0.98000 | 0.75744 | 400 |
| 20 | (87) | 50 | 1 | 49 | 0.97959 | 0.74198 | 316 |
| 21 | (39) | 51 | | | | | |
| 22 | (47) | 51 | 3 | 48 | 0.93750 | 0.69561 | 1333 |
| 23 | (55) | 51 | | | | ↓ | 956 |
| 24 | (36) | 54 | 1 | 45 | 0.97778 | 0.68015 | 1412 |
| 25 | (32) | 60 | 1 | 44 | 0.97727 | 0.66469 | 1659 |
| 26 | (73) | 63 | 1 | 43 | 0.97674 | 0.64923 | 540 |
| 27 | (13) | 64 | 1 | 42 | 0.97619 | 0.63378 | 2003 |
| 28 | (68) | 65 | 1 | 41 | 0.97561 | 0.61832 | 721 |
| 29 | (65) | 66 | 1 | 40 | 0.97500 | 0.60286 | 744 |
| 30 | (89) | 68 | 1 | 39 | 0.97436 | 0.58740 | 222 |
| 31 | (97) | 109* | 0 | (38) | 1.0 | ↓ | 109 |
| 32 | (11) | 127 | 1 | 37 | 0.97297 | 0.56153 | 2029 |
| 33 | (24) | 136 | 1 | 36 | 0.97222 | 0.55565 | 1719 |
| 34 | (53) | 147 | 1 | 35 | 0.97143 | 0.53977 | 963 |
| 35 | (94) | 161 | 1 | 34 | 0.97059 | 0.52390 | 195 |
| 36 | (96) | 166* | 0 | (33) | 1.0 | ↓ | 166 |

**Table 6.1** (continued)

| No. | (ID) | Follow-up time (in days) $t_i$ | No. of deaths $\delta_i$ | Risk set $R_i$ | $\hat{p}_i = \frac{R_i - \delta_i}{R_i}$ | $\hat{S}(t_i)$ | Time due for withdrawal |
|---|---|---|---|---|---|---|---|
| 37 | (67) | 228 | 1 | 32 | 0.96875 | 0.50753 | 682 |
| 38 | (93) | 236* | 0 | (31) | 1.0 | ↓ | 236 |
| 39 | (51) | 253 | 1 | 30 | 0.96667 | 0.49061 | 1072 |
| 40 | (16) | 280 | 1 | 29 | 0.96552 | 0.47369 | 1955 |
| 41 | (84) | 297 | 1 | 28 | 0.96429 | 0.45677 | 389 |
| 42 | (78) | 304* | 0 | (27) | 1.0 | ↓ | 304 |
| 43 | (58) | 322 | 1 | 26 | 0.96154 | 0.43921 | 900 |
| 44 | (88) | 338* | 0 | (25) | 1.0 | ↓ | 388 |
| 45 | (86) | 388* | 0 | (24) | 1.0 | ↓ | 388 |
| 46 | (81) | 438* | 0 | (23) | 1.0 | ↓ | 438 |
| 47 | (80) | 455* | 0 | (22) | 1.0 | ↓ | 455 |
| 48 | (76) | 498* | 0 | (21) | 1.0 | ↓ | 498 |
| 49 | (64) | 551 | 1 | 20 | 0.95000 | 0.41725 | 757 |
| 50 | (71) | 588* | 0 | (19) | 1.0 | ↓ | 588 |
| 51 | (72) | 591* | 0 | (18) | 1.0 | ↓ | 591 |
| 52 | (7) | 624 | 1 | 17 | 0.94118 | 0.39270 | 2038 |
| 53 | (69) | 659* | 0 | (16) | 1.0 | ↓ | 659 |
| 54 | (23) | 730 | 1 | 15 | 0.93333 | 0.36652 | 1813 |
| 55 | (63) | 814* | 0 | (14) | 1.0 | ↓ | 814 |
| 56 | (30) | 836 | 1 | 13 | 0.92308 | 0.33833 | 1670 |
| 57 | (59) | 837* | 0 | (12) | 1.0 | ↓ | 837 |
| 58 | (56) | 874* | 0 | (11) | 1.0 | ↓ | 874 |
| 59 | (50) | 897 | 1 | 10 | 0.90000 | 0.30449 | 1058 |
| 60 | (46) | 994 | 1 | 9 | 0.88889 | 0.27066 | 1175 |
| 61 | (21) | 1024 | 1 | 8 | 0.87500 | 0.23683 | 1877 |
| 62 | (49) | 1105* | 0 | (7) | 1.0 | ↓ | 1105 |
| 63 | (41) | 1263* | 0 | (6) | 1.0 | ↓ | 1263 |
| 64 | (14) | 1350 | 1 | 5 | 0.80000 | 0.18946 | 1982 |
| 65 | (40) | 1366* | 0 | (4) | 1.0 | ↓ | 1366 |
| 66 | (33) | 1535* | 0 | (3) | 1.0 | ↓ | 1535 |
| 67 | (34) | 1548* | 0 | (2) | 1.0 | ↓ | 1548 |
| 68 | (25) | 1774* | 0 | (1) | 1.0 | ↓ | 1774 |

*Withdrawn alive.

From Crowley, J. and Hu, M., Stanford University Technical Report No. 2, Oct. 15, 1974.

**Table 6.2a** Heart transplantation.
Exact time of death or withdrawal recorded

| $t_i$ to $t_{i+1}$ | $h_i$ | $N_i$ | $d_i$ | $w_i$ | $\tilde{N}_i'$ | $\tilde{q}_i$ | $\tilde{P}_i$ | $\tilde{A}_i$ | $\tilde{m}_i$ |
|---|---|---|---|---|---|---|---|---|---|
| 0–50 | 50 | 68 | 16 | 3 | 65.88 | 0.2429 | 1.0000 | 2876 | 0.005563 |
| 50–100 | 50 | 49 | 11 | 0 | 49.00 | 0.2245 | 0.7571 | 1993 | 0.005519 |
| 100–200 | 100 | 38 | 4 | 2 | 36.75 | 0.1088 | 0.5872 | 3446 | 0.001161 |
| 200–400 | 200 | 32 | 5 | 4 | 30.33 | 0.1649 | 0.5233 | 5446 | 0.000918 |
| 400–700 | 300 | 23 | 2 | 6 | 19.76 | 0.1012 | 0.4370 | 5704 | 0.000351 |
| 700–1000 | 300 | 15 | 4 | 3 | 13.42 | 0.2981 | 0.3928 | 3482 | 0.001149 |
| 1000–1300 | 300 | 8 | 1 | 2 | 7.23 | 0.1383 | 0.2757 | 1892 | 0.000529 |
| 1300–1600 | 300 | 5 | 1 | 3 | 2.83 | 0.3534 | 0.2376 | 899 | 0.001112 |
| 1600+ | — | 1 | 0 | 1 | | | 0.1536 | — | — |

For example, for the fourth interval (200–400) we have $h_i = 200$, $N_i = 32$, $d_i = 5$, $w_i = 4$, and the corresponding times of withdrawal are 236, 304, 338, 388. Thus from (6.11)

$$\tilde{N}_i' = (32-4) + \frac{36+104+138+188}{200} = 30.33,$$

so that $\tilde{q}_4 = 5/30.33 = 0.1649$.

We also estimate the central death rates. Here the exact times of the five deaths are: 228, 253, 280, 297, 322, and there are $N_{i+1} = 23$ patients who did not die and were exposed to risk for the full interval. The total amount of *person-days* in this interval is [from (6.3)] $\tilde{A}_4 = (28+53+80+97+122) + (36+104+138+188) + 23 \cdot 300 = 5446$, so that the observed central death rate [from (6.4)] is $\tilde{m}_4 = 5/5446 = 0.000918$ per day per person.

The estimated survival function is $\tilde{P}_i = \tilde{p}_0 \cdot \tilde{p}_1 \cdots \tilde{p}_{i-1}$.

(b) Similar calculations were carried out neglecting the information on exact times ($t_{ij}^0$'s) and using $N_i'$ [given by (6.14)], and $A_i$ [given by (6.9)] and

**Table 6.2b** Heart transplantation.
Exact time of death or withdrawal not utilized

| $t_i$ to $t_{i+1}$ | $N_i'$ | $\hat{q}_i^{(1)}$ | $\hat{p}_i^{(1)}$ | $\hat{P}_i^{(1)}$ | var($\hat{q}_i^{(1)}$) | var($\hat{P}_i^{(1)}$) | $A_i$ | $\hat{m}_i^{(1)}$ | $\hat{q}_i^{(2)}$ |
|---|---|---|---|---|---|---|---|---|---|
| 0–50 | 66.5 | 0.2406 | 0.7594 | 1.0000 | 0.0027 | | 2925 | 0.005470 | 0.2413 |
| 50–100 | 49 | 0.2245 | 0.7755 | 0.7594 | 0.0036 | 0.0027 | 2175 | 0.005057 | 0.2245 |
| 100–200 | 37 | 0.1081 | 0.8919 | 0.5889 | 0.0026 | 0.0037 | 3500 | 0.001143 | 0.1083 |
| 200–400 | 30 | 0.1667 | 0.8333 | 0.5253 | 0.0046 | 0.0038 | 5500 | 0.000909 | 0.1677 |
| 400–700 | 20 | 0.1000 | 0.9000 | 0.4377 | 0.0045 | 0.0040 | 5700 | 0.000351 | 0.1008 |
| 700–1000 | 13.5 | 0.2963 | 0.7037 | 0.3939 | 0.0154 | 0.0041 | 3450 | 0.001159 | 0.3023 |
| 1000–1300 | 7 | 0.1429 | 0.8571 | 0.2772 | 0.0175 | 0.0044 | 1950 | 0.000513 | 0.1445 |
| 1300–1600 | 3.5 | 0.2857 | 0.7143 | 0.2376 | 0.0583 | 0.0046 | 900 | 0.001111 | 0.3101 |
| 1600+ | 0.5 | 0.0000 | 1.0000 | 0.1697 | 0.0000 | 0.0056 | — | | 0.0000 |

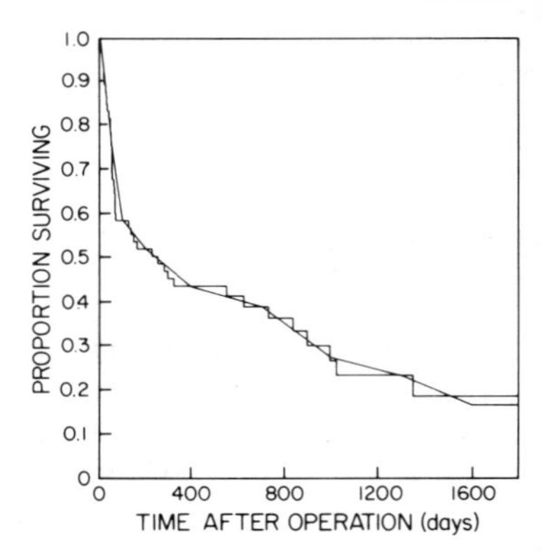


**Fig. 6.2** Heart transplant data. Product-limit and actuarial estimates of survival function.

estimating $q_i$ from (6.15). The results are presented in Table 6.2b. For the reasons explained in the next section, the actuarial estimator of $q_i$ is denoted by $\hat{q}_i^{(1)}$ and correspondingly, $\hat{P}_i^{(1)}$ and $\hat{m}_i^{(1)}$. The $\hat{P}_i^{(1)}$ function is displayed in Fig. 6.2 (continuous line).

There are no noticeable differences between the two sets of estimators given in Tables 6.2a and 6.2b respectively, except when $N_i'$s are very small. In such cases, neither of the two approximations is very reliable.

In Sections 6.3 and 6.4, we will discuss some other estimators of $q_i$. In fact, we will reach the conclusion that the actuarial estimator (6.15) is the most suitable for regular use. Nonstatisticians, who are willing to accept this conclusion may skip these sections. The material contained in them may, however, be found helpful by those feeling confused about which estimator it is 'best' to use.

## 6.3 SOME MAXIMUM LIKELIHOOD ESTIMATORS OF $q_i$

The concept of "effective number of exposed to risk" was introduced in actuarial work from practical considerations, not directly based on the statistical theory of estimation. Approaching the problem of estimating $q_i$ from a statistical point of view, we need to make some *assumptions* about the distribution of time at death and, possibly, about the distribution of time due for withdrawal. We now introduce some new estimators of $q_i$, distinguishing each by a superscript ($\hat{q}_i^{(k)}$). The actuarial estimator (6.15) will now be denoted by $\hat{q}_i^{(1)}$ (see Table 6.2b).

Generally, there are two ways of treating censored survival data:

1. One approach is to assume a certain form of the failure distribution, and treat the observed withdrawal time as an observed survival time

according to a given failure distribution. This is also our attitude in regard to censored observations (see also Chapter 13).

2. On the other hand, many authors introduce a hypothetical time due for withdrawal as a random variable having its own distribution. The observed time for individual $(j), t_{ij}^o$ $(t_i \leqslant t_{ij}^o < t_{i+1})$ is the minimum of failure time and withdrawal time, whichever occurs first. We have, in effect, a problem in competing risk theory, as we shall see in Chapter 9.

These two approaches are also used in follow-up studies. It appears that estimators based on models with certain distributional assumptions and yielding rather complicated expressions, are sometimes regarded as more "rigorous and scientific" as distinguished from the actuarial estimator that is attributed as "intuitive." In practice, the discrepancies are negligible when $q_i$ are small and $N_i$'s are large. In the opposite situation, none of these estimators is close to the true $q_i$. Nevertheless, there is some confusion as to which one to use; in this section we try to clarify some points related to this question.

### 6.3.1 Failure Time Alone Regarded as a Random Variable

In this section, we use the approach discussed in (1) to construct ML-estimators of $q_i$.

***Uniform Distribution of Time at Death.*** We now assume that the time at death is approximately *uniform* over $[t_i, t_{i+1})$. We first suppose that the individual times of exposure to risk, $t_{ij}^o$ (defined in (6.2)) or their fractions of $h_i, \theta_{ij}^o$ [defined in (6.6)] are recorded and utilized in each interval $[t_i, t_{i+1})$. Denoting the (conditional) survival function in $[t_i, t_{i+1})$ by $S_i(t|t_i)$, we have in this case

$$S_i(t|t_i) = 1 - \frac{t - t_i}{h_i} q_i = 1 - \theta_i(t) q_i \qquad \text{for } t_i \leqslant t < t_{i+1}, \tag{6.20}$$

where $\theta_i(t) = (t - t_i)/h_i$, and so the (conditional) PDF is

$$f_i(t|t_i) = \frac{1}{h_i} q_i.$$

The likelihood function (conditional on $N_i, w_i$)

$$L_i(q_i|N_i, w_i) \propto \left(\frac{1}{h_i} q_i\right)^{d_i} \left[\prod_j (1 - \theta_{ij}^* q_i)\right] (1 - q_i)^{N_{i+1}} \tag{6.21}$$

where $\theta_{ij}^*$ is as defined in (6.6).

The ML-estimator of $q_i$ can be obtained by iterative techniques [Elveback (1958)].

Often the individual times $t_{ij}^o$ are not recorded (or not utilized), but instead, $\theta_{ij}^* = \frac{1}{2}$ (i.e., withdrawal time at the midpoint of the interval) is used as an approximation for all individuals. Then (6.21) takes the form

$$L_i(q_i|N_i, w_i) \propto \left(\frac{1}{h_i} q_i\right)^{d_i} \left(1 - \tfrac{1}{2} q_i\right)^{w_i} (1 - q_i)^{N_{i+1}}. \tag{6.22}$$

Taking logarithms of both sides, differentiating with respect to $q_i$, and equating to zero, we obtain the maximum likelihood equation

$$\frac{\partial \log L_i}{\partial q_i} = \frac{d_i}{q_i} - \frac{w_i}{2 - q_i} - \frac{N_{i+1}}{1 - q_i} = 0. \tag{6.23}$$

Substituting $N_{i+1} = N_i - w_i - d_i$, we obtain the quadratic equation

$$N_i q_i^2 - (2N_i + d_i - w_i) q_i + 2d_i = 0,$$

leading to the ML-estimator of $q_i$ (denoted here by $\hat{q}_i^{(2)}$)

$$\hat{q}_i^{(2)} = \frac{(2N_i + d_i - w_i) - \sqrt{(2N_i + d_i - w_i)^2 - 8N_i d_i}}{2N_i}. \tag{6.24}$$

Values of $\hat{q}_i^{(2)}$, for the data of Example 6.1, are shown in the last column of Table 6.2b. In this model, the only random variable is $d_i$. Assuming that approximately $\mathcal{E}(d_i|N_i, w_i) \doteqdot (N_i - \frac{1}{2} w_i) q_i$, the large sample variance is approximately

$$\operatorname{var}\left(q_i^{(2)}|N_i, w_i\right) \doteqdot \left[\mathcal{E}\left(-\frac{\partial^2 \log L_i}{\partial q_i^2}\right)\right]^{-1} \doteqdot \left[\frac{N_i - \frac{1}{2} w_i}{p_i q_i} - \frac{(2 - q_i^2) w_i}{2p_i^2 (1 + p_i)^2}\right]^{-1} \tag{6.25}$$

(see Exercise 6.5).

Note that when $w_i$ is small, this variance is approximately $p_i q_i / (N_i - \frac{1}{2} w_i)$.

***Exponential Distribution of Time at Death.*** In this case, $S_i(t|t_i) = e^{-\mu_i(t - t_i)}$. If the individual times at death or withdrawal in $[t_i, t_{i+1})$ are utilized, the likelihood, *as a function of* $\mu_i$, is

$$L(\mu_i|N_i,w_i)=\left[\prod_j \mu_i e^{-\mu_i h_i \theta_{ij}}\right]\left[\prod_j e^{-\mu_i h_i \theta_{ij}^*}\right]e^{-\mu_i h_i N_{i+1}}. \tag{6.26}$$

The likelihood equation is

$$\frac{\partial \log L_i}{\partial \mu_i}=\frac{d_i}{\mu_i}-h_i\left[\sum_j \theta_{ij}+\sum_j \theta_{ij}^*+N_{i+1}\right]=0,$$

from which it is easy to see that $L_i(\hat{\mu}_i|N_i,w_i)$ is maximized by

$$\hat{\mu}_i=\frac{d_i}{h_i\left[\sum_j \theta_{ij}+\sum_j \theta_{ij}^*+N_{i+1}\right]}=\frac{\text{number of deaths}}{\text{amount of exposed to risk}}. \tag{6.27}$$

[See (6.4).] Note that $N_{i+1}=N_i-d_i-w_i$.

The corresponding maximum likelihood estimator of $p_i$ is

$$\hat{p}_i=e^{-\hat{\mu}_i h_i}, \tag{6.28}$$

where $\hat{\mu}_i$ is given by (6.27).

On the other hand, if all that is known that $d_i$ deaths occurred *at some time* in $[t_i,t_{i+1})$, and it is assumed that the *observed average* time of withdrawal is the midpoint $\frac{1}{2}(t_i+t_{i+1})$, the likelihood now, *as a function of* $p_i$, is

$$L_i(p_i|N_i,w_i)\propto(1-p_i)^{d_i}p_i^{w_i/2}p_i^{N_{i+1}}=(1-p_i)^{d_i}p_i^{N_i-w_i/2-d_i},$$

whence the maximum likelihood estimator of $q_i$ is

$$\hat{q}_i^{(3)}=\frac{d_i}{N_i-\frac{1}{2}w_i}, \tag{6.29}$$

*which is identical with the actuarial estimator* $\hat{q}_i^{(1)}$. This rather remarkable coincidence provides a link between the probabilistic and actuarial approach to $q_i$. From statisticians' viewpoint it may provide an aura of respectability for the actuarial estimator.

The variance of $\hat{q}_i^{(3)}$ (conditional on $N_i$, $w_i$) is given by (6.18).

### 6.3.2 Failure Time and Censoring Time Regarded as Random Variables

Let $T$ be the failure time (time at death) with SDF $S_i(t|t_i)$ in $[t_i, t_{i+1})$, and let $U$ denote the censoring time with PDF $g_i(u|t_i)$. Assume that $T$ and $U$ are *independent*.

It is often stressed that planned withdrawals have a different distribution than those randomly lost from the study. For simplicity, we assume that there are *no* random withdrawals.

Let $p_{(w)i}$ be the probability that an individual alive at $t_i$ will withdraw before he dies, in $[t_i, t_{i+1})$. We have

$$p_{(w)i} = \int_{t_i}^{t_{i+1}} S_i(y|t_i) g_i(y|t_i)\, dy \tag{6.30}$$

(see also Chapter 9—here death and withdrawal are considered as "competing causes" of exit; $p_{(w)i}$ is the crude probability of exit by withdrawal in the presence of mortality).

***Special Cases.***

1. Suppose $S_i(t|t_i) = 1 - \frac{1}{h_i}(t - t_i)q_i$ and $g_i(u|t_i) = \frac{1}{h_i}$, that is, $T$ and $U$ are both *uniform*. Then from (6.30),

$$p_{(w)i} = \frac{1}{h_i}\int_{t_i}^{t_{i+1}}\left[1 - \frac{1}{h_i}(y - t_i)q_i\right]dy \doteqdot 1 - \tfrac{1}{2}q_i. \tag{6.31}$$

2. Suppose that $S_i(t|t_i) = e^{-\mu_i(t - t_i)}$, and $g_i(u|t_i) = \frac{1}{h_i}$, that is, $T$ is *exponential* and $U$ *uniform*. Then from (6.30),

$$p_{(w)i} = \frac{1}{h_i}\int_{t_i}^{t_{i+1}} e^{-\mu_i(y - t_i)}\, dy$$

$$= \frac{1}{\mu_i h_i}(1 - e^{-\mu_i h_i}) \doteqdot -\frac{1 - p_i}{\log p_i}, \tag{6.32}$$

or approximately,

$$p_{(w)i} \doteqdot p_i^{1/2}. \tag{6.33}$$

Chiang (1968, p. 272) showed by numerical calculations that (6.33) is, indeed, a very good approximation to (6.32).

It has been further suggested that the $N_i$ individuals observed at $t_i$ should be separated into two groups: those who are due for withdrawal and those who are not, and that they constitute two *independent* random samples.

Let $c_i$ denote the number planned to be withdrawn in $[t_i, t_{i+1})$, and $d_i'$ the number of deaths among those $c_i$ in $[t_i, t_{i+1})$. In this group, an individual either is withdrawn or dies, so $q_{(w)i} = 1 - p_{(w)i}$ is the probability of not being withdrawn (i.e., of dying) before withdrawal takes place.

Among the $(N_i - c_i)$ individuals who are not due for withdrawal, there are $(d_i - d_i')$ deaths. Clearly, the likelihood

$$L_i \propto q_i^{d_i - d_i'}(1 - q_i)^{(N_i - c_i) - (d_i - d_i')} \times q_{(w)i}^{d_i'}(1 - q_{(w)i})^{c_i - d_i'}. \tag{6.34}$$

Substituting an approximate formula for $q_{(w)i}$ (in terms of $q_i$), we obtain an ML-estimator of $q_i$, the conditional probability of dying in $[t_i, t_{i+1})$ given alive at $t_i$.

It is clear from this section, how many assumptions have been made to obtain (6.34) and how artificial they appear to be. This brief outline however, has been given to provide some instructive background. The results in two special cases (of uniform and exponential distributions) are briefly as follows.

1. *T and U are both uniformly distributed over* $(t_i, t_{i+1})$. In this case [from (6.31)], $q_{(w)i} = \frac{1}{2}q_i$, and the likelihood (6.34) takes the form

$$L_i(q_i|N_i, c_i) \propto q_i^{d_i}(1 - q_i)^{N_{i+1}}(2 - q_i)^{w_i}, \tag{6.35}$$

where

$$N_{i+1} = (N_i - c_i) - (d_i - d_i')$$

and

$$w_i = c_i - d_i'.$$

This, of course, leads to the same likelihood equation as given in (6.23) yielding $\hat{q}_i^{(2)}$ given by (6.24). For this model, however, not only $d_i$, but also

$w_i$ are random variables. We have

$$\mathcal{E}(d_i - d_i' | N_i, c_i) = (N_i - c_i) q_i,$$

$$\mathcal{E}(d_i' | c_i) \doteq c_i \cdot \tfrac{1}{2} q_i,$$

so that

$$\mathcal{E}(d_i | N_i, c_i) \doteq \left(N_i - \tfrac{1}{2} c_i\right) q_i,$$

and

$$\mathcal{E}(w_i | c_i) \doteq c_i \left(1 - \tfrac{1}{2} q_i\right).$$

In this case the approximate variance

$$\operatorname{var}\left(\hat{q}_i^{(2)} | N_i, c_i\right) \doteq \left[ \mathcal{E}\left( - \frac{\partial^2 \log L_i}{\partial q_i^2} \right) \right]^{-1}$$

$$\doteq \frac{q_i(1-q_i)(2-q_i)}{N_i(2-q_i)-c_i} = \frac{p_i q_i}{N_i}\left[1 - \frac{c_i}{N_i(1+p_i)}\right]^{-1} \tag{6.36}$$

is different from that in (6.25). [See Elveback (1958) and Exercise 6.5.] When $c_i$ is small as compared with $N_i$, (6.36) is approximately $p_i q_i / N_i$.

2. *T is exponentially, and U uniformly distributed.* In this case [from (6.33)], the likelihood (6.34) takes the form

$$L_i(p_i | N_i, c_i) \propto (1-p_i)^{d_i - d_i'} p_i^{(N_i - c_i) - (d_i - d_i')} \left(1 - p_i^{1/2}\right)^{d_i} \left(p_i^{1/2}\right)^{c_i - d_i'}. \tag{6.37}$$

This leads to the quadratic maximum likelihood equation in $p_i^{1/2} = x$:

$$(2N_i - c_i)x^2 + d_i' x - (2N_i + w_i) = 0, \tag{6.38}$$

so that the ML-estimator of $q_i$ (denoted here by $\hat{q}_i^{(4)}$) is

$$\hat{q}_i^{(4)} = 1 - \left[ \frac{-d_i' + \sqrt{d_i'^2 + 4(2N_i - c_i)(2N_{i+1} + w_i)}}{2(N_i - c_i)} \right]^2. \tag{6.39}$$

[Note that $2N_{i+1}+w_i=(2N_i-c_i)-(2d_i-d_i')$.] This estimator was derived by Chiang (1961).

Its approximate variance is

$$\mathrm{var}\left[\hat{q}_i^{(4)}|N_i,c_i\right] \doteq \left[\frac{N_i-c_i}{p_iq_i}+\frac{c_i}{4p^{3/2}(1-p^{1/2})}\right]^{-1} \tag{6.40}$$

[see Chiang (1961), Drolette (1975) and Exercise 6.6]. Estimators $\hat{q}_i^{(2)}$ and $\hat{q}_i^{(4)}$ are biased.

## 6.4 SOME OTHER ESTIMATORS OF $q_i$

### 6.4.1 Moment Estimator of $q_i$

Assuming, as in Section 6.3.2 (1), that the time at death and time due for withdrawal are each *uniformly* distributed, we have

$$\mathcal{E}(d_i-d_i'|N_i,c_i)=(N_i-c_i)q_i \qquad \text{and} \qquad \mathcal{E}(d_i'|c_i)\doteq c_i\cdot\tfrac{1}{2}q_i,$$

so that

$$\mathcal{E}(d_i|N_i,c_i)=(N_i-c_i)q_i+q_i+\tfrac{1}{2}c_iq_i=\left(N_i-\tfrac{1}{2}c_i\right)q_i. \tag{6.41}$$

Equating the observed number of deaths to its expected value, we obtain the *moment* estimator of $q_i$ (denoted here by $\hat{q}_i^{(5)}$),

$$\hat{q}_i^{(5)}=\frac{d}{N_i-\frac{1}{2}c_i} \tag{6.42}$$

[see also Elveback (1958)]. This is similar to the actuarial estimator $\hat{q}_i^{(1)}$, but with $w_i$ replaced by $c_i$.

Clearly, $\hat{q}_i^{(5)}$ is an *unbiased* estimator of $q_i$. The variance of $d_i$ (conditional on $N_i$ and $c_i$) is

$$\begin{aligned}\mathrm{var}(d_i|N_i,c_i) &\doteq (N_i-c_i)q_i(1-q_i)+\tfrac{1}{2}c_iq_i\left(1-\tfrac{1}{2}q_i\right)\\ &=\left(N_i-\tfrac{1}{2}c_i\right)p_iq_i+\tfrac{1}{4}c_iq_i^2,\end{aligned} \tag{6.43}$$

so that

$$\operatorname{var}\left(q_i^{(5)} \mid N_i, c_i\right) = \left(N_i - \tfrac{1}{2}c_i\right)^{-2} \operatorname{var}(d_i \mid N_i, c_i)$$

$$\doteqdot \frac{p_i q_i}{N_i - \frac{1}{2}c_i} + c_i \left[ \frac{q_i}{2\left(N_i - \frac{1}{2}c_i\right)} \right]^2. \qquad (6.47)$$

When $q_i$ and $c_i$ are small, $\operatorname{var}(\hat{q}_i^{(5)} \mid N_i, c_i) \doteqdot p_i q_i / (N_i - \frac{1}{2}c_i)$.

### 6.4.2 Estimator of $q_i$ Based on Reduced Sample

Drolette (1975) pointed out that in practice $d_i' \ll c_i q_{(w)i}$ and recommended that in follow-up studies with the anniversary method only the number of exposed to risk for the *full* interval should be used in estimating $q_i$. In effect, this is acting *as if* all individuals due for withdrawal were removed at the beginning of the interval. The estimator proposed by Drolette (denoted here by $\hat{q}_i^{(6)}$) is

$$\hat{q}_i^{(6)} = \frac{d_i - d_i'}{N_i - c_i}, \qquad (6.48)$$

and its variance (conditional on $N_i, c_i$) is

$$\operatorname{var}\left(\hat{q}_i^{(6)} \mid N_i, c_i\right) = \frac{p_i q_i}{N_i - c_i}. \qquad (6.49)$$

Clearly, $\hat{q}_i^{(6)}$ is a binomial proportion, and so it is an unbiased estimator of $q_i$.

## 6.5 COMPARISON OF VARIOUS ESTIMATORS OF $q_i$

Much discussion of the relative merits of different estimators is essentially a discussion of the merits of different assumptions, since, despite the apparent dissimilarities in the formulae, the estimators produce nearly identical results, provided that $N_i$ is fairly large and $q_i$ small ($<0.3$, say) [Kuzma (1967), Drolette (1975), Elandt-Johnson (1977)].

In fact, the close agreement among the numerical values of $\hat{q}_i^{(1)}$ through $\hat{q}_i^{(5)}$ is not fortuitous; it has been shown [Johnson (1977)] that there are some close mathematical relationships among them, so that one can be approximated by another.

For convenience, the formulae for estimators $\hat{q}_i^{(1)}$ through $\hat{q}_i^{(6)}$, their variances, and the assumptions under which they were derived are summarized in Table 6.3.

**Table 6.3.** Estimators of $q_i$, assumptions, and properties

| Estimator | Method of estimation | Assumptions in $[t_i, t_{i+1})$: Distribution — Time at death | Distribution — Withdrawal time | Loses observed | Deaths $d_i'$ utilized separately | Properties of the estimator: Biasedness | Approximate variance (Conditional on $N_i$, $c_i$, or $w_i$) | References |
|---|---|---|---|---|---|---|---|---|
| (1) $\hat{q}_i^{(1)} - \dfrac{d}{N_i - \frac{1}{2}w_i}$ | Actuarial | Expected value $\frac{1}{2}(t_i + t_{i+1})$ | Average $\frac{1}{2}(t_i + t_{i+1})$ | Yes | No | Negatively biased | $\dfrac{p_i q_i}{N_i - \frac{1}{2}w_i}$ | Frost (1933), Berkson and Gage (1950) Elveback (1958), Cutler and Ederer (1958) Kuzma (1967), Gehan(1969) Drolette (1975) Elandt-Johnson (1977) |
| (2) $\hat{q}_i^{(2)} = \dfrac{(2N_i + d_i - w_i) - \sqrt{(2N_i + d_i - w_i)^2 - 8N_i d_i}}{2N_i}$ | Maximum likelihood | Uniform | All at $\frac{1}{2}(t_i + t_{i+1})$ | Yes | No | Biased | $\left[\dfrac{N_i - \frac{1}{2}w_i}{p_i q_i} - \dfrac{(2 - q_i^2)w_i}{2p_i^2(1 + p_i)^2}\right]^{-1}$ | Elandt-Johnson (1977) |
| | | | Uniform | No | Yes | Positively biased | $\dfrac{p_i q_i}{N_i}[1 - \dfrac{c_i}{N_i(1 + P_i)}]^{-1}$ | Elveback (1958), Drolette (1975) |
| (3) $\hat{q}_i^{(3)} = \dfrac{d_i}{N_i - \frac{1}{2}w_i}$ (the same as $q_i^{(1)}$) | Maximum likelihood | Exponential | Average $\frac{1}{2}(t_i + t_{i+1})$ | Yes | Yes | Negatively biased | $\dfrac{p_i q_i}{N_i - \frac{1}{2}w_i}$ | Elandt-Johnson (1977) |
| (4) $\hat{q}_i^{(4)} = 1 - \left[\dfrac{-d_i' + \sqrt{d_i'^2 + 4(2N_i - c_i)(2N_{i+1} + w_i)}}{2(2N_i - c_i)}\right]^2$ | Maximum likelihood | Exponential | Uniform | No | Yes | Negatively biased | $\left[\dfrac{N_i - c_i}{p_i q_i} + \dfrac{c_i}{4p_i^{3/2}(1 - p_i^{1/2})}\right]^{-1}$ | Chiang (1968), Kuzma (1967) Drolette (1975) |
| (5) $\hat{q}_i^{(5)} = \dfrac{d_i}{N_i - \frac{1}{2}c_i}$ | First moment | Uniform | Uniform | No | Yes | Unbiased | (Exact) $\dfrac{p_i q_i}{N_i - \frac{1}{2}c_i} + c_i\left[\dfrac{q_i}{2(N_i - \frac{1}{2}c_i)}\right]^2$ | Elveback (1958), Elandt-Johnson (1977) |
| (6) $q^{(6)} = \dfrac{d_i - d_i'}{N_i - c_i}$ | Maximum likelihood | Uniform | All at $t_i$ | No | Not at all | Unbiased | (Exact) $\dfrac{p_i q_i}{N_i - c_i}$ | Drolette (1975) |

$N_i$ = total number of living at $t_i$.
$d_i$ = total number of deaths in $[t_i, t_{i+1})$.
$N_{i+1} = (N_i - d_i - w_i)$ = survivors at the end of the interval $[t_i; t_{i-1})$.

$c_i$ = number due for withdrawal in $[t_i, t_{i+1})$.
$d_i'$ = number of deaths among $c_i$ in $[t_i, t_{i+1})$.
$w_i$ = number of actually withdrawn in $[t_i, t_{i+1})$.

In conclusion, we may say that for sufficiently large sample sizes and small $q_i$ ($<0.3$), one can use any of the estimators $\hat{q}_i^{(1)}$ through $\hat{q}_i^{(5)}$. Of course, the simplest is the actuarial estimator $\hat{q}_i^{(1)}$ defined by (6.15) and we would recommend it for use. It is a *good* and *robust* estimator of $q_i$.

## 6.6 ESTIMATION OF CURVE OF DEATHS

When data are incomplete, we can only construct a fitted histogram from the estimated survival function with width equal to $h_i$ and height equal to

$$\hat{f}(t_i^*)=\frac{\hat{P}_i-\hat{P}_{i+1}}{h_i}=\frac{\hat{P}_i\hat{q}_i}{h_i}. \tag{6.50}$$

This may also be used as an estimator of the density at the midpoint $t_i^*=t_i+\frac{1}{2}h_i$. Its approximate variance can be calculated from (5.27), replacing $N_i$ by $N_i'=N_i-\frac{1}{2}w_i$.

## 6.7 PRODUCT-LIMIT METHOD OF ESTIMATING THE SURVIVAL FUNCTION FROM INDIVIDUAL TIMES AT DEATH

We now consider a situation in which the exact time at death or withdrawal is recorded for each of all $N$ individuals who were enrolled at any time during the trial.

Let $t_0=0$ be the starting point and $t_e$ the ending point on the follow up scale. Imagine the period $(0,t_e]$ to be divided into $M$ (closed to the right) intervals $(0,t_1]$, $(t_1,t_2],\ldots,$ $(t_{M-1},t_e]$, so fine that the occurrence of more than one event (death or withdrawal) in the same interval has a negligible probability.

Let $R_i$ be the number of individuals alive in the trial at $t_i$, or in practice, just before $t_i$. These individuals are said to constitute the *risk set* $R_i$ at $t_i$ (just before $t_i$). Let

$$\phi_i=\begin{cases}1 & \text{if death occurs in } (t_{i-1},t_i],\\ 0 & \text{otherwise,}\end{cases}$$

and $q_{i-1}$ be the (conditional) probability of death in $(t_{i-1},t_i]$ given alive at $t_{i-1}$.

The likelihood

$$L_i\propto(q_{i-1})^{\phi_i}(p_{i-1})^{R_i-\phi_i} \tag{6.52}$$

leads to the *unbiased* ML-estimator

$$\hat{q}_{i-1} = \begin{cases} \dfrac{1}{R_i} & \text{if death occurs in } (t_{i-1}, t_i] \\ 0 & \text{otherwise.} \end{cases} \tag{6.53}$$

Since $(t_{i-1}, t_i]$ is as narrow as we please, $\hat{q}_{i-1} = 1/R_i$ if death occurs at $t_i$. Of course

$$\hat{p}_{i-1} = 1 - \hat{q}_{i-1} = \begin{cases} \dfrac{R_i - 1}{R_i} & \text{if death occurs at } t_i \\ 1 & \text{otherwise,} \end{cases} \tag{6.53a}$$

and the estimated survival function is

$$\hat{S}(t_i) = \hat{P}_i = \hat{p}_0 \cdot \hat{p}_1 \cdots \hat{p}_{i-1}, \tag{6.54}$$

with $\hat{S}(0) = \hat{P}_0 = 1$.

Clearly, if there is no death at $t_i$, then $p_{i-1} = 1$, so that division points with no death do not contribute to the estimate in (6.54). Therefore, for practical purposes, it is sufficient to use, as division points, the *ordered* times at death

$$t'_1 < t'_2 < \cdots < t'_j < \cdots < t'_K, \tag{6.55}$$

where $K$ denotes the number of deaths at *distinct* time points. $\mathcal{R}_j$ is now the risk set at $t'_j$, and

$$\hat{S}(t) = \hat{P}_i = \begin{cases} 1 & \text{for } t < t'_1, \\ \displaystyle\prod_{j=1}^{i} \frac{R_j - 1}{R_j} & \text{for } t'_i \leqslant t < t'_{i+1},\ i = 1, 2, \ldots, K-1, \\ \displaystyle\prod_{j=1}^{K} \frac{R_j - 1}{R_j} & \text{for } t \geqslant t'_K. \end{cases} \tag{6.56}$$

This is called the *product-limit* (PL) estimator of $S(t)$. It was derived by Kaplan and Meier (1958), and is often called the *Kaplan-Meier estimator* of $\hat{S}(t)$.

An approximate (Greenwood's) formula for variance [cf. (5.22)] is

$$\text{est. var}\left(\hat{P}_i \mid \{R_j\}\right) = \hat{P}_i^2 \sum_{j=1}^{i} \frac{\hat{q}_{j-1}}{R_j \hat{p}_{j-1}} = \hat{P}_i^2 \sum_{j=1}^{i} \frac{1}{R_j(R_j - 1)}. \tag{6.57}$$

If there are multiple deaths—$\delta_j$ deaths at $t'_j$—then $(R_j-1)/R_j$ in (6.56) is replaced by $(R_j-\delta_j)/R_j$. In this case, formula (6.57) takes the form

$$\text{est. var}(\hat{P}_i|\{R_j\}) = \hat{P}_i^2 \sum_{j=1}^{i} \frac{\delta_j}{R_j(R_j-\delta_j)}. \tag{6.58}$$

**Example 6.2** As an illustration of the product-limit method, we use the heart transplant data given in Table 6.1. Here, one death is observed at $t'_0=0+$, so that $\hat{S}(0+)=1$. We also have at $t'_i=1$, one death and one withdrawal [patients No. 2 (28) and No. 3 (100)]. In such cases the smaller rank is usually arbitrarily assigned to the death. There is also one case of multiplicity—of deaths at $t_i=51$, with $\delta_i=3$. The sizes of the risk sets, $R_i$, are given in column 5 of Table 6.1. The numbers in the parentheses are values of $R_i$ at times of observed withdrawals. The values of $\hat{S}(t'_i)$ are given in column 7. There is a graphical representation of $\hat{S}(t)$ in Fig. 6.2 (step function).

## 6.8 ESTIMATION OF SURVIVAL FUNCTION USING THE CUMULATIVE HAZARD FUNCTION

We notice that $\hat{q}_{i-1}$ in (6.53) can be obtained by supposing a *uniform* distribution of time at death in $(t_{i-1}, t_i]$ and then assuming that its length $t_i - t_{i-1} = h_{i-1}$ tends to zero. One may also use a similar argument but supposing that the time at death is *exponential* with constant $\lambda_i$ over $(t_{i-1}, t_i]$. Using the ordered times at death (6.55) as division points, we obtain

$$\hat{\lambda}_i = \frac{1}{\sum_{j\in R_i} h_{i-1}} = \frac{1}{h_{i-1}R_i} \qquad i=1,2,\ldots \tag{6.59}$$

and

$$\tilde{q}_{i-1} = 1-\exp(-\hat{\lambda}_i h_{i-1}) = 1-\exp\left(-\frac{1}{R_i}\right) \tag{6.60}$$

so that

$$\tilde{S}(t_i) = \exp\left[-\sum_{j=1}^{i} \frac{1}{R_j}\right] \tag{6.61}$$

[cf. (13.53)]. Note that

$$\sum_{j=1}^{i} \frac{1}{R_j} = \tilde{\Lambda}(t_i)$$

is the cumulative hazard function.

If there is multiplicity at $t_i$, either an arbitrary order can be assigned to the deaths, or $\lambda_i$ is estimated from the formula $\hat{\lambda}_i = \delta_i / h_{i-1} R_i$. The reader might estimate SDF for heart transplant data given in Table 6.1, using (6.61), and compare with that obtained by the Kaplan-Meier method.

## REFERENCES


Batten, R. W. (1978). *Mortality Table Construction*. Prentice-Hall, Englewood Cliffs, N. J.

Benjamin, B., and Haycocks, H. W. (1970). *The Analysis of Mortality and Other Actuarial Statistics*. Cambridge University Press, Cambridge, Chapter 2.

Berkson, J., and Gage, R. P. (1950). Calculation of survival rates for cancer. *Proc. Staff Meet. Mayo Clinic* **25**, 270–286.

Breslow, N., and Crowley, J. (1974). A large sample study of the life table and product limit estimates under random censorship. *Ann. Statist.* **2**, 437–453.

Chiang, C. L. (1961). A stochastic study of the life table and its application. III. The follow-up studies with the consideration of competing risks. *Biometrics* **17**, 57–78.

Chiang, C. L. (1968). *Introduction to Stochastic Processes in Biostatistics*. Wiley, New York, Chapter 12.

Crowley, J., and Hu, M. (1974). Covariance analysis of heart transplant survival data, Stanford University, Technical Report #2, pp. 1–35.

Cutler, S. J., and Ederer, F. (1958). Maximum utilization of the life table method in analyzing survival. *J. Chron. Dis.* **8**, 699–713.

Dorn, H. (1950). Methods of analysis for follow-up studies. *Human Biol.* **22**, 238–248.

Drolette, M. E. (1975). The effect of incomplete follow-up. *Biometrics* **31**, 135–144.

Elandt-Johnson, R. C. (1977). Various estimators of conditional probabilities of death in follow-up studies: Summary of results. *J. Chron. Dis.* **30**, 247–256.

Elveback, L. (1958). Actuarial estimation of survivorship in chronic disease. *J. Amer. Statist. Assoc.* **53**, 420–440.

Frost, W. H. (1933). Risk of persons in familial contact with pulmonary tuberculosis. *Amer. J. Publ. Health* **23**, 426–432.

Gehan, E. A. (1969). Estimating survival functions from the life table. *J. Chron. Dis.* **21**, 629–644.

Gershenson, H. (1961). *Measurement of Mortality*. Society of Actuaries, Chicago.

Kaplan, E. L., and Meier, P. (1958). Nonparametric estimation from incomplete observations. *J. Amer. Statist. Assoc.* **53**, 457–481.

Kuzma, J. W. (1967). A comparison of two life table methods. *Biometrics* **23**, 51–64.

Nelson, W. (1972). Theory and application of hazard plotting for censored failure data. *Technometrics* **14**, 945–966.

## EXERCISES

**6.1.** Suppose that the lifetime, $X$, of electric bulbs is exponential with PDF

$$f_X(x) = \lambda e^{-\lambda x}, \qquad x>0, \qquad \lambda>0.$$

A random sample of $n$ bulbs was selected for quality testing. The period for the whole experiment was $(0,\tau)$, but the individual bulbs were planned to be maintained for periods $t_1, t_2, \ldots, t_n$ ($t_j$ for the $j$th bulb) unless they failed previously. Assume that $t_j$'s are mutually independent random variables, each uniformly distributed over $(0,\tau)$, that is, with PDF

$$g_{T_j}(t) = \frac{1}{\tau}, \qquad 0<t<\tau,$$

for $j=1,2,\ldots,n$.

The total number of failures recorded over the interval $(0,\tau)$ is $D$.

(*a*) Find the conditional (given $t_1, t_2, \ldots, t_n$) expected value and variance of $D$.

(*b*) Show that the (unconditional) expected value of $D$ is

$$\mathcal{E}(D) = n\left[1 - \frac{1-e^{-\lambda\tau}}{\lambda\tau}\right].$$

(*c*) Find the variance of $D$.

**6.2.** Do Exercise 6.1, assuming that the PDF of maintenance time is exponential

$$g_{T_j}(t) = \mu e^{-\mu t}, \qquad \mu>0, \qquad t>0.$$

[Notice that the observation time is only over the period $(0,\tau)$.]

**6.3.** The data in Table E6.1 are reconstructed from data given in a paper by J. B. Storer ((1965) "Radiation resistance with age in normal and irradiated populations of mice," *Radiation Res.* **25**, 435–459).

The data represent mortality experiences of 1068 control, and 1047 irradiated (with dose 300 roentgens) mice of $BDF_1/J$ strain. The data are grouped in equal intervals, each of length $h=100$ days.

Here $N_i$ is the number of survivors at the beginning of the interval $[t_i, t_{i+1})$, and $d_i$ is the observed number of deaths in $[t_i, t_{i+1})$. At a prede-

**Table E6.1** Mortality experience of controlled and irradiated mice

| Age (in interval units $i$ to $i+1$) | Control | | | | | Irradiated (300 roentgens) | | | | |
|---|---|---|---|---|---|---|---|---|---|---|
| | Age (in days) | $N_i$ | $d_i$ | $w_i$ | $\tau_i^*$ (in days) | Age (in days) | $N_i$ | $d_i$ | $w_i$ | $\tau_i^*$ (in days) |
| 1–2 | 100–200 | 1068 | 4 | — | — | 100–200 | 1047 | 4 | — | — |
| 2–3 | 200–300 | 1064 | 7 | 142 | 206 | 200–300 | 1043 | 16 | 147 | 258 |
| 3–4 | 300–400 | 915 | 13 | — | — | 300–400 | 880 | 35 | — | — |
| 4–5 | 400–500 | 902 | 13 | — | — | 400–500 | 845 | 52 | — | — |
| 5–6 | 500–600 | 889 | 33 | 135 | 545 | 500–600 | 793 | 85 | 122 | 578 |
| 6–7 | 600–700 | 721 | 63 | — | — | 600–700 | 586 | 112 | 174 | 683 |
| 7–8 | 700–800 | 658 | 106 | 9 | 800 | 700–800 | 300 | 120 | 15 | 800 |
| 8–9 | 800–900 | 543 | 140 | 155 | 851 | 800–820 | 165 | 30 | 135 | 820 |
| 9–10 | 900–1000 | 248 | 149 | 17 | 950 | | | | | |
| 10–11 | 1000–1100 | 28 | 28 | — | — | | | | | |

termined (fixed) time point $\tau_i^*$ in the interval $[t_i, t_{i+1})$, $w_i$ mice were removed from the experiment and used in another study.

(*a*) Estimate the effective numbers of initial exposed to risk, and evaluate the survival function for each group. Draw the two estimated survival functions on the same graph.

(*b*) For convenience, an interval of 100 days as a *time unit* is to be used. For each group: calculate for each interval the amount of mouse-time units and evaluate the central rate per mouse per time unit.

(*c*) Assuming constant mortality rates within each interval, one can evaluate the force of mortality from the formula $\hat{\mu}(t_i^*) = (-1/h_i)\log \hat{p}_i$, where $(t_i^* = t_i + \frac{1}{2}h_i)$. [Notice that we have $(h_i = 1)$.] Of course, this also is an approximation to the central rate. Calculate $\hat{\mu}_i$'s from the formula given above and compare with the central rates calculated in (*b*).

**6.4.** The heart transplant data in Table 6.1 include (in the last column) time due for withdrawal. This is equal to termination time *minus* the time at entry (on a chronological scale). Arrange these in descending order of time due for withdrawal and calculate the (hypothetical) empirical distribution—that is, if there were no deaths.

*Hint:* see Section 5.2.

How does this distribution relate to the time at entry distribution?

**6.5.**

(*a*) Likelihood functions given in (6.22), Section 6.3.1, and in (6.35), Section 6.3.2 are identical, although they were obtained under different assumptions. State the assumptions in each case.

(*b*) Show that the second derivative of the log likelihood (6.22) is

$$\frac{\partial^2 \log L_i}{\partial q_i^2} = -\left[\frac{d_i}{q_i^2} + \frac{w_i}{(2-q_i)^2} + \frac{w_i}{(2-q_i)^2} + \frac{N_i - d_i - w_i}{(1-q_i)^2}\right]. \quad \text{(A)}$$

Note that conditionally on $N_i$ and $w_i$, the only random variable is $d_i$. Assuming $\mathcal{E}(d_i|N_i, w_i) \doteqdot (N_i - \frac{1}{2}w_i)q_i$ (for explanation, see Section 6.2.3), obtain the formula (6.25) for the approximate variance of $\hat{q}_i^{(2)}$ given in (6.24).

(*c*) In the situation of Section 6.3.2, both $d_i$ and $d_i'$ (or equivalently $w_i$) are random variables with $\mathcal{E}(d_i) \doteqdot (N_i - \frac{1}{2}c_i)q_i$ and $\mathcal{E}(w_i) \doteqdot c_i(1 - \frac{1}{2}q_i)$. Show that the variance of $\hat{q}_i^{(2)}$ (conditional on $N_i, c_i$) is given by (6.36).

**6.6.** Derive the conditional variance of $\hat{q}_i^{(4)}$ given in (6.40). Note the assumptions in this case. Compare with the situation discussed in Section 6.3.1, for the estimator $\hat{q}_i^{(3)}$ given by (6.29).

*Hint:* The approximate variance (6.40) can be obtained using

$$\mathcal{E}\left[-\frac{\partial^2 \log L_i}{\partial q_i^2}\right]^{-1}$$

or by using the statistical differentials formula (3.82).

**6.7.** The data in Table E6.2 are taken from Susarla and Van Ryzin ((1978), *Ann. Statist.* **6**, 740–754) and represent survival times (in weeks) of 81 patients with melanoma.

(*a*) Arrange the data in descending order and evaluate the empirical SDF (using the product-limit method) at the points:

$$t = 0, 16, 20, 40, 60, 80, 100, 120, 140, 160, 180, 200, 220, 240.$$

(*b*) Group the data into fixed intervals (e.g., of length 25 or 30 weeks) and evaluate the SDF from the grouped data.

**Table E6.2** Survival time of 81 patients with melanoma

| | | | | | | | | |
|---|---|---|---|---|---|---|---|---|
| 136 | 58 | 55+ | 181+ | 21 | 23 | 190+ | 65 | 234 |
| 194+ | 14 | 90 | 20 | 130 | 213+ | 215+ | 124 | 108+ |
| 54 | 98 | 193+ | 138 | 141 | 110 | 67+ | 50 | 26 |
| 103 | 59 | 134+ | 147+ | 162+ | 65 | 40 | 34 | 57 |
| 81+ | 152+ | 125+ | 151+ | 34 | 158 | 27 | 148+ | 27 |
| 132+ | 140+ | 32 | 130+ | 38 | 85 | 129+ | 100+ | 19 |
| 118 | 53 | 140+ | 66 | 46 | 37 | 50+ | 114+ | 124+ |
| 26 | 102 | 93+ | 80+ | 60 | 86+ | 21+ | 44+ | 23 |
| 80 | 73+ | 19 | 38 | 31 | 25 | 76+ | 13+ | 16+ |

+ Censored observations.

**6.8.** The data in Table E6.3 are taken from Hoel and Walburg [(1972), *J. Nat. Cancer Inst.* **49**, 361–372] and represent the ages at death of 96 conventional mice, classified by two causes of death: cause $C_1$—with lung tumor ($n_1=27$ mice), and cause $C_2$—without lung tumor ($n_2=69$ mice).

**Table E6.3** Ages at death of 96 conventional mice

| | |
|---|---|
| Lung tumor (27) | 381, 477, 485, 515, 539, 563, 565, 582, 603, 616, 624, 650, 651, 656, 659, 672, 679, 698, 702, 709, 723, 731, 775, 779, 795, 811, 839 |
| No lung tumor (69) | 45, 198, 215, 217, 257, 262, 266, 371, 431, 447, 454, 459, 475, 479, 484, 500, 502, 503, 505, 508, 516, 531, 541, 553, 556, 570, 572, 575, 577, 585, 588, 594, 600, 601, 608, 614, 616, 632, 632, 638, 642, 642, 642, 644, 644, 647, 647, 653, 659, 660, 662, 663, 667, 667, 673, 673, 677, 689, 693, 718, 720, 721, 728, 760, 762, 773, 777, 815, 886. |

(*a*) Order the 96 observations in descending order and calculate the overall survival function (without taking into account the cause of death). Note the mortality data are complete (with some multiplicities).

(*b*) Evaluate the empirical SDF for 69 deaths with no lung tumor [i.e., deaths from $C_2$ in *the presence* of deaths from $C_1$; see also Section 12.4.1(1)].

(*c*) Evaluate the empirical SDF from cause $C_2$, but treating the deaths from cause $C_1$ as withdrawals [i.e., deaths from $C_2$, in *the absence* of deaths from $C_1$; see also Section 12.4.1(3)].

(*d*) Represent the curves obtained in (*b*) and (*c*) on the same graph. Comment on the results.

**6.9.** Show that $\tilde{q}_i$ defined in (6.12) and $\tilde{\tilde{q}}_i$ defined in (6.13) are not identical.

**6.10.** Schwartz and Lazar [(1964). *Rev. Statist. Appl., Paris*, **12** (3), 15–28] obtained a formula ("formula G") for $\hat{q}_i$ based on the assumptions that

(*i*) probabilities of death in the *i*th period are the same among individuals due for withdrawal in that period as among all other individuals, and

(*ii*) among those individuals who are due for withdrawal in a period and also die during the same period (whether before or after the time due for withdrawal) there is a known fixed proportion $\omega$ (not depending on the period) who die before withdrawal, and so are observed to die.

The formula is obtained by eliminating $q_i'$ from the equations

$$\frac{d_i}{N_i} = \hat{q}_i(1-\omega q_i'),$$

$$\frac{w_i}{N_i} = q_i'\left[1-(1-\omega)\hat{q}_i\right].$$

What interpretation would you give to $q_i'$ and $\omega$? Explain how $\omega$ differs from $q_{w(i)}$. Obtain the formula for $\hat{q}_i$ and show that

$$N_i \operatorname{var}(\hat{q}_i) \doteqdot q_i(1-q_i)\left[1-\frac{1-\omega}{1-\omega q_i}q_i'\right]^{-1}.$$

CHAPTER 7

# Fitting Parametric Survival Distributions

## 7.1 INTRODUCTION

When fitting *parametric* survival functions, the choice of the form of survival function (i.e., the family to which it belongs) is supposed to have already been decided. Thus we try to fit a Gompertz distribution, or a Weibull distribution or a lognormal distribution, and so on. Our interest is then in *estimating values of the parameters* appearing in the mathematical formula for CDF (or SDF) of the family of distributions considered.

There are, indeed, situations when we *compare* the fits obtained by using two or more different families of distributions. In each separate fitting, however, we consider only members of the families being fitted, and the only unknown aspects are the values of the parameters. Statistical theory provides us with many clues for designing useful methods of fitting. The theory is, however, too closely based on assumptions of stochastic stability, to be applied directly and unquestioningly to most data encountered in survival analysis. For example, when fitting a Gompertz distribution, it is not always advisable to assume that there really is an underlying "true" Gompertz distribution, and seek to obtain estimated values of the parameters ($R$ and $a$) as "close," in some sense, to the "true" values as possible. What we usually wish to obtain is a Gompertz distribution giving survival probabilities as close as possible to the observed proportions, without taking too seriously assumptions about the way in which these proportions came to be observed.

Therefore, although we include accounts of methods of estimation—such as maximum likelihood—which are based on probabilistic models, we emphasize methods of estimation aimed at fitting observed survival proportions and advocate judging the methods according to how well they do this.

## 7.2 SOME METHODS OF FITTING PARAMETRIC DISTRIBUTION FUNCTIONS

We distinguish two broad groups of methods for fitting distributions—*graphical and analytical.*

Most *graphical* methods rely on plotting some functions of the hazard rate function (HRF plots), or cumulative hazard function (CHF plots), against some functions of $x$. The functions are usually so chosen that if the parametric form of the survival function is reasonably appropriate, an approximately straight-line plot will be obtained. The plotting can be facilitated by the use of special graph paper—*probability* paper or *hazard* paper. Especially useful in survival analysis are *cumulative hazard* papers on which $[x, \Lambda(x)]$ can be entered directly to give a straight line plot [Nelson (1972)]. These methods are discussed in more detail in Section 7.7. Although graphical methods give reasonably useful results in many (perhaps, most) cases, some people are rather disturbed by their subjective nature, and prefer an analytical approach.

Among *analytical* methods, we distinguish *ad hoc* methods, and *standard* (orthodox) methods. For example, the method described in Section 4.11 (fitting a Gompertz distribution to the right hand tail of a life table) is an *ad hoc* method. The method of *percentile points*, based on making the fitted SDF equal to the observed SDF, at a few selected points, is also an *ad hoc* method.

On the other hand, methods such as maximum likelihood, minimum Chi-square, or least squares are regarded as mathematically and statistically "sound," although the "soundness" depends on the accuracy of the assumptions on which they are based. Many computer programs are available for fairly general use of these methods.

## 7.3 EXPLOITATION OF SPECIAL FORMS OF SURVIVAL FUNCTION

We have already mentioned that straight-line relationships between functions of SDF and time (age) can be useful in fitting specified (parametric) forms of SDF's. In this section, we derive some of the relationships for a few common survival distributions (discussed in Section 3.10). For convenience, we use $X$ to denote both age and survival time. Furthermore, we denote the estimates calculated directly from empirical data by a superscript $o$ (for observed); a caret above a symbol will denote an estimate obtained from the fitted curve. For example, $\lambda_X^o(x)$ and $\hat{\lambda}_X(x)$ denote observed and fitted hazard rates at time $x$, respectively. (Note that when dealing with population life tables, force of mortality, $\mu_x$, is used instead of $\lambda_X(x)$ to conform to established tradition.)

***Least value Type 1 and Gompertz distributions.*** The hazard rate of these two distributions [from (3.49)] is

$$\lambda_X(x) = Re^{ax}, \tag{7.1}$$

so that

$$\log\lambda_X(x) = \log R + ax. \tag{7.2}$$

We recall that for *grouped* data, the observed hazard rate over the interval $[x_i, x_{i+1})$ of length $h_i$, evaluated at the midpoint of the interval, $x_i^* = x_i + \frac{1}{2}h_i$, is

$$\lambda_X^o(x_i^*) = -\frac{1}{h_i}\log p_i^o = -\frac{1}{h_i}\log\frac{S_X^o(x_i + h_i)}{S_X^o(x_i)}. \tag{7.3}$$

Thus, plotting $\lambda_X^o(x)$ against $x$ at the midpoints of the intervals, we should get [from (7.2)] approximately a straight line.

The SDF of *extreme value Type 1* is [from (3.48)]

$$S_X(x) = \exp\left(-\frac{R}{a}e^{ax}\right), \qquad -\infty < x < \infty, \tag{7.4}$$

so that the CHF is

$$\Lambda_X(x) = -\log S_X(x) = \frac{R}{a}e^{ax}, \tag{7.5}$$

and

$$\log\Lambda_X(x) = \log[-\log S_X(x)] = \log\frac{R}{a} + ax. \tag{7.6}$$

If we plot the observed values, $\log[-\log S_X(x)]$, against $x$, we should obtain an approximately straight line if model (7.4) is correct.

On the other hand, *Gompertz SDF*

$$S_X(x) = \exp\left[\frac{R}{a}(1 - e^{ax})\right], \qquad x > 0 \tag{7.7}$$

is, in fact, the truncated (at $x = 0$) SDF given by (7.4). There is no simple relationship between $x$ and $S_X(x)$, but as a first approximation we may use (7.6) to estimate $R$ and $a$.

For a *Makeham-Gompertz* SDF [from (3.53)]

$$\lambda_X(x) = A + Re^{ax}. \tag{7.8}$$

Since the constant $A$ is unknown, we cannot use the relationship

$$\log[\lambda_X(x) - A] = \log R + ax. \tag{7.9}$$

It may be worthwhile to proceed by trial and error. Using a few trial values of $A$, we plot $\log[\lambda_X(x) - A]$ against $x$, and continue, adjusting $A$ until the closest approach to a straight line is attained.

***Weibull Distribution.*** The HRF of the Weibull distribution with location parameter $\xi = 0$ is [from (3.55)]

$$\lambda_X(x) = \frac{c}{\theta^c} x^{c-1}, \qquad x > 0, \qquad \theta > 0, \tag{7.10}$$

so that

$$\log \lambda_X(x) = \log \frac{c}{\theta^c} + (c-1)\log x. \tag{7.11}$$

The SDF [from (3.54)] is

$$S_X(x) = \exp\left[-\left(\frac{x}{\theta}\right)^c\right], \qquad x > 0, \qquad \theta > 0, \tag{7.12}$$

and

$$\log[-\log S_X(x)] = \log \Lambda_X(x) = c \log x - c \log \theta. \tag{7.13}$$

When $x$ is replaced by $(x - \xi)$ and $\xi$ is unknown, we may use trial and error techniques to estimate also $\xi$.

***Logistic Distribution.*** The HRF [from (3.67)]

$$\lambda_X(x) = b[1 + \exp(a - bx)]^{-1}$$

does not lead to a simple straight line relation. However [from (3.69)], we have

$$\log \frac{F_X(x)}{S_X(x)} = bx - a. \tag{7.14}$$

For some other distributions (for instance, normal, lognormal, gamma) appropriate probability paper or hazard paper should be used, if it is available.

## 7.4 FITTING DIFFERENT DISTRIBUTION FUNCTIONS OVER SUCCESSIVE PERIODS OF TIME

Plotted data sometimes indicate that is is not possible to fit a single distribution function of a simple known mathematical form over the whole range of data. For example, cancer data often indicate rather high mortality rates in the first few years after diagnosis, but the patients who survive this critical period seem to experience subsequent mortality similar to that in the general population of the same age. In the general case, there might be more than two such periods.

Let $0<\xi_1<\xi_2<\cdots<\xi_k<\infty$ be $k$ *fixed* points at which changes of SDF might reasonably be expected. These points are often suggested by graphical displays, even if no additional information on the disease process is available. We denote the SDF's in the successive intervals by $S_0(x)$, $S_1(x),\ldots, S_k(x)$. Generally, $S_i(x)$ and $S_j(x)$ $(i\neq j)$ can even belong to different families; in practice, however, they are usually different members of the same family.

For convenience, let

$$p_i(x|\xi_i)=\frac{S_i(x)}{S_i(\xi_i)} \qquad \text{for } \xi_i\leqslant x<\xi_{i+1}, \tag{7.15}$$

$i=0,1,\ldots,k$, denote the conditional probability ${}_{x-\xi_i}p_{\xi_i}$ of surviving till time (age) $x$, given alive at time $\xi_i$. (We interpret $\xi_0=0$, and so $S_0(0)=1$, and $\xi_{k+1}=\infty$.)

In particular, we have

$$p_0(\xi_1|0)=S_0(\xi_1) \qquad \text{and} \qquad p_i(\xi_{i+1}|\xi_i)=\frac{S_i(\xi_{i+1})}{S_i(\xi_i)} \tag{7.16}$$

for $i=1,2,\ldots,k-1$.

The unconditional probability of surviving until time $x$, where $\xi_i\leqslant x<\xi_{i+1}$, is

$$p_0(\xi_1|0)\cdot p_1(\xi_2|\xi_1)\cdots p_{i-1}(\xi_i|\xi_{i-1})p_i(x|\xi_i)$$

$$=S_0(\xi_1)\cdot\frac{S_1(\xi_2)}{S_1(\xi_1)}\frac{S_2(\xi_3)}{S_2(\xi_2)}\cdots\frac{S_i(x)}{S_i(\xi_i)}. \tag{7.17}$$

Therefore the piecewise survival function, $S_X(x)$ over $k$ successive periods is

$$S_X(x)=\begin{cases} S_0(x) & \text{for } 0\leqslant x<\xi_1 \\ S_0(\xi_1)\cdot\dfrac{S_1(x)}{S_1(\xi_1)} & \text{for } \xi_1\leqslant x<\xi_2 \\ S_0(\xi_1)\cdot\dfrac{S_1(\xi_2)}{S_1(\xi_1)}\cdot\dfrac{S_2(x)}{S_2(\xi_2)} & \text{for } \xi_2\leqslant x<\xi_3 \\ -------------- & \\ S_0(\xi_1)\cdot\dfrac{S_1(\xi_2)}{S_1(\xi_1)}\cdot\dfrac{S_2(\xi_3)}{S_2(\xi_2)}\cdots\dfrac{S_k(x)}{S_k(\xi_k)} & \text{for } \xi_k\leqslant x. \end{cases} \tag{7.18}$$

The corresponding PDF is

$$f_X(x)=\begin{cases} f_0(x) & \text{for } 0\leqslant x<\xi_1 \\ S_0(\xi_1)\cdot\dfrac{f_1(x)}{S_1(\xi_1)} & \text{for } \xi_1\leqslant x<\xi_2 \\ S_0(\xi_1)\cdot\dfrac{S_1(\xi_2)}{S_1(\xi_1)}\cdot\dfrac{f_2(x)}{S_2(\xi_2)} & \text{for } \xi_2\leqslant x<\xi_3 \\ -------------- & \\ S_0(\xi_1)\cdot\dfrac{S_1(\xi_2)}{S_1(\xi_1)}\cdot\dfrac{S_2(\xi_3)}{S_2(\xi_2)}\cdots\dfrac{f_k(x)}{S_k(\xi_k)} & \text{for } \xi_k\leqslant x. \end{cases} \tag{7.19}$$

Of course, when the parametric form of $S_i(x)$ is specified, simple graphical methods discussed in Section 7.3 can be used to estimate parameters of each piece.

## 7.5 FITTING A 'PIECE-WISE' PARAMETRIC MODEL TO A LIFE TABLE: AN EXAMPLE

Approximation to the life table function, $l_x$, over fixed intervals $[x, x+n)$ by using uniform or exponential distributions with different parameters in each interval is, in fact, an application of the piece-wise fitting discussed above. We have applied this procedure to estimate a complete life table by fitting a Gompertz distribution (Section 4.11).

In this section we present an example of fitting a parametric model to a life table, in which the results of Section 7.4, as well as graphical fitting, are used.

**Example 7.1** We consider fitting Gompertz and Weibull SDF's to the United States (White Males) Life Table 1969–1971 (see Table 4.1).

Let $\mu_x^o = -\log p_{x-\frac{1}{2}}^o$ denote the force of mortality estimated at age $x$. Note that since we obtain values of $p_{x-\frac{1}{2}}^o$ at integer ages (e.g., 35, 36, 37,...), we have values of $\mu_x^o$ at midyears (35.5, 36.5, 37.5,...).

Fitting Gompertz distribution by HRF-plots, we plot $\log(-\log p_{x-\frac{1}{2}}^o)$ against $x$ (Fig. 7.1a), and fitting Weibull distribution by HRF-plots, we plot $\log(-\log p_{x-\frac{1}{2}}^o)$ against $\log x$ (Fig. 7.2a).

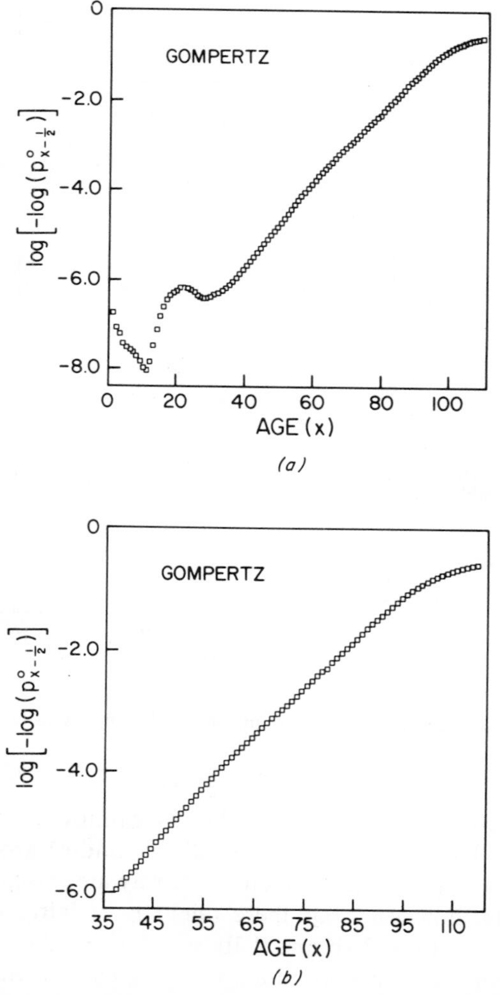


**Fig. 7.1** U.S. Life Tables 1969–1971 White Males. Estimated hazard plots for fitting.

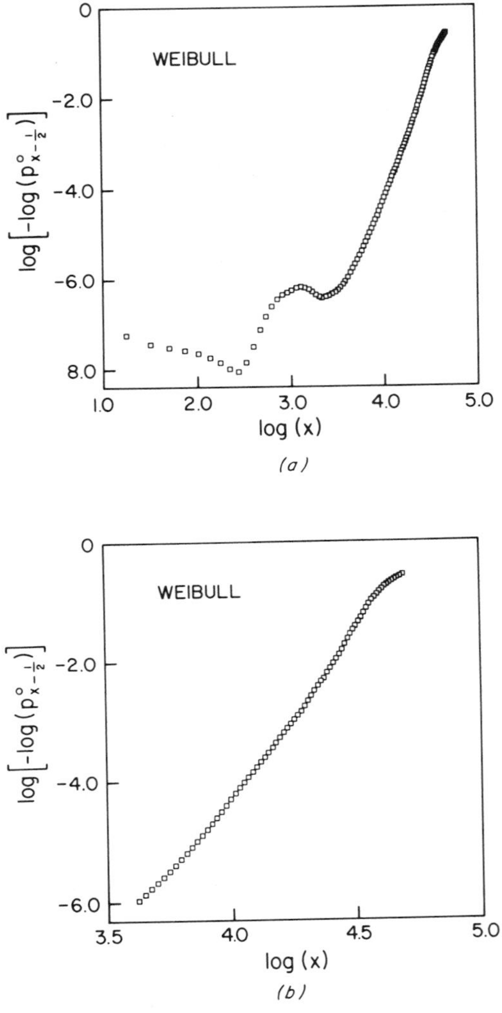


**Fig. 7.2** U.S. Life Tables 1969–1971 White Males. Estimated hazard plots for fitting.

It is clear from Figs. 7.1a and 7.2a that we cannot get a linear fit for the whole lifespan. There is a "hump" (local maximum) around ages 20–25, often ascribed to fatalities in accidents—mostly automobile accidents. The Figures indicate, however, that there might be a fairly good fit for ages above 35. (See also Figs. 7.1b and 7.2b, which show the parts of Figs. 7.1a and 7.2a, respectively, corresponding to ages greater than 35.)

1. Straight lines were fitted *by eye* to the plots in each of Figs. 7.1b and 7.2b by taking two arbitrarily chosen points on them and calculating the coefficients. The equations of these lines are:

*for Gompertz:* (Fig. 7.1b): $\log \mu_x = 0.0897x - 9.2292$; from which the estimates of $a$ and $R$ [in (7.2)] are: $\hat{a} = 0.0897$; $\hat{R} = 0.00009813$;

*for Weibull* (Fig. 7.2b): $\log \mu_x = 5.0964 \log x - 24.5302$ from which the estimates of $c$ and $\theta$ [in (7.11)] are: $\hat{c} = 6.0964$; $\hat{\theta} = 75.2059$.

Note that we are, in fact, fitting only *tails* of distributions corresponding to $x > 35$. We are *not* fitting (two parameter) Gompertz or (three parameter) Weibull SDF's, *starting* at $x = 35$. Indeed, for Weibull with $c > 1$ (as in the case here) this would imply that $\mu_{35} = 0$, which seems very unreasonable.

To calculate the fitted $l_x$ values, $l_x^{(1)}$, say, we have to use the formula (7.17)

$$l_x^{(1)} = l_{x_1} \cdot \frac{\hat{S}_1(x)}{\hat{S}_1(x_1)}, \qquad x > x_1, \tag{7.20}$$

where the age range is from $x_1$ upward (in our case $x_1 = 35$), $l_{x_1}$ is a life table value [corresponding $S_0(x)$], and $\hat{S}_1(x)$ is the fitted SDF [compare (4.66)].

Figures 7.3 and 7.5 show the estimated SDFs based on these fitted lines applied to values of $x$ greater than 35. The Gompertz fit does appear to be better than Weibull, but some improvement in the latter might be possible by adjustment of parameter values (see Exercise 7.12).

2. Figures 7.1b and 7.2b suggest that quite accurate results might be obtained by fitting *two successive* lines (one for younger and one for older

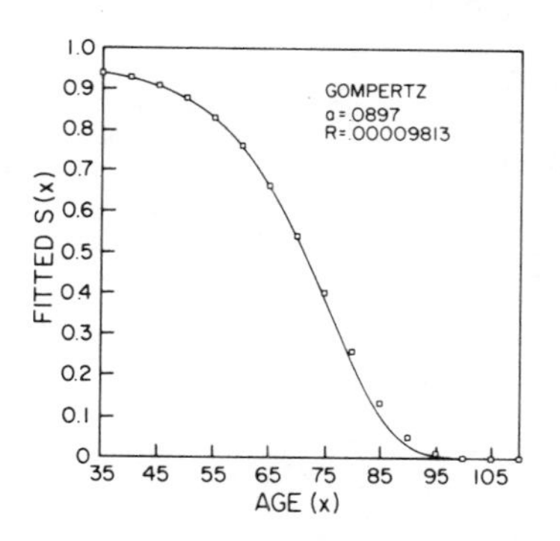


**Fig. 7.3**

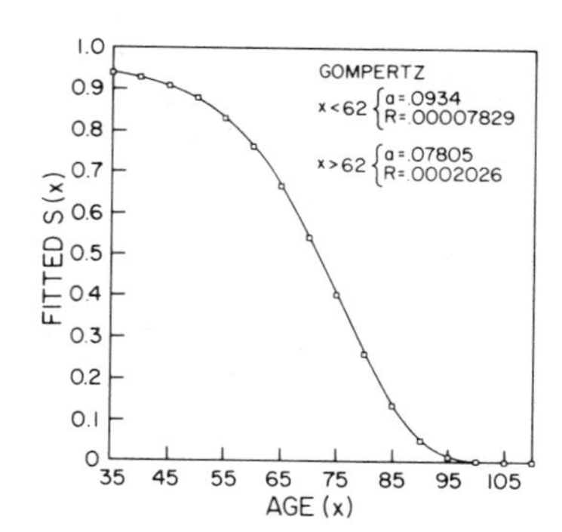


**Fig. 7.4**

U.S. Life Tables 1969–1971 White Males. Fitted survival function (age 35 and above).

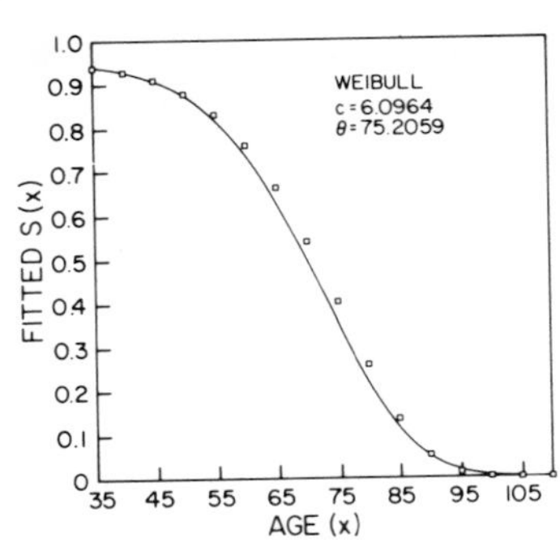


Fig. 7.5

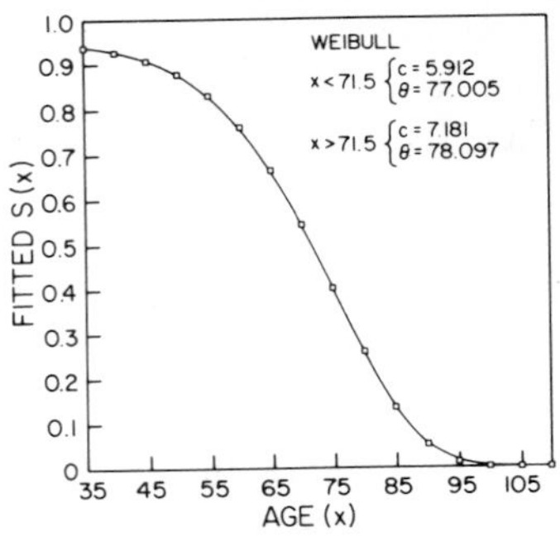

Fig. 7.6

U.S. Life Tables 1969–1971 White Males. Fitted survival function (age 35 and above).

ages in the range $x>35$). In fact, very good fits are so obtained, especially for the Weibull plot (Fig. 7.6).

The equations of these lines (fitted *by eye*) are:
*for Gompertz* (Fig. 7.4)

$$35<x<62: \qquad \log\mu_x = 0.0934x - 9.4451,$$

from which $\hat{a}_1 = 0.0934, \quad \hat{R}_1 = 0.00007829,$

and

$$x>62: \qquad \log\mu_x = 0.07805x - 8.5043,$$

from which $\hat{a}_2 = 0.07805, \quad \hat{R}_2 = 0.0002026.$

*for Weibull* (Fig. 7.6)

$$35<x<65: \qquad \log\mu_x = 4.8898\log x - 23.7965,$$

from which $\hat{c}_1 = 5.8898, \quad \hat{\theta}_1 = 76.8116,$

and

$$x>65: \qquad \log\mu_x = 6.3273\log x - 29.7973,$$

from which $\hat{c}_2 = 7.3273, \quad \hat{\theta}_2 = 76.5864.$

Note that in calculating $l_x^{(1)}$ for the lower age ranges ($x_1<x<x_2$), we use formula (7.20). For the upper age ranges ($x>x_2$), we use formula

$$l_x^{(2)} = l_{x_1}\cdot\frac{\hat{S}_1(x_2)}{\hat{S}_1(x_1)}\cdot\frac{\hat{S}_2(x)}{\hat{S}_2(x_2)} = l_{x_2}^{(1)}\cdot\frac{\hat{S}_2(x)}{\hat{S}_2(x_2)}, \tag{7.21}$$

where $S_2(x)$ is the fitted SDF for ages $x > x_2$.

Since a population life table does not represent directly an actual mortality experience, although it is usually *derived* from such an experience, there are no actually observed numbers of deaths, and so it is inappropriate to apply significance tests of goodness of fit in the present context.

The life table used in Example 7.1 has already been smoothed and the parameters of the fitted curves, estimated by eye provide a very good fit. When we deal with experimental data, the observed points often do not lie so closely on a straight line. Still, quick estimates by eye can be fairly good for use as initial estimates when fitting by an analytical method (e.g., by maximum likelihood).

## 7.6 MIXTURE DISTRIBUTIONS

Groups of individuals who are subjected to lifetime distribution studies are often not from a single homogeneous source. If it is possible to assign each individual to a well defined source and numbers for each such source are sufficiently large, it is preferable to analyze data from each source separately, leaving possible synthesis later. Often, it is not possible to identify sources in this way. We may then seek to make allowance for possible heterogeneity of sources by fitting a *mixture* of distributions.

We confine our discussion here to a *two-component* mixture, each belonging to the same known location-scale family. The first step is to plot the observed SDF, $S^o(x)$ or a function of it against $x$, or a function of $x$. (Of course, we may use an appropriate probability paper and plot $S^o(x)$ against $x$.) From such a plot we may get some indication whether we need a second component.

At this point we insert a word of caution against regarding plots like those in Figures 7.4 or 7.6 (where two consecutive straight lines appear to give a good fit) as evidence pointing toward a mixture. What was actually fitted in Fig. 7.4 was *not a mixture* of two Gompertz distributions, but one Gompertz for age $35 < x < 62$, followed by another for $x > 62$. It is also possible that a mixture would fit quite well, but the following discussion shows that this is not necessarily the case.

Suppose that there are *really* two components and their location parameters differ sufficiently (as compared with the scale parameters). Thus the SDF for lower (higher) values should reflect mainly variation of the component with lower (higher) value of location parameter. Denoting the SDF's of the two components by $S_j(x)$ $(j=1,2)$, the mixture SDF is

$$S_X(x)=\phi S_1(x)+(1-\phi)S_2(x), \tag{7.22}$$

where $\phi$ $(0<\phi<1)$ is the proportion of the first component in the mixture. Under the stated assumptions, $S_1(x)$ corresponds overall to much shorter lifetimes than $S_2(x)$. For small values of $x$, we then have $S_2(x)\doteqdot 1$, and

$$S_X(x)\doteqdot \phi S_1(x)+(1-\phi) \tag{7.23}$$

or

$$F_X(x)\doteqdot \phi[1-S_1(x)]=\phi F_1(x). \tag{7.23a}$$

On the other hand, for large values of $x$, $S_1(x)\doteqdot 0$, and so

$$S_X(x)\doteqdot (1-\phi)S_2(x). \tag{7.24}$$

The observed proportions, $S^o(x)$ and $F^o(x)$, should follow similar patterns.

If $F^o(x)$ (for small $x$) or $S^o(x)$ (for large $x$), are plotted against $x$ on probability paper (appropriate to the family), we would *not*, in general, expect to get a straight line because of the multipliers, $\phi$ and $(1-\phi)$, respectively.

By taking trial values of $\phi$, it is possible to proceed, by trial and error, continuing until an adequate approach to a straight line plot is obtained.

Plotting the hazard function leads to similar situations. The HRF of the mixture is

$$\lambda_X(x)=\frac{\phi f_1(x)+(1-\phi)f_2(x)}{\phi S_1(x)+(1-\phi)S_2(x)} \tag{7.25}$$

(using an obvious notation). Under the conditions described earlier in this section [i.e., when $S_1(x)$ and $S_2(x)$ are sufficiently separated], it is possible to use a trial and error method analogous to that described for SDF.

Much of the above argument also applies when $S_1(x)$ and $S_2(x)$ are from different families. Of course, it is necessary to use different probability papers when dealing with the two tails in such cases.

**Example 7.2** The data in Table 7.1 are from Hoel (1972) and represents time at death of 99 irradiated mice. There are three causes of death: *thymic lymphoma*, (cause $C_1$), *reticulum cell sarcoma* (cause $C_2$), and *other causes*

**Table 7.1** Time (in days) and cause of death of 99 RFM male mice treated with radiation dose of 300 r

| | | | | | | | | | |
|---|---|---|---|---|---|---|---|---|---|
| Thymic | 159, | 189, | 191, | 198, | 200, | 207, | 220, | 235, | 245, |
| lymphoma | 250, | 256, | 261, | 265, | 266, | 280, | 343, | 356, | 383, |
| (22) | 403, | 414, | 428, | 432 | | | | | |
| Reticulum cell | 317, | 318, | 399, | 495, | 525, | 536, | 549, | 552, | 554, |
| sarcoma | 557, | 558, | 571, | 586, | 594, | 596, | 605, | 612, | 621, |
| (38) | 628, | 631, | 636, | 643, | 647, | 648, | 649, | 661, | 663, |
| | 666, | 670, | 695, | 697, | 700, | 705, | 712, | 713, | 738, |
| | 748, | 753, | | | | | | | |
| Other | 40, | 42, | 51, | 62, | 163, | 179, | 206, | 222, | 228, |
| causes | 249, | 252, | 282, | 324, | 333, | 341, | 366, | 385, | 407, |
| (39) | 420, | 431, | 441, | 461, | 462, | 482, | 517, | 517, | 524, |
| | 564, | 567, | 586, | 619, | 620, | 621, | 622, | 647, | 651, |
| | 686, | 761, | 763 | | | | | | |

(cause $C_3$). For illustrative purposes, we use only the data for causes: $C_1$ with $n_1 = 22$ observations and $C_2$ with $n_2 = 38$ observations. (The data for $C_3$ will be analyzed later, in Example 7.3.)

We now analyze the data *as if we had no information other than time of death*. (Information on cause of death will be ignored in this analysis.) Suppose that we are interested in fitting two parameter *Weibull distributions*. We first plot $\log[-\log S^o(x)]$ against $\log x$ [see (7.11)], and obtain Fig. 7.7. There appears to be a change of slope at about $x = 450$

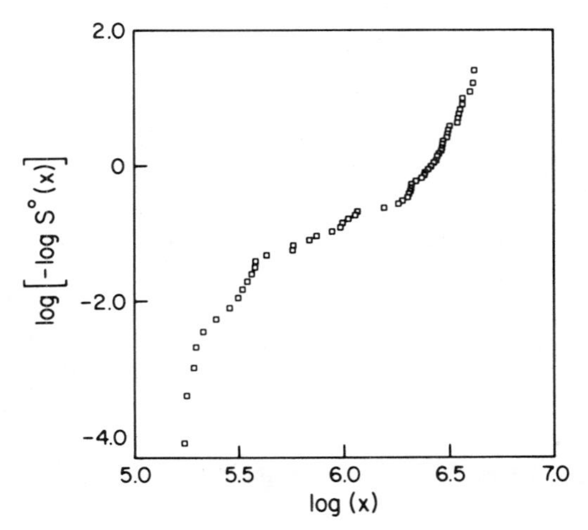


**Fig. 7.7** Empirical log $[-\log S^o(x)]$ versus $\log(x)$ plot. ($N = 60$)

(log $x \doteq 6.1$). According to formula (7.24) we cannot estimate $S_2(x)$ simply by fitting a straight line to the right-hand part of the plot. From (7.24), we have

$$\hat{S}_2(x) \doteq \frac{S^o(x)}{1-\phi}, \tag{7.26}$$

and to estimate $\hat{S}_2(x)$, we have to plot

$$\log\{-\log[1-\phi)^{-1}S^o(x)]\} = \log[\log(1-\phi) - \log S^o(x)]$$

against $\log x$.

We use a series of trial values of $\log(1-\phi)$ until we get the nearest approach (in our judgment) to a straight line. This can be done quite simply using visual display computer output or even by hand-plotting. The result provides: (1) an estimate of $\phi$, and (2) estimates of parameters, $\hat{c}_2$ and $\hat{\theta}_2$, say, of $S_2(x)$.

In the present case, we obtain

$$\hat{\phi} = 0.5, \qquad \hat{c}_2 = 11.6, \qquad \hat{\theta}_2 = 672.7$$

(see also Figs. 7.8a, b, c).

The lower part of the plot of $\log[-\log S^o(x)]$ will be estimated by $\hat{S}_1(x)$. From (7.23a)

$$\hat{S}_1(x) \doteq 1 - \frac{F^o(x)}{\phi}. \tag{7.27}$$

Plotting $\log\{-\log[1-\phi^{-1}F^o(x)]\}$ against $\log x$ we would hope to get a nearly straight-line plot. In the present case, this is so, and the plot can be used to estimate the parameters of $\hat{S}_1(x)$, giving $\hat{c}_1 = 5.0$, $\hat{\theta}_1 = 301.2$ (see Figs. 7.9a, b, c).

A more symmetrical approach would be to obtain simultaneously plots of:

$$\log\{-\log[(1-\phi)^{-1}S^o(x)]\} \text{ against } \log x \text{ for } x > 450, \text{ for example,}$$

and

$$\log\{-\log[1-\phi^{-1}F^o(x)]\} \text{ against } \log x \text{ for } x < 300, \text{ for example,}$$

and try to make *both* as nearly linear as possible by varying $\phi$. If

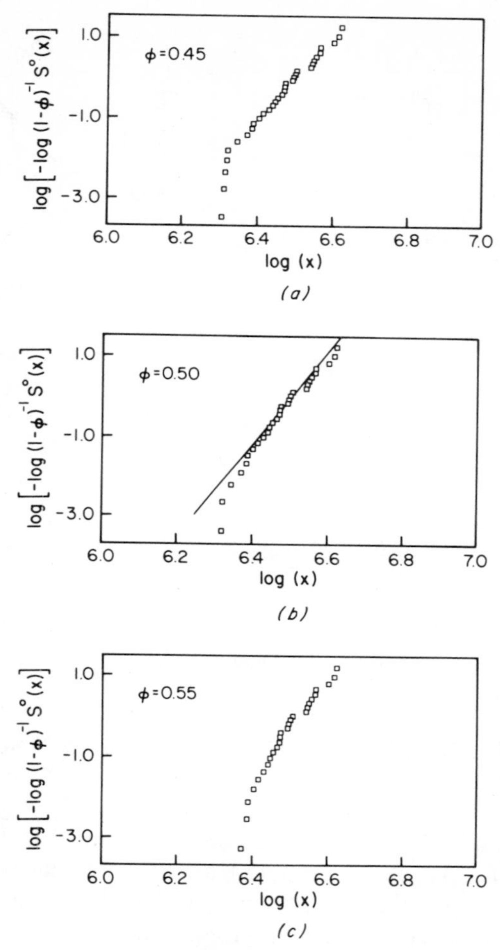


**Fig. 7.8** Trial plot—upper part.

satisfactory closeness to linearity in both plots is not attainable simultaneously (i.e., for the same value of $\phi$), it would be an indication that the model is not adequate.

If there are more than two components or if there is a considerable overlap between two components, the problem is more difficult. Maximum likelihood estimation can be used, but with caution (see remarks in Section 7.8). Even with three components, if each is well separated from the others, it is possible to extend the methods described above in a natural way.

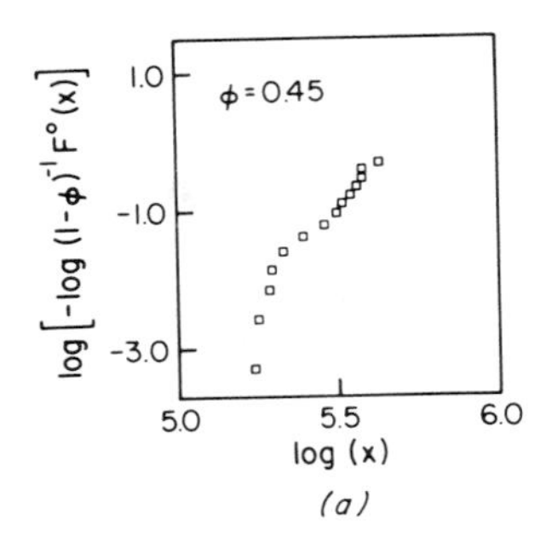


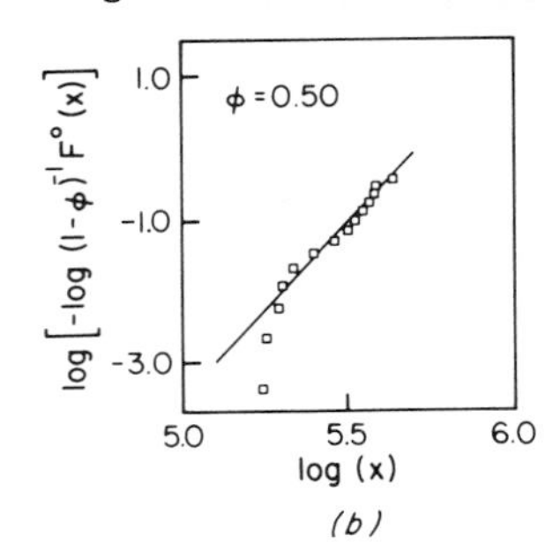


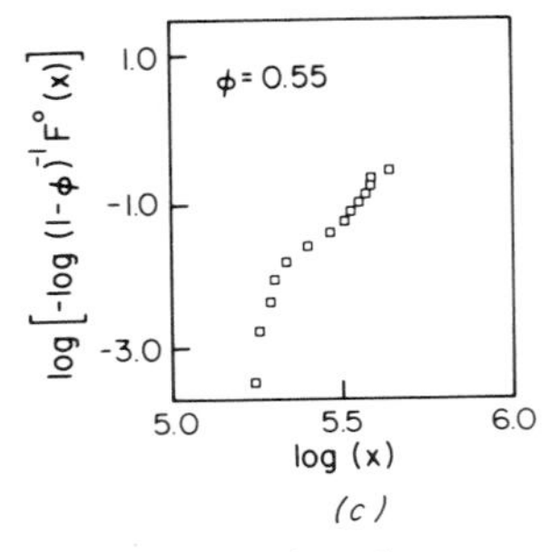


**Fig. 7.9** Trial plot—lower part.

## 7.7 CUMULATIVE HAZARD FUNCTION PLOTS—NELSON'S METHOD FOR UNGROUPED DATA

In Sections 7.2 and 7.3 we have discussed a graphical method of plotting cumulative hazard, $\Lambda_X(x) = -\log S_X(x)$, against $x$, or more generally, some functions of these quantities to obtain a straight line. The method can be used regardless of whether data are grouped or ungrouped. However, for ungrouped data representing individual (ordered) times at death, some simple techniques can be used [Nelson (1972)].

### 7.7.1 Complete Data

Let $x_1' < x_2' < \cdots < x_N'$ represent $N$ (distinct) ordered times of death. Then from (5.2a)

$$S^o(x) = \frac{N-i}{N} = \frac{N-1}{N} \cdot \frac{N-2}{N-1} \cdots \frac{N-i}{N-i+1} \tag{7.28}$$

$$= \left(1 - \frac{1}{N}\right)\left(1 - \frac{1}{N-1}\right) \cdots \left(1 - \frac{1}{N-i+1}\right) \tag{7.28a}$$

for $x_i' \leqslant x < x_{i+1}'$. Hence

$$\Lambda^o(x) = -\log S^o(x) = -\sum_{j=1}^{i} \log\left(1 - \frac{1}{N-j+1}\right) \tag{7.29}$$

for $x_i' \leqslant x < x_{i+1}'$.

However, for sufficiently large $(N-j+1)$ we may use an approximation

$$\log\left(1 - \frac{1}{N-j+1}\right) \doteqdot -\frac{1}{N-j+1}, \tag{7.30}$$

and substituting this into (7.29) we obtain

$$\Lambda^o(x) = -\log S^o(x) \doteqdot \sum_{j=1}^{i} \frac{1}{N-j+1}$$

$$= \left(\frac{1}{N} + \frac{1}{N-1} + \frac{1}{N-2} + \cdots + \frac{1}{N-i+1}\right). \tag{7.31}$$

Note that the terms in (7.31) represent the reciprocals of reverse rank orders of times at death.

Let $\Lambda^o(x_i')$ denote the cumulative function at the (ordered) time point $x_i'$. Clearly, from (7.31)

$$0 < \Lambda^o(x_1') < \Lambda^o(x_2') < \cdots < \Lambda^o(x_i'). \tag{7.32}$$

It can also be shown [Nelson (1972)] that

$$\mathcal{E}\left[\Lambda^o(X_i')\right] = \overline{\Lambda}^o(X_i') = \frac{1}{N} + \frac{1}{N-1} + \cdots + \frac{1}{N-i+1}. \tag{7.33}$$

Therefore, instead of (7.29), Nelson proposed to use (7.31), and stated simple rules for calculating an empirical CHF, namely:

1. Write times at death ($x_i'$) in decreasing order.
2. Write the ranks of observations in the reverse (decreasing) order. (If two observations are recorded at the same time point, assign their orders arbitrarily.)
3. Calculate the reciprocals of the reverse rank orders $(N-i+1)^{-1}$, and the approximate empirical CHF from (7.31) at each $x_i'$.
4. Plot an appropriate function of the empirical CHF so obtained against an appropriate function of $x$ to obtain a straight line.

Since many values of $\Lambda^o(x)$ are between 0 and 1 (and so $\log \Lambda^o(x)$ is negative), Nelson suggested using $\Lambda^o(x) \cdot 10^2$ rather than $\Lambda^o(x)$. Of course, this is not essential, and we do not use this transformation in our example.

We notice that for small values of $(N-i+1)$ approximation (7.30) is not good. However, we are often seeking only a rough estimation of a useful survival function model. Also estimation of the slope and intercept may cause some problems, because the $\Lambda^o(x)$ (and so their functions) are not independent. Again, the estimates are often needed only as initial estimates when using a computer program for calculating the ML-estimates. Estimating by eye or unweighted regression method (although incorrect) should be satisfactory for this purpose.

**Example 7.3** The method is applied to observations representing the time at death of 39 mice who died from other causes (see Table 7.2). Note that there were two deaths at time 517 days, the order of these deaths was assigned arbitrarily.

We wish to fit a Gompertz distribution. But because the relationship between $\Lambda_X(x)$ and $x$ for Gompertz distribution is not simple, we fit, instead, the extreme value Type 1 (see Section 7.3). For extreme value Type 1 we have the linear relationship [cf. (7.2)]

$$\log \Lambda^o(x_i') = y_i = \log R + a x_i' + e_i, \tag{7.34}$$

where $e_i$ is a random error.

For comparison, we use two ways of calculating empirical $\Lambda^o(x_i')$.

1. We first calculate $S^o(x_i')$ from (7.28), $-\log S^o(x_i') = \Lambda^o(x_i')$ from (7.29), and finally $\log[-\log S^o(x_i')] = y_i$. These results are given in columns (3)–(5) of Table 7.2. Fitting linear (unweighted) linear regression (7.34), we obtain the estimates for $R$ and $a$: $\check{R} = 3.316 \cdot 10^{-4}$, and $\check{a} = 5.693 \cdot 10^{-3}$.

2. We also calculate approximate values of $\Lambda^o(x_i')$ (denoted by $\overline{\Lambda}^o(x_i')$ in this context) by Nelson's method, that is using the approximation (7.31). The results are given in columns (8)–(9) of Table 7.2. The estimated values

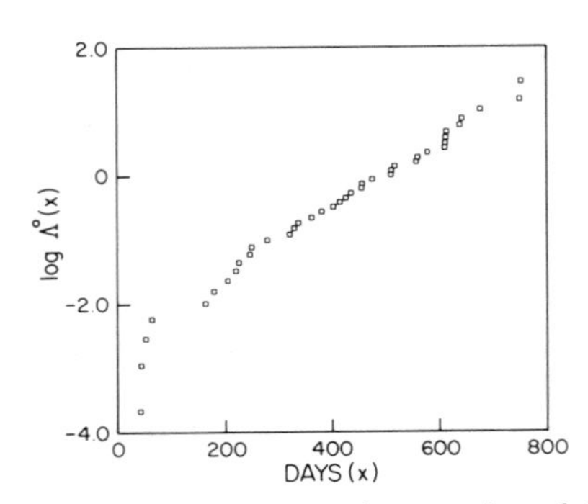


**Fig. 7.10** Cumulative hazard versus time plot.

**able 7.2** Fitting extreme value Type 1 distribution by cumulative hazard unction method [Mice data from: Hoel (1972)]

| $i$ | $x_i'$ | $S^o(x_i')$ | $-\log S^o(x_i')$ | $\log[-\log S^o(x_i')]$ | $N-i+1$ | $\frac{1}{N-i+1}$ | $\overline{\Lambda}^o(x_i')$ | $\log \overline{\Lambda}^o(x_i')$ |
|---|---|---|---|---|---|---|---|---|
| 1 | 40 | 0.9744 | 0.0260 | −3.65 | 39 | 0.0256 | 0.0256 | −3.66 |
| 2 | 42 | 0.9487 | 0.0526 | −2.94 | 38 | 0.0263 | 0.0520 | −2.96 |
| 3 | 51 | 0.9231 | 0.0800 | −2.53 | 37 | 0.0270 | 0.0790 | −2.54 |
| 4 | 62 | 0.8974 | 0.1082 | −2.22 | 36 | 0.0278 | 0.1068 | −2.24 |
| 5 | 163 | 0.8718 | 0.1372 | −1.99 | 35 | 0.0286 | 0.1353 | −2.00 |
| 6 | 179 | 0.8462 | 0.1671 | −1.79 | 34 | 0.0294 | 0.1647 | −1.80 |
| 7 | 206 | 0.8205 | 0.1978 | −1.62 | 33 | 0.0303 | 0.1950 | −1.63 |
| 8 | 222 | 0.7949 | 0.2296 | −1.47 | 32 | 0.0312 | 0.2263 | −1.49 |
| 9 | 228 | 0.7692 | 0.2624 | −1.34 | 31 | 0.0323 | 0.2586 | −1.35 |
| 0 | 249 | 0.7436 | 0.2963 | −1.22 | 30 | 0.0333 | 0.2919 | −1.23 |
| 1 | 252 | 0.7179 | 0.3314 | −1.10 | 29 | 0.0345 | 0.3264 | −1.12 |
| 2 | 282 | 0.6923 | 0.3677 | −1.00 | 28 | 0.0357 | 0.3621 | −1.02 |
| 3 | 324 | 0.6667 | 0.4055 | −0.90 | 27 | 0.0370 | 0.3991 | −0.92 |
| 4 | 333 | 0.6410 | 0.4447 | −0.81 | 26 | 0.0385 | 0.4376 | −0.83 |
| 5 | 341 | 0.6154 | 0.4855 | −0.72 | 25 | 0.0400 | 0.4776 | −0.74 |
| 6 | 366 | 0.5896 | 0.5281 | −0.64 | 24 | 0.0417 | 0.5193 | −0.66 |
| 7 | 385 | 0.5641 | 0.5725 | −0.56 | 23 | 0.0435 | 0.5627 | −0.57 |
| 8 | 407 | 0.5385 | 0.6190 | −0.48 | 22 | 0.0455 | 0.6082 | −0.50 |
| 9 | 420 | 0.5128 | 0.6678 | −0.40 | 21 | 0.0476 | 0.6558 | −0.42 |
| 0 | 431 | 0.4872 | 0.7191 | −0.33 | 20 | 0.0500 | 0.7058 | −0.35 |
| 1 | 441 | 0.4615 | 0.7732 | −0.26 | 19 | 0.0526 | 0.7584 | −0.28 |
| 2 | 461 | 0.4359 | 0.8303 | −0.19 | 18 | 0.0556 | 0.8140 | −0.21 |
| 3 | 462 | 0.4103 | 0.8910 | −0.12 | 17 | 0.0588 | 0.8728 | −0.14 |
| 4 | 482 | 0.3846 | 0.9555 | −0.05 | 16 | 0.0625 | 0.9353 | −0.07 |
| 5 | 517 | 0.3590 | 1.0245 | 0.02 | 15 | 0.0667 | 1.0020 | 0.002 |
| 6 | 517 | 0.3333 | 1.0986 | 0.09 | 14 | 0.0714 | 1.0734 | 0.07 |
| 7 | 524 | 0.3077 | 1.1787 | 0.16 | 13 | 0.0769 | 1.1503 | 0.14 |
| 8 | 564 | 0.2821 | 1.2657 | 0.24 | 12 | 0.0833 | 1.2337 | 0.21 |
| 9 | 567 | 0.2564 | 1.3610 | 0.31 | 11 | 0.0909 | 1.3246 | 0.28 |
| 0 | 586 | 0.2308 | 1.4663 | 0.38 | 10 | 0.1000 | 1.4246 | 0.35 |
| 1 | 619 | 0.2051 | 1.5841 | 0.46 | 9 | 0.1111 | 1.5357 | 0.43 |
| 2 | 620 | 0.1795 | 1.7177 | 0.54 | 8 | 0.1250 | 1.6607 | 0.51 |
| 3 | 621 | 0.1538 | 1.8718 | 0.63 | 7 | 0.1429 | 1.8035 | 0.59 |
| 4 | 622 | 0.1282 | 2.0541 | 0.72 | 6 | 0.1667 | 1.9702 | 0.68 |
| 5 | 647 | 0.1026 | 2.2773 | 0.82 | 5 | 0.2000 | 2.1702 | 0.77 |
| 6 | 651 | 0.0769 | 2.5649 | 0.94 | 4 | 0.2500 | 2.4202 | 0.88 |
| 7 | 686 | 0.0513 | 2.9704 | 1.09 | 3 | 0.3333 | 2.7535 | 1.01 |
| 8 | 761 | 0.0256 | 3.6636 | 1.30 | 2 | 0.5000 | 3.2535 | 1.18 |
| 9 | 763 | 0.0 | — | — | 1 | 1.0000 | 4.2535 | 1.45 |

of the parameters obtained by fitting a regression line of type (7.34) are $\tilde{R}=3.192\cdot 10^{-4}$, $\tilde{a}=5.687\cdot 10^{-3}$.

The results from (1) and (2) do not differ much. Figure 7.10 represents the CHF-plots when Nelson's method (2) was used.

### 7.7.2 Incomplete Data

CHF-plotting can also be used when the data are *incomplete*, that is, when we observe withdrawals. We recall (from Section 6.7), the product-limit estimator of $S(x)$,

$$S^{o}(x)=\prod_{j=1}^{i}\left(1-\frac{1}{R_j}\right) \qquad \text{for } x_i' \leqslant x < x_{i+1}' \tag{7.35}$$

[compare (6.56)]. Hence

$$\Lambda^{o}(x_i')=-\sum_{j=1}^{i}\log\left(1-\frac{1}{R_j}\right), \tag{7.36}$$

or approximately

$$\Lambda^{o}(x_i')\doteqdot\sum_{j=1}^{i}\frac{1}{R_j}. \tag{7.37}$$

Practical rules for calculating $\Lambda^{o}(x_i')$ are similar to those for complete data, but here we need to know the number of individuals available "just before" time $x_i'$, that is, the number $R_i$ in the risk set $\mathfrak{R}_i$.

## 7.8 CONSTRUCTION OF THE LIKELIHOOD FUNCTION FOR SURVIVAL DATA: SOME EXAMPLES

The basic concepts and definitions of likelihood function, and maximum likelihood estimators (MLE) of parameters (with two examples) have been briefly discussed in Section 3.15. Some other examples have been given in Section 6.3. Here we give yet further examples, typical of those occurring in the analysis of survival data.

**Example 7.4.** *Simple random sample—exact times at death.*

We wish to find the MLE's, $\hat{a}$ and $\hat{R}$, for a Gompertz distribution using the mice data given in Table 7.2, and discussed in Example 7.3. Since there are

$N=39$ observations in the random sample, the likelihood is of the form

$$L=\prod_{j=1}^{39} f(x_j;R,a), \tag{7.38}$$

where

$$f(x;R,a)=Re^{ax}\exp\left[\frac{R}{a}(1-e^{ax})\right] \tag{7.39}$$

$x>0$, $R>0$, $a>0$.

In this case, computer program "MAXLIK" [Kaplan and Elston (1973)] was used. As initial estimates, for $a$ and $R$, we used those obtained in Example 7.3 for extreme value Type 1 distribution, and obtained from CHF-plotting (using Nelson's method). The MLEs are $\hat{a}=4.675\cdot 10^{-3}$, and $\hat{R}=5.221\cdot 10^{-4}$. The corresponding fitted SDF to the data is displayed in Fig. 7.11.

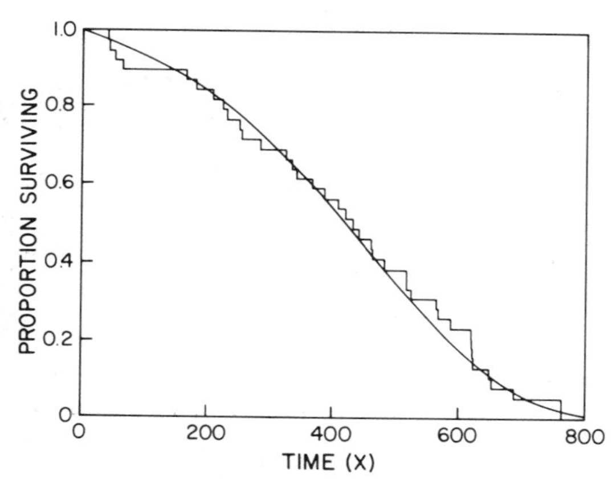


**Fig. 7.11** Fitted Gompertz distribution to mice data. [Hoel (1972); Other causes, N=39]

**Example 7.5.** *Mixtures of distributions.*

We fit a *mixture* of Weibull distributions to the mice data discussed in Example 7.2. We note again that we analyze these data as if we had no information on causes of deaths, they just represent $N=60$ individual times at death. We wish to fit a *mixture* of Weibull distributions.

The likelihood is

$$L=\prod_{j=1}^{60}\left\{\phi c_1\theta_1^{-c_1}x_j^{c_1-1}\exp\left[-\left(\frac{x_j}{\theta_1}\right)^{c_1}\right]+(1-\phi)c_2\theta^{-c_2}x_j^{c_2-1}\exp\left[-\left(\frac{x_j}{\theta_2}\right)^{c_2}\right]\right\}. \tag{7.40}$$

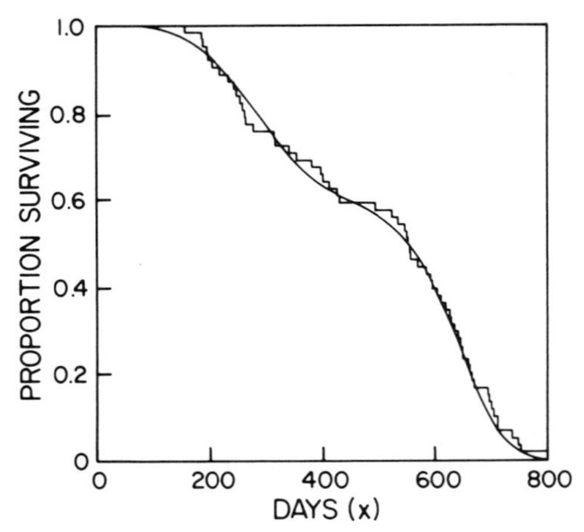


**Fig. 7.12** Fitted mixture of two Weibull distributions.

The results are: $\hat{\phi}=0.3967$, $\hat{c}_1=3.7643$, $\hat{\theta}_1=313.4$, and $\hat{c}_2=3.7643$, $\hat{\theta}_2=697.12$. These estimates are fairly close to the estimates obtained by graphical methods. The fitted mixture of Weibull SDF's is presented in Fig. 7.12.

It should be emphasized that if the information on causes of death (as given in Table 7.1) were to be utilized, we would have a different problem. The likelihood function would be different (see Section 9.9 and Example 12.2).

**Example 7.6.** *Data with recorded times of entry and departure.*

Let $X$ denote time since a certain initial event (e.g., since the start of experiment, since birth, etc.).

Let $S(x;\boldsymbol{\theta})$ denote the survival function where $\boldsymbol{\theta}'=(\theta_1,\ldots,\theta_s)$ is a vector of $s$ parameters, which we wish to estimate from the data. Suppose that for individual ($j$) we have records of time *at entry*, $\xi_j$, into the experience and time of *departure*— failure or withdrawal, $x_j$.

1. If failure was observed at time $x_j$, then the contribution to likelihood from the individual ($j$) is

$$L_j=-\frac{S'(x_j;\boldsymbol{\theta})}{S(\xi_j;\boldsymbol{\theta})}=\frac{f(x_j;\boldsymbol{\theta})}{S(\xi_j;\boldsymbol{\theta})}, \tag{7.41}$$

where

$$-S'(x_j;\boldsymbol{\theta})=-\left.\frac{dS(x;\boldsymbol{\theta})}{dx}\right|_{x=x_j}=f(x_j;\boldsymbol{\theta})$$

is the probability density function, and $S(\xi_j;\boldsymbol{\theta})=\Pr\{X>\xi_j|\boldsymbol{\theta}\}$ is the probability of surviving until time $\xi_j$.

2. On the other hand, if $x_j$ is the time at which $(j)$ was lost from observation (withdrawn), then the contribution of $(j)$ to the likelihood function is

$$L_j(\boldsymbol{\theta})=L_j=\frac{S(x_j;\boldsymbol{\theta})}{S(\xi_j;\boldsymbol{\theta})}. \tag{7.42}$$

The likelihood function for the whole set of $N$ observations is

$$L=\prod_{j=1}^{N} L_j \tag{7.43}$$

or equivalently

$$\log L=\sum_{j=1}^{N}\log L_j. \tag{7.44}$$

The set of $s$ likelihood equations

$$\frac{\partial \log L}{\partial \theta_r}=0, \qquad r=1,2,\ldots,s \tag{7.45}$$

may sometimes yield explicit formulae for the estimators of $\theta$'s. In general, it is necessary, and often simpler, to use a computer program that maximizes $L$ directly.

To illustrate the construction of the likelihood function we use the data in Table 7.3. These give: (1) time at entry; (2) time of departure; (3) mode of entry ($F$=failure; $W$=withdrawal). The survival distribution to be fitted is denoted by $S(x,\theta)$; in a particular situation it would take a specified form. The contributions to the likelihood function are given in the last (fourth) column of Table 7.3. The product of these terms constitute the likelihood function of these 18 individuals. There could be a fifth column showing values of $\log L_j$, which would be summed to give the log-likelihood function.

**Example 7.7.** *Quantal response data*

Consider the following special situation. All individuals are observed from time zero, but for each individual ($j=1,2,\ldots,N$) the only observation is whether a specified event (which may be, but need not be, failure or death) occurs *before* a specified time $\tau_j$. The analysis follows the same lines as described in this chapter. For the $j$th individual the contribution to the

**Table 7.3.** Construction of likelihood

| (1) Entry time | (2) Departure time | (3) Mode of departure | (4) Likelihood | |
|---|---|---|---|---|
| 0 | 70 | W | $S(70;\boldsymbol{\theta})$ | |
| 0 | 141 | F | $-S'(141;\boldsymbol{\theta})$ | |
| 0 | 120 | F | $-S'(120;\boldsymbol{\theta})$ | |
| 0 | 110 | W | $S(110;\boldsymbol{\theta})$ | |
| 0 | 98 | F | $-S'(98;\boldsymbol{\theta})$ | |
| 0 | 172 | F | $-S'(172;\boldsymbol{\theta})$ | |
| 0 | 180 | F | $-S'(180;\boldsymbol{\theta})$ | |
| 0 | 102 | F | $-S'(102;\boldsymbol{\theta})$ | |
| 0 | 41 | W | $S(41;\boldsymbol{\theta})$ | |
| 0 | 152 | F | $-S'(152;\boldsymbol{\theta})$ | |
| 20 | 182 | F | $-S'(182;\boldsymbol{\theta})$ | $\times[S(20;\boldsymbol{\theta}]^{-1}$ |
| 20 | 190 | F | $-S'(190;\boldsymbol{\theta})$ | |
| 40 | 144 | F | $-S'(144;\boldsymbol{\theta})$ | $\times[S(40;\boldsymbol{\theta}]^{-1}$ |
| 40 | 107 | W | $S(107;\boldsymbol{\theta})$ | |
| 60 | 177 | F | $S'(177;\boldsymbol{\theta})$ | $\times[S(60;\boldsymbol{\theta}]^{-1}$ |
| 60 | 125 | F | $S'(125;\boldsymbol{\theta})$ | |
| 80 | 120 | W | $S(120;\boldsymbol{\theta})$ | $\times[S(80;\boldsymbol{\theta}]^{-1}$ |
| 80 | 137 | F | $S'(137;\boldsymbol{\theta})$ | |

(total) likelihood is

$$L_j = \begin{cases} S(\tau_j;\boldsymbol{\theta}) & \text{if the event } \textit{does not} \text{ occur before or at } \tau_j, \\ F(\tau_j;\boldsymbol{\theta}) & \text{if the event occurs before or at } \tau_j. \end{cases} \tag{7.46}$$

The method of maximum likelihood is directly applicable to the estimation of $\theta$'s. Special techniques for calculating MLE when $S(t;\boldsymbol{\theta})$ is normal (probit analysis), logistic (logit analysis) or rectangular (rankit analysis) have been evolved [e.g., Finney (1964)]. These are, in fact, specific applications of a general iterative method of calculating the MLE's for location-scale families from binary response data developed by R. A. Fisher (1935) [see Bliss (1935) and Cox (1970)].

These techniques are sometimes called *dosage mortality* techniques. If (presumably) random chosen animals are given a prespecified dosage ($y$) of a drug and the *response* (occurrence or nonoccurrence of event) is recorded, we then have the situation described above, with the *dosage* ($y$) playing the part of the *elapsed time* ($t$). The description *dosage mortality* is apt when the recorded event is death. Similar data arise in industrial testing when an item is subjected to a shock of prespecified amount (the "dosage"—such as the height from which the item is dropped) and the recorded event is some kind of failure (e.g., breakage).

If there are $k$ dosages, $y_1, y_2, \ldots, y_k$, and $n_i$ animals are tested at dosage $y_i$ with $d_i$ among them dying ($i = 1, 2, \ldots, k$) then the likelihood function is

$$L = \prod_{i=1}^{k} \binom{n_i}{d_i} [S(y_i;\boldsymbol{\theta})]^{n_i - d_i} [F(y_i;\boldsymbol{\theta})]^{d_i}. \tag{7.47}$$

The log-likelihood is

$$\log L = \sum_{i=1}^{k} \binom{n_i}{d_i} + \sum [(n_i - d_i)\log S(y_i;\boldsymbol{\theta}) + d_i \log F(y_i;\boldsymbol{\theta})]. \tag{7.48}$$

The maximum likelihood equations are

$$\frac{\partial \log L}{\partial \theta_u} = \sum_{i=1}^{k} \left[ \frac{n_i - d_i}{S(y_i;\boldsymbol{\theta})} - \frac{d_i}{F(y_i;\boldsymbol{\theta})} \right] \frac{\partial S(y_i;\boldsymbol{\theta})}{\partial \theta_u} = 0 \tag{7.49}$$

for $u = 1, 2, \ldots, s$ (provided $k > s$).

**Example 7.8.** *Likelihood when fitting k successive (different) distribution functions.*

Suppose that we have records on deaths and withdrawals (alive) of all $N$ individuals ever involved in a follow-up study. Suppose that HRF (or CHF) plots indicate that it might be appropriate to split the data into $k$ groups at fixed points $0 < \xi_1 < \xi_2 < \ldots < \xi_k < \infty$, and fit a different member of a family of SDFs in each interval $\xi_i \leqslant x \leqslant \xi_{i+1}$. Then the piecewise SDF and PDF are given by (7.18) and (7.19), respectively.

Let $x_{ij}$ be the *observed* time (age) of death or withdrawal of $j$th individual in $[\xi_i, \xi_{i+1})$. If he dies, his contribution to the likelihood is [from (7.19)]

$$S_0(\xi_1) \cdot \frac{S_1(\xi_2)}{S_1(\xi_1)} \cdots \frac{f_i(x_{ij})}{S_i(\xi_i)} = S_0(\xi_1) \cdot \frac{S_1(\xi_2)}{S_1(\xi_1)} \cdots \frac{S_i(x_{ij})}{S_i(\xi_i)} \lambda_i(x_{ij}), \tag{7.50}$$

where $\lambda_i(x_{ij}) = f_i(x_{ij}) / S_i(x_{ij})$.

If $(ij)$ withdraws at $x_{ij}$, his contribution is

$$S_0(\xi_1) \cdot \frac{S_1(\xi_2)}{S_1(\xi_1)} \cdots \frac{S_i(x_{ij})}{S_i(\xi_i)}. \tag{7.51}$$

If there are $d_i$ deaths and $w_i$ withdrawals, with $d_i + w_i = n_i$ in $[\xi_i, \xi_{i+1})$ then the contribution to the overall (unconditional) likelihood is

$$L_i = \left[ \frac{S_0(\xi_1)}{S_1(\xi_1)} \frac{S_1(\xi_2)}{S_2(\xi_2)} \cdots \frac{S_{i-1}(\xi_i)}{S_i(\xi_i)} \right]^{n_i} \prod_{j=1} \left[ \lambda_i(x_{ij}) \right]^{\delta_{ij}} S_i(x_{ij}), \tag{7.52}$$

with

$$\delta_{ij} = \begin{cases} 1 & \text{if } (ij) \text{ dies,} \\ 0 & \text{otherwise.} \end{cases}$$

The overall likelihood is

$$L = \prod_{i=0}^{k} L_i. \tag{7.53}$$

It is left to the reader to construct the conditional (on the sets of survivors at $\xi_i$, $i = 1, 2, \ldots, k$) likelihood and show that it is the same as the unconditional likelihood (7.52).

Construction of the likelihood when the SDF is 'piecewise' exponential is discussed by Aroian and Robison (1966).

**Example 7.9.** *Data grouped in fixed intervals.*

1. Suppose that times at death observed on a cohort of $N$ individuals (*complete* data) are grouped in $M$ fixed intervals, $[x_i, x_{i+1})$. Let $d_i$ be the number of deaths in $[x_i, x_{i+1})$, with

$$d_0 + d_1 + \ldots + d_{M-1} = N.$$

Let $S(x; \boldsymbol{\theta})$ be the (unknown) parametric SDF, which we wish to fit to the data. The (unconditional) probability of death in $[x_i, x_{i+1})$ is

$$\pi_i(\boldsymbol{\theta}) = S(x_i; \boldsymbol{\theta}) - S(x_{i+1}; \boldsymbol{\theta}). \tag{7.54}$$

The likelihood function is the *multinomial* probability regarded as a function of $\theta$'s, that is

$$L(\boldsymbol{\theta}) = N! \prod_{i=0}^{M-1} \left\{ \frac{[\pi_i(\boldsymbol{\theta})]^{d_i}}{d_i!} \right\}. \tag{7.55}$$

The MLE's $\hat{\boldsymbol{\theta}}$ maximize (7.55). (Compare Section 5.4 and Exercise 5.4 for nonparametric estimation, where the $\pi_i$'s themselves are the parameters that are estimated.)

2. When data are *censored* (incomplete), it is often assumed that the $w_i$ withdrawals occur at the midpoint of the interval $x_i^* = x_i + \frac{1}{2}h_i$, $(h_i = x_{i+1} - x_i)$, unless otherwise specified. *Conditional* on surviving till $x_i$, we have, for the interval $[x_i, x_{i+1})$:

(a) Probability of death

$$q_i(\boldsymbol{\theta}) = \frac{S(x_i;\boldsymbol{\theta}) - S(x_{i+1};\boldsymbol{\theta})}{S(x_i;\boldsymbol{\theta})}. \tag{7.56a}$$

(b) Probability of surviving until $x_i^*$

$$p_i^*(\boldsymbol{\theta}) = \frac{S(x_i;\boldsymbol{\theta}) - S(x_i^*;\boldsymbol{\theta})}{S(x_i;\boldsymbol{\theta})}. \tag{7.56b}$$

(c) Probability of surviving full interval

$$p_i(\boldsymbol{\theta}) = 1 - q_i(\boldsymbol{\theta}) = \frac{S(x_{i+1};\boldsymbol{\theta})}{S(x_1;\boldsymbol{\theta})}. \tag{7.56c}$$

The *conditional* likelihood is

$$L_i(\boldsymbol{\theta}) \propto [q_i(\boldsymbol{\theta})]^{d_i}[p_i^*(\boldsymbol{\theta})]^{w_i}[p_i(\boldsymbol{\theta})]^{N_i - d_i - w_i}, \tag{7.57}$$

where $N_i$ is the number of individuals alive at $x_i$.

The overall likelihood is

$$L = \prod_{i=0}^{M-1} L_i.$$

(Compare Section 6.3.1. Notice similarities and differences.)

**Example 7.10.** *Estimation for location-scale families of distributions.*

There is a remarkable set of general properties of MLE's of the location and scale parameters ($\xi$ and $\theta$) in location-scale families of distributions (see Section 3.9) which facilitates construction of tests and confidence intervals relating to values of these parameters. The properties are the following.

If $\hat{\xi}$ and $\hat{\theta}$ are MLE's of the parameters $\xi$ and $\theta$, respectively, based on a random sample of size $n$, then the distribution of the following statistics

$$T_1 = \frac{\hat{\theta}}{\theta}, \qquad T_2 = \frac{\hat{\xi} - \xi}{\theta}, \qquad \text{and} \qquad T_3 = \frac{\hat{\xi} - \xi}{\hat{\theta}} \tag{7.58}$$

*do not depend* on the values of the parameters themselves, although, of course, they do depend on sample size and the population distribution. The result is very general; in particular, it does not require the MLE's to be solutions of the ML equations. A proof [given by Antle and Bain (1969)] is not difficult.

These properties can be exploited in a number of ways, provided we can determine the distributions of $T_1$, $T_2$, and $T_3$. Analytic derivation is usually rather difficult, but simulation has been used to construct practically useful tables of percentage points for many important population distributions [Bain (1978)].

As an example of application of the general property to a special family, let us consider the two-parameter Weibull family with SDF

$$S_X(x,\beta,c)=\exp\left[-\left(\frac{x}{\beta}\right)^c\right]=\exp\{-\exp[c(\log x-\log\beta)]\},$$

$$x>0, \qquad \beta>0, \qquad c>0. \quad (7.59)$$

The distribution of $\log X$ belongs to a location-scale family with $\xi=\log\beta$ and $\theta=1/c$. It follows that if $\hat{c}$ and $\hat{\beta}$ are MLE's of $c$ and $\beta$, respectively, based on a random sample, then the distributions of

$$T_1=\frac{\hat{c}}{c}, \qquad T_2=c\log\frac{\hat{\beta}}{\beta}, \qquad \text{and} \qquad T_3=\hat{c}\log\frac{\hat{\beta}}{\beta}$$

depend on the sample size $n$, but *not* on the values of $c$ or $\beta$.

***Some Special Features of the Maximum Likelihood Method.***
It is necessary to bear in mind:

1. The maximum likelihood estimators *may not satisfy likelihood equations* $\partial\log L/\partial\theta=0$; for example, when the likelihood is maximized by parameter values, some of which are at the boundaries of their ranges of possible values.
2. It is necessary to check that a set of solutions, likelihood equations corresponds to the absolute *maximum*, and not a local maximum or (local) minimum of $L$.
3. There may be more than one set of solutions, each corresponding to a *local maximum*.
4. There are, indeed, cases wherein the likelihood can be *infinite* for sets of values of $(\theta_1,\ldots,\theta_s)$. These include situations that are, by no means, pathological. For example, when fitting a mixture of two normal distributions with these parameters—$\phi$ is the proportion of the first distribution;

$\mu_1$, $\sigma_1^2$ are the mean and variance of the first distribution; $\mu_2$, $\sigma_2^2$ are the mean and variance of the second distribution—we can make $L$ as large as desired by putting $\mu_1$ equal to any of the observed values, and letting $\sigma_1^2$ tend to zero keeping $\phi$ ($>0$), $\mu_2$, and $\sigma_2^2$ fixed. Nevertheless, such situations are not, overall, of frequent occurrence.

5. On the assumptions that the survival distribution really does belong to the family fitted, the sampling is random, and the sample size is sufficiently large, approximate variances and covariances can be evaluated using formulae (3.95).

The books by Bain (1978) and Gross and Clark (1975) contain useful detailed information on fitting parametric survival distributions.

## 7.9 MINIMUM CHI-SQUARE AND MINIMUM MODIFIED CHI-SQUARE

1. When *grouped* data are *complete* (as in Example 7.9), the approximate $X^2$ statistic

$$X^2 = X^2(\boldsymbol{\theta}) \doteq \sum_{i=0}^{M-1} \frac{[d_i - N\pi_i(\boldsymbol{\theta})]^2}{N\pi_i(\boldsymbol{\theta})}, \tag{7.60}$$

where $\pi_i(\boldsymbol{\theta})$ is given by (7.54), can be used in estimation. Minimizing (7.60) with respect to the $\theta$'s, we obtain the *minimum Chi-square estimators*.

2. Technically it is usually much easier to minimize (7.60) with $N\pi_i(\boldsymbol{\theta})$ in the denominator replaced by $d_i$, that is

$$X'^2 = X'^2(\boldsymbol{\theta}) = \sum_{i=0}^{M-1} \frac{[d_i - N\pi_i(\boldsymbol{\theta})]^2}{d_i}. \tag{7.61}$$

By minimizing (7.61), we obtain *minimum modified Chi-square* estimators of $\theta$'s.

The same type of program (e.g., MAXLIK) used for calculating ML-estimators can be used to calculate minimum and minimum modified Chi-square estimators. For sufficiently large samples, the results are usually close to the ML-estimators.

## 7.10 LEAST SQUARES FITTING

1. Here we briefly describe fitting by least squares, when the dependent variable is hazard rate or some function thereof, and we have *complete* mortality data, *grouped* in fixed intervals.

If $N_i$ is the number of survivors at the beginning of the interval $[x_i, x_{i+1})$ and $d_i$ is the number of deaths, then $p_i^o = (N_i - d_i)/N_i$ is the observed (i.e., estimated) conditional probability of surviving over $[x_i, x_{i+1})$ given alive at $x_i$ [compare (5.10)]. The notation $p_i^o$, used in this context, was explained in Section 7.3. Analogously to (5.30), the observed hazard rate at $x_i^* = x_i + \frac{1}{2}h_i$ is

$$\lambda_X^o(x_i^*) \doteq -\frac{1}{h_i}\log p_i^o, \tag{7.62}$$

and for the Gompertz distribution, we have

$$\lambda_X(x_i) = Re^{ax_i},$$

or

$$\log[\lambda_X(x_i^*)] = \log R + ax_i^* = b + ax_i^*. \tag{7.63}$$

Writing $\log\lambda_X^o(x_i^*) = y_i^o$, we have the linear regression function

$$y_i^o = b + ax_i^* + e_i, \qquad i = 0, 1, \ldots, M-1, \tag{7.64}$$

where $e_i$ is a random error.

To estimate slope $a$ and intercept $b$, we can minimize the *weighted* sum of squares

$$SS = \sum_{i=0}^{M-1} c_i(y_i^o - b - ax_i^*)^2, \tag{7.65}$$

where $c_i$ is a suitably chosen weight. The estimators of $a$ and $b$ (and so of $R$) so obtained are the *least-square estimators*.

From evidence presented by Gehan and Siddiqui (1973) (with three kinds of weights), and from our own experience, it appears that the *unweighted* (i.e., taking $c_i = 1$ for all $i$) least-square estimates are, on the whole, closest to the true values of the parameters although their variances are somewhat higher than those obtained from weighted least squares. Anyway, when these estimates are used as initial estimates for fitting by maximum likelihood, unweighted regression will be the simplest to use in practice.

For *Weibull* distribution, $[\log\lambda_X^o(x_i^*)]$ will be used versus $\log x_i^*$ [compare (7.11)].

2. The method can also be used for *incomplete* (censored) data, estimating $p_i^o$ from formula (6.15) [i.e., $p_i^o = 1 - q_i^o$, where $q_i^o = d_i/(N_i - \frac{1}{2}w_i)$] and applying techniques described in (1) for complete data.

One remark needs to be made on the techniques described in this section: the estimators $p_i^o$ and $p_j^o$ $(i \neq j)$ are not independent, but they are uncorrelated, and *conditionally on $N_i$ (and $w_i$)* they are independent. This fact should be taken into account when calculating standard errors of the estimates of the parameters. We do not discuss this here any further.

The method often gives good approximate results, and the estimates so obtained are good initial values for fitting by the maximum likelihood method as described in Example 7.9.

A full (unconditional) treatment of survival data by the least-square method using the estimators of $S(x_i)$, $i=1,2,\ldots,N$ and their variance-covariance structure has been given by Grizzle et al. (1969).

## 7.11 FITTING A GOMPERTZ DISTRIBUTION TO GROUPED DATA: AN EXAMPLE

We have already described some methods of fitting parametric models to grouped data, including the method of maximum likelihood (Example 7.9), minimum Chi-square (Section 7.9) and fitting a linear regression of hazard rate on time (Section 7.10). Here we illustrate these methods on some complete data.

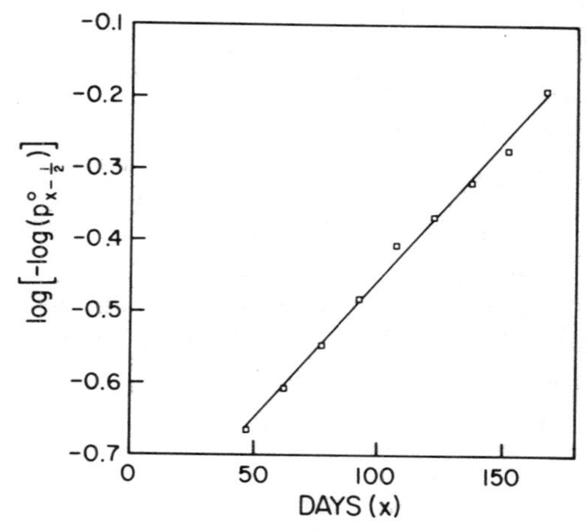


**Fig. 7.13** Fitted regression line.

**Example 7.11** We use the data from Furth et al. (1959), already given in Table 5.4 (grouped in Table 5.5), and discussed in Example 5.2. In Table 7.4, these data are again grouped into intervals different than in Example 5.2 including the interval [0,40) with observed number of deaths equal to zero.

*1. Empirical survival function.* The empirical survival function denoted here by $S^o(x_i)$ (which corresponds to $\hat{P}_i$ in terms of notation of

**Table 7.4** Estimation of survival function.
Mice data from Fürth et al. (1959)

| $i$ | $[x_i, x_{i+1})$ | $d_i$ | $N_i$ | $S^o(x_i)$ | $p_i^o$ | $-\frac{1}{h_i}\log p_i^o = y_i^o$ | $x_i^*$ |
|---|---|---|---|---|---|---|---|
| — | 0–40 | 0 | 208 | 1.00000 | 1.00000 | — | — |
| 0 | 40–55 | 4 | 208 | 1.00000 | 0.98077 | 0.00129 | 47.5 |
| 1 | 55–70 | 7 | 204 | 0.98077 | 0.96569 | 0.00233 | 62.5 |
| 2 | 70–85 | 12 | 197 | 0.94711 | 0.93909 | 0.00419 | 77.5 |
| 3 | 85–100 | 21 | 185 | 0.88942 | 0.88649 | 0.00803 | 92.5 |
| 4 | 100–115 | 37 | 164 | 0.78846 | 0.77439 | 0.01705 | 107.5 |
| 5 | 115–130 | 40 | 127 | 0.61058 | 0.68504 | 0.02522 | 122.5 |
| 6 | 130–145 | 40 | 87 | 0.41827 | 0.54023 | 0.04105 | 137.5 |
| 7 | 145–160 | 29 | 47 | 0.22596 | 0.38298 | 0.06398 | 125.5 |
| 8 | 160–175 | 16 | 18 | 0.08654 | 0.11111 | 0.14648 | 167.5 |
| 9 | 175+ | 2 | 2 | 0.00962 | 0.00000 | — | — |
| Total | | 208 | | | | | |

Section 5.3) was evaluated using formula (5.8). It is represented in Fig. 7.14, and the frequency distribution by the histogram in Fig. 7.15.

We now wish to fit a *Gompertz* distribution to these data over the range $(0, \infty)$, using the method of maximum likelihood. To use the computer program MAXLIK, we need initial estimates of the parameters. We obtain these by the method of least squares described in Section 7.10.

2. *Fitting least square regression of the "observed" hazard rates on age.* The appropriate data for this method are given in the last two columns of Table 7.4. Note that here $p_i^o$ corresponds to $\hat{p}_i$ from (5.10).

By minimizing (7.65) with $c_i = 1$, and using standard techniques for fitting a linear regression, we obtain $\hat{a}^{(1)} = 0.03841$, and $\hat{b}^{(1)} = -8.4204$, so

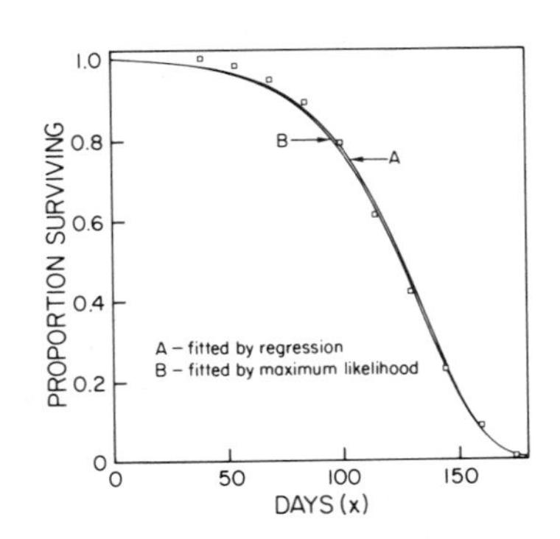


**Fig. 7.14** Fitted survival functions.

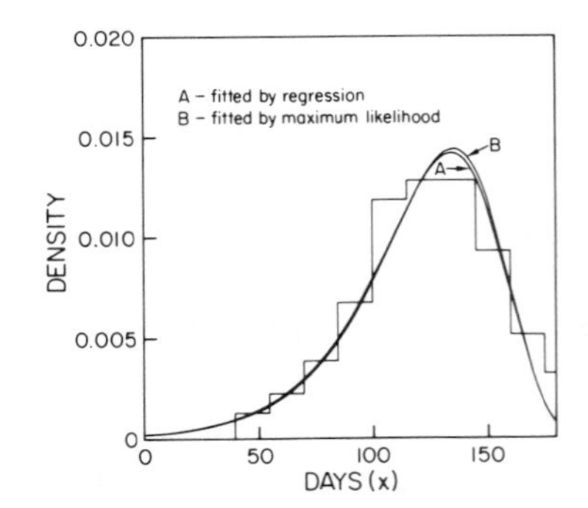


**Fig. 7.15** Fitted probability density functions.

that $\hat{R}^{(1)} = \exp(\hat{b}^{(1)}) = 0.0002203$. Figure 7.13 represents the fitted line.

Using these values, we also calculate estimated values of the SDF, denoted here by $\hat{S}^{(1)}(x_i)$. (See Table 7.5 and Fig. 7.14).

3. *Fitting by the method of maximum likelihood.* We wish to fit the (untruncated) Gompertz SDF over the range $(0, \infty)$, so that $S(0) = S(\infty) = 1$. We notice that if $d_i = 0$, then $[\pi_i(\boldsymbol{\theta})]^o = 1$, so that there is no contribution from such cells to the likelihood.

On the other hand, the method of minimum modified Chi-square cannot be used, since we would have a denominator equal to zero [see (7.61)]. Since we wish to compare these methods, we impose an arbitrary regrouping of the data, combining the two first intervals into one $[0, 55)$, and so use only the information that the four deaths occurred at some times in $[0, 55)$, putting $\pi_0(a, R) = S(0; a, R) - S(55; a, R)$. The last interval is taken as $[175, \infty)$, with $d_9 = 2$, so that $\pi_9(a, R) = S(175; a, R) - S(\infty; a, R) = S(175; a, R)$. This regrouping has no particular merits, except that it does not affect the results greatly, since the numbers of deaths involved are small. It must be kept in mind, however, that the results would have been different if we had used other groupings. There is room for some subjective judgment in these matters, and the reader is advised to repeat the method for other groupings (e.g., using the grouping as in Table 5.5).

We obtain the ML-estimates, $\hat{a}^{(2)} = 0.03901$ and $\hat{R}^{(2)} = 0.0001998$. The appropriate values of SDF, $\hat{S}^{(2)}(x_i)$ are given in Table 7.5 and in Fig. 7.14.

**Table 7.5** Comparison of Gompertz SDF's fitted to Fürth (1959) mice data by different methods

| $i$ | $[x_i, x_{i+1})$ | $d_i$ | $S^{(o)}$ | $\hat{S}^{(1)}$ | $\hat{S}^{(2)}$ | $\hat{S}^{(3)}$ | $\hat{S}^{(4)}$ | $X^2$ [for $\hat{S}^{(2)}$] |
|---|---|---|---|---|---|---|---|---|
| | 0–40 | 0 | 1.0000 | 1.0000 | 1.0000 | 1.0000 | 1.0000 | |
| 0 | 40–55 | 4 | 1.0000 | 0.9793 | 0.9809 | 0.9797 | 0.9829 | 2.1563 |
| 1 | 55–70 | 7 | 0.9808 | 0.9592 | 0.9621 | 0.9599 | 0.9656 | 0.0001 |
| 2 | 70–85 | 12 | 0.9471 | 0.9244 | 0.9292 | 0.9256 | 0.9347 | 0.0004 |
| 3 | 85–100 | 21 | 0.8894 | 0.8655 | 0.8729 | 0.8674 | 0.8810 | 0.1198 |
| 4 | 100–115 | 37 | 0.7885 | 0.7700 | 0.7802 | 0.7725 | 0.7908 | 1.8119 |
| 5 | 115–130 | 40 | 0.6106 | 0.6253 | 0.6379 | 0.6283 | 0.6492 | 0.0000 |
| 6 | 130–145 | 40 | 0.4183 | 0.4318 | 0.4443 | 0.4348 | 0.4531 | 0.3013 |
| 7 | 145–160 | 29 | 0.2260 | 0.2234 | 0.2322 | 0.2254 | 0.2350 | 0.4539 |
| 8 | 160–175 | 16 | 0.0865 | 0.0692 | 0.0724 | 0.0699 | 0.0710 | 0.5786 |
| 9 | 175+ | 2 | 0.0096 | 0.0086 | 0.0089 | 0.0087 | 0.0080 | 0.0016 |
| | | | | | | | | 5.4239 |

$(o)$ = empirical; (1) = fitted regression on hazard; (2) = maximum likelihood; (3) = minimum chi-square; (4) = minimum modified chi-square.
Note: $\hat{S}^{(k)} \equiv \hat{S}^{(k)}(x_i)$, $k = 1, 2, 3, 4$.

Note that $\hat{S}^{(1)}(x)$ and $\hat{S}^{(2)}(x)$ are very close. The same impression is given by Fig. 7.15 (where the PDF's are exhibited), although here we can see the discrepancies from the observed data (represented by the histogram) rather more clearly.

Estimates of Gompertz parameters, obtained by using two other methods are: (1) minimum Chi-square estimates: $\hat{a}^{(3)}=0.03856$, $\hat{R}^{(3)}=0.0002115$ (2) minimum modified Chi-square estimates: $\hat{a}^{(4)}=0.04009$, $\hat{R}^{(4)}=0.0001740$. The corresponding fitted SDFs are given in Table 7.5.

The reader interested in further examples of comparisons of various methods of fitting will find some in Barnett (1951), Cramer and Wold (1935), and de Vylder (1975). Berkson (1955, 1956) gives an interesting discussion of several methods of fitting.

## 7.12 SOME TESTS OF GOODNESS OF FIT

It has been pointed out in Section 7.1, that the essential feature of fitting parametric SDF's is *estimation* rather than *testing hypotheses*. Nevertheless, formal significance tests (and less formal but related techniques) can contribute useful evidence, where a decision on acceptability of a fitted distribution has to be made. In this section, we describe some more commonly used techniques that are applicable to *single observation* data on time of death.

### 7.12.1 Graphical "Test"

1. Consider first situations, in which we have a hypothesized SDF, $S_0(x)$, say, and want to judge the agreement of an observed mortality experience, $S^o(x)$, say, with it. For example, $S_0(x)$ may represent an established life table, perhaps a regional life table, and the mortality experience that of a well-defined group of individuals living within the region (inhabitants of certain counties, students at certain schools, life insurance policy holders with a certain company, etc.). In such cases there is no fitting procedure—our $S_0(x)$ is *predetermined*; but we do have the problem of assessing goodness of fit of the predetermined $S_0(x)$ to the empirical $S^o(x)$.

When the data are *complete*, the resultant graph will fall entirely within the square with diagonal points $(0,0)$, $(1,1)$ as shown in Fig. 7.16. If the fit is good, the points $[S_0(x), S^o(x)]$ should be close to the diagonal line $OP$ on which $S^o(x)=S_0(x)$. Such a graphical "test" is useful as a quick check to see whether it is worthwhile to apply some further analytical test(s).

When the data are *truncated*, the same method can be used over the restricted range. However, it seems rather dangerous to extrapolate the

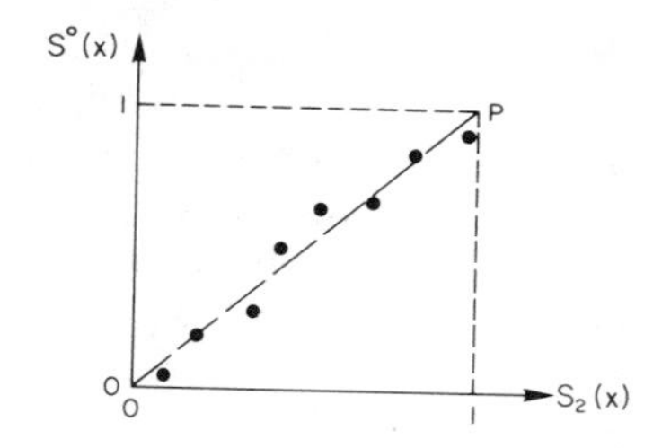


**Fig. 7.16** Empirical versus hypothesized SDF plot.

results for making judgments about what would happen if the data were complete.

For *incomplete* (with withdrawals) data, the same technique can be used; the empirical $S^o(x)$ is calculated in this case by the product-limit method.

2. The same kind of graphical test can be used when the hypothetical $S_0(x)$ is *not known*, although the *family* to which it belongs is specified. In this case, we may use the fitted $\hat{S}(x)$. Of course, in neither case [whether $S_0(x)$ or $\hat{S}(x)$] is a formal significance level used.

### 7.12.2 Kolmogorov-Smirnov Statistic. Limiting Distribution

There are many nonparametric methods for testing goodness of fit when the hypothesized $S_0(x)$ is predetermined. These can be found in standard textbooks on nonparametric methods [e.g., Hollander and Wolfe (1973)].

In this section, we discuss briefly the well-known Kolmogorov-Smirnov (briefly, K-S) statistic. To conform with other textbooks, we present the problem in terms of $F(x)$ rather than $S(x)$. It is not difficult to express the formulae in terms of SDF's.

***(1) Complete Data.*** The Kolmogorov-Smirnov statistic is defined as

$$D_N = \max_x |F^o(x) - F_0(x)|. \tag{7.66}$$

It is used in testing the hypothesis $H_0: F(x) = F_0(x)$ against the alternative $H_1: F(x) \neq F_0(x)$ (two-sided test), and assuming that $N$ single measurements (in this context, a *complete* set of times at death in a cohort of $N$ individuals) are available.

Clearly, the distribution of $D_N$ depends on the sample size $N$. Approximate tables of percentage points of $D_N$, for small $N$, can be found in some textbooks [e.g., Hollander and Wolfe (1973), Table A22].

For large $N$ ($>50$, say), we can use the limiting distribution

$$\lim_{N \to \infty} \Pr\{\sqrt{N} D_N > y\} = 2 \sum_{j=1}^{\infty} (-1)^{j+1} \exp(-2j^2 y^2). \tag{7.67}$$

Let $\alpha$ denote the significance level, and $y_{1-\alpha}$ such a value of $y$ for which the right-hand side of (7.67) is equal to $1-\alpha$. Then if $\sqrt{N}\,D_N > y_{1-\alpha}$ we reject $H_0$, and if $\sqrt{N}\,D_N \leqslant y_{1-\alpha}$ we do not reject $H_0$.

Some values of $y_{1-\alpha}$ for different $\alpha$ are given below.

Approximate upper percentage points of $\sqrt{N}\,D_N$

| $\alpha$ | 0.10 | 0.05 | 0.025 | 0.01 |
|---|---|---|---|---|
| $y_{1-\alpha}$ | 1.2238 | 1.3581 | 1.4802 | 1.6276 |

(See also the last column, i.e., $\phi = 1.0$ in Table 7.6.)

Better approximation is obtained by taking $(\sqrt{N} + 0.12 + 0.11/\sqrt{N})D_N$ in place of $\sqrt{N}\,D_N$ [Stephens (1974)].

To construct *one-sided* tests, we can use the statistics

$$D_N^+ = \max_x \left[ F^o(x) - F_0(x) \right] \tag{7.68}$$

[for alternatives $F(x) > F_0(x)$] or

$$D_N^- = \max_x \left[ F_0(x) - F^o(x) \right] \tag{7.69}$$

[for alternatives $F(x) < F_0(x)$].

For large samples, approximate limits for $D_N^+$, $D_N^-$ are the same as those for $D_N$ with the significance level doubled, that is, use $y_{1-2\alpha}$ in place of $y_{1-\alpha}$.

The formulae

$$D_N^+ = \max_i \left[ \frac{i}{N} - F_0(x_i') \right] \tag{7.70}$$

and

$$D_N^- = \max_i \left[ F_0(x_i') - \frac{i-1}{N} \right], \tag{7.71}$$

where $x_1' < x_2' < \cdots < x_N'$ are ordered times at death, are useful for calculating $D_N^+$ and $D_N^-$.

We can also calculate $D_N$ from

$$D_N = \max(D_N^+, D_N^-). \tag{7.72}$$

***(2) Censored and truncated data.*** In survival analysis or in life testing, the experiment is often terminated before all $N$ individuals (items) fail.

One of the two schemes is usually followed:

1. Continue the experiment only until the $r$ smallest failure times out of $N$ are observed (*censored* sample). We define the K-S *censored* statistic as

$$D_N\left(\frac{r}{N}\right) = D_N(\phi) = \max_{x < x_r'} |F^o(x) - F_0(x)|, \tag{7.73}$$

where $\phi = r/N$. In this case, $x_r'$ is a random variable.

2. Continue the experiment until predetermined time $\tau$ (truncated sample). We define the K-S *truncated* statistic

$$D_N(\tau) = \max_{x < \tau} |F^o(x) - F_0(x)|. \tag{7.74}$$

Generally, the exact distributions of $D_N(\phi)$ and $D_N(\tau)$ are different. Extensive tables of these two distributions are given by Barr and Davidson (1973).

However, as $N \to \infty$ the limiting distributions in both situations are the same if $F_0(\tau) = \phi$ [Conover (1967)].

Koziol and Byar (1975) calculated values of $y_{1-\alpha}(\phi)$ such that

$$\lim_{N\to\infty} \Pr\left\{\sqrt{N}\, D_N(\phi) > y_{1-\alpha}(\phi)\right\} = \alpha, \tag{7.75}$$

for different $\phi$ and $\alpha$. Independently, a program for this purpose was also written in the Department of Biostatistics, University of North Carolina at Chapel Hill by Anna Colosi, and extensive tables calculated. Table 7.6 is a part of our larger tables.

**Table 7.6** Critical values of $y_{1-\alpha}(\phi)$ for truncated distribution of $\sqrt{N}D_N(\phi)$ statistic

| | $\phi$ | | | | | | | | | |
|---|---|---|---|---|---|---|---|---|---|---|
| $\alpha$ | 0.01 | 0.02 | 0.03 | 0.04 | 0.05 | 0.06 | 0.07 | 0.08 | 0.09 | 0.10 |
| 0.10 | 0.1953 | 0.2753 | 0.3360 | 0.3867 | 0.4308 | 0.4703 | 0.5062 | 0.5392 | 0.5699 | 0.5985 |
| 0.05 | 0.2233 | 0.3147 | 0.3839 | 0.4417 | 0.4920 | 0.5569 | 0.5577 | 0.6152 | 0.6500 | 0.6825 |
| 0.025 | 0.2488 | 0.3505 | 0.4276 | 0.4918 | 0.5477 | 0.5975 | 0.6428 | 0.6844 | 0.7230 | 0.7589 |
| 0.01 | 0.2796 | 0.3938 | 0.4803 | 0.5523 | 0.6149 | 0.6707 | 0.7214 | 0.7679 | 0.8110 | 0.8512 |
| 0.005 | 0.3011 | 0.4240 | 0.5171 | 0.5946 | 0.6619 | 0.7219 | 0.7764 | 0.8264 | 0.8726 | 0.9157 |
| 0.001 | 0.3466 | 0.4880 | 0.5950 | 0.6840 | 0.7613 | 0.8303 | 0.8927 | 0.9500 | 1.0029 | 1.0523 |

| | $\phi$ | | | | | | | | | |
|---|---|---|---|---|---|---|---|---|---|---|
| $\alpha$ | 0.10 | 0.20 | 0.30 | 0.40 | 0.50 | 0.60 | 0.70 | 0.80 | 0.90 | 1.00 |
| 0.10 | 0.5985 | 0.8155 | 0.9597 | 1.0616 | 1.1334 | 1.1813 | 1.2094 | 1.2216 | 1.2238 | 1.2238 |
| 0.05 | 0.6825 | 0.9268 | 1.0868 | 1.1975 | 1.2731 | 1.3211 | 1.3471 | 1.3568 | 1.3581 | 1.3581 |
| 0.025 | 0.7989 | 1.0282 | 1.2024 | 1.3209 | 1.3997 | 1.4476 | 1.4717 | 1.4794 | 1.4802 | 1.4802 |
| 0.01 | 0.8512 | 1.1505 | 1.3419 | 1.4696 | 1.5520 | 1.5996 | 1.6214 | 1.6272 | 1.6276 | 1.6276 |
| 0.005 | 0.9157 | 1.2361 | 1.4394 | 1.5735 | 1.6582 | 1.7056 | 1.7258 | 1.7306 | 1.7308 | 1.7308 |
| 0.001 | 1.0523 | 1.4171 | 1.6456 | 1.7931 | 1.8828 | 1.9292 | 1.9464 | 1.9494 | 1.9495 | 1.9495 |

Note that $y_{1-\alpha}(1) = y_{1-\alpha}$.

***(3) Modified Kolomogorov-Smirnov Statistic.*** If only the family of the distributions is specified, but the parameters of the distribution are *not known* and are estimated from the sample, the critical values $y_{1-\alpha}$ are not correct.

Stephens (1974), using some Monte Carlo simulations, obtained some modified limits, $y'_{1-\alpha}$, for K-S (and several other) statistics, which can be used when the underlying distribution is normal or exponential, and when the unknown parameters are estimated from the sample. Some of his results are given in Table 7.7.

**Table 7.7** Modified K-S statistics and critical values $y'_{1-\alpha}$

| Modified statistic | Significance level $\alpha$ | | | |
|---|---|---|---|---|
| | 0.10 | 0.05 | 0.025 | 0.01 |
| | $S(x)$ normal with $\mu$ and $\sigma$ estimated | | | |
| $D\left(\sqrt{N}-0.01+\frac{0.85}{\sqrt{N}}\right)$ | 0.519 | 0.895 | 0.955 | 1.035 |
| | $S(x)$ exponential with $\lambda$ estimated | | | |
| $\left(D-\frac{0.2}{N}\right)\left(\sqrt{N}+0.26+\frac{0.5}{\sqrt{N}}\right)$ | 0.990 | 1.094 | 1.190 | 1.308 |

**Example 7.12** We wish to test, whether the Gompertz curve fits the mice data discussed in Examples 7.2 and 7.3. Of course, the true $S_0(x)$ is not known and we use in testing the fitted (by maximum likelihood) $\hat{S}(x)$. The estimated parameters are $\hat{a}=0.004615$, $\hat{R}=0.0005221$. The graphical test (not shown here) of $S^o(x)$ versus $S(x)=S(x;\hat{a},\hat{R})$, discussed in Section 7.12.1 seems to indicate that fit should be not too bad.

The calculated K-S statistic is $D=0.074436$. Since $N=39$, then $D\sqrt{N}=0.464855$.

Although the correct values of $y'_{1-\alpha}$ for Gompertz distribution with estimated parameters are not known, the observed value (0.465) is much smaller than the $y_{0.95}=1.358$ which would be the significance limit if the parameters were known in advance. Therefore, we may conclude that the fitted Gompertz SDF is acceptable.

### 7.12.3 Anderson-Darling $A^2$ Statistic

Studies by Stephens (1974) indicate that the Anderson-Darling (1954) statistic

$$A^2=N\int_{-\infty}^{\infty}\left[S_0(x)F_0(x)\right]^{-1}\left[S^o(x)-S_0(x)\right]^2 dF_0(x)$$

$$=-N-\frac{1}{N}\sum_{j=1}^{n}(2j-1)\left\{\log\left[F_0(x'_j)S_0(x'_{N-j+1})\right]\right\}, \qquad (7.76)$$

where $N$ is the sample size, provides a rather more powerful test than the $D$ statistic in many important situations. For all $N \geqslant 4$ upper percentage points for $A^2$ are given below [from Stephens (1974)].

Upper percentage points, $y_{1-\alpha}$, for Anderson-Darling $A^2$

| $\alpha$ | 0.10 | 0.05 | 0.025 | 0.01 |
|---|---|---|---|---|
| $y_{1-\alpha}$ | 1.933 | 2.242 | 3.070 | 3.857 |

When parameters are *estimated* from data, the percentage points are smaller as they are also for $D$.

Stephens (1974, 1977) introduced some modified $A^2$ statistics and evaluated the corresponding percentage points, using Monte Carlo simulation. Some of his results are shown in Table 7.8. Approximation is good, even for $N$ as small as 5.

**Table 7.8.** Modified upper percentage points for $A^2$ statistic

| Modified statistic | Significance level $\alpha$ | | | |
|---|---|---|---|---|
| | 0.10 | 0.05 | 0.025 | 0.01 |
| | $S(x)$ normal with $\mu$ and $\sigma$ estimated | | | |
| $A^2(1+4N^{-1}-25N^{-2})$ | 0.656 | 0.787 | 0.918 | 1.092 |
| | $S(x)$ exponential with $\lambda$ estimated | | | |
| $A^2(1+0.6N^{-1})$ | 1.078 | 1.341 | 1.606 | 1.957 |
| | $S(x)$ extreme value with $R$ and $a$ estimated | | | |
| $A^2(1+0.2N^{-1/2})$ | 0.637 | 0.757 | 0.877 | 1.038 |

### 7.12.4 Chi-Square Test for Grouped Data

For *complete grouped data*, the well known Pearson $X^2$-statistic can be used as a test criterion of goodness of fit. The observed frequencies, $O_i$, in $[x_i, x_{i+1})$, are the numbers of deaths, $d_i$. The expected frequencies, $E_i$, are calculated from $E_i = N\pi_i(\hat{\boldsymbol{\theta}})$, where $\pi_i(\hat{\boldsymbol{\theta}})$ is given by formula (7.54).

If there are $M$ groups, and $s$ parameters are estimated, the statistic

$$X^2 = \sum_{i=0}^{M-1} \frac{(O_i - E_i)^2}{E_i} \tag{7.77}$$

is approximately distributed as $\chi^2$ with $M-1-s$ degrees of freedom; (provided the type of SDF fitted is really appropriate).

**Example 7.13** The last column in Table 7.5 gives the individual values of $(O_i - E_i)^2/E_i$, where the ML-estimates $\hat{a}^{(2)} = 0.03856, \hat{R}^{(2)} = 0.0002115,$

calculated for Fürth et al. (1959), are used. The overall $X^2 \doteqdot 5.42$. The number of degrees of freedom is $10-1-2=7$. The 5% critical value is $X^2_{0.95;7}=14.07$. Since $5.42<14.07$, it seems reasonable to suppose that the Gompertz distribution fits these data. Of course, this does not imply that other models are excluded. One might also obtain a good fit with Weibull or lognormal distribution, for example.

## REFERENCES


Anderson, T. W. and Darling, D. A. (1954). A test of goodness of fit. *J. Amer. Statist. Assoc.* **49**, 765–769.

Antle, C. E. and Bain, L. J. (1969). A property of maximum likelihood estimators of location and scale parameters. *SIAM Rev.* **11**, 445–460.

Aroian, L. A. and Robison, D. E. (1966) Sequential life tests for the exponential distribution with changing parameter. *Technometrics*, **8**, 217–227.

Bain, L. J. (1978). *Statistical Analysis of Reliability and Life-Testing Models: Theory and Methods*, Marcel Dekker, New York.

Barnett, H. A. R. (1951). Graduation tests and experiments. *J. Inst. Actu. London* **77**, 15–54.

Barr, D. R. and Davidson, T. (1973). A Kolmogorov-Smirnov test for censored samples. *Technometrics* **15**, 739–757.

Berkson, J. (1955). Maximum likelihood and minimum $\chi^2$ estimates of the logistic function. *J. Amer. Statist. Assoc.* **50**, 130–162.

Berkson, J. (1956). Estimation by least squares and by maximum likelihood. *Proc. 3rd. Berkeley Symp.* **1**, 270–286.

Bliss, C. I. (1935). The calculation of the dosage-mortality curve. *Ann. Appl. Biol.* **22**, 134–167.

Cohen, A. C. (1965). Maximum likelihood estimation in the Weibull distribution based on complete and on censored samples. *Technometrics* **7**, 579–588.

Cox, D.R. (1970). *The Analysis of Binary Data*, Methuen, London

Cramer, H. and Wold, H. (1935). Mortality variations in Sweden. *Skand. Aktuar.* **18**, 161–241.

De Vylder, F. (1975). Maximum de vraisemblance et moindres carrés ponderés dans l'ajustement des tables de mortalité. *Assoc. Roy. Actu. Belges, Bull.* **70**, 35–41.

Finney, D. J. (1964). *Statistical Method in Biological Assay*, Griffin, London; Hafner, New York.

Fisher, R. A. (1935). The case of zero survivors. *Ann. Appl. Biol.* **22**, 164–165.

Fürth, J., Upton, A. C., and Kimball, A. W. (1959). Late pathologic effects of atomic detonation and their pathogenesis. *Radiat. Res., Suppl.* **1**, 243–264.

Gehan, E. A. and Siddiqui, M. M. (1973). Simple regression methods for survival time studies. *J. Amer. Statist. Assoc.* **68**, 848–856.

Grizzle, J. E., Starmer, C. F., and Koch, G. G. (1969). Analysis of categorical data by linear models. *Biometrics* **25**, 489–504.

Gross, A. J. and Clark, V. A. (1975). *Survival Distributions: Reliability Applications in the Biomedical Sciences*. Wiley, New York.

Hoel, D. G. (1972). A representation of mortality data by competing risks. *Biometrics* **28**, 475–488.

Hollander, M. and Wolfe, D. A. (1973). *Nonparametric Statistical Methods*. Wiley, New York, Chapters 4 and 10.

Johnson, N. L. and Kotz, S. (1970). *Distributions in Statistics: Continuous Univariate Distributions, -1 and 2*. Wiley, New York, Chapters 14, 17, 18, 20–23.

Kaplan, E. B. and Elston, R. C. (1973). A subroutine package for maximum likelihood estimation (MAXLIK). *University of North Carolina, Institute of Statistics Mimeo Series No. 823*.

Kimball, A. W. (1960). Estimation of mortality intensities in animal experiments. *Biometrics* **16**, 505–521.

Kodlin, D. (1967). A new response time distribution. *Biometrics* **23**, 227–239.

Koziol, J. A. and Byar, D. P. (1975). Percentage points of the asymptotic distributions of one- and two sample K-S statistics for truncated or censored data. *Technometrics* **17**, 507–510.

Krane, S. A. (1963). Analysis of survival data by regression techniques. *Technometrics* **5**, 161–174.

Kulldorf, G. (1961). *Estimation from Grouped and Partially Grouped Samples*. Wiley, New York.

Nelson, W. A. (1972). Theory and application of hazard plotting for censored failure data. *Technometrics* **14**, 945–966.

Rényi, A. (1953). On the theory of order statistics. *Acta. Math., Hungar. Acad. Sci.* **4**, 151–231.

Stephens, M. A. (1974). EDF statistics for goodness of fit and some comparisons. *J. Amer. Statist. Assoc.* **69**, 730–737.

Stephens, M. A. (1977). Goodness of fit for the extreme value distribution. *Biometrika* **64**, 583–588.

## EXERCISES

**7.1.** Explain the difference between: ($a$) a piecewise exponential SDF and ($b$) a mixture of exponential SDFs. Give graphical methods of fitting each kind of SDF. Also, describe graphical aids for deciding which of ($a$) and ($b$) is likely to be better to use with a given set of data.

**7.2.** Suppose that $N$ individuals are observed from time zero until time $k$ (an integer). In each unit interval $(j,j+1)$, the number of individuals failing, $n_j$, is recorded. The number of individuals surviving at time $k$ is $m$, so that

$$N=\sum_{j=0}^{k-1} n_j+m.$$

It is desired to fit an exponential SDF

$$S(x)=\exp(-\lambda x), \qquad \lambda, x>0,$$

to the data. Obtain equations satisfied by

(*i*) the maximum likelihood estimator of $\lambda$,
(*ii*) the minimum $\chi^2$ estimator of $\lambda$,
(*iii*) the modified minimum $\chi^2$ estimator of $\lambda$.

**7.3.** In Exercise 7.2, suppose that during the interval $(j, j+1)$, $r_j$ individuals are lost, so that all that is known of them is that they survived until time $j$. For $m$ to retain its meaning we must have

$$N = \sum_{j=0}^{k-1} (n_j + r_j) + m.$$

Obtain an equation satisfied by the maximum likelihood estimator of $\lambda$. What difficulties are encountered in trying to determine the minimum, or modified minimum $\chi^2$ estimator in this situation?

**7.4.** Suppose that the survival function, SDF, is represented by a set of successive distribution functions as defined in (7.18). Let $N_i$ be the number of individuals alive at $\xi_i$. Suppose that $d_i$ individuals died and $w_i$ were withdrawn alive in $(\xi_i, \xi_{i+1})$ and assume that individual times $(x_{ij})$ at death or withdrawal are known.

(*a*) Show that *conditionally* (on $N_i$), the contribution to overall likelihood from $(\xi_i, \xi_{i+1})$ is

$$L_i = \left[ \prod_j f_i(x_{ij}) \right] \left[ \prod_l S_i(x_{il}) \right] \left[ S_i(\xi_{i+1}) \right]^{N_i - d_i - w_i}$$

and the overall likelihood conditional on $\{N_i\}$ is

$$L_{\text{cond}} = \prod_{i=1}^{k} L_i.$$

(*b*) Show that $L_{\text{cond}}$ is identical with the unconditional likelihood [see (7.52) and (7.53)].

**7.5.** A mortality experience provides numbers of (central) exposed to risk age $x$ lbd for $x = 20\ (1)\ 59$, with corresponding deaths over a two-year period. Give directions for fitting the formula

$$q_x = \alpha q_x^{(1)} + (1-\alpha) q_x^{(2)},$$

where $q_x^{(j)}$ $(j = 1, 2)$ corresponds to one of two standard life tables, and $\alpha$ is an unknown parameter. How would you test whether this is a suitable model?

**7.6.** It has been suggested that the model in Exercise 7.5 corresponds to the assumption that the population from which the mortality experience has been obtained is a mixture, in proportions $\alpha:(1-\alpha)$ of two populations, $(\Pi_1, \Pi_2)$, with mortality in $\Pi_j$ following the $j$th standard life table ($j=1,2$).

(*a*) Explain why this is not so, and give an appropriate formula for $q_x$, in terms of the two standard life tables, if it were so.

(*b*) Under what circumstances might the model in Exercise 7.5 be appropriate?

**7.7.** "If a test of goodness-of-fit is used, that is especially sensitive to a particular kind of departure from the hypothesis, for example, an excess of actual over conjectured force of mortality which increases linearly with age, then a nonsignificant result should be interpreted *only* as a lack of evidence for *that particular* form of departure, and not as a confirmation of the conjectured set of values for force of mortality. However, if a general test (such as a Chi-square test) is used, which is not constructed to be especially sensitive to specific departures, nonsignificance can be so interpreted."

Comment on this passage, with special reference to the opinion in the second paragraph.

**7.8.** It has been suggested (see, for instance, H. L. Seal (1939) *J. Inst. Actu. London* **71**) that small observed values of Chi-squares, as well as large, should be regarded as evidence against goodness-of-fit; according to the scheme

| $\Pr[\chi^2 > \text{observed value}]$ | Verdict |
|---|---|
| $>0.999$ | much too probable |
| $>0.99,\ \leqslant 0.999$ | too probable |
| $>0.95,\ \leqslant 0.99$ | rather too probable |
| $\geqslant 0.01,\ <0.05$ | of doubtful improbability |
| $\geqslant 0.001,\ <0.01$ | improbable |
| $<0.001$ | very improbable |

Show that this interpretation would not be appropriate if the alternative hypotheses simply specified that the true mortality probabilities differ from those fitted.

Nevertheless there might be intuitive feeling supporting the proposed verdicts. Describe a model that might be used to support this intuition.

**Table E7.1** Cumulative probability of business failure in Poughkeepsie, 1844–1926

| | Type of business | | | |
|---|---|---|---|---|
| Age in years | Retail | Manufacture | Craft | Service |
| 0–1 | 0.296 | 0.231 | 0.307 | 0.327 |
| 1–2 | 0.438 | 0.346 | 0.454 | 0.457 |
| 2–3 | 0.532 | 0.469 | 0.551 | 0.551 |
| 3–4 | 0.594 | 0.547 | 0.607 | 0.618 |
| 4–5 | 0.643 | 0.602 | 0.660 | 0.669 |
| 5–6 | 0.684 | 0.655 | 0.697 | 0.708 |
| 6–7 | 0.715 | 0.678 | 0.727 | 0.743 |
| 7–8 | 0.741 | 0.702 | 0.753 | 0.769 |
| 8–9 | 0.762 | 0.726 | 0.772 | 0.792 |
| 9–10 | 0.782 | 0.746 | 0.791 | 0.812 |

From Lomax (1954); see Exercise 3.11

**7.9.** Table E7.1, is taken from Lomax (1954) (see Exercise 3.11). Using maximum likelihood, estimate the parameters of distribution (*ii*) of Exercise 3.11, to each of the four sets of data.

(Note, that the *numbers* of businesses are not given. Why does this not affect the MLE's?)

**7.10.** As Exercise 7.9, but fitting distribution (*i*) of Exercise 3.11. What special problems arise in this case?

**7.11.** T. N. Thiele ((1872) *J. Inst. Actu. London* **16**, 313–329) proposed the following formula for force of mortality at age $x$:

$$\mu_x = b_1 c_1^{-(x-d)} + b_2 c_2^{x-a} + b_3 c_3^{-(x-a)^2}$$

(all parameters positive).

(*i*) Show that this includes the Gompertz and Makeham-Gompertz laws as special cases.

(*ii*) The title of the paper starts "On a mathematical formula to express the rate of mortality throughout the whole of life...." How might this description be justified?

(*iii*) Outline a method for graphical fitting of the SDF associated with this law, assuming that in old age a Gompertz law gives a good representation of the mortality pattern.

**7.12.** Adjust the parameter values of the fitted Weibull SDF shown in Fig. 7.5 by trial and error, to improve the fit, especially over the age range 60–80, using graphical methods.

CHAPTER 8

# Comparison of Mortality Experiences

## 8.1 INTRODUCTION

We now come to the comparison of mortality experienced by two (or more) groups. Our primary concern is in describing in what ways (if at all) the mortality patterns in different groups (populations) appear to be *distinct*. Choice of methods of comparison depends heavily on the specific questions posed by the investigator, and on the type of mortality data available for the analysis. Essentially, we will be concerned with four types (classes) of survival data:

1. Comparison of two populations (two life tables).
2. Comparison of mortality experience with a population life table.
3. Comparison of two mortality experiences—complete data.
4. Comparison of two mortality experiences in follow-up and longitudinal studies (clinical trials).

For each type, appropriate methods of analysis will be presented, often extended to comparisons among more than two groups.

## 8.2 COMPARISON OF TWO LIFE TABLES

In this case, we do not have direct records of mortality experiences, but essentially a set of probabilities derived from (usually rather extensive) experiences. Of course, *no formal testing of hypotheses* is involved here. Many standard descriptive techniques can be used. A few, which seem to be especially useful and relevant, are discussed below.

### 8.2.1 Graphical Displays

The most straightforward methods seem to be graphical.

1. We may plot one SDF against the other, as in Section 7.12.1, and observe whether the points deviate from the diagonal (see Fig. 7.16).
2. We may plot the *survival functions* $(l_x/l_0)$; *histograms* representing the distributions of death, $d_x$; and *forces of mortality*, $\mu_x$, reflecting in different ways changes in mortality rates with age (see Example 8.1).

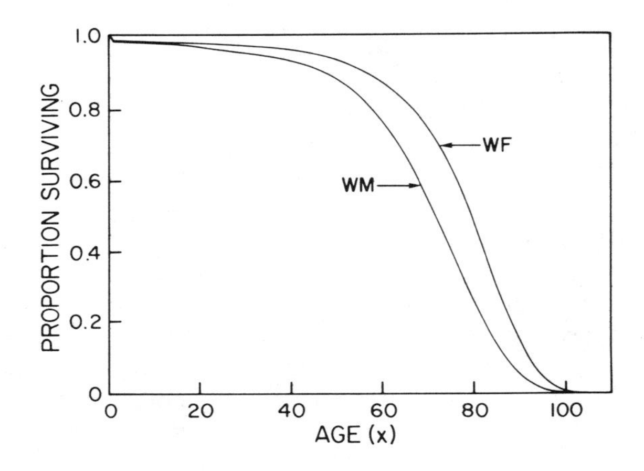


**Fig. 8.1a** Survival functions.

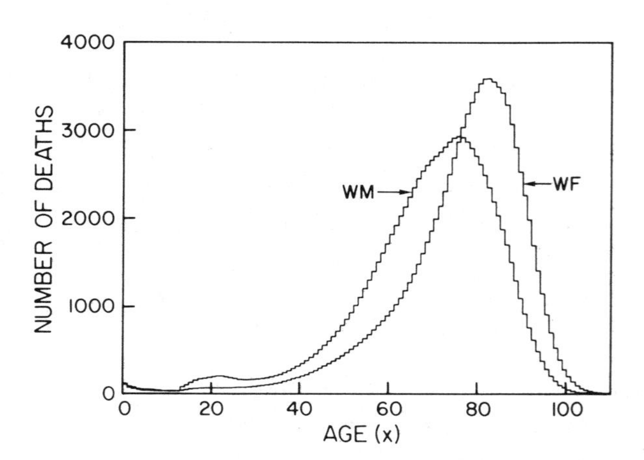


**Fig. 8.1b** Histograms of death distributions.

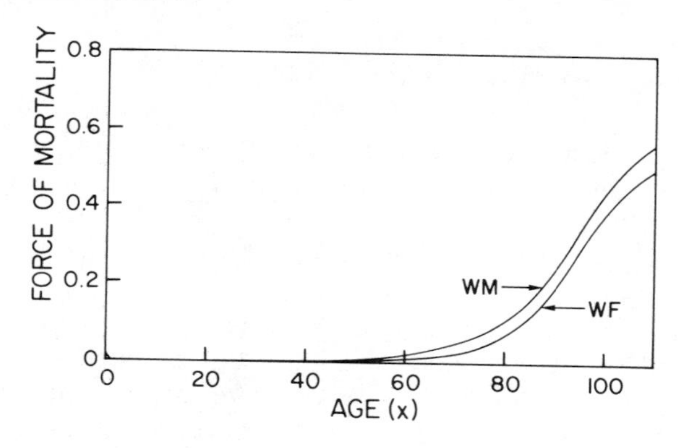


**Fig. 8.1c** Forces of mortality.

**Example 8.1** Figures 8.1a, b, and c represent life table functions for U. S. Males (WM) and White Females (WF), 1969–1971. As can be seen from these figures, each of the three functions yields a different kind of comparison.

### 8.2.2 Conditional Probabilities $q_x$

Although the PDF and HRF plots are useful in detecting differences and showing where they occur, it should be remembered that the main interest is, how well do the *death probabilities* calculated from different life tables agree with each other, and in what respects do they differ?

If the results of comparison of two life tables for each of a number of age intervals exhibit notable common features [for example, ${}_nq_x^{(1)}$ for life table of population 1 is always higher than the corresponding probability ${}_nq_x^{(2)}$ for life table of population 2, the ratio ${}_nq_x^{(1)}/{}_nq_x^{(2)}$, as a function of $x$, exhibits a quasilinear trend], it is useful to summarize such features in as concise a form as possible. For example, one might say that ${}_nq_x^{(1)}$ is substantially greater than ${}_nq_x^{(2)}$ in all age intervals, or ${}_5q_{30}^{(1)}$ is approximately 1.5 times as great as ${}_5q_{30}^{(2)}$ but their ratio decreases roughly linearly with increasing age, until at age 80 ${}_5q_{80}^{(1)}$ is no more than 1.2 times as great as ${}_5q_{80}^{(2)}$, and so forth.

### 8.2.3 Conditional Expectations and Median of Future Lifetime

Some authors like to use the expected future lifetime $\mathring{e}_x$ [defined by (4.12) or (4.20)] or the median future lifetime $Me_x$ (discussed in Example 4.3). These are often displayed graphically as functions of $x$.

## 8.3 COMPARISON OF MORTALITY EXPERIENCE WITH A POPULATION LIFE TABLE

This class of problems arises when the current mortality of a *group* is under investigation, to see whether it is sufficiently well described by a population life table. The group may be employees in an industrial plant exposed to occupational hazard (e.g., asbestos factory workers), patients who have undergone a certain treatment, and so on.

We should bear in mind that, regardless whether such a group is small or large, it does not represent a *random* sample; it is especially selected part of the whole population and its survival data are included in calculation of the population life table. Therefore any formal test that assumes independence of the total and study population data are, strictly speaking, inappropriate. However, the contribution of the study subpopulation to the resulting overall life table is usually small, and for practical purpose, we may use the population life table as a *standard* (*model*) with which we compare the mortality experience of the study population. Therefore, methods of comparison described in Chapter 7 might be approximately used here. Also, if the study population is large enough (over 10,000 exposed to risk, say), then methods of comparison of two life tables, described in Section 8.2 might be appropriate.

Two further methods are suggested below.

### 8.3.1 Test Based on Median of Future Lifetime

Let $x' = x + \phi$ (where $x$ is an integer and $0 \leqslant \phi < 1$) denote the exact present age, and $y_{x+\phi}$—the age up to which a person aged $(x+\phi)$ has 50% chance of surviving, according to the standard life table. We have $l_{y_{x+\phi}} = \frac{1}{2} l_{x+\phi}$. Following the method described in Example 4.3, letting $y_{x+\phi} = a + \theta$, where $a$ is an integer and $0 \leqslant \theta < 1$, we calculate $y_{x+\phi}$ from the approximate formula

$$y_{x+\phi} \doteqdot a + \frac{l_a - \frac{1}{2} l_{x+\phi}}{l_a - l_{a+1}}. \tag{8.1}$$

The (estimated) *expected median future lifetime* for an individual of present age $(x+\phi)$, is

$$Me_{x+\phi} = y_{x+\phi} - (x+\phi). \tag{8.2}$$

For each individual in the study, we record the actual time elapsed between the event of interest and death (e.g., between onset of disease and death), and compare it with $Me_{x+\phi}$. We score

$$1 \text{ if actual survival time} \geqslant Me_{x+\phi},$$

$$0 \text{ if actual survival time} < Me_{x+\phi}.$$

Note that data need not be complete, but each survivor should have passed his median $Me_{x+\phi}$ (and is known to have score 1).

Under the null hypothesis that there is no deviation from the standard life table, we would expect that the total score, $R$, say, for $N$ individuals, would have a binomial distribution with parameters, $N$, $\frac{1}{2}$, so that

$$\Pr\{R=r\}=\binom{N}{r}\left(\frac{1}{2}\right)^{N}. \tag{8.3}$$

Unless $N$ is small ($N \leqslant 10$, say), we may use the normal approximation

$$\Pr\{R \leqslant r\} \doteqdot \Phi\left[\frac{\left(r+\frac{1}{2}\right)-\frac{1}{2}N}{\frac{1}{2}\sqrt{N}}\right]=\Phi\left(\frac{2r+1-N}{\sqrt{N}}\right). \tag{8.4}$$

On this basis, we can apply a test of significance.

It is also possible, in a similar way, to use other percentiles (compare Example 4.3) as bases for test of significance.

### 8.3.2 Test Based on Expected Future Lifetime

When data are *complete*, comparison is also possible using the expectation ($\mathring{e}_x$) and variance ($V_x$) of future lifetime [see (4.12) and (4.25), respectively]. Of course, if we observe exact age $x'=x+\phi$, some further interpolation is required.

For $N$ deaths in the study population, we denote:

$T_N$ = sum of actual times elapsed between event and death;

$E_N$ = sum of corresponding values of $\mathring{e}_x$;

$V_N$ = sum of variances $V_x$.

Thus

$$Z=\frac{T_N-E_N}{\sqrt{V_N}} \tag{8.5}$$

is approximately distributed as unit normal, when the null hypothesis (that the expectations of future lifetimes in study and standard populations are equal) is valid.

*Remark 1.* If a study group is composed of individuals from different populations (e.g., males, females, blacks, whites, etc.), then for each individual an appropriate standard life table should be used.

*Remark 2.* The median test is usually to be preferred to the test based on expectations; $\mathring{e}_x$ calculated from abridged life table is often not very precise.

**Example 8.2** We use a part of the data from Table 5.1 on survival time since diagnosis of dementia for Swiss females aged $x=74$. (We assume that this is approximately the exact age—for explanation, see Example 5.1.) The Geneva life table 1968–1972 for females, part of which is exhibited in Table 8.1, was used as standard life table.

**Table 8.1** Part of life table for Geneva female population, 1968–72

| Age $x$ to $x+1$ | $l_x$ | $d_x$ | $L_x$ | $\mathring{e}_x$ | $V_x$ |
|---|---|---|---|---|---|
| 65–66 | 86,194 | 1,006 | 85,691 | 17.22 | 59.79 |
| 66–67 | 85,188 | 1,117 | 84,630 | 16.42 | 57.15 |
| 67–68 | 84,071 | 1,244 | 83,449 | 15.63 | 54.50 |
| 68–69 | 82,827 | 1,388 | 82,133 | 14.86 | 51.83 |
| 69–70 | 81,439 | 1,547 | 80,666 | 14.10 | 49.13 |
| 70–71 | 79,892 | 1,723 | 79,030 | 13.36 | 46.43 |
| 71–72 | 78,169 | 1,911 | 77,214 | 12.65 | 43.72 |
| 72–73 | 76,258 | 2,110 | 75,203 | 11.95 | 41.02 |
| 73–74 | 74,148 | 2,318 | 72,989 | 11.28 | 38.34 |
| 74–75 | 71,830 | 2,532 | 70,564 | 10.62 | 35.70 |
| 75–76 | 69,298 | 2,757 | 67,920 | 9.99 | 33.11 |
| 76–77 | 66,541 | 2,968 | 65,058 | 9.39 | 30.59 |
| 77–78 | 63,574 | 3,178 | 61,985 | 8.80 | 28.14 |
| 78–79 | 60,396 | 3,383 | 58,704 | 8.24 | 25.79 |
| 79–80 | 57,013 | 3,577 | 55,224 | 7.70 | 23.54 |
| 80–81 | 53,436 | 3,753 | 51,559 | 7.18 | 21.40 |
| 81–82 | 49,683 | 3,905 | 47,730 | 6.69 | 19.37 |
| 82–83 | 45,778 | 4,024 | 43,766 | 6.21 | 17.46 |
| 83–84 | 41,754 | 4,102 | 39,703 | 5.76 | 15.67 |
| 84–85 | 37,652 | 4,132 | 35,586 | 5.34 | 14.00 |
| 85–86 | 33,520 | 4,106 | 31,467 | 4.93 | 12.46 |
| 86–87 | 29,414 | 4,018 | 27,406 | 4.55 | 11.04 |
| 87–88 | 25,397 | 3,865 | 23,465 | 4.19 | 9.74 |
| 88–89 | 21,532 | 3,646 | 19,709 | 3.86 | 8.55 |
| 89–90 | 17,886 | 3,367 | 16,203 | 3.54 | 7.48 |

From this table $Me_{74} = 10.42$ years. For $x = 74$, we have $N = 25$, and $r = 2$. Thus from (8.4)

$$\Pr\{R > 2\} \doteq 1 - \Phi\left(\frac{4+1-25}{5}\right) = 1 - \phi(-4.00) \doteq 1.$$

Clearly, dementia is a disease associated with accelerated aging.

Applying the test based on $\mathring{e}_x$'s, we have $T_N = 112.82, \mathring{e}_{74} = 10.62$, so that $E_N = 25 \cdot 10.62 = 265.50, V_{74} = 35.70$, so that $V_N = 25 \cdot 35.70 = 892.50$. Thus

$$z \doteq \frac{112.82 - 265.50}{(892.50)^{1/2}} = -5.11.$$

If $Z$ is unit normal, $\Pr\{Z < z\} = \Phi(-5.11) \doteq 0$. This also gives a clearly significant result.

In the remaining sections we are concerned with comparisons among two or more mortality experiences.

## 8.4 SOME DISTRIBUTION-FREE METHODS FOR UNGROUPED DATA

We now wish to compare two mortality experiences. In other words, we consider two sets of observations *based on samples* (supposedly random), from two populations. It is necessary to take into account random variation in both sets of data.

In this section, we assume that the individual times at death (or withdrawal) are available, and we wish to construct some *distribution-free* (nonparametric) tests for comparison of two mortality experiences. As has been mentioned before, survivorship is a continuous process, and comparisons based on single statistics are not really appropriate. Nevertheless, to obtain some idea of how a study should progress, a preliminary nonparametric test is occasionally of some use. There are many tests of this kind; in this section we discuss two tests applicable to data from two *independent* random samples: the Kolmogorov-Smirnov test and the Wilcoxon test with some modifications. Comparisons will be made of failure rather than survival distribution functions.

### 8.4.1 Two Sample Kolmogorov-Smirnov Test

Let

$$t'_{g1} < t'_{g2} < \dots < t'_{gN_g}, \qquad g = 1, 2, \tag{8.6}$$

be the *ordered* failure times in the $g$th group, so that $F_g^o(t)=j/N_g$ for $t'_{gj} \leqslant t < t'_{g,j+1}$ is an empirical failure function for the $g$th group, as defined in (5.2). We assume that the sets $\{t'_{1j}\}$ and $\{t'_{2j}\}$ are *independent*.

The *two sample* Kolmogorov-Smirnov (briefly K-S) test statistic for testing $H_0: F_1(x)=F_2(x)$ against two-sided alternative $H_1: F_1(x) \neq F_2(x)$ is

$$D_{N_1,N_2} = \max_t |F_1^o(t) - F_2^o(t)| = \max_t |S_1^o(t) - S_2^o(t)|. \tag{8.7}$$

Large values of $D_{N_1,N_2}$ are regarded as significant differences between two population failure distributions.

For smaller sample sizes, $N_1$, $N_2$, there are various tables from which significance of $D_{N_1,N_2}$ can be assessed [e.g., Kim and Jennrich (1970) and Hollander and Wolfe (1973), Tables A.22 and A.23]. For sufficiently large values of $N_1$, $N_2$ ($>100$, say), the approximate upper percentage point limits $\sqrt{N_1 N_2/(N_1+N_2)}\, D_{N_1,N_2}$, given in Section 7.12.2 can be used. If

$$\sqrt{\frac{N_1 N_2}{N_1+N_2}}\; D_{N_1,N_2} > y_{1-\alpha}\,.$$

we reject $H_0: F_1(x)=F_2(x)$, against the alternative $H_1: F_1(x) \neq F_2(x)$ with the significance level $\alpha$.

If there are *ties* at time $t'_k$, say, we may calculate the *upper* and *lower* limits for $|F_1^o(t'_k) - F_2^o(t'_k)|$ in the following way. For each tied pair assign: (1) lower rank to all individuals from group 1; (2) lower rank to all individuals from group 2; then (3) find the corresponding two values of $|F_1^o(t'_k) - F_2^o(t'_k)|$—the greater value of the two gives the upper limit at this point; (4) apply the K-S test to the upper (and if the result is significant also the lower) limit.

**Example 8.3** We use data on dementia given in Table 5.1, to compare the survival after diagnosis for ages at diagnosis 73 and 74. These data and calculation of $D_{N_1,N_2}$ are given in Table 8.2.

We observe four ties. The value of $D_{N_1,N_2}$ is 0.1435 if the first of the two deaths at recorded time 1.67 years is from the first group (age at diagnosis, 73); otherwise it is 0.1294. We have

$$\sqrt{\frac{N_1 N_2}{N_1+N_2}}\; D_{N_1,N_2} = \sqrt{\frac{17\cdot 25}{42}} \times 0.1435 = 0.456.$$

From Section 7.12.2(1), we find, the approximate 5% limit is 1.358. From more extensive tables [Kim and Jennrich (1970)] the value is 1.295. Of

**Table 8.2** Comparison of empirical distributions of survival times for two groups of patients with dementia [ages of diagnosis: 73 ($g=1$), and 74 ($g=2$)]

| Rank | $g=1$ $t'_{1j}$ | $g=2$ $t'_{2j}$ | $F_1^o(t)$ | $F_2^o(t)$ | $\lvert F_1^o - F_2^o \rvert$ |
|---|---|---|---|---|---|
| 1 | | 0.50 | 0.0000 | 0.0400 | 0.0400 |
| 2 | 0.58 | | 0.0588 | | 0.0188 |
| 3 | | 1.00 | | 0.0800 | 0.0212 |
| 4 | 1.08 | | 0.1171 | | 0.0371 |
| 5 | | | | | 0.0965* |
| | 1.25 | 1.25 | 0.1765 | 0.1200 | 0.0029† |
| 6 | | | | | 0.0565 |
| 7 | | 1.41 | | 0.1600 | 0.0165 |
| 8 | | 1.42 | | 0.2000 | 0.0235 |
| 9 | | 1.58 | | 0.2400 | 0.0635 |
| 10 | | 1.66 | | 0.2800 | 0.1035 |
| 11 | | | | | 0.0447* |
| | 1.67 | 1.67 | 0.2353 | 0.3200 | 0.1435† |
| 12 | | | | | 0.0847 |
| 13 | 2.00 | | 0.2951 | | 0.0249 |
| 14 | 2.08 | | 0.3529 | | 0.0329 |
| 15 | 2.17 | | 0.4118 | | 0.0918 |
| 16 | | 2.25 | | 0.3600 | 0.0518 |
| 17 | | 2.33 | | 0.4000 | 0.0118 |
| 18 | | 2.92 | | 0.4400 | 0.0282 |
| 19 | | 3.08 | | 0.4800 | 0.0682 |
| 20 | | | | | 0.0094* |
| | 3.75 | 3.75 | 0.4706 | 0.5200 | 0.1082† |
| 21 | | | | | 0.0494 |
| 22 | | 3.92 | | 0.5600 | 0.0894 |
| 23 | | 4.17 | | 0.6000 | 0.1294 |
| 24 | 4.25 | | 0.5294 | | 0.0706 |
| 25 | 4.58 | | 0.5882 | | 0.0118 |
| 26 | | 4.67 | | 0.6400 | 0.0518 |
| 27 | 4.92 | | 0.6471 | | 0.0071 |
| 28 | | 5.00 | | 0.6800 | 0.0329 |
| 29 | | | | | 0.0259* |
| | 5.25 | 5.25 | 0.7059 | 0.7200 | 0.0729† |
| 30 | | | | | 0.0141 |
| 31 | | 5.83 | | 0.7600 | 0.0541 |
| 32 | 6.75 | | 0.7647 | | 0.0047 |
| 33 | | 7.25 | | 0.8000 | 0.0353 |
| 34 | 7.83 | | 0.8236 | | 0.0236 |
| 35 | 7.84 | | 0.8824 | | 0.0824 |
| 36 | | 8.50 | | 0.8400 | 0.0424 |
| 37 | 9.17 | | 0.9412 | | 0.1012 |
| 38 | | 9.33 | | 0.8800 | 0.0612 |
| 39 | | 10.33 | | 0.9200 | 0.0212 |
| 40 | | 11.25 | | 0.9600 | 0.0188 |
| 41 | 11.50 | | 1.0000 | | 0.0400 |
| 42 | | 12.50 | | 1.0000 | 0.0000 |

*Assuming death in group 1 first.
†Assuming death in group 2 first.

course, the observed value is not significant. When the alternative is $H_1: F_1(x) > F_2(x)$, the statistic $D^+_{N_1,N_2} = [F_1^o(x) - F_2^o(x)]$ is used. The critical value for significance level $\alpha$, will now correspond to $y_{1-2\alpha}$.

### 8.4.2 Two Sample Wilcoxon Test for Complete Data

If we are interested in comparing *average* lifetimes—that is, judging whether the $t'_{1j}$'s appear to be greater, or less, than the $t'_{2j}$'s—it is natural to consider using the *Wilcoxon test* based on ranks of ordered $t'_{gj}$'s. To apply this test:

1. Arrange all the $(N_1 + N_2)$ recorded times to failure in ascending order;
2. Assign rank 1 to the least, rank 2 to the next least, and so on until rank $(N_1 + N_2)$ is assigned to the greatest recorded value.
3. For one of the populations [conventionally the one—$\Pi_1$, say—with the smaller sample size $(N_1 \leqslant N_2)$], we add up the ranks for all individuals from that population.

Call the total $W_1$, say, the total score. If the SDF's in the two populations are *identical*, the $N_1$ ranks for individuals from $\Pi_1$ are equally likely to be any one of the $\binom{N_1+N_2}{N_1}$ possible sets of integers chosen from $1, 2, \ldots, (N_1 + N_2)$, and so the distribution of $W_1$, can be evaluated straightforwardly. Tables [e.g., Table 22 of Pearson and Hartley (1952)] are available giving lower percentage points of this null distribution of $W_1$ for $N_1 \leqslant N_2 = 2(1)25$.

Upper percentage points can be derived using the fact that the null distribution of $[N_1(N_1 + N_2 + 1) - W_1]$ is the same as that of $W_1$.

The moments of the distribution of $W_1$ are

$$\mathcal{E}(W_1) = \tfrac{1}{2}N_1(N_1 + N_2 + 1), \tag{8.9a}$$

$$\operatorname{var}(W_1) = \tfrac{1}{12}N_1 N_2(N_1 + N_2 + 1). \tag{8.9b}$$

For larger samples, the statistic

$$Z = \frac{W_1 - \mathcal{E}(W_1)}{\sqrt{\operatorname{var}(W_1)}} \tag{8.10}$$

is approximately distributed $N(0,1)$ and can be used as test statistic.

If lifetimes in $\Pi_1$ tend to be longer (shorter) than those in $\Pi_2$, one would expect to get rather large (small) values of $W_1$ as judged by the null distribution.

The statistic $W_1$ can be calculated in another way. Take every possible pair of observed values, one from $\Pi_1$, and one from $\Pi_2$, $N_1N_2$ pairs in all. Assign a score of 1 (or $-1$) when the value from $\Pi_1$ (or $\Pi_2$) is the greater. Then

$$W_1 = \tfrac{1}{2}\left[U_1 + N_1(N_1 + N_2 + 1)\right], \tag{8.11}$$

where $U_1$ is the sum of scores.

It should be clearly understood that the null distribution of $W_1$ corresponds to the assumption that $S_1(t)$ and $S_2(t)$ are *identical*, not just that the mean (expected) lifetime or median lifetime is the same in the two populations. It is true that the $W_1$ statistic is aimed at detecting differences in *means* of the SDF's, but differences in *shape* might counteract the effect of differences in means to some extent. On the other hand, differences in shape of SDF's may lead to misleading evidence of differences in expectations of life.

When there are *ties*, the formulae for mean and variance of $W_1$ [given in (8.9a) and (8.9b), respectively], no longer apply. There are formulae correcting for ties, but if these make much difference it is likely that inferences will be of dubious value, anyway.

Alternative approaches might be as follows: (1) Assign a rank to any $\Pi_1$ individual equal to the *average* of the ranks shared by all individuals with the same failure time (see Example 8.4 below). (2) Assume that for any two tied values the failure time from $\Pi_1$ ($\Pi_2$) comes *first*. Calculate $W_1$ for each extreme case, and observe if there is any serious difference. If there is serious disagreement, this should be regarded as letting us know that the recording technique is not adequately sensitive.

**Example 8.4** We return to Table 8.2. The value of $W_1$ can be calculated by adding the entries in column (1) ('ranks') corresponding to entries in column (2) (individuals from $\Pi_1$). We calculate

$$W_1 = 2 + 4 + 5\tfrac{1}{2} + 11\tfrac{1}{2} + \cdots + 41 = 370.$$

From tables [Pearson and Hartley (1972)], with $N_1 = 17, N_2 = 25$, we find that the lower 10% point of $W_1$ is 314, and the upper 10% point of $W_1$ is therefore $17(17 + 25 + 1) - 314 = 418$. Since $314 < 370 < 418$, the observed value of $W_1$ is *not significant*.

Using the normal approximation, we obtain

$$\mathcal{E}(W_1)=\tfrac{1}{2}\cdot 17\cdot(17+25+1)=365.5;$$

$$\mathrm{SD}(W_1)=\sqrt{\mathrm{var}(W_1)}=\sqrt{\tfrac{1}{12}\cdot 17\cdot 25(17+25+1)}=39.0.$$

Hence

$$z=\frac{370-365.5}{39.0}=0.115.$$

Again this is *not a significantly* large deviation from zero.

### 8.4.3 Modified Wilcoxon Tests for Incomplete Data

***Limits for $W_1$.*** When data are *incomplete* neither the Kolmogorov-Smirnov nor the Wilcoxon test presented above, can be applied. It is, of course, possible to obtain *limits* between which the value of $W_1$ must necessarily lie, if complete data were available. If very nearly the same conclusion would be reached on the basis of any observed value of $W_1$ over the range so obtained then we can, in effect, use the Wilcoxon test as if the data *were* complete.

**Example 8.5** For illustrative purposes, we use the following data:

Failure times (in hours)

| | | | | | | | | | | | |
|---|---|---|---|---|---|---|---|---|---|---|---|
| from $\Pi_1$: | 2, | 3, | 5, | 5, | 5*, | 7, | 8*, | 8*, | 10 | | $(N_1=9)$ |
| from $\Pi_2$: | 2*, | 4, | 4, | 7, | 7*, | 8, | 9, | 9, | 10*, | 11 | $(N_2=10)$ |

The recorded value 5*, for example, means "still alive after 5 hours, when observation ceased."

The *upper limit* is obtained by assigning greatest possible ranks to the individuals from $\Pi_1$. This is achieved by assuming all unobserved failure times in $\Pi_1$ to be as large as necessary, and all in $\Pi_2$ as small as possible. This is done by the following procedure:

1. in $\Pi_1$ data, regard 5*, 8*, 8* as if each is greater than 11;
2. in $\Pi_2$ data, regard 2* as also less than 3,
   7* as also less than 6,
   10* as also less than 11.

The resulting ranking is set out below:

| Rank: | 1 | 2 | 3 | 4 | 5 | 6 | 7 | 8 | 9 | 10 | 11 | 12 | 13 | 14 | 15 | 16 | 17 | 18 | 19 |
|---|---|---|---|---|---|---|---|---|---|---|---|---|---|---|---|---|---|---|---|
| from $\Pi_1$: | 2 | | 3 | | | 5 | 5 | 7 | | | | | | 10 | | | 5* | 8* | 8* |
| from $\Pi_2$: | | 2* | | 4 | 4 | | | | 7 | 7* | 8 | 9 | 9 | | 10* | 11 | | | |

This leads to

$$W_1^{\text{upper}} = 1+3+6+7+8\tfrac{1}{2}+14+17+18+19 = 93\tfrac{1}{2}.$$

The tie (7,7) is resolved by averaging—taking each rank as $8\frac{1}{2}$.

Similarly, applying the same procedure on data from $\Pi_2$, we obtain the *lower limit* for $W_1$. The results are

| Rank: | 1 | 2 | 3 | 4 | 5 | 6 | 7 | 8 | 9 | 10 | 11 | 12 | 13 | 14 | 15 | 16 | 17 | 18 | 19 |
|---|---|---|---|---|---|---|---|---|---|---|---|---|---|---|---|---|---|---|---|
| from $\Pi_1$: | 2 | 3 | | | 5 | 5 | 5* | 7 | | | 8* | 8* | | | 10 | | | | |
| from $\Pi_2$: | | | 4 | 4 | | | | | 7 | 8 | | | 9 | 9 | | 11 | 2* | 7* | 10* |

This leads to

$$W_1^{\text{lower}} = 1+2+5+6+7+8\tfrac{1}{2}+11+12+15 = 67\tfrac{1}{2}.$$

On the null hypothesis (of identical failure time distributions) and neglecting ties, the expected value of $W_1$ is $\frac{1}{2}\cdot 9\cdot(9+10+1) = 90$, and the standard deviation is $[\frac{1}{12}\cdot 9\cdot 10\cdot(9+10+1)]^{\frac{1}{2}} = 12.25$ approx. Although the $W_1^{\text{lower}}$ value deviates from the expected value by nearly two standard deviations, this would not be regarded as evidence of difference in the two failure time distributions. It must be remembered that this is the *lowest possible* value of $W_1$ consistent with the observations.

***Gilbert and Gehan's test.*** If the procedure described above does not give satisfactory results, one may try to guess a value for the Wilcoxon statistic which would have been obtained, had the data been complete.

The approach leading to (8.11) can be extended straightforwardly, by assigning the score 0 when it is impossible to tell whether the score should be $+1$ or $-1$ [Gilbert (1962), Gehan (1965)].

Let $t_{gj}^o$ denote the recorded (observed) time for $j$th individual in $g$th group. Let

$$t_{gj}^o = \begin{cases} t_{gj} & \text{if } j\text{th individual is observed to die;} \\ t_{gj}^* & \text{if } j\text{th individual is censored.} \end{cases} \tag{8.12}$$

In terms of observed time to failure, scoring is according to the scheme set out in Table 8.3.

**Table 8.3** Scores for data in Example 8.5

| Data | Score: 1 if | Score: 0 if | Score: −1 if |
|---|---|---|---|
| $t_{1j}, t_{2k}$ | $t_{1j} > t_{2k}$ | $t_{1j} = t_{2k}$ | $t_{1j} < t_{2k}$ |
| $t_{1j} > t^*_{1j}, t_{2k}$ | $t^*_{1j} > t_{2k}$ | $t^*_{1j} \leqslant t_{2k}$ | — |
| $t_{1j}, t_{2k} > t^*_{2k}$ | — | $t_{1j} \geqslant t^*_{2k}$ | $t_{1j} < t^*_{2k}$ |
| $t_{1j} > t^*_{1j}, t_{2k} > t^*_{2k}$ | — | always | — |

**Example 8.6** To illustrate the scoring procedure, we use the data of Example 8.5. The scores for the $9 \times 10 = 90$ possible pairs are set out in Table 8.4. (It is not ordinarily necessary to set out a table of this kind. It is given here merely to aid the explanation.)

**Table 8.4** Scores for data in Example 8.5

| $t^o_{1j}$ values from $\Pi_1$ | $t^o_{2j}$ values from $\Pi_2$: 2* | 4 (twice) | 7 | 7* | 8 | 9 (twice) | 10* | 11 | Total |
|---|---|---|---|---|---|---|---|---|---|
| 2 | −1 | −1 | −1 | −1 | −1 | −1 | −1 | −1 | −10 |
| 3 | 0 | −1 | −1 | −1 | −1 | −1 | −1 | −1 | −9 |
| 5 | 0 | +1 | −1 | −1 | −1 | −1 | −1 | −1 | −5 |
| 5* (twice) | 0 | +1 | 0 | 0 | 0 | 0 | 0 | 0 | +2 |
| 7 | 0 | +1 | 0 | −1 | −1 | −1 | −1 | −1 | −4 |
| 8* (twice) | 0 | +1 | +1 | 0 | +1 | 0 | 0 | 0 | +4 |
| 10 | 0 | +1 | +1 | 0 | +1 | +1 | 0 | −1 | +5 |
| Total | −1 | 5 | 0 | −4 | −1 | −3 | −4 | −5 | −11 |

We find from (8.11), $W_1 = \frac{1}{2}(-11 + 9 \cdot 20) = 84.5$.

***Assessing Significance of $W_1$ from Randomization Distribution.*** Clearly, the values of mean and standard deviation appropriate to $W_1$ calculated from complete data cannot be used here. When it is reasonable to suppose that on the null hypothesis, the $(N_1 + N_2)$ recorded times to failure, $t^o_{gj}$, can be regarded as a random allocation of $N_1$ values to $\Pi_1$, and the remaining $N_2$ values to $\Pi_2$, we can use the randomization distribution so generated in assessing the significance of the calculated $W_1$ value. The distribution has $\binom{N_1 + N_2}{N_1}$ equally likely (though not necessarily all distinct) values, each corresponding to a different allocation.

For this randomization distribution, Mantel (1967) has shown that we still have

$$\mathcal{E}(W_1)=\tfrac{1}{2}N_1(N_1+N_2+1),$$

[cf. (8.9a)].

For our data from Example 8.5, $\mathcal{E}(W_1)=\frac{1}{2}\cdot 9\cdot 20=90$. However, the variance is calculated from a formula different from that in (8.9b).

Let $G_r$ denote the sum of scores for the $r$th observed value $[r=1,2,\ldots,(N_1+N_2)]$, with respect to all $(N_1+N_2-1)$ other observed values. We call $G_r$ the *total score* for the $r$th individual (observation).

In our Example 8.5, we have $N_1+N_2=19$. The value 2 is the smallest, so naturally yields a score of $-1$ with respect to each of the remaining 18 observations. Therefore, the total score for this observation is $-18$. On the other hand, observation 2* is given score $+1$ with observation 2, and 0 with all remaining. Hence, the total score for this observation is $+1$, and so on. The total scores for Example 8.6 are shown in Table 8.5.

**Table 8.5** Total scores $G_r$

| Rank | Obs. value | Total score | Rank | Obs. value | Total score | Rank | Obs. value | Total score |
|---|---|---|---|---|---|---|---|---|
| 1 | 2 | −18 | 7 | 5* | +5 | 13 | 8* | +8 |
| 2 | 2* | +1 | 8 | 5* | +5 | 14 | 8* | +8 |
| 3 | 3 | −15 | 9 | 7 | −4 | 15 | 9 | +5 |
| 4 | 4 | −12 | 10 | 7 | −4 | 16 | 9 | +5 |
| 5 | 4 | −12 | 11 | 7* | +7 | 17 | 10 | +8 |
| 6 | 5 | −9 | 12 | 8 | 0 | 18 | 10* | +11 |
| | | | | | | 19 | 11 | +11 |

Then the randomization variance of the (modified) $W_1$ is

$$\operatorname{var}(W_1)=\frac{1}{4}\,\frac{N_1N_2}{(N_1+N_2)(N_1+N_2-1)}\sum_{r=1}^{N_1+N_2}G_r^2. \tag{8.13}$$

For our data,

$$\operatorname{var}(W_1)=\frac{1}{4}\cdot\frac{90}{342}\left[(-18)^2+(1)^2+\cdots+(11)^2\right]=100.92,$$

so that $\mathrm{SD}(W_1)=10.05$. A rough test may be based on the assumption that the null hypothesis distribution of $W_1$ is normal. In our case

$$z = \frac{84.5-90}{10.05} = -0.55,$$

giving a nonsignificant result.

The assumptions on which the randomization distribution is based are not always valid. The null hypothesis used here implies that withdrawals (as well as failures) will occur at the same times regardless of the population to which individuals are assigned. There is usually no way of checking whether this is so. Estimated distributions of *withdrawal* times in the populations, obtained by applying the product limit method, with failures and withdrawals interchanged, may be compared with a view to checking whether there are gross discrepancies among the distributions. In the absence of such indications one would feel more confident in using the randomization distribution approach.

## 8.5 SPECIAL PROBLEMS ARISING IN CLINICAL TRIALS AND PROGRESSIVE LIFE TESTING

One of the purposes of clinical trials is assessment of the effects of various treatments (e.g., operations, drugs) on mortality (or other phenomena under consideration). Many difficulties arise in reaching a decision. As we have repeatedly emphasized, there is no universal statistical test that on its own, suffices for a final decision whether a certain treatment 'is' beneficial or harmful. A few tests are presented in the following sections, which might sometimes be of help in clarifying difficulties associated with the nature of clinical trials.

### 8.5.1 Early Decision

In most clinical trials, the sample size for each treatment is fixed in advance. If, however, it becomes apparent in the course of the trial that one treatment is much better or much worse than any others, ethical considerations require that the trials should not be further continued without modification. In fact, in planning long-term clinical trials, it is almost always required to include rules specifying at which stage(s) of the trial the data should be examined. Broadly speaking, these rules are of two types:

1. Examine the data at certain fixed time point(s) (*truncation*);
2. Examine the data after specified numbers (or proportion) of deaths have occurred (*censoring*).

A test based on the modified Kolmogrov-Smirnov statistic (also referred to as Tsao-Conover test) is sometimes useful in early decision problems. This test is presented in Section 8.6.

### 8.5.2 Multistage Testing

Beneficial effects of a certain drug in the early states of a trial may be followed by harmful effects in a later stage. *Multistage* testing is desirable to help assess preference for one treatment over the other. Of course, if information from significance tests at earlier and later stages is used, the earlier and later tests are *not* independent. It is, indeed, not easy to obtain a precise significance level in these situations.

One possibility (although we do not recommend it highly) is to assess the nominal significance level *as if* the tests were *independent*, as shown below.

***Probability of at Least One Significant Result.*** Suppose that we apply $k$ *independent* tests of the same hypothesis $H_0$. Let

$$\alpha^* = \Pr\{\text{reject } H_0 | H_0 \text{ true}\}$$

be the *nominal* significance level for each test. Then the probability of at least one significant result in $k$ independent tests is

$$1 - \Pr\{\text{none significant}\} = 1 - (1 - \alpha^*)^k. \tag{8.14}$$

Let $\alpha$ be the *overall* significance level, in which we, in fact, are interested, such as, for example, $\alpha = 0.05$. Then we have

$$\alpha = 1 - (1 - \alpha^*)^k, \tag{8.15}$$

or solving for $\alpha^*$,

$$\alpha^* = 1 - (1 - \alpha)^{1/k}. \tag{8.15a}$$

For small $\alpha$ and $k$, (8.15a) can be approximated by $\alpha/k$. For example, for $k = 10$ and $\alpha = 0.05$, we obtain $\alpha^* = 0.00512$; this means that each of the 10 tests should be set up at the nominal level $\alpha^* \doteqdot 0.005$. An interesting discussion of this matter has been given by Tukey (1977).

***Repeated Experiments.*** Often the same type of experiment is carried out in different clinics, hospitals, countries, or even repeated several times in the same institution. Here the assumption that the experiments are *independent* is less dubious.

For a given significance level $\alpha$, some experiments might give significant results, others not. How might such results be combined to reach general conclusions?

Let $T_i$ be a test criterion for $H_0$, and $T_i^o$ an actually observed value of $T_i$, in the $i$th experiment. We calculate the *observed* significance level, $\alpha_i^o$, say, (often denoted by $P$ and called the "$P$-value"), that is

$$\Pr\{T_i > T_i^o | H_0\} = \alpha_i^o. \tag{8.16}$$

If $H_0$ is valid, the statistic

$$U = -2 \sum_{i=1}^{k} \log \alpha_i^o \tag{8.17}$$

is distributed as Chi-square with $2k$ degrees of freedom. Large values of $U$ are significant.

Of course, the adjustment of the significance level discussed in the previous section is also applicable here.

### 8.5.3 Testing for Trends in Mortality Patterns

In long-term experiments our main concern is whether the differences between two treatments have the same direction in mortality patterns—are they consistent over a longer period? An application of a series of (conditional) Pearson Chi-square tests (see Section 8.7), when data are grouped in fixed intervals, may be useful in such a situation. Also the Mantel-Haenszel and the log rank tests (see Section 8.8) are of this nature.

### 8.5.4 Staggered Entries and Withdrawals

In longitudinal studies and most clinical trials, patients often cannot all enter at the same time nor can they be followed until the death of the last member of the group. We then have so called "planned" withdrawal as well as "lost" patients (see Chapter 6). Most of the tests discussed in this chapter are primarily devised for cohort (complete) data. Therefore some adjustments need to be made.

For *ungrouped* data, it may be possible to calculate *limits* between which the value of the test statistic based on (unavailable) complete data must lie. One extreme would result from assessing that all individuals leaving before the end of the trial survived until, at least, the termination date; the other extreme might result from not counting these individuals at all. If the bounds are not too far apart, useful conclusions may still be drawn.

For *grouped* data, we usually replace the number of survivors at the beginning of the interval $[t_i, t_{i+1})$ in the $g$th group, $N_{gi}$, by the effective number of initial exposed to risk, $N'_{gi}$ (cf. Section 6.2.2, and also see Examples 8.10 and 8.11). There is a good discussion of the uses of significance tests in the analysis of data from clinical trials in Biometrics Seminar (1966). Meier (1975) also discusses this matter, together with the important topic of randomization.

## 8.6 CENSORED KOLMOGOROV-SMIRNOV (OR TSAO-CONOVER) TEST

***Exact Test.*** Let $x'_{11} < x'_{12} < \cdots < x'_{1N_1}$ and $y'_{21} < y'_{22} < \cdots < y'_{2N_2}$ be ordered times at death in two groups. Suppose that we want to test the hypothesis $H_0: F_1(x) = F_2(x)$ against the alternative $H_1: F_1(x) \neq F_2(x)$, and wish to construct tests based on the following three kinds of censoring. Observation is continued until:

1. *Exactly r deaths are observed in either group*, whichever comes first. This means, we continue observation until $\min(x'_r, y'_r)$ is reached. The statistic [Conover (1967)]

$$d_{N_1,N_2}(r) = \max_{x < \min(x'_r, y'_r)} |F_1^o(x) - F_2^o(x)| \tag{8.18a}$$

can be used for this purpose.

2. *At least r deaths are observed in both groups.* This means that the observation is continued until $\max(x'_r, y'_r)$ is reached. The suggested statistic [Tsao (1953), Conover (1967)] is

$$d'_{N_1,N_2}(r) = \max_{x \leqslant \max(x'_r, y'_r)} |F_1^o(x) - F_2^o(x)|. \tag{8.18b}$$

3. *Exactly r deaths are observed in the first group*, that is, we observe $x'_r$. The proposed statistic [Tsao (1953), Conover (1967)] is

$$d''_{N_1,N_2}(r) = \max_{x \leqslant x'_r} |F_1^o(x) - F_2^o(x)|. \tag{8.18c}$$

(Note that our notation $d$, $d'$, $d''$ corresponds to Tsao-Conover notation $d''$, $d'$, $d$, respectively.)

For $N_1 = N_2 = N = 3(1)10(5)20(10)40$, Tsao (1953) gives tables of $\Pr\{Nd''_{N,N}(r) \leqslant c\}$ and $\Pr\{d'_{N,N}(r) \leqslant c\}$, where $c$ is a positive integer. As

**Table 8.6** Values of $\Pr\{Nd_{N,N}(r) \geqslant c\}$

| | | $c$ | | | | | | | | |
|---|---|---|---|---|---|---|---|---|---|---|
| $N$ | $r$ | 2 | 3 | 4 | 5 | 6 | 7 | 8 | 9 | 10 |
| 20 | 2 | 0.48717 | | | | | | | | |
| | 3 | 0.73666 | 0.23076 | | | | | | | |
| | 4 | 0.86457 | 0.40748 | 0.10602 | | | | | | |
| | 5 | 0.93023 | 0.54292 | 0.21369 | 0.04712 | | | | | |
| | 6 | 0.96399 | 0.64680 | 0.30950 | 0.10652 | 0.02020 | | | | |
| | 7 | 0.98137 | 0.72655 | 0.39291 | 0.16637 | 0.05044 | 0.00832 | | | |
| | 8 | 0.99034 | 0.78784 | 0.46529 | 0.22285 | 0.08432 | 0.02265 | 0.00328 | | |
| | 9 | 0.99498 | 0.83499 | 0.52807 | 0.27487 | 0.11859 | 0.04022 | 0.00962 | 0.00123 | |
| | 10 | 0.99738 | 0.87129 | 0.58247 | 0.32229 | 0.15165 | 0.05907 | 0.01799 | 0.00384 | 0.00044 |
| 30 | 2 | 0.49152 | | | | | | | | |
| | 3 | 0.74130 | 0.23728 | | | | | | | |
| | 4 | 0.86830 | 0.41793 | 0.11239 | | | | | | |
| | 5 | 0.93291 | 0.55551 | 0.22626 | 0.05218 | | | | | |
| | 6 | 0.96580 | 0.66034 | 0.32721 | 0.11808 | 0.02372 | | | | |
| | 7 | 0.98255 | 0.74026 | 0.41491 | 0.18464 | 0.05953 | 0.01054 | | | |
| | 8 | 0.99109 | 0.80122 | 0.49086 | 0.24767 | 0.10005 | 0.02902 | 0.00457 | | |
| | 9 | 0.99545 | 0.84774 | 0.55662 | 0.30605 | 0.14156 | 0.05213 | 0.01368 | 0.00194 | |
| | 10 | 0.99767 | 0.88327 | 0.61356 | 0.35965 | 0.18225 | 0.07756 | 0.02614 | 0.00624 | 0.00080 |
| 40 | 2 | 0.49367 | | | | | | | | |
| | 3 | 0.74355 | 0.24050 | | | | | | | |
| | 4 | 0.87006 | 0.42298 | 0.11556 | | | | | | |
| | 5 | 0.93414 | 0.56147 | 0.23235 | 0.05474 | | | | | |
| | 6 | 0.96661 | 0.66659 | 0.33564 | 0.12379 | 0.02555 | | | | |
| | 7 | 0.98306 | 0.74642 | 0.42513 | 0.19346 | 0.06414 | 0.01174 | | | |
| | 8 | 0.99140 | 0.80706 | 0.50244 | 0.25939 | 0.10786 | 0.03237 | 0.00531 | | |
| | 9 | 0.99564 | 0.85314 | 0.56918 | 0.32039 | 0.15261 | 0.05828 | 0.01593 | 0.00236 | |
| | 10 | 0.99778 | 0.88816 | 0.62681 | 0.37637 | 0.19672 | 0.08690 | 0.03057 | 0.00765 | 0.00103 |

Conover (1967) pointed out, there are simple relations among these probabilities. For example,

$$\Pr\{Nd_{N,N}(r) \leqslant c\} = \begin{cases} \Pr\{Nd'_{N,N}(r-c) \leqslant c\} & \text{for } c < r, \\ 1 & \text{for } c \geqslant r. \end{cases} \tag{8.19}$$

Using (8.19), we have evaluated the values of $\Pr\{Nd_{N,N}(r) \geqslant c\}$ given in Table 8.6.

It appears that the question posed in (1) is more frequently asked than (2) or (3), so that it is more useful to tabulate these probabilities.

**Example 8.7** Two lots of electric bulbs, each lot of a different brand and containing $N=30$ items, were put on a life test. It was decided to try to reach a conclusion as soon as $r=10$ bulbs fail in either of the two lots. Suppose that the following results (based only on the *order* in which failures occur) are observed.

| Ordering: | $x'_1$ | $<x'_2$ | $<x'_3$ | $<y'_1$ | $<y'_2$ | $<x'_4$ | $<x'_5$ | $<x'_6$ | $<x'_7$ | $<y'_3$ | $<x'_8$ | $<x'_9$ | $<y'_4$ | $<y'_5$ | $<x'_{10}$ |
|---|---|---|---|---|---|---|---|---|---|---|---|---|---|---|---|
| Ranks $NF_1^o(t)$ | 1 | 2 | 3 | 3 | 3 | 4 | 5 | 6 | 7 | 7 | 8 | 9 | 9 | 9 | 10 |
| $NF_2^o(t)$ | 0 | 0 | 0 | 1 | 2 | 2 | 2 | 2 | 3 | 3 | 3 | 3 | 4 | 5 | 5 |
| $N\|F_1^o(t)-F_2^o(t)\|$ | 1 | 2 | 3 | 2 | 1 | 2 | 3 | 4 | 5 | 4 | 5 | 6 | 5 | 4 | 5 |

Here we have $N=30$. We obtain $30d_{30,30}(10) = 30|F_1^o(x'_{10}) - F_2^o(x'_{10})| = 10-4=6$. From Table 8.6, we find for $N=30, r=10$

$$\Pr\{30d_{30,30}(10) \geqslant 6\} = 0.18225.$$

This test provides insufficient evidence to prefer one brand to the other, if the $\alpha=0.05$ level of significance is being used.

We note that the maximum possible value of $N|F_1^o(x) - F_2^o(x)|$ over the completed experiment would be 25 (if all the remaining 20 lamps in the first group failed before any in the second group), and the minimum possible value would be 20.

***Limiting distribution of $d_{N_1,N_2}(r)$.*** The limiting distribution, as $N_1, N_2 \to \infty$ with $r/N_g = \phi_g$, $g=1,2$ being finite, of the statistic

$$\sqrt{\frac{N_1N_2}{N_1+N_2}}\, d_{N_1,N_2}(r) \tag{8.20}$$

is the same as the limiting distribution of $\sqrt{N}\, D_N(\phi)$ given in Section 7.12.2 (2) with $\phi=\min(\phi_1,\phi_2)$. Moreover, the statistics

$$\sqrt{\frac{N_1N_2}{N_1+N_2}}\; d'_{N_1,N_2}(r) \qquad \text{and} \qquad \sqrt{\frac{N_1N_2}{N_1+N_2}}\; d''_{N_1,N_2}(r)$$

have also the same limiting distribution, with $\phi=\max(\phi_1,\phi_2)$ and $\phi=\phi_1$, respectively.

**Example 8.8** Suppose that we have the same situation as in Example 8.7, except that the lot sizes are $N_1=200, N_2=250$. Then we have $\phi=\min\left(\frac{10}{200},\frac{10}{250}\right)=\frac{10}{250}=0.04$. The observed value of (8.18) is

$$\sqrt{\frac{200\cdot 250}{450}}\times\frac{60}{250}=0.253.$$

From Table 7.6 with $\phi=0.04$ and $\alpha=0.10$, we obtain $y_{0.90}(0.04)=0.3867$. The observed value (0.253) is not significant, even at the $\alpha=0.10$ level.

***Early decision.*** For *any* procedure (censored or not) early decision is possible as soon as the final outcome of the test to be used is determined and *cannot be changed* by further data from the experiment, as planned.

**Example 8.9** Suppose $N_1=N_2=N=250$, and we plan to stop the experiment as soon as there are 25 deaths in either of the two groups (whichever comes first). Thus $\phi=25/250=0.10$. Suppose that we wish to test the hypothesis $H_0: F_1(x)=F_2(x)$ against the alternative $H_1: F_1(x)\neq F_2(x)$ using the significance level $\alpha=0.05$. Using Table 7.6, we calculate the upper 5% point of the null hypothesis distribution

$$y_{0.95}(0.10)\sqrt{\frac{N_1+N_2}{N_1N_2}}=0.6825\sqrt{\frac{1}{125}}=0.0610.$$

If the observed value of $D_{N_1,N_2}(0.10)$ exceeds 0.0610 the null hypothesis will be rejected. Since here $N_1=N_2=N$, we can also express this in terms of the difference between numbers of deaths in two groups. We have $250\cdot 0.0610=15.25\approx 16$. Therefore, as soon as the number of deaths in the two groups differs by 16 or more the experiment can be stopped and an *early decision to reject $H_0$ can be reached.* With some adjustment of the significance level $\alpha$ (as discussed in Section 8.5.2), early decision can also be reached at some further stages.

The experiment could also be stopped, and a *nonsignificant* verdict recorded, as soon as there are at least 10 deaths (note that $(0.10–0.0610)\cdot 250 = 9.75 \approx 10$, in *each* group, without there having been a difference of 16 or more between the numbers of deaths at any time, because the maximum future difference arising subsequently in the experiment as planned would be $|25-10| = 15 < 16$.

The method just described is well-known; in quality control work it is called *curtailed sampling*.

## 8.7 TRUNCATED DATA. PEARSON'S CONDITIONAL $X^2$ TEST

### 8.7.1 Two Sample Problem

As in Section 8.6, we first consider two groups of patients, each subjected to a different treatment. We suppose that the planned period of observation is divided into $m$ *fixed* intervals $[t_0 = 0, t_1)$, $[t_1, t_2), \ldots, [t_{m-1}, t_m)$, and that a statistical test is based on the data available at the *end* of each interval.

For example, in long-term clinical trials, the data may be analyzed every 3, 6, or 12 months; or there might be an initial analysis after 12 months, followed by one every 6 months subsequently. Many variations are possible, but analysis always takes place at the *end* of an interval.

We first assume, also, that *if* there are any losses, withdrawals or new entries, (if any) occur at the beginning of an interval, and not within any interval.

We now consider one particular interval, the $(i+1)$th one, $[t_i, t_{i+1})$. We use $N_{gi}\,(g = 1, 2)$ to denote the number of patients in group $g$ at time $t_i$ (including any new entries, but excluding losses and withdrawals), and by $d_{gi}$ the number of deaths among the $N_{gi}$ patients in $[t_i, t_{i+1})$. These data can be arranged in a $2\times 2$ table (see Table 8.7).

**Table 8.7** Survival data for interval $[t_i, t_{i+1})$

| | Died | Survived | Total |
|---|---|---|---|
| Group 1 | $d_{1i}$ | $N_{1i} - d_{1i}$ | $N_{1i}$ |
| Group 2 | $d_{2i}$ | $N_{2i} - d_{2i}$ | $N_{2i}$ |
| Total | $d_{1i} + d_{2i}$ | $(N_{1i} + N_{2i})$<br>$-(d_{1i} + d_{2i})$ | $N_{1i} + N_{2i}$ |

The observed proportion of deaths in group $j$ during the period $t_i$ to $t_{i+1}$ is

$$\hat{q}_{gi} = \frac{d_{gi}}{N_{gi}}, \qquad g = 1,2. \tag{8.21}$$

The overall proportion of deaths in the two groups is

$$\hat{q}_i = \frac{d_{1i} + d_{2i}}{N_{1i} + N_{2i}} = \frac{N_{1i}\hat{q}_{1i} + N_{2i}\hat{q}_{2i}}{N_{1i} + N_{2i}}. \tag{8.22}$$

*Conditionally* on $N_{gi}$, the number of deaths, $d_{gi}$, has a binomial distribution (see Section 5.7). If the mortality probabilities $q_{1i}$, $q_{2i}$ in groups 1,2, respectively, over the interval $[t_i, t_{i+1})$ are the same ($q_{1i} = q_{2i} = q_i$) under the null hypothesis, then the estimated expected number of deaths in group $g$ is

$$E_{gi} = N_{gi}\hat{q}_i. \tag{8.23}$$

It follows that the expected number of survivors is

$$N_{gi} - E_{gi} = N_{gi}(1 - \hat{q}_i) \tag{8.23a}$$

Note that $(N_{gi} - d_{gi}) - (N_{gi} - E_{gi}) = -(d_{gi} - E_{gi})$.

It is natural to use the well-known Pearson's Chi-square statistic with 1 degree of freedom,

$$X_i^2(1) = \sum_{g=1}^{2}\left(\frac{1}{E_{gi}} + \frac{1}{N_{gi} - E_{gi}}\right)(d_{gi} - E_{gi})^2 = \frac{1}{1 - \hat{q}_i}\sum_{g=1}^{2}\frac{(d_{gi} - E_{gi})^2}{E_{gi}}, \tag{8.24}$$

to test for difference between $q_{1i}$ and $q_{2i}$.

If $q_{1i} = q_{2i}$, then $X_i^2(1)$ is approximately distributed as Chi-square with 1 degree of freedom. Of course, if for a given significance level $\alpha$, $X_i^2(1) > \chi_{1-\alpha}^2(1)$, the difference $|\hat{q}_{1i} - \hat{q}_{2i}|$ is regarded as significant. One can construct a sequence of such statistics, $X_0^2(1), X_1^2(1), \ldots, X_{m-1}^2(1)$, one for each interval $[t_0, t_1), [t_1, t_2), \ldots, [t_{m-1}, t_m]$, respectively.

*Conditional* on the set of $N_{gi}$'s ($i = 0, 1, \ldots, m-1$) and supposing the null hypothesis $H_0: q_{1i} = q_{2i}$ for all $i$ to be valid, it is reasonable to regard the

$X_i^2(1)$ as mutually *independent*, so that

$$X^2(m) = \sum_{i=0}^{m-1} X_i^2(1) \tag{8.25}$$

is approximately distributed as Chi-square with $m$ degrees of freedom. Using a specified (approximate) significance level $\alpha$, we can base a test on the rule of rejecting the null hypothesis if $X^2(m)$ exceeds the upper $100\alpha\%$ point of the distribution of Chi-square with $m$ degrees of freedom.

***Progressive Censoring.*** When (as usually is the case), losses and planned withdrawals occur *within* intervals, we may make use of the concept of effective *number of exposed to risk* (see Section 6.2.2). Although the $d_{gi}$'s can no longer be regarded as binomial variables, in practice, we usually regard the $q_{gi}$'s as binomial proportions with parameters, $N'_{gi}$, $q_{gi}$, where $N'_{gi}$ is the effective number of initial exposed to risk at $t_i$ in the $g$th group.

We calculate the $X_i^2(1)$ using formula (8.24), and $X^2(m)$, using formula (8.25) but with $N_{gi}$ replaced by $N'_{gi}$ (see Example 8.10).

### 8.7.2 Extension to $k$ Treatments

This is straightforward. Instead of the $2\times2$ table (Table 8.7), we will have a $k\times2$ table, one row for each treatment. The statistic

$$X_i^2(k-1) = \frac{1}{1-\hat{q}_i}\sum_{j=1}^{k}\frac{(d_{gi}-E_{gi})^2}{E_{gi}}, \tag{8.26}$$

where, in an obvious notation

$$\hat{q}_i = \frac{d_{1i}+d_{2i}+\cdots+d_{ki}}{N_{1i}+N_{2i}+\cdots+N_{ki}}, \qquad E_{gi} = N_{gi}\hat{q}_i \tag{8.27}$$

is used for testing the null hypothesis $q_{1i} = q_{2i} = \cdots = q_{ki}$. Under the null hypothesis, (8.26) is distributed as Chi-square with $k-1$ degrees of freedom.

**Example 8.10** Table 8.8 represents survival data in four groups of adult-onset diabetic patients under various drug treatments. The first group

**Table 8.8** Survival data in four treatment groups of diabetic patients in the UGDP-study

| Year of follow-up | $i$ | Placebo ($g=1$) | | | Tolbutamide ($g=2$) | | | Insulin standard ($g=3$) | | | Insulin variable ($g=4$) | | |
|---|---|---|---|---|---|---|---|---|---|---|---|---|---|
| | | $N_{1i}$ | $w_{1i}$ | $d_{1i}$ | $N_{2i}$ | $w_{2i}$ | $d_{2i}$ | $N_{3i}$ | $w_{3i}$ | $d_{3i}$ | $N_{4i}$ | $w_{4i}$ | $d_{4i}$ |
| 1 | 0 | 205 | — | — | 204 | — | — | 210 | — | 2 | 204 | — | 4 |
| 2 | 1 | 205 | — | 5 | 204 | — | 5 | 208 | — | 2 | 200 | — | 3 |
| 3 | 2 | 200 | — | 4 | 199 | — | 5 | 206 | — | 2 | 197 | — | 3 |
| 4 | 3 | 196 | 4 | 4 | 194 | 5 | 5 | 204 | 2 | 2 | 194 | 5 | 1 |
| 5 | 4 | 188 | 23 | 4 | 184 | 24 | 5 | 198 | 25 | 4 | 188 | 21 | — |
| 6 | 5 | 161 | 43 | 3 | 155 | 41 | 4 | 171 | 48 | 2 | 167 | 38 | 4 |
| 7 | 6 | 115 | 50 | 1 | 110 | 47 | 5 | 120 | 56 | 3 | 125 | 60 | 1 |
| 8 | 7 | 64 | 36 | — | 58 | 33 | 1 | 60 | 32 | 4 | 64 | 35 | 2 |
| Total | | | | 21 | | | 30 | | | 21 | | | 18 |

*From *Diabetes* (1970) **19**, Suppl. 2, 218.

(placebo) is a control group, whereas the remaining three are treatment groups with hypoglycemic agents (tolbutamide and two systems of applications of insulin). The data are from UGDP (1970), where also a description of the study can be found.

In this Example, we compare the mortality of tolbutamide and placebo groups at the end of each year, using conditional Pearson's $X^2$-tests. Since there were withdrawals (starting in the fourth year), the 'effective numbers of initial exposed to risk' were calculated from the (approximate) formula: $N'_{gi} = N_{gi} - \frac{1}{2} w_{gi}$, where $w_{gi}$ is the number of withdrawals in $[t_i, t_{i+1})$ (6.14).
The calculations of individual $X_i^2(1)$'s is shown in Table 8.9. [For the last (eighth) year, it was inappropriate to calculate $X_7^2(1)$ since $E_{g7} < 1$. If one does so, the value would be 1.14].

Clearly, the results for individual years as well as the overall result are nonsignificant. One can compare tolbutamide with insulin, or can compare all four groups jointly by calculating $X_i^2(3)$'s from (8.26). It appears that none of these results were significant (see Exercise 8.9). Nevertheless, the UGDP recommended that tolbutamide should be discontinued. We notice that there is a consistent (although small) effect indicating that the tolbutamide group exhibits higher mortality than in the remaining groups (total number of deaths was 30 in the tolbutamide group, whereas only 18–21 in the other groups). The decision was mainly based on this quite important observation rather than on formal (approximate) tests. Some more discussion on this will be given in Example 8.11.

## 8.8 TESTING FOR CONSISTENT DIFFERENCES IN MORTALITY. MANTEL-HAENSZEL AND LOGRANK TESTS

### 8.8.1 Mantel-Haenszel Test

The tests discussed in Section 8.7 do not take into account the *signs of differences* among the estimates $\hat{q}_{gi}$'s. We are, however, usually interested in the consistency of differences along the time scale (i.e., over different values of $i$).

Forms of Chi-square tests appropriate to this interest have been suggested by Mantel and Haenszel (1959; 1966) and by Peto and Peto (1972).

***Two Treatments.*** We again consider the situation described in Section 8.7 with the notation used in Table 8.8. Mantel and Haenszel (1959; 1966) suggested that one can consider the *conditional* distribution of $d_{1i}$ (for example) given $N_{1i}, N_{2i}$ *and* the total number of deaths $d_{1i}+d_{2i}$ as *fixed* (so that $\hat{q}_i, E_{1i}$, and $E_{2i}$ are also fixed). On the hypothesis of no mortality difference between group 1 and group 2, the *conditional* distribution of $d_{1i}$ is *hypergeometric*.

This is the basis of the so called Fisher's (1937) exact test. Of course, the assumption that $d_{1i}+d_{2i}$ is fixed is not really appropriate, in physical terms, but this does not invalidate the use of a distribution restricted by the condition $d_{1i}+d_{2i}$ constant as a device to control the overall significance level [Barnard (1947), Pearson (1947)].

The (conditional) expected value of $d_{1i}$ (under the null hypothesis) is

$$E_{1i}=N_{1i}\hat{q}_i=N_{1i}\frac{d_{1i}+d_{2i}}{N_{1i}+N_{2i}} \tag{8.28}$$

as defined in (8.23). The (conditional) variance of $d_{1i}$ (and also of $d_{1i}-E_{1i}$) is

$$\operatorname{var}(d_{1i})=V_{1i}=\frac{N_{1i}N_{2i}(d_{1i}+d_{2i})[(N_{1i}+N_{2i})-(d_{1i}+d_{2i})]}{(N_{1i}+N_{2i})^2(N_{1i}+N_{2i}-1)},$$

$$=\frac{N_{1i}N_{2i}}{N_{1i}+N_{2i}-1}\hat{q}_i(1-\hat{q}_i). \tag{8.29}$$

**Table 8.9.** Comparison of tolbutamide with placebo. Pearson's and Mantel-Haenszel Chi-square tests

| | Placebo | | | Tolbutamide | | | | Pearson's | Mantel-Haenszel Test | | |
|---|---|---|---|---|---|---|---|---|---|---|---|
| $i$ | $N'_{1i}$ | $d_{1i}$ | $E_{1i}$ | $N'_{2i}$ | $d_{2i}$ | $E_{2i}$ | $\hat{q}_i$ | $X^2$ | $\sum_h^i d_{1h} - \sum_h^i E_{1h}$ | $\sum_h^i V_{1h}$ | $X^2$ |
| 0 | 205 | — | — | 204 | — | — | — | — | | | |
| 1 | 205 | 5 | 5.01 | 204 | 5 | 4.99 | 0.02445 | 0.00 | −0.01 | 2.445 | 0.00 |
| 2 | 200 | 4 | 4.51 | 199 | 5 | 4.49 | 0.02256 | 0.12 | −0.52 | 4.650 | 0.06 |
| 3 | 194 | 4 | 4.53 | 191.5 | 5 | 4.47 | 0.02335 | 0.13 | −1.05 | 6.853 | 0.16 |
| 4 | 176.5 | 4 | 4.56 | 172 | 5 | 4.44 | 0.02582 | 0.14 | −1.61 | 9.050 | 0.29 |
| 5 | 139.5 | 3 | 3.56 | 134.5 | 4 | 3.44 | 0.02559 | 0.18 | −2.17 | 10.764 | 0.44 |
| 6 | 90 | 1 | 3.06 | 86.5 | 5 | 2.94 | 0.03399 | 2.93 | −4.23 | 12.221 | 1.46 |
| 7 | 46 | — | 0.53 | 41.5 | 1 | 0.47 | 0.01143 | — | −4.76 | 12.470 | 1.82 |
| Sum | | 21 | | | 30 | | | 3.50 | | | |

On the null hypothesis

$$X_{1i}^2 = \frac{(d_{1i} - E_{1i})^2}{V_{1i}} \tag{8.30}$$

is approximately distributed as Chi-square with 1 degree of freedom, provided neither $N_{1i}$ or $N_{2i}$ is too small.

We often would like our test to be especially sensitive to *consistent* differences between death probabilities in groups 1 and 2, that is, such that the death probabilities are greater (or less) in group 1 than in group 2 in each of the intervals $[0, t_1)$, $[t_1, t_2), \ldots, [t_{m-1}, t_m)$. Formally, the null and alternative hypotheses are specified as $H_0: q_{1i} = q_{2i}$ and $H_1: q_{1i} > q_{2i}$ (or $H_1: q_{1i} < q_{2i}$) for $i = 0, 1, \ldots, m-1$, respectively.

It is natural to use the test statistic

$$T_1 = \sum_{i=0}^{m-1} (d_{1i} - E_{1i}) = \sum_{i=1}^{m-1} d_{1i} - \sum_{i=0}^{m-1} E_{1i}. \tag{8.31}$$

Conditionally on fixed values of $N_{gi}$'s and the $(d_{1i} + d_{2i})$'s, the null hypothesis variance of $T_1$ is

$$\text{var}(T_1) = V_1 = \sum_{i=0}^{m-1} V_{1i} = \sum_{i=0}^{m-1} \frac{N_{1i} N_{2i}}{N_{1i} + N_{2i} - 1} \hat{q}_i (1 - \hat{q}_i). \tag{8.32}$$

Note that we can define

$$T_2 = \sum_{i=0}^{m-1} d_{2i} - \sum_{i=0}^{m-1} E_{2i} \tag{8.33}$$

in a similar manner, and since $T_1 + T_2 = 0$, the variance of $T_2$ is the same as that of $T_1$.

Mantel and Haenszel (1959) suggested using the statistic

$$X_1^2 = \frac{T_1^2}{V_1} \tag{8.34}$$

to test for consistent (in sign) differences in mortality probabilities. On the null hypothesis, $X_1^2$ would be approximately distributed as Chi-square with 1 degree of freedom.

*Withdrawals* can be treated similarly as in Section 8.7. We replace $N_{gi}$ by $N'_{gi}$.

Mantel and Haenszel also considered *multistage testing*, in which fresh values of $X^2_{1i}$ and $X^2_1$ are calculated from time to time, as more data become available. They also emphasized, as we have done, the importance of checking for possible differences in detail in patterns of mortality, as well as overall differences.

In this connection it is important to note that if treatment 1 is better (i.e., leads to lower mortality) in some subgroups (e.g., age or time intervals) and worse in others, the value of $T_1$ is affected by the relative amounts of data available for the different groups. If the subgroups in which treatment 1 is worse, happen to contain more data than those in which it is better, there will be an apparently worse (i.e., greater) overall value for $T_1$, and conversely. It can, therefore, be misleading to speak of consistent (or even on the whole) superiority or inferiority on the basis of the test using $X^2_1$ without taking into account the values of individual $(d_{1i}-E_{1i})$'s.

***Extension to k treatments.*** Suppose that we wish to test for consistently higher (or lower) mortality with treatment $g$ as compared with the *average* of all $k$ treatments. We can reduce our procedure to the $k=2$ treatment problem, by considering treatment $g$ as the first treatment, and the $(k-1)$ remaining as the second treatment.

Let

$$N_{\cdot i}=\sum_{g=1}^{k} N_{gi} \qquad \text{and} \qquad \hat{q}_i=\frac{\sum_{g=1}^{k} d_{gi}}{N_{\cdot i}}. \tag{8.35}$$

Thus the expected number of deaths in group $g$ is $E_{gi}=N_{gi}\hat{q}_i$, and in the remaining group of $(k-1)$ treatments, it is $(N_{\cdot i}-N_{gi})\hat{q}_i$. Also, the variance of $d_{gi}$ is

$$V_{gi}=\frac{N_{gi}(N_{\cdot i}-N_{gi})}{N_{\cdot i}-1}q_i(1-q_i). \tag{8.36}$$

We calculate

$$T_g=\sum_{i=0}^{m-1}(d_{gi}-E_{gi})=\sum_{i=0}^{m-1}d_{gi}-\sum_{i=0}^{m-1}E_{gi}, \tag{8.37}$$

and its variance (conditional on $N_{gi}$'s and $\Sigma_j d_{gi}$'s)

$$V_g = \sum_{i=0}^{m-1} V_{gi} = \sum_{i=0}^{m-1} \frac{N_{gi}(N_{\cdot i} - N_{gi})}{N_{\cdot i} - 1} \hat{q}_i(1 - \hat{q}_i), \tag{8.38}$$

Thus under null hypothesis of no treatment effect, the statistic

$$X_g^2 = \frac{T_g^2}{V_g} \tag{8.39}$$

is approximately distributed as Chi-square with 1 degree of freedom.

We also notice that (under null hypothesis) the statistic

$$Z_g = \frac{T_g}{\sqrt{V_g}} \tag{8.40}$$

is approximately distributed as unit normal, and can be used as test statistic for consistent difference. This test has the advantage that it also gives the direction (sign + or −) of deviation.

Methods taking account of withdrawals also follow the same lines as for the case of two treatments. The comments at the end of the discussion of the two-treatment case apply with equal force here.

**Example 8.11** We illustrate Mantel-Haenszel techniques, using the data given in Table 8.8 and discussed in Example 8.10. Here we test whether the effect of each of the four treatments is consistently higher (or lower) as compared with the *average* of the four treatments. Calculations of $T_g$'s are shown in Tables 8.10a, b and the variances $V_{gi}$ are given in Table 8.10c.

The only statistic ($Z_g$ or $X_g^2$), which, on its own, might be regarded as significant (with significance level 0.02 for one-sided, or 0.04 for two-sided test) would be that for tolbutamide ($g=2$) against the remainder. The sign + indicates some excess mortality for tolbutamide.

***Special Case of the Mantel-Haenszel Test.*** Suppose that the successive periods of time are so short that there is no more than *one death in each period*. This means that (in the case of $k$ treatments) we have $\Sigma_{g=1}^{k} d_{gi} = 0$ or 1, and this holds for all $i$. Then for each period with no deaths, we have

**Table 8.10a.** Calculations for Mantel-Haenszel tests

| $i$ | Placebo | | | Tolbutamide | | | Insulin standard | | | Insulin variable | | | | | |
|---|---|---|---|---|---|---|---|---|---|---|---|---|---|---|---|
| | $N'_{1i}$ | $d_{1i}$ | $E_{1i}$ | $N'_{2i}$ | $d_{2i}$ | $E_{2i}$ | $N'_{3i}$ | $d_{3i}$ | $E_{3i}$ | $N'_{4i}$ | $d_{4i}$ | $E_{4i}$ | $\sum_g N'_{gi}$ | $\sum_g d_{gi}$ | $\hat{q}_i$ |
| 0 | 205 | — | 1.49 | 204 | — | 1.49 | 210 | 2 | 1.53 | 204 | 4 | 1.49 | 823 | 6 | 0.00729 |
| 1 | 205 | 5 | 3.76 | 204 | 5 | 3.75 | 208 | 2 | 3.82 | 200 | 3 | 3.67 | 817 | 15 | 0.01836 |
| 2 | 200 | 4 | 3.49 | 199 | 5 | 3.47 | 206 | 2 | 3.60 | 197 | 3 | 3.44 | 802 | 14 | 0.01746 |
| 3 | 194 | 4 | 3.48 | 191.5 | 5 | 3.44 | 203 | 4 | 3.64 | 191.5 | 1 | 3.44 | 780 | 14 | 0.01795 |
| 4 | 176.5 | 4 | 2.73 | 172 | 5 | 2.66 | 185.5 | 2 | 2.87 | 177.5 | — | 2.74 | 711 | 11 | 0.01546 |
| 5 | 139.5 | 3 | 3.43 | 134.5 | 4 | 3.31 | 147 | 3 | 3.62 | 148 | 4 | 3.64 | 569 | 14 | 0.02460 |
| 6 | 90 | 1 | 2.72 | 86.5 | 5 | 2.62 | 92 | 4 | 2.78 | 95 | 1 | 2.87 | 363 | 11 | 0.03026 |
| 7 | 46 | — | 0.78 | 41.5 | 1 | 0.70 | 44 | — | 0.74 | 46.5 | 2 | 0.78 | 178 | 3 | 0.01685 |
| Total | | 21 | 21.88 | | 30 | 21.44 | | 19 | 22.60 | | 18 | 22.07 | | | |

**Table 8.10b**

| $i$ | Placebo | | | Tolbutamide | | | Insulin standard | | | Insulin variable | | |
|---|---|---|---|---|---|---|---|---|---|---|---|---|
| | $\sum_h^i d_{1h}$ | $\sum_h^i E_{1h}$ | Diff. | $\sum_h^i d_{2h}$ | $\sum_h^i E_{2h}$ | Diff. | $\sum_h^i d_{3h}$ | $\sum_h^i E_{3h}$ | Diff. | $\sum_h^i d_{4h}$ | $\sum_h^i E_{4h}$ | Diff. |
| 0 | — | 1.49 | −1.49 | — | 1.49 | −1.49 | 2 | 1.53 | 0.47 | 4 | 1.49 | 2.51 |
| 1 | 5 | 5.25 | −0.25 | 5 | 5.24 | −0.24 | 4 | 5.35 | −1.65 | 7 | 5.16 | 1.84 |
| 2 | 9 | 8.74 | 0.26 | 10 | 8.71 | 1.29 | 6 | 8.95 | −2.95 | 10 | 8.60 | 1.40 |
| 3 | 13 | 12.22 | 0.78 | 15 | 12.15 | 2.85 | 10 | 12.59 | −2.59 | 11 | 12.04 | −1.04 |
| 4 | 17 | 14.95 | 2.05 | 20 | 14.81 | 5.19 | 12 | 15.46 | −3.46 | 11 | 14.78 | −3.78 |
| 5 | 20 | 18.38 | 1.62 | 24 | 18.12 | 5.88 | 15 | 19.08 | −4.08 | 15 | 18.42 | −3.42 |
| 6 | 21 | 21.10 | −0.10 | 29 | 20.74 | 8.26 | 19 | 21.86 | −3.86 | 16 | 21.29 | −5.29 |
| 7 | 21 | 21.88 | −0.88 | 30 | 21.44 | 8.56 | 19 | 22.60 | −3.60 | 18 | 22.07 | −4.07 |
| | | | $(=T_1)$ | | | $(=T_2)$ | | | $(=T_3)$ | | | $(=T_4)$ |

**Table 8.10c.** Values of $V_{gi}$

| $i$ | $g$ | | | |
|---|---|---|---|---|
| | 1 | 2 | 3 | 4 |
| 0 | 1.115 | 1.112 | 1.133 | 1.112 |
| 1 | 2.771 | 2.762 | 2.798 | 2.726 |
| 2 | 2.579 | 2.570 | 2.630 | 2.553 |
| 3 | 2.023 | 2.550 | 2.651 | 2.550 |
| 4 | 2.023 | 1.988 | 2.090 | 2.031 |
| 5 | 2.531 | 2.469 | 2.621 | 2.632 |
| 6 | 1.990 | 1.940 | 2.022 | 2.065 |
| 7 | 0.568 | 0.530 | 0.552 | 0.572 |
| $V_g=\sum_{i=0}^{7} V_{gi}$ | 16.15 | 15.921 | 16.497 | 16.241 |
| $\sqrt{V_g}$ | 4.02 | 3.99 | 4.06 | 4.03 |
| $T_g/\sqrt{V_g}$ | −0.22 | +2.15 | −0.89 | −1.01 |
| $X^2$ | 0.048 | 4.622 | 0.792 | 1.020 |

$E_{gi}=V_{gi}=0$, and for each period with one death, we have from (4.44)

$$q_i=\frac{1}{N_{\cdot i}},\qquad \text{and so}\qquad E_{gi}=N_{gi}q_i=\frac{N_{gi}}{N_{i\cdot}},\tag{8.41}$$

where $N_{i\cdot}=N_{1i}+\cdots+N_{ki}$.

Also from (8.36)

$$\begin{aligned}V_{gi}&=\frac{N_{gi}(N_{\cdot i}-N_{gi})}{N_{\cdot i}-1}\,\frac{1}{N_{\cdot i}}\cdot\frac{N_{\cdot i}-1}{N_{\cdot i}}\\&=\frac{N_{gi}}{N_{\cdot i}}\left(1-\frac{N_{gi}}{N_{\cdot i}}\right)=E_{gi}(1-E_{gi}).\end{aligned}\tag{8.42}$$

Then, we have from (8.37)

$$T_g=\sum_{i=0}^{m-1} d_{gi}-\sum_{i=0}^{m-1} E_{gi}=d_g-\sum_h E_{gh},\tag{8.43}$$

where $d_g=\sum_{i=0}^{m-1} d_{gi}$ is the total number of deaths up to time $t_m$, and $h$ is summed over those values for which there is a death in the corresponding interval (and so $E_{gh}\neq 0$).

We also have

$$V_g = \sum_h E_{gh}(1 - E_{gh}). \tag{8.44}$$

The statistic

$$Z_g = \frac{T_g}{\sqrt{V_g}} \tag{8.45}$$

is approximately distributed as unit normal, and

$$Z_g^2 = X_g^2 = \frac{T_g^2}{V_g} = \frac{\left(d_g - \sum_h E_{gh}\right)^2}{\sum_h E_{gh}(1 - E_{gh})} \tag{8.46}$$

is approximately distributed as Chi-square with 1 degree of freedom. Note that this method takes care of withdrawals. Ties can be treated similarly as for the Kolmogorov-Smirnov test (Section 8.4.1) by calculating upper and lower limits.

### 8.8.2 Logrank test

In case of $k=2$ groups, the statistic (8.46) evaluated for group 1 ($g=1$) is

$$X_1^2 = \frac{T_1^2}{V_1} = \frac{\left(d_1 - \sum_h E_{1h}\right)^2}{\sum_h E_{1h}(1 - E_{1h})}. \tag{8.47}$$

This statistic is used in testing for a consistent discrepancy in mortality between groups 1 and 2. (Note that $X_2^2 = X_1^2$).

Peto and Peto (1972) suggested a somewhat different statistic also based on sums of observed and expected deaths and called it the *logrank statistic*. For $k=2$ groups, it is defined as

$$X'^2(1) = \frac{\left(d_1 - \sum_h E_{1h}\right)^2}{\sum_h E_{1h}} + \frac{\left(d_2 - \sum_h E_{2h}\right)^2}{\sum_h E_{2h}}, \tag{8.48}$$

and is supposed to be approximately distributed as Chi-square with 1 degree of freedom.

In the case when there are no ties, (8.48) can be written in the form

$$X'^2(1)=\frac{\left(d-\sum_h E_{1h}\right)^2}{(1/d)\left(\sum_h E_{1h}\right)\left(d-\sum_h E_{1h}\right)},\tag{8.48a}$$

where $d=d_1+d_2$ is the total number of deaths up to time $t_m$.

$$\text{Since}\quad \frac{1}{d}\left(\sum E_{1h}\right)\left(d-\sum E_{1h}\right)=\sum E_{1h}-d\bar{E}_{1h}^2,\qquad \text{where } \bar{E}_{1\cdot}=\frac{1}{d}\sum E_{1h},$$

$$=\sum E_{1h}(1-E_{1h})+\sum\left(E_{1h}-\bar{E}_{1\cdot}\right)^2,$$

$$\geqslant \sum E_{1h}(1-E_{1h}),\tag{8.49}$$

then we can see that the logrank statistic (8.48a) is never greater than the statistic (9.31). Relatively, the logrank test will be more conservative (less likely to establish significance) than $X_1^2$ defined in (8.47) [see Peto and Pike (1973) and Haybittle and Freedman (1979)].

**Example 8.12** We calculate the Mantel-Haenszel criterion $X_1^2$ [defined in (8.47)] and logrank criterion for survival data in dementia given in Table 8.2. Here the data are complete; we calculate these statistics using all 42 deaths. Since there are *ties* at four times of death, we calculate the extremes of the possible range of values for $\sum E_{jh}$.

We obtain the minimum possible value by assuming the death from group 1 to occur first in each of the four ties. The minimum value is

$$\min \sum E_{1h}=\frac{17}{42}+\frac{17}{41}+\frac{16}{40}+\frac{16}{39}+\frac{15}{38}+\frac{14}{37}+\frac{14}{36}+\cdots+\frac{1}{3}+\frac{1}{2}=16.392.$$

The corresponding value of $\sum E_{1i}(1-E_{1i})=9.632$.

The maximum value is

$$\max \sum E_{1h}=\frac{17}{42}+\cdots+\frac{15}{38}+\frac{15}{37}+\frac{14}{36}+\cdots+\frac{1}{3}+\frac{1}{2}$$

$$=\min \sum E_{1h}+\frac{1}{37}+\frac{1}{31}+\frac{1}{22}+\frac{1}{13}=16.574.$$

The corresponding value of $\Sigma E_{1h}(1-E_{1h})=9.660$. The total number of deaths in group 1 is 17.

1. For the Mantel-Haenszel statistic $X_1^2$ defined in (8.47) and $Z_1=\sqrt{X_1^2}$, we obtain

$$\frac{(17-16.392)^2}{9.632}=0.0384, \quad \text{and} \quad \sqrt{0.0384} \doteq 0.20 \quad \text{using min} \sum E_{1h},$$

$$\frac{(17-16.574)^2}{9.660}=0.0188, \quad \text{and} \quad \sqrt{0.0188} \doteq 0.14 \quad \text{using max} \Sigma E_{1h}.$$

2. For the logrank statistic defined in (8.48), we obtain

$$\frac{42(17-16.392)^2}{16.392 \cdot 25.608}=0.0369 \quad \text{using min} \sum E_{1h}$$

$$\frac{42(17-16.574)^2}{16.574 \cdot 25.426}=0.0181 \quad \text{using max} \sum E_{1h}.$$

In either case the test gives a nonsignificant result.

### 8.8.3 Extension to *r* Experiments (Classes)

In Mantel-Haenszel tests as described in the previous section, where there were two treatments, the division into subgroups was according to elapsed time. This, of course, may be combined (in a cross-classification) with division according to another type of classification, giving subsubgroups (sub$^2$-groups). Such a classification may represent data from different experiments, from different types of jobs in a specific occupation, from different demographic groups, and so on, or from groups classified according to sets of initial values of especially chosen concomitant variables (risk factors).

For the $s$th such set, we can calculate from (8.39)

$$^{s}X_g^2=\frac{^{s}T_g^2}{^{s}V_g}, \qquad s=1,2,\ldots,r. \tag{8.50}$$

By summing $^{s}T_g$'s and $^{s}V_g$'s over $s=1,2,\ldots,r$ we obtain a combined Chi-square

$$X_{g\ \text{comb}}^2(1)=\frac{\left(\sum_{s=1}^{r} {}^{s}T_g\right)^2}{\left(\sum_{s=1}^{r} {}^{s}V_g\right)}, \tag{8.51}$$

which (on the null hypothesis) is approximately distributed as Chi-square with 1 degree of freedom. Sequences (in time) of such tests can be used in testing for consistent differences in mortality in time over the sets of risk factors. Modifications for logrank tests are straightforward.

## 8.9 PARAMETRIC METHODS

In those cases, where the *parametric* form of the failure distribution in each population is firmly established, one may utilize this knowledge to construct tests comparing the *parameter* values, and so, indirectly, the CDF's (SDF's) in the different populations. It is, however, essential to bear in mind that the ultimate aim of comparison (whatever techniques are used) is to compare survival probabilities, and parameter values are only of interest insofar as they are used to evaluate these probabilities. It is then recommended that before any comparison of two (or more) parametric SDF's is made, we should first try *to fit* parametric forms to each distribution and test for goodness of fit, using appropriate methods described in Chapter 7. For specified distributions, such as exponential, Weibull, normal, and so on, exact tests have been derived. Many such tests and their applications in reliability theory are given in the book by Mann et al. (1974). Some of these tests are also discussed by Gross and Clark (1975), Chapter 7.

*Likelihood ratio tests* can be used when testing the hypothesis that the parameters of two distributions are the same—that is $H_0: \boldsymbol{\theta}_1 = \boldsymbol{\theta}_2 = \boldsymbol{\theta}$—against alternatives ($H_1$) that $\theta_{1j} \neq \theta_{2j}$ for at least one $j = 1, 2, \ldots, s$.

Let $\hat{\boldsymbol{\theta}} = (\hat{\theta}_1, \ldots, \hat{\theta}_s)$ be the vector of the ML-estimators of $\theta_j$'s when $H_0$ is true, and $L(\hat{\boldsymbol{\theta}})$ is the maximized value of the likelihood function. Similarly, let $\hat{\boldsymbol{\theta}}_1 = (\hat{\theta}_{11}, \ldots, \hat{\theta}_{1s})$ and $\hat{\boldsymbol{\theta}}_2 = (\hat{\theta}_{21}, \ldots, \hat{\theta}_{2s})$ be the ML-estimators without restriction by $H_0$, and $L(\hat{\boldsymbol{\theta}}_1; \hat{\boldsymbol{\theta}}_2)$ the corresponding value of maximized likelihood. Then

$$-2 \log \frac{L(\hat{\boldsymbol{\theta}})}{L(\hat{\boldsymbol{\theta}}_1; \hat{\boldsymbol{\theta}}_2)} \tag{8.52}$$

is (under the null hypothesis) approximately distributed as Chi-square with $s$ degrees of freedom (cf. Section 3.15.4).

## 8.10 SEQUENTIAL METHODS

We have already, in Section 8.6, encountered situations in which early decisions can be reached without awaiting completion of a trial as originally planned. Much more elaborate and comprehensive methods for arriving at early decisions, while controlling probabilities of erroneous decision at acceptable levels, have been worked out. They are included in the subject of *sequential analysis*. A useful account of some of these methods in applications to clinical trials is contained in a book by Armitage (1975).

Since application of these methods is primarily associated with the *planning* of trials, we will not discuss this topic further.

## REFERENCES

Armitage, P. (1965). *Sequential Medical Trials*. Blackwell, Oxford; Halstead Press, New York.

Barnard, G. A. (1947). Significance tests for 2×2 tables. *Biometrika* **34**, 123–138.

Biometrics Seminar, S. J. Cutler (Chairman) (1966). The role of hypothesis testing in clinical trials. *J. Chron. Dis.* **19**, 857–882.

Conover, W. J. (1967). The distribution functions of Tsao's truncated Smirnov tests. *Ann. Math. Statist.* **38**, 1208–1215.

Fisher, R. A. (1937). *The Design of Experiments*. Oliver and Boyd, Edinburgh.

Gehan, E. A. (1965). A generalized Wilcoxon test for comparing arbitrarily singly-censored data. *Biometrika* **52**, 203–223.

Gilbert, G. P. (1962). Random censorship, Ph.D. Thesis, University of Chicago.

Gross, A. J. and Clark, V. A. (1975). *Survival Distributions: Reliability Applications in the Biomedical Sciences*. Wiley, New York, Chapter 8.

Haybittle, J. L. and Freedman, L. S. (1979). Some comments on the logrank test statistic in clinical trial applications. *The Statistician, London* **28**, 199–208.

Kim, P. J. and Jennrich, R. I. (1970). Tables of the exact sampling distribution of the two-sample Kolmogorov-Smirnov criterion $D_{m,n}(mn)$. *Selected Tables in Mathematical Statistics*, American Mathematical Society, Providence, Rhode Island. Vol. I, pp. 79–170.

Mann, N. R., Schafer, R. D., and Singpurwalla, N. D. (1974). *Methods for Statistical Analysis of Reliability and Life Data*. Wiley, New York.

Mantel, N. (1966). Evaluation of survival data and two new rank order statistics arising in its consideration. *Cancer Chemother. Rep.* **50**, 163–170.

Mantel, N. (1967). Ranking procedures for arbitrary restricted observation. *Biometrics* **23**, 65–78.

Mantel, N. and Haenszel, W. (1959). Statistical aspects of the analysis of data from retrospective studies of disease. *J. Natl. Cancer Inst.* **22**, 719–748.

Meier, P. (1975). Statistics and medical experimentation. *Biometrics* **31**, 511–529.

Pearson, E. S. (1947). The choice of statistical tests illustrated on the interpretation of data classed in a 2×2 table. *Biometrika* **34**, 134–167.

Pearson, E. S. and Hartley, H. O. (1972). *Biometrika Tables for Statisticians*, Vol. II, Table 22. Cambridge University Press, Cambridge.

Peto, R. (1972). Rank tests of maximal power against Lehman-type alternatives. *Biometrika* **59**, 472–474.

Peto, R. and Peto, J. (1972). Asymptotically efficient rank invariant test procedures (with discussion). *J. Roy. Statist. Soc. Ser. A* **135**, 185–198.

Peto, R. and Pike, M. C. (1973). Conservatism of the approximation $\Sigma(O-E)^2/E$ in the logrank test for survival data or tumor incidence data, *Biometrics* **29**, 579–584.

Peto, R., Pike, M. C., Armitage, P., Breslow, N. E., Cox, D. R., Howard, S. V., Mantel, N., McPherson, K., Peto, J. and Smith, P. G. (1977). Design and analysis of randomized elimination trials requiring prolonged observation of each patient. II. Analysis and examples, *Br. J. Cancer* **35**, 1–39.

University Group Diabetes Program. A study of the effects of hypoglycemic agents on vascular complications in patients with adult-onset diabetes, *Diabetes* **19** (Suppl. 2), Appendix A, 816–830. [UGDP (1970)]

Tsao, C. K. (1954). An extension of Massey's distribution of the maximum deviation between two-sample cumulative step functions. *Ann. Math. Statist.* **25**, 587–592.

Tukey, J. (1977). Some thoughts on clinical trials, especially problems of multiplicity. In: "Medical Research, Statistics and Ethics", *Science*, **198**, 679–684.

## EXERCISES

**8.1.** Compare the USA and USSR life tables for males (females) given in Table E8.1, using different methods described in Section 8.2.

**Table E8.1.** USA and USSR life tables

| Age group | Males | | | | Females | | | |
|---|---|---|---|---|---|---|---|---|
| | USA,* 1960 | | USSR,† 1958–1959 | | USA, 1960 | | USSR, 1958–1959 | |
| $x$ to $x+n$ | ${}_nq_x$ | $l_x$ | ${}_nq_x$ | $l_x$ | ${}_nq_x$ | $l_x$ | ${}_nq_x$ | $l_x$ |
| 0–1 | 0.02970 | 100,000 | 0.04424 | 100,000 | 0.02267 | 100,000 | 0.03677 | 100,000 |
| 1–5 | 0.00475 | 97,030 | 0.01619 | 95,576 | 0.00391 | 97,733 | 0.01557 | 96,323 |
| 5–10 | 0.00280 | 96,569 | 0.00642 | 94,029 | 0.00206 | 97,351 | 0.00478 | 94,823 |
| 10–15 | 0.00275 | 96,299 | 0.00491 | 93,425 | 0.00163 | 97,150 | 0.00345 | 94,370 |
| 15–20 | 0.00653 | 96,034 | 0.00777 | 92,966 | 0.00270 | 96,992 | 0.00471 | 94,044 |
| 20–25 | 0.00894 | 95,407 | 0.01164 | 92,244 | 0.00349 | 96,730 | 0.00628 | 93,601 |
| 25–30 | 0.00864 | 94,554 | 0.01453 | 91,170 | 0.00443 | 96,392 | 0.00724 | 93,013 |
| 30–35 | 0.00993 | 93,737 | 0.01766 | 89,845 | 0.00606 | 95,965 | 0.00856 | 92,340 |
| 35–40 | 0.01432 | 92,806 | 0.02191 | 88,258 | 0.00896 | 95,383 | 0.01088 | 91,550 |
| 40–45 | 0.02284 | 91,477 | 0.02865 | 86,324 | 0.01397 | 94,528 | 0.01433 | 90,554 |
| 45–50 | 0.03740 | 89,388 | 0.03872 | 83,851 | 0.02102 | 93,207 | 0.01870 | 89,256 |
| 50–55 | 0.06081 | 86,045 | 0.05764 | 80,604 | 0.03161 | 91,248 | 0.02714 | 87,587 |
| 55–60 | 0.08875 | 80,813 | 0.08565 | 75,958 | 0.04577 | 88,364 | 0.03973 | 85,210 |
| 60–65 | 0.13368 | 73,641 | 0.12086 | 69,452 | 0.07224 | 84,320 | 0.06049 | 81,825 |
| 65–70 | 0.18755 | 63,797 | 0.16604 | 61,058 | 0.10729 | 78,229 | 0.09937 | 76,875 |
| 70–75 | 0.25918 | 51,832 | 0.23164 | 50,920 | 0.16882 | 69,836 | 0.16127 | 69,236 |
| 75–80 | 0.35452 | 38,398 | 0.31951 | 39,125 | 0.26577 | 58,046 | 0.25156 | 58,070 |
| 80–85 | 0.49792 | 24,785 | 0.43495 | 26,624 | 0.41906 | 42,619 | 0.37297 | 43,462 |
| 85+ | 1.00000 | 12,444 | 1.00000 | 15,044 | 1.00000 | 24,759 | 1.00000 | 27,252 |

*From Preston, S. et al. (1972). *Causes of Death: Life Tables for National Populations*. Seminar Press, New York.

†From *Demographic Year Book—Mortality Statistics* (1974), 26th issue, pp. 1058–1059, United Nations, New York.

**8.2.** The Pearson's Chi-square test given by (8.25) and the Mantel-Haenszel Chi-square test given by (8.34) are both used as tests of homogeneity (i.e., for heterogeneity). What aspects of heterogeneity does each test take into account? Formulate in each case the null hypothesis and the alternative against which the test is especially aimed.

**8.3.** Given $f_X(x;\theta)=(1/\theta)e^{-x/\theta}$, $x>0$, $\theta>0$ (the PDF of exponential distribution):

(*a*) Prove that, in fact, $X$ is distributed as $\frac{1}{2}\theta\chi_2^2$, where $\chi_2^2$ means "Chi-square with 2 degrees of freedom."

(*b*) Let $x_1'<x_2'<\cdots<x_N'$ denote the (ordered) failure times of a cohort consisting of $N$ individuals. Show that the ML-estimator of $\theta$ is

$$\hat{\theta}=\frac{1}{N}\sum_{j=1}^{N} x_j'=\bar{x}.$$

(*c*) Show that the statistic

$$\frac{2N\hat{\theta}}{\theta}=\frac{2N\bar{x}}{\theta}$$

is distributed as $X_{2N}^2$ (i.e., Chi-square with $2N$ d.f.).

*Hint*: Use results of (*a*) in this Exercise.

(*d*) Consider two independent exponential populations with PDF $f_{X_i}(x_i;\ \theta_i)$, $i=1,2$. Let $N_1$ and $N_2$ be the corresponding number of deaths in random samples, from populations 1 and 2, respectively. Show that the statistic $F=\bar{x}_1/\bar{x}_2$ is distributed as $(\theta_1/\theta_2)F_{2N_1,2N_2}$.

**8.4.** Let $S_i(x_i)=\exp[-u_i(x_i)/\theta_i]$ $x_i>0$, $\theta_i>0$, where $u_i(x_i)$ is a monotonic increasing function of $x_i$ such that $u_i(x_i)\to 0$ as $x_i\to 0$ and $u_i(x_i)\to\infty$ as $x_i\to\infty$, $i=1,2$. Using the results of Exercise 3.4 and Exercise 8.3, construct a test for testing hypothesis $H_0:\theta_1=\theta_2$.

**8.5.** As a rule of thumb it has been suggested that to obtain reasonably sound comparisons among a number of different groups in respect of mortality, dates should be available of sufficient extent to include at least $m$ deaths in each group. Suggested values of $m$ range from 20 to 50.

(*i*) On the assumption of Poisson distributions of number of deaths in each group, examine the adequacy of this suggestion.

(*ii*) In your answer to (*i*), you will work, at least in part, in terms of *expected* numbers of deaths. It has been suggested that a minimum *expected* number of deaths, rather than a minimum *observed* number of deaths, should be used as a criterion of adequacy. Comment on the practicality of this suggestion.

(*iii*) On the basis of the first suggested criterion, one might plan to continue a trial until each group experiences a specified minimum number of deaths, and then stop. Since usually, the last death will occur in just one group, it is proposed that comparisons among the other groups can be carried out ignoring the stopping rule. Comment on this proposal, and also on the feasibility of ignoring the stopping rule, even in comparisons involving the group containing the last death.

**8.6.** $X_1,\ldots,X_m; Y_1,\ldots,Y_n$ are independent random variables corresponding to observed values in random samples of size $m,n$ respectively from two exponential populations. Each of $X_1,\ldots,X_m$ has SDF

$$\exp\left(-\frac{x}{\theta}\right), \qquad x>0, \qquad \theta>0,$$

and each of $Y_1,\ldots,Y_n$ has SDF

$$\exp\left(-\frac{y}{\phi}\right), \qquad y>0, \qquad \phi>0.$$

Obtain a likelihood ratio tests of the hypothesis $\theta=\phi$, against alternative $\theta\neq\phi$ based on:

(*i*) the complete set of $(m+n)$ sample values,

(*ii*) all observed values less than some prespecified number $\zeta$,

(*iii*) all observed less than or equal to the $r$th smallest among the $X$'s, or among the $Y$'s $[r\leqslant \min(m,n)]$, whichever of these two values is the larger,

(*iv*) the $r$ smallest values among the $X$'s and the $r$ smallest values among the $Y$'s.

In each case explain, in simple terms, the method of data collection which would lead to use of the test you construct.

*Hint*: See Exercise 8.3.

**8.7.** (*a*) Show how to apply the results of Exercise 8.6 to test for differences between two Weibull SDF's with

(*i*) a known (common) value of the slope parameter and

(*ii*) known (different) values of the slope parameter.

(Note that in case (*ii*) the two SDF's cannot be *identical*, and consider what kinds of differences you might find of interest.)

(*b*) Suggest a practical procedure for use in each of (*a*) (*i*) and (*a*) (*ii*) when the slope parameter(s) is (are) not known exactly, but can reasonably be supposed to be between 0.8 and 1.25.

**8.8.** Extend your answers to Exercises 8.6 and 8.7 to situations in which there are samples from more than two populations to be compared.

**8.9.** For the survival data given in Table 8.6, construct a Pearson's $X^2$-test to compare:

(*a*) Tolbutamide with insulin treatment (standard and variable combined).

(*b*) Four groups (placebo, tolbutamide, insulin standard, and insulin variable).

# Part 3

# MULTIPLE TYPES OF FAILURE

CHAPTER 9

# Theory of Competing Causes: Probabilistic Approach

## 9.1 CAUSES OF DEATH: BASIC ASSUMPTIONS

In Part 2 we discussed a number of univariate failure distributions and pointed out that these distributions can be used in many situations, where failure is not necessarily death, but can be any nonrepetitive, all-or-none phenomenon.

In the context of mortality data, we may have information beyond the simple facts of death or survival. In particular, death certificates record not only the time of death, but also cause(s) of death. This last information enables a classification of death by causes to be made. Similar situations can arise in other contexts. For example failure of a metal strip under test conditions can occur in a number of different ways—by cracking, buckling, shearing, etc. In this text, we use the terminology of mortality and survival functions, because this is our primary interest; clearly, the methods and ideas to be discussed can be applied in broader fields, just as in Part 2.

What is really "the cause" of death? Present day death certificates in many countries list morbid conditions and impairments (insofar as these are known to the physician) regarded as contributing to the death of an individual. In the United States, the medical statement on death certificates can specify four kinds of "causes" of death (more precisely, conditions that contribute to death). These are *immediate*, *intervening*, *underlying*, and *contributory causes of death*. Definitions and further details on the coding of these conditions can be found in *Vital Statistics* U.S. 1955, Supplement (1965).

However, only one of these—the *underlying* cause—has been commonly utilized in the analysis of multiple cause mortality data. The underlying cause is defined as "the disease or injury which *initiated* the train of morbid conditions leading directly to death" [Vital Statistics Instruction Manual (1962)]. Therefore, analysis of mortality by cause of death is based on assumption A1:

A1. *Each death is due to a single cause.*

This assumption is, indeed, a consequence of our lack of knowledge of what "the cause of death" really means; a joint event of death from two or more causes simultaneously has, as yet, not been clearly defined for living organisms, although this might be not true for mechanical devices. Models for which A1 does not hold will be discussed in Section 15.5.

Another tacit assumption, although not often spelled out, is assumption A2:

A2. *Each individual in a given population is liable to die from any of the causes operating in this population.*

Heterogeneous populations in which A2 does not hold will be discussed in Section 9.9.

## 9.2 SOME BASIC PROBLEMS

Consider a population in which there are $k$ causes of death, $C_1, C_2, \ldots, C_k$, say, operating simultaneously. When a member of this population dies, the *age* at death and the (underlying) *cause* of death are recorded. These records, called the *mortality data*, are used in analysis. Methods appropriate for analysis of such data are known as competing causes analysis, or more often as *competing risk analysis*.

We ask the following questions:

1. *What is the structure (probability distribution) of age at death from different causes acting simultaneously in a given population*? We want to find expected proportions of deaths from different causes in various age groups, and evaluate changes in these proportions with time. These problems are concerned with probability distributions of age at death from various causes in the *whole population*; we refer to them as *public* probabilities. An answer to question (1) can be obtained from multiple decrement life tables; their construction will be discussed in Chapter 10.

2. *What is the probability that a newborn individual will die after age $x$ from cause $C_\alpha$*? Of course, here is a tacit assumption that he does not die

from any other cause before age $x$. This is a kind of *private* probability, since it concerns an individual's chance of dying from a particular cause. This problem will be discussed in Chapter 11.

3. *How might the mortality pattern change if a certain cause, $C_\alpha$, were "eliminated"?* This question has been asked for over 200 years. Although attempts have been made to answer this question by constructing single decrement life tables, the problem of evaluating the effect on mortality of a chronic disease is, at present, beyond our capacity (compare Section 11.2).

Various methods of estimating public and private probabilities from cohort and clinical-type mortality data are presented in Chapter 12. In the present chapter, we describe a probabilistic approach to competing risk theory. The succeeding three Chapters (10–12) exploit this approach.

## 9.3 "TIMES DUE TO DIE"

In the formal development of the theory, it is convenient to use the concept of a "*time due to die*" for each competing cause. It is supposed that each individual is endowed—presumably at birth—with a set of such times, one for each competing cause. The actual observed time at death is, of course, the minimum of these times. It is the *only* one of them that can actually be observed.

It must be emphasized that this concept is introduced *purely for mathematical convenience*. It is not claimed that "times due to die" exist in any real sense—merely that their introduction facilitates development of the theory.

Let $X_1, X_2, \ldots, X_k$ denote the hypothetical (potential) times due to die and define their joint survival distribution function (SDF) as

$$S_{1\ldots k}(x_1,\ldots,x_k)=\Pr\left\{\bigcap_{\alpha=1}^{k}(X_\alpha>x_\alpha)\right\}. \tag{9.1}$$

It is convenient (although not essential) to assume that $S_{1\ldots k}(x_1,\ldots,x_k)$ is a *proper* distribution, as defined in (3.8). This implies that also each of the *marginal* survival functions

$$S_\alpha(x_\alpha)=S_{1\ldots k}(0,\ldots,x_\alpha,\ldots,0) \tag{9.2}$$

for $\alpha=1,2,\ldots,k$ is a proper survival distribution function. Note that this property is satisfied, if assumption A2 holds.

We also define the *multivariate hazard rate component with respect to* $x_\alpha$ at the point $\boldsymbol{x}=(x_1,\dots,x_k)$ as

$$h_\alpha(x_1,\dots,x_k)=h_\alpha(\boldsymbol{x})=h_\alpha(\cdot)$$

$$=\lim_{h\to 0}\frac{1}{h}\Pr\left\{\bigcap_{\beta\neq\alpha}^{k}(X_\beta>x_\beta)\cap(x_\alpha<X_\alpha\leqslant x_\alpha+h)\,|\,\bigcap_{\beta=1}^{k}(X_\beta>x_\beta)\right\}$$

$$=-\frac{1}{S_{1\dots k}(\cdot)}\cdot\frac{\partial S_{1\dots k}(\cdot)}{\partial x_\alpha}=-\frac{\partial\log S_{1\dots k}(\cdot)}{\partial x_\alpha}. \tag{9.3}$$

The $(1\times k)$ vector

$$\mathbf{h}(\cdot)=[h_1(\cdot),h_2(\cdot),\dots,h_k(\cdot)] \tag{9.4}$$

is the *multivariate hazard vector* as defined by Johnson and Kotz (1973).

## 9.4 THE OVERALL AND 'CRUDE' SURVIVAL FUNCTIONS

### 9.4.1 The Overall Survival Function

In view of assumption A1, that each death is due to a single cause, we cannot observe $(X_1,\dots,X_k)$ jointly. Instead, we observe time at death

$$X=\min(X_1,\dots,X_k). \tag{9.5}$$

The *overall* (i.e., from any cause) survival function is

$$S_X(x)=\Pr\{X>x\}=\Pr\left\{\bigcap_{\alpha=1}^{k}(X_\alpha>x)\right\}=S_{1\dots k}(x,x,\dots,x). \tag{9.6}$$

This is, of course, a function of the single variable $x$. Its hazard rate is

$$h_X(x)=-\frac{1}{S_X(x)}\cdot\frac{dS_X(x)}{dx}=-\frac{d\log S_X(x)}{dx} \tag{9.7}$$

$$=-\frac{d\log S_{1\dots k}(x,x,\dots,x)}{dx}. \tag{9.7a}$$

### 9.4.2 The Crude and Net Hazard Rates

We can also identify the *causes of death*. The hazard rate at age $x$ for cause $C_\alpha$, in the *presence* of all other causes acting simultaneously in a population is defined as

$$h_\alpha(x)=\lim_{h\to 0}\frac{1}{h}\Pr\left\{(x<X_\alpha<x+h)\bigcap_{\substack{\beta=1\\ \beta\neq\alpha}}^{k} k(X_\beta>x)|\bigcap_{\beta=1}^{k}(X_\beta>x)\right\}$$

$$=-\frac{1}{S_X(x)}\frac{\partial S_{1\ldots k}(\cdot)}{\partial x_\alpha}\Bigg|_{\{x_\beta=x\}}=-\frac{\partial\log S_{1\ldots k}(\cdot)}{\partial x_\alpha}\Bigg|_{\{x_\beta=x\}}, \qquad (9.8)$$

where $\{x_\beta=x\}$ means $\cap_{\beta=1}^{k}(x_\beta=x)$.

Note that $h_\alpha(x)$ is equal to the multivariate hazard rate component defined in (9.3), evaluated at the point $(x,x,\ldots,x)$. In our notation

$$h_\alpha(x)=h_\alpha(x,x,\ldots,x). \qquad (9.9)$$

This is called a *crude hazard rate* (or crude force of mortality). It describes the (instantaneous) rate of dying from cause $C_\alpha$ at age $x$ when all causes are acting simultaneously. [cf. (3.4), also (9.24), (9.25)]

Since

$$\frac{d\log S_{1\ldots k}(x,\ldots,x)}{dx}=\sum_{\alpha=1}^{k}\frac{\partial\log S_{1\ldots k}(x_1,\ldots,x_k)}{\partial x_\alpha}\Bigg|_{\{x_\beta=x\}}, \qquad (9.10)$$

it follows that

$$h_X(x)=h_1(x)+h_2(x)+\cdots+h_k(x). \qquad (9.11)$$

Note that the *additivity property* of crude hazard rates stated in (9.11) is the consequence of assumption A1, that is, that deaths from different causes are mutually exclusive events.

On the other hand, the hazard rate of the marginal SDF, defined in (9.2) sometimes called the *net hazard rate* (net force of mortality) is

$$\lambda_\alpha(x)=-\frac{d\log S_\alpha(x)}{dx}=-\frac{d\log S_{1\ldots k}(0,\ldots,x,\ldots 0)}{dx}. \qquad (9.12)$$

It is the instantaneous rate of dying associated with the hypothetical time due to death, $X_\alpha$; it is sometimes claimed, although, in general, unjustifiably that this represents the instantaneous death rate for cause $C_\alpha$ alone [cf. Section 11.2.1 (1)].

Note that in general,

$$\lambda_\alpha(x)\neq h_\alpha(x).$$

### 9.4.3 The Crude Probability Distribution for Cause $C_\alpha$

Define a random index

$$I=\begin{cases}\alpha & \text{if } X_\alpha = X \text{ (if death is from } C_\alpha\text{)},\\ 0 & \text{otherwise,}\end{cases} \tag{9.13}$$

where $X=\min(X_1,\dots,X_k)$.

If we observe the pair of values $(X,I)$, then time at death and the cause of death are identified; it is customary to speak about an *identified minimum*. If, on the other hand, only $X$ is observed, then it is referred to as a *nonidentified minimum* [Nádas (1971), Basu and Ghosh (1978)]. In Section 9.4.1, we have obtained the distribution of $X=\min(X_1,\dots,X_k)$ [see (9.6)] for a given joint SDF. In this section, we derive the distribution of the random pair $(X,I)$.

The conditional probability of death from $C_\alpha$ in an interval $(x,x+dx)$, given alive at age $x$, and in the presence of all other causes acting simultaneously in a population, is approximately $h_\alpha(x)\,dx$. The unconditional probability of death from $C_\alpha$ in $(x,x+dx)$ is then $S_X(x)h_\alpha(x)\,dx$. Hence, the crude (in the presence of all causes) probability distribution of time at death for $C_\alpha$ is

$$Q_\alpha^*(x)=\Pr\{(X\leqslant x)\cap(I=\alpha)\}=\int_0^x h_\alpha(t)S_X(t)\,dt. \tag{9.14}$$

The symbol * [as in $Q_\alpha^*(x)$] will be used to denote that the failure or survival (parametric) distribution for cause $C_\alpha$ is regarded *in the presence* of all causes acting simultaneously in a population.

Let $\pi_\alpha$ denote the expected proportion of deaths from $C_\alpha$. We have

$$\pi_\alpha=\Pr\{(X<\infty)\cap(I=\alpha)\}=\Pr\{I=\alpha\}=\int_0^\infty h_\alpha(t)S_X(t)\,dt=Q_\alpha^*(\infty), \tag{9.15}$$

with

$$\pi_1+\pi_2+\cdots+\pi_k=1. \tag{9.16}$$

The crude probability of eventually dying from $C_\alpha$, at an age greater than $x$, is

$$P_\alpha^*(x)=\Pr\{(X>x)\cap(I=\alpha)\}=\int_x^\infty h_\alpha(t)S_X(t)\,dt. \tag{9.17}$$

Of course,

$$\pi_\alpha = Q_\alpha^*(\infty) = P_\alpha^*(0). \tag{9.18}$$

Clearly, we also have

$$\Pr\{X \leqslant x\} = F_X(x) = Q_1^*(x) + Q_2^*(x) + \cdots + Q_k^*(x), \tag{9.19}$$

and

$$\Pr\{X > x\} = S_X(x) = P_1^*(x) + P_2^*(x) + \cdots + P_k^*(x). \tag{9.20}$$

Note that

$$F_X(x) + S_X(x) = 1,$$

but

$$Q_\alpha^*(x) + P_\alpha^*(x) \leqslant 1. \tag{9.21}$$

The crude probability functions, $Q_\alpha^*(x)$ and $P_\alpha^*(x)$, are not proper distributions in the sense of (3.8). We can, however, define a proper CDF or SDF in terms of these functions. For example, the proper SDF associated with cause $C_\alpha$, and denoted by $S_\alpha^*(x)$, is

$$\begin{aligned} S_\alpha^*(x) &= \Pr\{\text{age at death} > x \mid \text{death from } C_\alpha \text{ sometime}\} \\ &= \Pr\{X > x \mid I = \alpha\} = \frac{P_\alpha^*(x)}{P_\alpha^*(0)} = \frac{1}{\pi_\alpha} P_\alpha^*(x) \\ &= \frac{1}{\pi_\alpha} \int_x^\infty h_\alpha(t) S_X(t)\, dt. \end{aligned} \tag{9.22}$$

Thus

$$F_\alpha^*(x) = 1 - S_\alpha^*(x) \tag{9.22a}$$

represents the actual (observable) distribution of time at death among those who die from cause $C_\alpha$ *in the presence* of all causes.

The corresponding PDF is then

$$f_\alpha^*(x) = \frac{1}{\pi_\alpha} h_\alpha(x) S_X(x) = -\frac{1}{\pi_\alpha} \frac{dP_\alpha^*(x)}{dx}. \tag{9.23}$$

Hence

$$h_\alpha(x) = -\frac{1}{S_X(x)}\frac{dP_\alpha^*(x)}{dx}. \tag{9.24}$$

On the other hand, the hazard rate of the distribution $S_\alpha^*(x)$ is

$$\lambda_\alpha^*(x) = -\frac{d\log S_\alpha^*(x)}{dx} = \frac{f_\alpha^*(x)}{S_\alpha^*(x)} = \frac{h_\alpha(x)S_X(x)}{\int_x^\infty h_\alpha(t)S_X(t)\,dt},$$

$$= -\frac{1}{P_\alpha^*(x)}\cdot\frac{dP_\alpha^*(x)}{dx}. \tag{9.25}$$

Comparing (9.24) and (9.25), we notice that

$$\frac{h_\alpha(x)}{\lambda_\alpha^*(x)} = \frac{P_\alpha^*(x)}{S_X(x)} = \pi_\alpha(x). \tag{9.26}$$

This gives the conditional probability of death after age $x$ from cause $C_\alpha$, given alive at age $x$. (See also Sections 9.7 and 10.4.)

We may also obtain another expression for $h_\alpha(x)$. Substituting for $S_X(x)$ from (9.20) into (9.24), we obtain

$$h_\alpha(x) = -\frac{1}{S_X(x)}\frac{dP_\alpha^*(x)}{dx} = -\frac{dP_\alpha^*(x)/dx}{\sum_{\beta=1}^{k} P_\beta^*(x)}. \tag{9.27}$$

Further, taking into account the relationships (9.23) and (9.25), we can express $h_\alpha(x)$ in terms of $S_\alpha^*(x)$'s and $\lambda_\alpha^*(x)$.

We then have

$$h_\alpha(x) = \frac{\pi_\alpha f_\alpha^*(x)}{\sum_{\beta=1}^{k}\pi_\beta S_\beta^*(x)} = \frac{\pi_\alpha\lambda_\alpha^*(x)S_\alpha^*(x)}{\sum_{\beta=1}^{k}\pi_\beta S_\beta^*(x)}. \tag{9.28}$$

(See also Examples 9.4 and 9.5.)

## 9.5 CASE WHEN $X_1, \ldots, X_k$ ARE INDEPENDENT

Of special interest is the situation when $X_1, X_2, \ldots, X_k$ are *mutually independent*. In this case, the joint SDF is of the form

$$S_{1\ldots k}(x_1,\ldots,x_k) = S_1(x_1)S_2(x_2)\ldots S_k(x_k), \tag{9.29}$$

where $S_\alpha(x_\alpha)$ is the marginal SDF with respect to $X_\alpha$, and its (net) hazard rate is $\lambda_\alpha(x) = -[d\log S_\alpha(x)]/dx$.

Note that, in this case we do have

$$\lambda_\alpha(x) = h_\alpha(x), \qquad \alpha = 1,2,\ldots,k, \tag{9.30}$$

where $h_\alpha(x)$ is the crude hazard rate defined in (9.8). Since time due to die from cause $C_\alpha$, $X_\alpha$, is independent of any other $X_\beta$, $S_\alpha(x)$ may be thought of as a SDF from cause $C_\alpha$ acting *alone* in a population in which all individuals die from only that cause. From (9.30) we see that, in this case, the crude (i.e., in the presence of all causes) rate, $h_\alpha(x)$, and the net (for cause $C_\alpha$ alone) rate, $\lambda_\alpha(x)$, are the same. However, we should emphasize that this is not true for the crude and net probability distributions, because the (proper) SDF in the presence of all causes is as given in (9.22), that is,

$$S_\alpha^*(x) = \frac{1}{\pi_\alpha}\int_x^\infty h_\alpha(t) S_X(t)\,dt,$$

whereas the marginal SDF is

$$S_\alpha(x) = \exp\left[-\int_0^\infty \lambda_\alpha(t)\,dt\right] = \exp\left[-\int_0^x h_\alpha(t)\,dt\right].$$

Clearly,

$$S_\alpha^*(x) \neq S_\alpha(x). \tag{9.31}$$

## 9.6 EQUIVALENCE AND NONIDENTIFIABILITY THEOREMS IN COMPETING RISKS

For the purpose of this section, we define a *family* of distributions as a set of distributions of the same mathematical form; each member of the family is determined by the values of the parameters.

Two basic problems arise:

1. *Given the distribution of the identified minimum* $(X,I)$ *is it possible to identify the family of the joint, and marginal SDF's?* The answer, in the general case, is "no". The reasons for this answer are given by Equivalence Theorem. (See Section 9.6.1.)
2. *Assuming that the joint SDF belongs to a specified family, under what conditions can we identify the parameters of this SDF from the distribution of* $(X,I)$? This problem is discussed in Section 9.6.2, and then in Section 12.3.1.

### 9.6.1 Equivalent Models of Survival Distribution Functions

Consider two sets of random variables $(X_1,\ldots,X_k)$ and $(X'_1,\ldots,X'_k)$, with corresponding joint SDF's, $S_{1\ldots k}(x_1,\ldots,x_k)$ and $S'_{1\ldots k}(x_1,\ldots,x_k)$. Let

$$X=\min(X_1,\ldots,X_k) \qquad \text{and} \qquad X'=\min(X'_1,\ldots,X'_k).$$

We define equivalence of survival models as follows: *Two survival models are equivalent if*

$$S_X(x)=S'_{X'}(x) \qquad \text{and} \qquad P^*_\alpha(x)=P'^*_\alpha(x) \text{ for all } \alpha. \tag{9.32}$$

We now prove two useful theorems: the Factorization Theorem and the Equivalence Theorem.

First, we recall from (9.11)

$$h_X(x)=\sum_{\alpha=1}^{k} h_\alpha(x). \tag{9.11}$$

Integrating both sides of (9.11) over the interval $(0,x)$, we obtain

$$\int_0^x h_X(t)\,dt=\sum_{\alpha=1}^{k}\int_0^x h_\alpha(t)\,dt. \tag{9.33}$$

It follows that

$$\exp\left[-\int_0^x h_X(t)dt\right]=\prod_{\alpha=1}^{k}\exp\left[-\int_0^x h_\alpha(t)\,dt\right],$$

or

$$S_X(x)=S_{1\ldots k}(x,x,\ldots,x)=\prod_{\alpha=1}^{k} G_\alpha(x), \tag{9.34}$$

where

$$G_\alpha(x)=\exp\left[-\int_0^x h_\alpha(t)\,dt\right]. \tag{9.35}$$

Note that $G_\alpha(x)$ represents a distribution associated with cause $C_\alpha$ alone, assuming that the hazard rate of this distribution is $h_\alpha(x)$. We summarize the result (9.34) in the *Factorization Theorem*:

**Factorization Theorem.** *Any joint SDF, $S_{1\ldots k}(x_1,\ldots,x_k)$, can be factorized at $x_1=x_2=\cdots x_k=x$ into $k$ terms as is* (9.34); *the $\alpha$th term is given by* (9.35).

We make two remarks. First, the overall SDF in (9.34) must represent a proper distribution. It follows that at least one $G_\alpha(x)$ must be proper, that is, we must have, for *at least* one $X_\alpha$,

$$\lim_{x\to\infty}\int_0^x h_\alpha(t)\,dt=\infty, \qquad \alpha=1,2,\ldots,k, \tag{9.36}$$

[see Miller (1977)]. Second, if $X_1,\ldots,X_k$ are mutally independent, relation (9.34) holds; the reverse may not necessarily be true [Gail (1975)]. A counter example has been given by Hakulinen and Rahiala (1977).

We now construct *another* model of the joint survival function. Let $X_1'$, $X_2',\ldots,X_k'$ be another set of *independent* random variables with SDF's

$$S_\alpha'(x_\alpha)=\exp\left[-\int_0^{x_\alpha}h_\alpha(t)\,dt\right]=G_\alpha(x_\alpha). \tag{9.37}$$

The joint SDF is

$$S'_{1\ldots k}(x_1,\ldots,x_k)=\prod_{\alpha=1}^{k}G_\alpha(x_\alpha). \tag{9.38}$$

Since $X'=\min(X_1',\ldots,X_k')$,

$$S'_{X'}(x)=S'_{1\ldots k}(x,\ldots,x)=\prod_{\alpha=1}^{k}G_\alpha(x)=S_X(x). \tag{9.39}$$

Since $X_1',\ldots,X_k'$ are independent, the crude and marginal (net) hazard rates are the same, and from (9.37) equal to $h_\alpha(t)$. Hence

$$P_\alpha'^*(x)=\int_x^\infty h_\alpha'(t)S'_{X'}(t)\,dt=\int_x^\infty h_\alpha(t)S_X(t)\,dt=P_\alpha^*(x). \tag{9.40}$$

We have proved the *Equivalence Theorem*:

**Equivalence Theorem.** *For each joint SDF with dependent times due to die there always exists an equivalent model with independent times due to die.* [See Cox (1962), and also Tsiatis (1975).]

It can be shown, by contradiction, that the independent model is unique. The probability function, $P_\alpha^*(x)$, cannot be affected by any transformation such that $\min(X_1,\ldots,X_k)$ remains unchanged provided values of the other variables $X_\alpha$ [$>\min(X_1,\ldots,X_k)$] remain greater than $\min(X_1,\ldots,X_k)$. Therefore, there is an infinity of models with dependent times due to die equivalent [in terms of our definition given by (9.32)] to a given joint SDF. The given joint SDF (usually of fairly well-known form) will be called a *base* SDF, while the unique SDF with independent times due to die, defined in (9.38), will be called the *core* SDF [see Elandt-Johnson (1979)].

### 9.6.2 Nonidentifiability of the Member of a Parametric Family of Distributions

Suppose that for some reasons, we decide to fit a parametric SDF to multiple cause mortality data. In other words, we select a distribution model that is supposed to represent a *base* SDF. Could we identify the parameters of this SDF [question (2) at the beginning of Section 9.6]? The answer depends on the particular family concerned.

For example, Nádas (1971) has shown that if the SDF is bivariate normal, then its parameters can be uniquely determined from the distribution of $(X,I)$; Basu and Ghosh (1978) have discussed some other examples and conditions for identifiability of parameters of some other bivariate distributions. (See also Example 9.2 and discussion in Section 12.3.1.)

Formal consequences of equivalence and nonidentifiability will become apparent in Chapters 11 and 12.

## 9.7 PROPORTIONAL HAZARD RATES

Suppose that the ratio $h_\alpha(x)/h_X(x)$ does not depend on $x$, that is

$$h_\alpha(x)=c_\alpha h_X(x), \tag{9.41}$$

where $c_\alpha$ is a constant between 0 and 1. In this case, $h_\alpha(x)$ and $h_X(x)$ are said to be *proportional*.

We now derive some useful properties of various survival distributions, when the crude hazard rates are proportional.

1. From (9.17), we have

$$P_\alpha^*(x)=\int_x^\infty h_\alpha(t)S_X(t)\,dt=c_\alpha\int_x^\infty h_X(t)S_X(t)\,dt=c_\alpha S_X(x), \tag{9.42}$$

and so

$$P_\alpha^*(0)=c_\alpha=\pi_\alpha. \tag{9.43}$$

Hence

$$S_\alpha^*(x)=\frac{1}{\pi_\alpha}P_\alpha^*(x)=S_X(x). \tag{9.44}$$

We have obtained an important and useful result:

*If the crude rate, $h_\alpha(x)$, is proportional to the overall hazard rate $h_X(x)$, then the (conditional) crude SDF among those who eventually die from cause $C_\alpha$, $S_\alpha^*(x)$, is the same as the overall survival function, $S_X(x)$.*

It is of interest to study the ratio given in (9.26), $P_\alpha^*(x)/S_X(x)$. This is, in fact, the conditional probability of death after age $x$ from cause $C_\alpha$ in a population, given alive at age $x$. If this probability remains fairly stable, then the distribution of deaths from $C_\alpha$ in the presence of all causes, $S_\alpha^*(x)$, has the same pattern as the overall distribution, $S_X(x)$ (cf. also Section 10.4).

Note that this result does not require the times due to die to be independent [Elandt-Johnson (1976)]. Moreover, it can be shown that it also does not require validity of assumption A1 (that different causes are mutually exclusive events).

If (9.41) holds for each $\alpha = 1,2,\ldots,k$, and assumption A1 holds, then we also have

$$c_1 + c_2 + \cdots + c_k = 1. \tag{9.45}$$

In this case, it is easily seen that

$$\frac{h_\alpha(x)}{h_\beta(x)} = \frac{c_\alpha}{c_\beta} = c_{\alpha\beta}, \tag{9.46}$$

and so

$$\frac{P_\alpha^*(x)}{P_\beta^*(x)} = c_{\alpha\beta}, \tag{9.47}$$

for any $\alpha$, $\beta$.

2. There is another interesting property that is implied by the assumption of proportionality of hazard rates. This is associated with the marginal of the core SDF, $G_\alpha(x)$, defined in (9.35).

If (9.41) holds, we have from (9.35)

$$G_\alpha(x) = \exp\left[-c_\alpha \int_0^x h_X(t)\,dt\right] = \left\{\exp\left[-\int_0^x h_X(t)\,dt\right]\right\}^{c_\alpha} = [S_X(x)]^{c_\alpha}. \tag{9.48}$$

This means that if the hazard rates $h_\alpha(x)$ and $h_X(x)$ are proportional, the marginal of the core SDF is a fractional power (with the power exponent equal to the coefficient of proportionality, $c_\alpha$) of $S_X(x)$.

As we will see in Chapter 11, $G_\alpha(x)$ is not directly observable, but it is estimable. If we can find empirically that the crude hazard rates are proportional, then $G_\alpha(x)$ can easily be estimated from (9.48).

## 9.8 EXAMPLES

We illustrate the results obtained in Sections 9.2–9.7 by examples.

**Example 9.1** Consider a bivariate Farlie-Gumbel-Morgenstern family of distributions,

$$S_{12}(x_1,x_2)=S_1(x_1)S_2(x_2)\left[1+\theta F_1(x_1)F_2(x_2)\right], \qquad |\theta|\leqslant 1 \qquad (9.49)$$

[Johnson and Kotz (1972), pp. 262–263]. Clearly, the marginal SDF's are $S_1(x)$ and $S_2(x)$, respectively.

Let $S_\alpha(t)=e^{-x_\alpha}$, $\alpha=1,2$, be both standard exponential distributions. Then

$$S_{12}(x_1,x_2)=e^{-(x_1+x_2)}\left[1+\theta(1-e^{-x_1})(1-e^{-x_2})\right]. \qquad (9.50)$$

Hence, the overall survival function defined in (9.6) is

$$S_X(x)=S_{12}(x,x)=e^{-2x}\left[1+\theta(1-e^{-x})^2\right], \qquad (9.51)$$

and its hazard rate is

$$h_X(x)=-\frac{d\log S_X(x)}{dx}=2\left[1-\theta\frac{e^{-x}(1-e^{-x})}{1+\theta(1-e^{-x})^2}\right]. \qquad (9.52)$$

The crude hazard rate, $h_1(x)$, is from (9.8)

$$h_1(x)=-\left.\frac{d\log S_{12}(x_1,x_2)}{dx_1}\right|_{x_1=x_2=x}=1-\theta\frac{e^{-x}(1-e^{-x})}{1+\theta(1-e^{-x})^2}=\frac{1}{2}h_X(x), \qquad (9.53)$$

and, by symmetry, we also have $h_2(x)=h_1(x)$.

We note that the hazard rates are proportional with $c_1=c_2=\frac{1}{2}$. It follows [from (9.43)] that the crude SDF's of the base SDF given by (9.49) are

$$P_1^*(x)=P_2^*(x)=\tfrac{1}{2}S_X(x)=\tfrac{1}{2}e^{-2x}\left[1+\theta(1-e^{-x})^2\right], \qquad (9.54)$$

The marginal SDF's of the core SDF at $x_1=x_2=x$, are (in view of proportionality of hazards),

$$G_\alpha(x)=\left[S_X(x)\right]^{\frac{1}{2}}=e^{-x}\left[1+\theta(1-e^{-x})^2\right]^{\frac{1}{2}} \qquad (9.55)$$

so that the core SDF is [from (9.38)]

$$S'_{12}(x_1,x_2)=G_1(x_1)G_2(x_2)$$

$$=e^{-(x_1+x_2)}\left[1-\theta(1-e^{-x_1})^2\right]^{\frac{1}{2}}\left[1-\theta(1-e^{-x_2})^2\right]^{\frac{1}{2}}. \quad (9.56)$$

**Example 9.2** We now consider a bivariate Gumbel Type A survival function in the form

$$S_{12}(x_1,x_2)=S_1(x_1)S_2(X_2)\exp\left\{\rho\left[\frac{1}{-\log S_1(x_1)}+\frac{1}{-\log S_2(x_2)}\right]^{-1}\right\}, \quad (9.57)$$

with $0\leqslant\rho\leqslant 1$. Again the marginal SDFs are $S_1(x_1)$ and $S_2(x_2)$.

In fact, Gumbel (1962) confined his definition to the cases, where $S_1(x_1)$, $S_2(x_2)$ belong to one of the three types of limiting extreme value distributions. Of course, this restriction is not necessary to define (9.57).

We recall from (3.5) the cumulative hazard function

$$\Lambda_\alpha(x_\alpha)=\int_0^{x_\alpha}\lambda_\alpha(t)\,dt=-\log S_\alpha(x_\alpha).$$

Thus, (9.57) can be rewritten in the form

$$S_{12}(x_1,x_2)=\exp\left\{-\left[\Lambda_1(x_1)+\Lambda_2(x_2)\right]+\rho\frac{\Lambda_1(x_1)\Lambda_2(x_2)}{\Lambda_1(x_1)+\Lambda_2(x_2)}\right\}. \quad (9.58)$$

The overall SDF from (9.6) is

$$S_X(x)=S_{12}(x,x)=\exp\left\{-\left[\Lambda_1(x)+\Lambda_2(x)\right]+\rho\frac{\Lambda_1(x)\Lambda_2(x)}{\Lambda_1(x)+\Lambda_2(x)}\right\}, \quad (9.59)$$

with the hazard rate

$$h_X(x)=\left[\lambda_1(x)+\lambda_2(x)\right]-\rho\frac{\lambda_1(x)\Lambda_2^2(x)+\lambda_2(x)\Lambda_1^2(x)}{\left[\Lambda_1(x)+\Lambda_2(x)\right]^2}. \quad (9.60)$$

The crude hazard rates from (9.8) are

$$h_1(x)=\lambda_1(x)\left\{1-\rho\left[\frac{\Lambda_2(x)}{\Lambda_1(x)+\Lambda_2(x)}\right]^2\right\},$$

$$h_2(x)=\lambda_2(x)\left\{1-\rho\left[\frac{\Lambda_1(x)}{\Lambda_1(x)+\Lambda_2(x)}\right]^2\right\}. \quad (9.61)$$

Of course,

$$h_X(x)=h_1(x)+h_2(x).$$

In the general case, $P_\alpha^*(x)$'s and $G_\alpha(x)$'s ($\alpha=1,2$) cannot be obtained in explicit forms. This can be, however, sometimes done, when $S_\alpha(x_\alpha)$'s are specified.

**Example 9.3** Consider again the bivariate Gumbel Type A distribution defined in (9.57), and suppose that $S_\alpha(x_\alpha)=\exp(-\lambda_\alpha x_\alpha)$, $\lambda_\alpha>0$, $x_\alpha>0$, so that $\Lambda_\alpha(x_\alpha)=\lambda_\alpha x_\alpha$, $\alpha=1,2$.

Then the joint SDF (9.57) (base SDF) is of the form

$$S_{12}(x_1,x_2)=\exp\left\{-\left[(\lambda_1 x_1+\lambda_2 x_2)-\rho\frac{\lambda_1\lambda_2}{\lambda_1 x_1+\lambda_2 x_2}x_1x_2\right]\right\}. \tag{9.62}$$

Now the overall SDF (defined in (9.59)) is

$$S_X(x)=S_{12}(x,x)=\exp\left\{-\left[(\lambda_1+\lambda_2)-\rho\frac{\lambda_1\lambda_2}{\lambda_1+\lambda_2}\right]x\right\}. \tag{9.63}$$

Notice that in this case, $S_X(x)$ is also exponential with the parameter (which is, of course, also its hazard rate)

$$(\lambda_1+\lambda_2)-\rho\frac{\lambda_1\lambda_2}{\lambda_1+\lambda_2}=\gamma,\text{ say.} \tag{9.64}$$

The crude hazard rates defined in (9.61) are

$$h_1(x)=\lambda_1\left[1-\rho\left(\frac{\lambda_2}{\lambda_1+\lambda_2}\right)^2\right]=\gamma_1,\text{ say,}$$

and (9.65)

$$h_2(x)=\lambda_2\left[1-\rho\left(\frac{\lambda_1}{\lambda_1+\lambda_2}\right)^2\right]=\gamma_2,\text{ say.}$$

The hazard rates are proportional with

$$\frac{h_\alpha(x)}{h_X(x)}=\frac{\gamma_\alpha}{\gamma}=c_\alpha \qquad \text{and} \qquad \gamma_1+\gamma_2=\gamma. \tag{9.66}$$

We now evaluate $P^*_\alpha(x)$ (from (9.17)

$$P^*_\alpha(x)=\int_x^\infty \gamma_\alpha e^{-\gamma t}\,dt=\frac{\gamma_\alpha}{\gamma}e^{-\gamma x},\qquad \alpha=1,2. \tag{9.67}$$

[Note that (9.67) can be obtained directly from (9.42) as a consequence of the proportionality of hazard rates.]

Clearly,

$$S^*_\alpha(x)=\frac{P^*_\alpha(x)}{P^*_\alpha(0)}=e^{-\gamma x},\qquad \alpha=1,2. \tag{9.68}$$

Similarly [from (9.35) or (9.48)]

$$G_\alpha(x)=\exp\left[-\int_0^x \gamma_\alpha dt\right]=e^{-\gamma_\alpha x}$$

$$=[S_X(x)]^{\gamma_\alpha/\gamma}=S'_\alpha(x),\qquad \alpha=1,2. \tag{9.69}$$

These are the marginal SDF's of the core SDF

$$S'_{12}(x_1,x_2)=e^{-\gamma_1 x_1}\cdot e^{-\gamma_2 x_2}. \tag{9.70}$$

Clearly, $\gamma_1$ and $\gamma_2$ (and so $\gamma$) can be identified from (9.67), but we cannot identify, $\lambda_1$, $\lambda_2$, and $\rho$. There are only two equations (9.65) relating these three parameters to $\gamma_1$ and $\gamma_2$. The additional condition, that $X_1$ and $X_2$ are independent, (that is $\rho=0$) is necessary to identify the joint SDF. [For a formal proof, see Basu and Ghosh (1978).]

**Example 9.4** In the previous examples, we started with a known joint SDF and derived $S^*_\alpha(x)$ and $G_\alpha(x)$. In practice, however, we only observe the $S^*_\alpha(x)$'s. In this example, we assume that the $S^*_\alpha(x)$'s are known. We can then derive the $G_\alpha(x)$'s, and so the core joint SDF, but not the base SDF.

For simplicity, consider only two causes, $C_1$ and $C_2$.

Let

$$S^*_\alpha(x)=\exp\left[-\int_0^x \lambda^*_\alpha(t)\,dt\right]=\exp[-\Lambda^*_\alpha(t)] \tag{9.71}$$

be the observable SDF defined in (9.22), and $\lambda^*_\alpha(t)$ is its hazard rate [cf. (9.25)]. For convenience, assume $\lambda^*_1(x)>\lambda^*_2(x)$.

The crude hazard rate, $h_1(x)$, is from (9.28)

$$h_1(x) = \frac{\pi_1 \lambda_1^*(x) S_1^*(x)}{\pi_1 S_1^*(x) + \pi_2 S_2^*(x)} = \frac{\pi_1 \lambda_1^*(x) \exp[-\Lambda_1^*(x)]}{\pi_1 \exp[-\Lambda_1^*(x)] + \pi_2 \exp[-\Lambda_2^*(x)]},$$

$$= \lambda_1^*(x) \cdot \frac{c \cdot \exp[-\Lambda^*(x)]}{1 + c \cdot \exp[-\Lambda^*(x)]}, \tag{9.72}$$

where

$$\Lambda^*(x) = \Lambda_1^*(x) - \Lambda_2^*(x) > 0 \qquad \text{and} \qquad c = \pi_1/\pi_2. \tag{9.73}$$

By a similar argument, and noticing that $\Lambda_2^*(x) - \Lambda_1^*(x) = -\Lambda^*(x)$ and $\pi_2/\pi_1 = 1/c$, we obtain

$$h_2(x) = \lambda_2^*(x) \frac{\exp[\Lambda^*(x)]}{c + \exp[\Lambda^*(x)]}. \tag{9.74}$$

Of special interest is the case, when $\lambda_1^*(x)/\lambda_2^*(x) = \theta$, $0 < \theta \leqslant 1$, that is, the the hazard rates are proportional. In this case,

$$\Lambda^*(x) = (1-\theta)\Lambda_1^*(x) = \frac{1-\theta}{\theta} \Lambda_2^*(x), \tag{9.75}$$

and so

$$h_1(x) = \frac{c\lambda_1^*(x) \exp[-(1-\theta)\Lambda_1^*(x)]}{1 + c \exp[-(1-\theta)\Lambda_1^*(x)]}. \tag{9.76}$$

Substituting $1 + c \exp[-(1-\theta)\Lambda_1^*(t)] = z$ and remembering that $\pi_1 + \pi_2 = 1$, we obtain

$$-\int_0^x h_1(t)\,dt = \log\left[\{\pi_2 + \pi_1 \exp[-(1-\theta)\Lambda_1^*(x)]\}^{1/(1-\theta)}\right]. \tag{9.77}$$

Hence

$$G_1(x) = \exp\left[-\int_0^x h_1(t)\,dt\right] = \{\pi_2 + \pi_1 \exp[-(1-\theta)\Lambda_1^*(x)]\}^{1/(1-\theta)}. \tag{9.78}$$

We have $G_1(0) = 1$, and as $x \to \infty$, $G_1(x) \to (\pi_2)^{1/(1-\theta)}$, so that $G_1(x)$ is an improper SDF.

Similarly,

$$h_2(x)=\lambda_2^*(x)\frac{\exp\left[\frac{1-\theta}{\theta}\Lambda_2^*(x)\right]}{c+\exp\left[\frac{1-\theta}{\theta}\Lambda_2^*(x)\right]}, \tag{9.79}$$

and

$$G_2(x)=\left\{\pi_1+\pi_2\exp\left[\frac{1-\theta}{\theta}\Lambda_2^*(x)\right]\right\}^{-\theta/(1-\theta)} \tag{9.80}$$

We have $G_2(0)=1$, and as $x\to\infty$ $G_2(x)\to 0$, so that $G_2(x)$ is a proper distribution.

Note that although $\lambda_1^*(x)$ and $\lambda_2^*(x)$ are proportional, $h_1(x)$ and $h_2(x)$ are not proportional, unless $\lambda_1^*(x)=\lambda_2^*(x)$.

**Example 9.5** The results obtained in Example 9.4 are now illustrated assuming that $S_1^*(x)$ and $S_2^*(x)$ are Gompertz distributions with proportional hazard rates, that is

$$S_\alpha^*(x)=\exp\left[\frac{R_\alpha}{a}(1-e^{ax})\right], \qquad \alpha=1,2.$$

We have

$$\lambda_\alpha^*(x)=R_\alpha e^{ax},$$

so that

$$\frac{\lambda_2^*(x)}{\lambda_1^*(x)}=\frac{R_2}{R_1}=\theta, \qquad 1-\theta=\frac{R_1-R_2}{R_1}=\frac{R}{R_1}, \quad \text{and} \quad \frac{1-\theta}{\theta}=\frac{R}{R_2}.$$

Also

$$(1-\theta)\Lambda_1^*(x)=-\frac{R}{a}(1-e^{ax}),$$

$$\frac{1-\theta}{\theta}\Lambda_2^*(x)=-\frac{R}{a}(1-e^{ax}).$$

Substituting appropriate expressions into (9.78) and (9.80), we obtain

$$G_1(x)=\left\{\pi_2+\pi_1\exp\left[\frac{R}{a}(1-e^{ax})\right]\right\}^{R_1/R}, \tag{9.81}$$

and

$$G_2(x)=\left\{\pi_1+\pi_2\exp\left[\frac{R}{a}(e^{ax}-1)\right]\right\}^{-R_2/R}, \tag{9.82}$$

respectively.

For review and further examples, see Chiang (1968), Gail (1975), David and Moeschberger (1978), Birnbaum (1979), Elandt-Johnson (1979).

## 9.9 HETEROGENEOUS POPULATIONS: MIXTURE OF SURVIVAL FUNCTIONS

So far, we have discussed models associated with well-behaved proper joint survival functions. Such functions arise when there is no differential susceptibility to dying from different causes in a population (assumption A2).

If, however, causes of death are closely related to diseases that are genetically controlled, the population can be regarded as a *mixture* with respect to risk (liability) of dying from different causes. Many models can be constructed for such *heterogeneous* population, and it would be difficult to present a general treatment of this problem. But to give some idea, how the model can be constructed, we present a simple model for three causes.

Let $C_1$ and $C_2$ be two specific causes of death, and let $C_3$ denote all other causes, except $C_1$ and $C_2$. Since everybody must eventually die, we assume that everybody is liable to die from other causes ($C_3$). For the specific causes, we assume that only a proportion $\phi_1$ is liable to die solely from $C_1$ (and $C_3$), the proportion $\phi_2$, solely from $C_2$ (and $C_3$); and the proportion $\phi_{12}$ from $C_1$ and $C_2$ (and $C_3$) in a given population.

Let $S_\alpha^+(x_\alpha)$ be a (hypothetical) SDF for cause $C_\alpha$ ($\alpha=1,2$) alone in an isolated population, and let $S_3(x_3)$ be the SDF for $C_3$. Assume that times due to die from different causes are independent.

For a population that is a mixture of individuals liable to die from the three different causes we have

$$\begin{aligned}S_{123}(x_1,x_2,x_3)=\big[&\phi_1S_1^+(x_1)+\phi_2S_2^+(x_2)+\phi_{12}S_1^+(x_1)S_2^+(x_2)\\&+(1-\phi_1-\phi_2-\phi_{12})\big]S_3(x_3).\end{aligned} \tag{9.83}$$

We notice that the marginal SDFs are

$$\begin{aligned}S_1(x_1)&=S_{123}(x_1,0,0)=[1-(\phi_1+\phi_{12})][1-S_1^+(x_1)],\\S_2(x_2)&=S_{123}(0,x_2,0)=[1-(\phi_2+\phi_{12})][1-S_2^+(x_2)],\end{aligned} \tag{9.84}$$

and for all other causes ($C_3$), the marginal SDF is $S_3(x_3)$.

Notice that $S_1(x_1)$ and $S_2(x_2)$ are improper distributions. Also, in general,

$$S_{123}(x_1,x_2,x_3) \neq S_1(x_1)S_2(x_2)S_3(x_3), \tag{9.85}$$

so that the (hypothetical) times due to die in the population are *not independent*.

Evaluation of various kinds of survival functions can be done in the ways described in Sections 9.2–9.7, and is illustrated in Example 9.6.

**Example 9.6** Let $S_\alpha^+(x_\alpha) = \exp(-\lambda_\alpha x_\alpha)$, $\alpha = 1, 2$, and also $S_3(x_3) = \exp(-\lambda_3 x_3)$. Assume in model (9.83) that $\phi_{12} = 0$, that is, the mechanisms that control the liability of dying from $C_1$ or $C_2$ are in some way mutually exclusive. The joint SDF defined in (9.83) takes the form

$$S_{123}(x_1,x_2,x_3) = [\phi_1 \exp(\lambda_1 x_1) + \phi_2 \exp(-\lambda_2 x_2) + (1-\phi_1-\phi_2)]\exp(-\lambda_3 x_3). \tag{9.86}$$

We now evaluate various SDF's.

1. The overall SDF is

$$\begin{aligned} S_X(x) &= S_{123}(x,x,x) \\ &= [\phi_1 \exp(-\lambda_1 x) + \phi_2 \exp(-\lambda_2 x) + (1-\phi_1-\phi_2)]\exp(-\lambda_3 x) \end{aligned} \tag{9.87}$$

—this is, of course, a mixture of exponential survival functions.

2. We now find the crude probability function $P_1^*(x)$. We have [from (9.3)]

$$-\left.\frac{\partial S_{123}(\cdot)}{\partial x_1}\right|_{\{x_\beta = x\}} = \phi_1 \lambda_1 \exp[-(\lambda_1+\lambda_3)x],$$

so that from (9.14) and (9.3), we have

$$\begin{aligned} P_1^*(x) &= \int_x^\infty \left[ -\left.\frac{\partial S_{123}(\cdot)}{\partial x_1}\right|_{\{x_\beta = t\}} dt = \phi_1 \lambda_1 \int_x^\infty \exp[-(\lambda_1+\lambda_3)t]\,dt \right. \\ &= \phi_1 \frac{\lambda_1}{\lambda_1+\lambda_3} \exp[-(\lambda_1+\lambda_3)x], \end{aligned} \tag{9.88}$$

and so

$$P_1^*(0)=\pi_1=\phi_1\frac{\lambda_1}{\lambda_1+\lambda_3} \tag{9.89}$$

—this is the proportion expected to die from $C_1$ *in the population*. The (proper) SDF, given that an individual dies from $C_1$, is

$$S_1^*(x)=\frac{P_1^*(x)}{P_1^*(0)}=\exp[-(\lambda_1+\lambda_3)x]. \tag{9.90}$$

This is exponential with hazard rate $(\lambda_1+\lambda_3)$, which is higher than that when $C_1$ acts alone. This implies that in the population, in which all causes are acting, there is a greater proportion of deaths at younger ages. This is because some individuals who would have died of $C_1$ at more advanced ages, had $C_1$ been acting alone, will die, in fact, from $C_3$ at younger ages.

3. We now find the marginal survival function $S_1(x)$. We have

$$S_1(x)=S_{123}(x,0,0)=1-\phi_1[1-\exp(-\lambda_1 x)]=1-\phi_1 F_1^+(x), \tag{9.91}$$

which is an improper distribution.

The hazard rate of $S_1(x)$ is

$$\lambda_1(x)=-\frac{d\log S_1(x)}{dx}=\frac{\phi_1\lambda_1\exp(-\lambda_1 x)}{1-\phi_1[1-\exp(-\lambda_1 x)]}=\frac{\lambda_1\exp(-\lambda_1 x)}{\dfrac{1-\phi_1}{\phi_1}+\exp(-\lambda_1 x)}, \tag{9.92}$$

while the hazard rate of $S_1^+(x)$ is $\lambda_1$. Of course, $\lambda_1(x)\leqslant\lambda_1$.

From (9.91), we also derive the hypothetical distribution for cause $C_1$ alone:

$$F_1^+(x)=\frac{1-S_1(x)}{\phi_1} \quad \text{and} \quad S_1^+(x)=1-F_1^+(x). \tag{9.93}$$

4. We now find the core SDF associated with the base SDF given by (9.83).

The crude hazard rate $h_1(x)$ is [from (9.3)]

$$h_1(x)=\frac{\phi_1\lambda_1\exp[-(\lambda_1+\lambda_3)x]}{S_X(x)}, \tag{9.94}$$

where $S_X(x)$ is given by (9.87). Then from (9.35), we have

$$G_1(x_1) = S_1'(x_1) = \exp\left[-\int_0^{x_1} h_1(t)\,dt\right]. \tag{9.95}$$

Unfortunately, the integral (9.95), with $h_1(t)$ given by (9.94), cannot be evaluated in explicit form. We may obtain its value by numerical integration.

In a similar way, we obtain $S_2'(x_2) = G_2(x_2)$, and, of course $S_3'(x_3) = S_3(x_3)$. Then the core SDF is

$$S_{123}'(x_1, x_2, x_3) = S_1'(x_1) S_2'(x_2) S_3(x_3). \tag{9.96}$$

The hypothetical times due to die of this SDF, $X_1'$, $X_2'$, $X_3'$ are mutually independent. Writing $X' = \min(X_1', X_2', X_3')$, we recall (from the Equivalence Theorem in Section 9.6.1) that $S_{X'}(x)$ and $P_1'^*(x)$ are identical with $S_X(x)$ given by (9.87) and $P_1^*(x)$ given by (9.88), respectively. The same applies to $P_2'^*(x)$ and $P_3'^*(x)$.

## REFERENCES


Basu, A. P. and Ghosh, J. K. (1978). Identifiability of the multinormal and other distributions under competing risk model. *J. Mult. Anal.* **8**, 413–429.

Berman, S. M. (1963). Note on extreme values, competing risks and semi-Markov processes. *Ann. Math. Statist.* **34**, 1104–1106.

Birnbaum, Z. W. (1979). On the mathematics of competing risks. *DHEW Publication No. (PHS) 79-1351*. U.S. Department of Health, Education, and Welfare, pp. 1–58.

Chiang, C. L. (1968). *Introduction to Stochastic Processes*. Wiley, New York, Chapter 11.

Cox, D. R. (1962). *Renewal Theory*. Methuen, London, Chapter 10.

David, H. A. and Moeschberger, M. L. (1978). *The Theory of Competing Risks*. Griffin's Statistical Monographs and Courses No. 39, MacMillan, New York.

Elandt-Johnson, R. C. (1976). Conditional failure time distributions under competing risk theory with dependent failure times and proportional hazard rates. *Scand. Actu. J.*, 37–51.

Elandt-Johnson, R. C. (1978). Some properties of bivariate Gumbel Type A distributions with proportional hazard rates. *J. Mult. Anal.* **8**, 244–254.

Elandt-Johnson, R. C. (1979). Equivalence and nonidentifiability in competing risks: A review and critique. *Institute of Statistics Mimeo Series No. 1222*, Department of Biostatistics, Univ. of North Carolina, Chapel Hill, N.C. pp. 1–19.

Gail, M. (1975). A review and critique of some models used in competing risk analysis. *Biometrics* **31**, 209–222.

Gumbel, E. J. (1962). Multivariate extremal distributions. *Bull. Inst. Int. Stat.* **39**, Part 2, 469–475. Paris.

Hakulinen, T. and Rahiala, M. (1977). An example of the risk dependence and additivity of intensities in the theory of competing risks. *Biometrics* **33**, 557–559.

Johnson, N. L. and Kotz, S. (1972). *Distributions in Statistics: Continuous Multivariate Distributions*. Wiley, New York.

Johnson, N. L. and Kotz, S. (1975). A vector-valued multivariate hazard rate. *J. Mult. Anal.* **5**, 53–66.

Miller, D. R. (1977). A note on independence of multivariate lifetimes in competing risks, *Ann. Statist.* **5**, 576–579.

Nádas, A. (1971). The distribution of the identified minimum of a normal pair determines the distribution of the pair. *Technometrics* **13**, 201–202.

Tsiatis, A. (1975). A nonidentifiability aspect of the problem of competing risks. *Proc. Natl. Acad. Sci. USA* **72**, 20–22.

*Vital Statistics Instruction Manual, Part II. Procedures for Coding Multiple Causes of Deaths Occurring in 1955.* (1962). National Office of Vital Statistics, Washington, D.C.

## EXERCISES

**9.1.** If the hypothetical times due to die from $k$ causes, $X_1,\ldots,X_k$, are mutually *independent*, why are the events: "Death is from cause $C_\alpha$, $\alpha=1,2,\ldots,k$," *not independent?*

**9.2.** Consider a joint survival function

$$S_{12}(x_1,x_2)=S_1(x_1)S_2(x_2),$$

and assume that $S_1(x_1),S_2(x_2)$ are proper SDF's.

(*a*) Define the crude hazard rate, $h_\alpha(x)$, and the marginal hazard rate $\lambda_\alpha(x)$, $\alpha=1,2$. Show that $\lambda_\alpha(x)=h_\alpha(x)$.

(*b*) Derive the distribution of time at death from cause $C_1$ *in the presence* of cause $C_2$, among those who would eventually die from $C_1$, $S_1^*(x)$. Using the definition of hazard rate [given in (3.4)], give the formula for hazard rate corresponding to $S_1^*(x)$, denoting it by $\lambda_1^*(x)$, say. Explain, why $\lambda_1^*(x)\neq h_1(x)$? Does the result $\lambda_\alpha^*(x)\neq h_\alpha(x)$ hold generally?

(*c*) Suppose that the crude hazard rates $h_1(x)$, and $h_2(x)$ are proportional to the overall hazard rate $h(x)$, that is, $h_\alpha(x)/h_X(x)=c_\alpha$, $\alpha=1,2$. Calculate $\lambda_\alpha^*(x)$, $\alpha=1,2$, in this case.

(*d*) Calculate $\lambda_1^*(x)$, and $\lambda_2^*(x)$, when $h_1(x)/h_2(x)=c$.

**9.3.** Let $h_\alpha(x)$ and $\lambda_\alpha(x)$ be the crude and net hazard rates, respectively, for $\alpha=1,2,\ldots,k$, and let $X_1,\ldots,X_k$ denote the times due to fail. Which of the following statements are true and which are false?

(*a*) If $X_1, X_2, \ldots, X_k$ are *mutually independent*, then
 (*i*) $h_1(x)+h_2(x)+\cdots+h_k(x)=h_X(x)$;
 (*ii*) $\lambda_1(x)+\lambda_2(x)+\cdots+\lambda_k(x)=h_X(x)$.

(*b*) If $X_1, X_2, \ldots, X_k$ are *not independent*, then
 (*i*) $h_1(x)+h_2(x)+\cdots+h_k(x)=h_X(x)$;
 (*ii*) $\lambda_1(x)+\lambda_2(x)+\cdots+\lambda_k(x)=h_X(x)$.

**9.4.** What might be an interpretation of the survival function $G_\alpha(x)=\exp[-\int_0^x h_\alpha(t)\,dt]$ defined in (9.35)?

**9.5.** Discuss the problem of nonidentifiability presented on Sections 9.6.1 and 9.6.2. Which aspects of nonidentifiability are discussed in each section?

**9.6.** Let $S_{12}(x_1,x_2)$ represent a joint SDF (defined in (9.1) for $k=2$), and let $S_{12}^*(x_1,x_2)=G_1(x_1)G_2(x_2)$ be the joint SDF of independent times due to die (defined in (9.38)).

(*a*) Under what conditions are these two models equivalent?

Suppose they are equivalent, and that a goodness of fit test indicates that $S_{12}(x_1,x_2)$ gives good fit to some fit to some mortality data from two causes.

(*b*) Does model $S_{12}^*(x_1,x_2)$ fit these data in exactly the same way?

(*c*) On what basis might you prefer one model over the other?

CHAPTER 10

# Multiple Decrement Life Tables

## 10.1 MULTIPLE DECREMENT LIFE TABLES: NOTATION

In Chapter 9, we have used a probabilistic approach to introduce various distributions associated with failures from causes, $C_1, C_2, \ldots, C_k$. We pointed out that the only observable survival functions are $S_X(x)$, which is the SDF from all causes, and $P_\alpha^*(x)$, which is the crude probability function of failure after age $x$ from specified cause $C_\alpha$ for each $\alpha = 1, 2, \ldots, k$. Estimation of these survival functions from population mortality data results in construction of *multiple decrement life tables* (briefly, MDLT), which are a part of life tables arranged by causes of death (for example, the U. S. Life Tables by Causes of Death, 1969–1971).

In this chapter, we discuss the construction of MDLT from current population (cross-sectional) mortality data. Using these tables, we will try to answer question (1), asked in the Introduction to Chapter 9, about mortality structure from different causes of death acting simultaneously in a population.

First about notation. All life table functions in MDLT will have the prefix $a$ (denoting: in the presence of *all* causes). For example, $al_x$, $aq_x$, $am_x$, $ad_x$. Since the MDLT are usually abridged, we shall also use the prefix $n$, that is, ${}_naq_x$, ${}_nam_x$, ${}_nad_x$. For cause $C_\alpha$, we then have ${}_naq_{\alpha x}$, ${}_nam_{\alpha x}$, ${}_nad_{\alpha x}$, and so on. [In English Life Tables the corresponding notation is $(al)_x$, ${}_n(aq)_x$, ${}_n(aq)_x^{(\alpha)}$, ${}_n(am)_x^{(\alpha)}$, etc.]

## 10.2 DEFINITIONS OF THE MDLT FUNCTIONS

For convenience, we recall definitions of life tables functions for all causes (see Chapter 4), using the modified notation with the prefix $a$.

$al_x$ denotes the expected number of survivors at exact age $x$ out of $al_0$ starters. We usually take $al_0 = 10{,}000{,}000$ (to get some reasonable numbers of expected deaths for some not common causes, for certain ages).
${}_n aq_x$ is the conditional death probability in $[x, x+n)$ given alive at age $x$;
${}_n ad_x$ is the total expected number of deaths in age interval $[x, x+n)$;
${}_n aL_x$ is the amount of person-years in $[x, x+n)$, and finally,
${}_n am_x$ is the overall central rate per year per person;

Clearly, $al_x / al_0$ is the analogue of the overall (from all causes) survival function, $S_X(x)$.

We introduce some further notation.

$al_{\alpha x}$ is the number living at exact age $x$ who ultimately are expected to die from cause $C_\alpha$.

Clearly, the ratio $al_{\alpha x} / al_0$ is an analogue of $P_\alpha^*(x)$ defined in (9.17), and $al_{\alpha 0} / al_0 = \pi_\alpha$ is the expected proportion of deaths from cause $C_\alpha$. It follows that $al_{\alpha x} / al_{\alpha 0}$ is an analogue of $S_\alpha^*(x)$ defined in (9.22), and $(1 - al_{\alpha x} / al_{\alpha 0})$ is an analogue of $F_\alpha^*(x)$, the distribution of time at death in the population, among those who die from cause $C_\alpha$.

We also introduce the notation:

${}_n ad_{\alpha x}$ the expected number of deaths between age $x$ and $x+n$ from cause $C_\alpha$, among $al_x$ living at age $x$.

Since we assume that deaths from different causes are mutually exclusive events, we have

$$ad_x = ad_{1x} + ad_{2x} + \cdots + ad_{kx}, \tag{10.1}$$

and

$$al_x = al_{1x} + al_{2x} + \cdots + al_{kx}. \tag{10.2}$$

## 10.3 RELATIONSHIPS AMONG FUNCTIONS OF MULTIPLE DECREMENT LIFE TABLE

We also define

${}_n aq_{\alpha x}$ the crude conditional probability of death from cause $C_\alpha$ between age $x$ and $x+n$, in the presence of all other causes, given alive at exact age $x$.

We have

$$ {}_n aq_x = \frac{{}_n ad_x}{al_x}, \tag{10.3} $$

and

$$ {}_n aq_{\alpha x} = \frac{{}_n ad_{\alpha x}}{al_x}. \tag{10.4} $$

Hence, from (10.1) we have

$$ {}_n aq_x = {}_n aq_{1x} + {}_n aq_{2x} + \cdots + {}_n aq_{kx}. \tag{10.5} $$

Also, from (10.3) and (10.4), we obtain an important formula

$$ {}_n aq_{\alpha x} = \frac{{}_n ad_{\alpha x}}{{}_n ad_x} {}_n aq_x. \tag{10.6} $$

We also define

${}_n am_{\alpha x}$ the central rate (per year per person) associated with cause $C_\alpha$.

We have

$$ {}_n am_{\alpha x} = \frac{{}_n ad_{\alpha x}}{{}_n aL_x}. \tag{10.7} $$

Clearly,

$$ {}_n am_x = {}_n am_{1x} + {}_n am_{2x} + \cdots + {}_n am_{kx}, \tag{10.8} $$

and

$$ {}_n am_{\alpha x} = \frac{{}_n ad_{\alpha x}}{{}_n ad_x} \cdot {}_n am_x. \tag{10.9} $$

Using *linear* interpolation on $al_x$, we have

$$ {}_n aL_x \doteq n\left(al_x - \tfrac{1}{2}\,{}_n ad_x\right), $$

and so

$$ al_x \doteq \frac{1}{n}\left({}_n aL_x + \frac{n}{2}\,{}_n ad_x\right), $$

so that (10.7) can be written in the form

$$ {}_{n}am_{\alpha x}=\frac{{}_{n}ad_{\alpha x}}{n\left(al_{x}-\frac{1}{2}{}_{n}ad_{x}\right)}=\frac{{}_{n}aq_{\alpha x}}{n\left(1-\frac{1}{2}{}_{n}aq_{x}\right)} \tag{10.10}$$

(cf. also Section 4.6.2.).

Hence ${}_{n}aq_{\alpha x}$ can also be written in the form

$$ {}_{n}aq_{\alpha x}\doteqdot\frac{{}_{n}am_{\alpha x}}{\frac{1}{n}\left(1+\frac{n}{2}\cdot{}_{n}am_{x}\right)}, \tag{10.11}$$

or multiplying the numerator and denominator by ${}_{n}aL_{x}$,

$$ {}_{n}aq_{\alpha x}\doteqdot\frac{{}_{n}ad_{\alpha x}}{\frac{1}{n}\left({}_{n}aL_{x}+\frac{n}{2}\cdot{}_{n}ad_{x}\right)}. \tag{10.11a}$$

Note that the numerator in (10.11) is the age- and cause-specific central rate, ${}_{n}am_{\alpha x}$, whereas the denominator includes the overall age-specific rate, ${}_{n}am_{x}$.

We may use an improved formula, replacing $\frac{1}{2}n$ by $n(1-{}_{n}f_{x})$ as discussed in Section 4.8. Then (10.11) takes the modified form

$$ {}_{n}aq_{\alpha x}\doteqdot\frac{{}_{n}am_{\alpha x}}{\frac{1}{n}\left[1+n(1-{}_{n}f_{x}){}_{n}am_{x}\right]}. \tag{10.12}$$

[cf. (4.59)]

## 10.4 CRUDE FORCES OF MORTALITY

In life table terminology, the phrase "force of mortality" is used rather than "hazard rate." Also the symbols $a\mu_{x}$ and $a\mu_{\alpha x}$ are used for $h_{X}(x)$ and $h_{\alpha x}$, respectively.

The overall force of mortality is then

$$ a\mu_{x}=-\frac{d\log(al_{x})}{dx}=-\frac{1}{al_{x}}\cdot\frac{d(al_{x})}{dx} \tag{10.13}$$

(cf. Section 4.3).

The analogue of $h_{\alpha}(x)$ defined in (9.27) is

$$ a\mu_{\alpha x}=-\frac{1}{al_{x}}\frac{d(al_{\alpha x})}{dx}. \tag{10.14}$$

Of course,

$$a\mu_x = a\mu_{1x} + a\mu_{2x} + \cdots + a\mu_{kx}. \tag{10.15}$$

When the force of mortality for $C_\alpha$ is *proportional* to the overall force of mortality, that is, when

$$\frac{{}_n a\mu_{\alpha x}}{{}_n a\mu_x} = c_\alpha \quad \text{for all } x, \tag{10.16}$$

then, from (9.44), we also have

$$\frac{al_{\alpha x}}{al_x} = c_\alpha = \pi_\alpha. \tag{10.17}$$

Note that, in general, the ratio

$$\frac{al_{\alpha x}}{al_x} = \pi_{\alpha x} \tag{10.18}$$

is not constant. Note that this is analogous to (9.26). In fact (10.18) represents the conditional probability of death from cause $C_\alpha$ after age $x$, given alive at age $x$. If this ratio is constant over all intervals, it means, that the survival probability for cause $C_\alpha$ in the population, $al_{\alpha x}/al_{\alpha 0}$ [the analogue of $S_\alpha^*(x)$] is the same as the survival probability for all causes $al_x/al_0$ [the analogue of $S_X(x)$] (cf. Section 9.7).

## 10.5 CONSTRUCTION OF MULTIPLE DECREMENT LIFE TABLES FROM POPULATION (CROSS-SECTIONAL) MORTALITY DATA

### 10.5.1 Mortality Data: Evaluation of ${}_n aq_x$ and ${}_n aq_{\alpha x}$

Mortality data are published in *Vital Statistics*. They usually represent the observed numbers of deaths in age intervals 0–1, 1–5, 5–10, etc., until age 85+, from any specific causes.

We recall (from Section 4.9.4) the notation:

${}_nK_x$ the midyear (or census) population in age group $x$ to $x+n$;
${}_nf_x$ the expected fraction of the last $n$ years of life of individuals present age $x$;
${}_nD_x$ the *observed* number of all deaths in $[x, x+n)$;
${}_nM_x = {}_nD_x/{}_nK_x$, the age specific death rate.

We introduce similar notation for cause $C_\alpha$:

${}_nD_{\alpha x}$ the *observed* number of deaths from cause $C_\alpha$ in $[x, x+n)$, and
${}_nM_{\alpha x} = {}_nD_{\alpha x}/{}_nK_{\alpha x}$, the observed age and cause specific death rate.

Clearly, we have

$$ {}_nD_x = {}_nD_{1x} + {}_nD_{2x} + \cdots + {}_nD_{kx}, \tag{10.19} $$

and

$$ {}_nM_x = {}_nM_{1x} + {}_nM_{1x} + {}_nM_{2x} + \cdots + {}_nM_{kx}. \tag{10.20} $$

Our purpose is to estimate ${}_naq_{\alpha x}$'s from the population mortality data. Similarly, as in Section 4.9, we make the assumption that the central rates of the life table are approximately equal to the corresponding *observed* age specific (central) rates, that is

$$ {}_nam_x \doteq {}_nM_x \qquad \text{and} \qquad {}_nam_{\alpha x} \doteq {}_nM_{\alpha x}, \qquad \alpha = 1, 2, \ldots, k. \tag{10.21} $$

We then estimate ${}_naq_x$ from the formula

$$ {}_n\widehat{aq}_x = \frac{{}_nM_x}{\dfrac{1}{n}\left[1 + n(1 - {}_nf_x){}_nM_x\right]}. \tag{10.22} $$

Introducing the effective number of lives at exact age $x$,

$$ N'_x = \frac{1}{n}\left[{}_nK_x + n(1 - {}_nf_x){}_nD_x\right], \tag{10.23} $$

formula (10.22) takes the form

$$ {}_n\widehat{aq}_x = \frac{{}_nD_x}{N'_x}. \tag{10.24} $$

The crude probabilities ${}_naq_{\alpha x}$'s, can be estimated from formulae

$$ {}_n\widehat{aq}_{\alpha x} = \frac{{}_nD_{\alpha x}}{N'_x}, \tag{10.25} $$

or using the relation (10.24), from

$$ {}_n\widehat{aq}_{\alpha x} = \frac{{}_nD_{\alpha x}}{{}_nD_x}\,{}_n\widehat{aq}_x \qquad \text{for } \alpha = 1, 2, \ldots, k. \tag{10.26} $$

### 10.5.2 Construction of MDLT

Once ${}_n\widehat{aq}_x$'s and the ${}_n\widehat{aq}_{\alpha x}$'s are calculated, we assume that

$$ {}_naq_x \doteq {}_n\widehat{aq}_x \quad \text{and} \quad {}_naq_{\alpha x} \doteq {}_n\widehat{aq}_{\alpha x}. \tag{10.27}$$

First, we construct the regular life table for all causes as described in Section 4.9. We take a radix $al_0$, and then calculate progressively

$$al_{x+n} = al_x(1 - {}_naq_x), \tag{10.28}$$

and

$${}_nad_x = al_x \cdot {}_naq_x. \tag{10.29}$$

The next step is to calculate the ${}_nad_{\alpha x}$'s. These can be obtained from the formula

$${}_nad_{\alpha x} = al_x \cdot {}_naq_{\alpha x}, \tag{10.30}$$

or combining (10.26) and (10.29), from the formula

$${}_nad_{\alpha x} = \frac{{}_nD_{\alpha x}}{{}_nD_x} \cdot {}_nad_x. \tag{10.31}$$

The U. S. Life Tables by Causes of Death provide tables that include columns for $al_x$, ${}_nad_x$, and ${}_nad_{\alpha x}$ (see, for example, the 1969–1971 Tables 6–10 on pp. 40–47). Some causes in these tables are subsets of other cases (e.g. Malignant Neoplasms of Digestive System are included in All Malignant Neoplasms); only deaths from mutually exclusive causes add to the total deaths.

Next, we calculate the $al_{\alpha x}$ functions by adding together all deaths from cause $C_\alpha$ which occur *after* age $x$, that is

$$al_{\alpha x} = {}_nad_{\alpha x} + {}_nad_{\alpha, x+n} + \cdots + {}_\infty ad_{\alpha \omega'}, \tag{10.32}$$

$$= {}_naq_{\alpha x} \cdot al_x + {}_naq_{\alpha, x+n} \cdot al_{x+n} + \cdots + al_{\alpha\omega'}, \tag{10.32a}$$

where $\omega'$ is the last recorded age (e.g., $\omega' = 85$).

The U.S. Life Tables do not give $al_{\alpha x}$'s explicitly, but they give the conditional probabilities $\pi_{\alpha x} = al_{\alpha x}/al_x$. In the U. S. Life Tables 1969–1971, these probabilities are given in sets of Tables No. 11–15 on pp. 48–57.

To illustrate the techniques and discuss what one might learn about mortality models from these tables, we give an example.

## 10.6 SOME MAJOR CAUSES OF DEATH: AN EXAMPLE OF CONSTRUCTING THE MDLT

Table 10.1 represents the 1970 Census population (${}_nK_x$) and the observed numbers of deaths for four major causes of death for White Males: $C_1$, Malignant neoplasms (140–209); $C_2$, Diseases of the circulatory system (390–448); $C_3$, Accidents (external) (E800–E999); and $C_4$, All remaining causes. The numbers in parentheses are code numbers according to the Eighth Classification (1967).

**Table 10.1** Deaths from four major causes. U.S. 1970 census population of White Males

| Age group $x$ to $x+n$ | Population size ${}_nK_x$ | Total deaths ${}_nD_x$ | Malignant neoplasms ${}_nD_{1x}$ | Circulatory system ${}_nD_{2x}$ | Accidents (external) ${}_nD_{3x}$ | Other causes ${}_nD_{4x}$ | Fraction of last interval lived ${}_nf_x$ |
|---|---|---|---|---|---|---|---|
| 0–1 | 1,501,250 | 31,725 | 65 | 267 | 935 | 30,458 | 0.129 |
| 1–5 | 5,873,083 | 4,910 | 501 | 163 | 2,119 | 2,127 | 0.430 |
| 5–10 | 8,633,093 | 4,099 | 708 | 134 | 2,119 | 1,138 | 0.470 |
| 10–15 | 9,033,725 | 4,382 | 526 | 163 | 2,708 | 985 | 0.597 |
| 15–20 | 8,291,270 | 12,200 | 742 | 319 | 9,628 | 1,511 | 0.549 |
| 20–25 | 6,940,820 | 13,812 | 867 | 443 | 10,834 | 1,668 | 0.502 |
| 25–30 | 5,849,792 | 9,897 | 800 | 595 | 7,051 | 1,451 | 0.488 |
| 30–35 | 4,925,069 | 9,130 | 941 | 1,202 | 5,246 | 1,741 | 0.517 |
| 35–40 | 4,784,375 | 12,459 | 1,608 | 3,104 | 5,059 | 2,688 | 0.535 |
| 40–45 | 5,194,497 | 21,819 | 3,392 | 7,848 | 5,608 | 4,971 | 0.537 |
| 45–50 | 5,257,619 | 35,992 | 6,463 | 15,817 | 5,955 | 7,757 | 0.535 |
| 50–55 | 4,832,555 | 53,092 | 10,893 | 25,843 | 5,782 | 10,574 | 0.535 |
| 55–60 | 4,310,921 | 76,502 | 17,133 | 38,853 | 5,791 | 14,725 | 0.529 |
| 60–65 | 3,647,243 | 98,781 | 22,505 | 52,365 | 5,145 | 18,766 | 0.522 |
| 65–70 | 2,807,974 | 113,614 | 24,691 | 63,340 | 4,404 | 21,179 | 0.515 |
| 70–75 | 2,107,552 | 122,829 | 24,317 | 72,177 | 3,890 | 22,445 | 0.508 |
| 75–80 | 1,437,628 | 124,979 | 21,468 | 77,556 | 3,495 | 22,460 | 0.496 |
| 80–85 | 805,564 | 101,556 | 14,260 | 67,265 | 2,746 | 17,285 | 0.479 |
| 85+ | 486,957 | 90,339 | 8,630 | 63,998 | 2,469 | 15,242 | — |
| Sum | 86,720,987 | 942,117 | 160,510 | 491,452 | 90,984 | 199,171 | |

From these data, ${}_nM_{\alpha x}$'s and ${}_naq_{\alpha x}$'s were calculated as described in Section 10.4.1. The values of ${}_nf_x$ from Table 4.4, column 5 were used.

Table 10.2 includes values of ${}_naq_x$ and ${}_naq_{\alpha x}$'s, and Table 10.3 exhibits the expected numbers of deaths from different causes.

Then, using the ${}_nad_{\alpha x}$ columns, the values of $al_{\alpha x}$ from (10.32), and the probabilities $\pi_{\alpha x}$ from (10.18) were obtained (see Tables 10.4 and 10.5, respectively).

It is worthwhile noticing that for malignant neoplasms, the ratio $\pi_{1x} = al_{1x}/al_x$ is fairly constant over a rather wide range of ages (0–65). It is, on the average, approximately 0.175. This indicates that also the ratios of forces of mortality, $a\mu_{1x}/a\mu_x$, are approximately constant. This is not true for diseases of circulatory system. The ratio $\pi_{2x} = al_{2x}/al_x$ increases with age from 0.55 at age $x=0$ to 0.71 at age $x=85$.

**Table 10.2.** Conditional probabilities of death from specific causes, ${}_naq_{\alpha x}$, in the population U. S. White Males, 1970

| Age group $x$ to $x+n$ | Total ${}_naq_x$ | Malignant neoplasms ${}_naq_{1x}$ | Circulatory system ${}_naq_{2x}$ | Accidents (external) ${}_naq_{3x}$ | Other causes ${}_naq_{4x}$ |
|---|---|---|---|---|---|
| 0–1 | 0.0207505 | 0.0000425 | 0.0001746 | 0.0006116 | 0.0199218 |
| 1–5 | 0.0033377 | 0.0003406 | 0.0001108 | 0.0014404 | 0.0014459 |
| 5–10 | 0.0023711 | 0.0004095 | 0.0000775 | 0.0012258 | 0.0006583 |
| 10–15 | 0.0024230 | 0.0002909 | 0.0000901 | 0.0014974 | 0.0005446 |
| 15–20 | 0.0073328 | 0.0004460 | 0.0001917 | 0.0057869 | 0.0009082 |
| 20–25 | 0.0099008 | 0.0006215 | 0.0003176 | 0.0077661 | 0.0011957 |
| 25–30 | 0.0084228 | 0.0006808 | 0.0005064 | 0.0060007 | 0.0012349 |
| 30–35 | 0.0092276 | 0.0009511 | 0.0012148 | 0.0053021 | 0.0017596 |
| 35–40 | 0.0129363 | 0.0016696 | 0.0032229 | 0.0052528 | 0.0027910 |
| 40–45 | 0.0207998 | 0.0032336 | 0.0074814 | 0.0053460 | 0.0047388 |
| 45–50 | 0.0336922 | 0.0060500 | 0.0148063 | 0.0055745 | 0.0072614 |
| 50–55 | 0.0535635 | 0.0109897 | 0.0260725 | 0.0058334 | 0.0106679 |
| 55–60 | 0.0851710 | 0.0190745 | 0.0432557 | 0.0064472 | 0.0163936 |
| 60–65 | 0.1271860 | 0.0289764 | 0.0674228 | 0.0066245 | 0.0241623 |
| 65–70 | 0.1842297 | 0.0400374 | 0.1027084 | 0.0071413 | 0.0343426 |
| 70–75 | 0.2548624 | 0.0504562 | 0.1497627 | 0.0080715 | 0.0465720 |
| 75–80 | 0.3565581 | 0.0612471 | 0.2212629 | 0.0099710 | 0.0640771 |
| 80–85 | 0.4745087 | 0.0666281 | 0.3142880 | 0.0128304 | 0.0807622 |
| 85+ | 1.0000000 | 0.0955291 | 0.7084205 | 0.0273304 | 0.1687200 |

A trend in the opposite direction is observed for accidental deaths ($C_3$); the ratio $\pi_{3x}$ decreases from 0.079 to 0.027. It should be noticed that $C_3$ covers a large class of accidental deaths, including poisonings, suicides, homicides, and injuries resulting from war.

We also obtained the (proper) age at death distributions—the analogues of $F_\alpha^*(x)$'s defined in Section 9.4.3; we denote them here by $F_{\alpha x}^*$'s. We have $F_{\alpha x}^* = 1 - al_{\alpha x}/al_{\alpha 0}$. These distributions are given in Table 10.6.

Figure 10.1 gives graphical representation of these results. Above age 85 the functions are not known, so that the curves are smoothed by hand (dotted lines). Figure 10.2 gives curves of death from various causes acting simultaneously in the population. They are, in fact, the approximate

**Table 10.3.** Life table deaths from specified causes. U. S. White Males, 1970

| Age group $x$ to $x+n$ | Conditional probability of death ${}_naq_x$ | Number living at beginning of age interval $al_x$ | Total deaths ${}_nad_x$ | Malignant neoplasms ${}_nad_{1x}$ | Circulatory system ${}_nad_{2x}$ | Accidents (external) ${}_nad_{3x}$ | Other causes ${}_nad_{4x}$ |
|---|---|---|---|---|---|---|---|
| 0–1 | 0.0207505 | 10,000,000 | 207,505 | 425 | 1,746 | 6,116 | 199,218 |
| 1–5 | 0.0033377 | 9,792,495 | 32,684 | 3,335 | 1,085 | 14,105 | 14,159 |
| 5–10 | 0.0023711 | 9,759,811 | 23,141 | 3,997 | 756 | 11,963 | 6,425 |
| 10–15 | 0.0024230 | 9,736,670 | 23,592 | 2,832 | 877 | 14,580 | 5,303 |
| 15–20 | 0.0073328 | 9,713,078 | 71,224 | 4,332 | 1,862 | 56,209 | 8,821 |
| 20–25 | 0.0099008 | 9,641,854 | 95,462 | 5,992 | 3,062 | 74,880 | 11,528 |
| 25–30 | 0.0084228 | 9,546,392 | 80,407 | 6,499 | 4,834 | 57,285 | 11,789 |
| 30–35 | 0.0092276 | 9,465,985 | 87,348 | 9,003 | 11,499 | 50,190 | 16,656 |
| 35–40 | 0.0129363 | 9,378,637 | 121,325 | 15,659 | 30,226 | 49,264 | 26,176 |
| 40–45 | 0.0207998 | 9,257,312 | 192,550 | 29,934 | 69,258 | 49,490 | 43,868 |
| 45–50 | 0.0336922 | 9,064,762 | 305,412 | 54,842 | 134,216 | 50,531 | 65,823 |
| 50–55 | 0.0535635 | 8,759,350 | 469,181 | 96,262 | 228,378 | 51,097 | 93,444 |
| 55–60 | 0.0851710 | 8,290,169 | 706,082 | 158,131 | 358,597 | 53,448 | 135,906 |
| 60–65 | 0.1271860 | 7,584,087 | 964,590 | 219,760 | 511,340 | 50,241 | 183,249 |
| 65–70 | 0.1842297 | 6,619,497 | 1,219,508 | 265,027 | 679,878 | 47,272 | 227,331 |
| 70–75 | 0.2548624 | 5,399,989 | 1,376,254 | 272,463 | 808,717 | 43,586 | 251,488 |
| 75–80 | 0.3565581 | 4,023,735 | 1,434,695 | 246,442 | 890,303 | 40,121 | 257,829 |
| 80–85 | 0.4745087 | 2,589,040 | 1,228,522 | 172,503 | 813,704 | 33,218 | 209,097 |
| 85+ | 1.0000000 | 1,360,518 | 1,360,518 | 129,969 | 963,819 | 37,183 | 229,547 |

**Table 10.4.** Expected number of individuals present age $x$, who ultimately die from cause $C_\alpha$, $al_{\alpha x}$ U. S. White Males, 1970

| Age group $x$ to $x+n$ | Total $al_x$ | Malignant neoplasms $al_{1x}$ | Circulatory system $al_{2x}$ | Accidents (external) $al_{3x}$ | Other causes $al_{4x}$ |
|---|---|---|---|---|---|
| 0–1 | 100,000 | 16,974 | 55,142 | 7,908 | 19,976 |
| 1–5 | 97,925 | 16,970 | 55,124 | 7,847 | 17,984 |
| 5–10 | 97,598 | 16,936 | 55,113 | 7,706 | 17,843 |
| 10–15 | 97,367 | 16,896 | 55,106 | 7,586 | 17,779 |
| 15–20 | 97,131 | 16,868 | 55,097 | 7,440 | 17,726 |
| 20–25 | 96,418 | 16,825 | 55,078 | 6,878 | 17,637 |
| 25–30 | 95,464 | 16,765 | 55,048 | 6,129 | 17,522 |
| 30–35 | 95,660 | 16,700 | 54,999 | 5,557 | 17,404 |
| 35–40 | 93,786 | 16,610 | 54,884 | 5,054 | 17,238 |
| 40–45 | 92,573 | 16,453 | 54,582 | 4,562 | 16,976 |
| 45–50 | 90,648 | 16,154 | 53,890 | 4,067 | 16,537 |
| 50–55 | 87,594 | 15,606 | 52,547 | 3,562 | 15,879 |
| 55–60 | 82,902 | 14,643 | 50,264 | 3,051 | 14,944 |
| 60–65 | 75,841 | 13,062 | 46,678 | 2,516 | 13,585 |
| 65–70 | 66,195 | 10,864 | 41,564 | 2,014 | 11,753 |
| 70–75 | 54,000 | 8,214 | 34,765 | 1,541 | 9,480 |
| 75–80 | 40,237 | 5,489 | 26,678 | 1,105 | 6,965 |
| 80–85 | 25,890 | 3,025 | 17,775 | 704 | 4,386 |
| 85+ | 13,605 | 1,300 | 9,638 | 372 | 2,295 |

**Table 10.5.** Conditional probabilities $[\pi_{\alpha x} = al_{\alpha x}/al_x]$ of eventually dying from specified causes alive at age $x$. U. S. White Males, 1969–1971

| Exact age $x$ | Malignant neoplasms $\pi_{1x}$ | Circulatory system $\pi_{2x}$ | Accidents (external) $\pi_{3x}$ | Other causes $\pi_{4x}$ |
|---|---|---|---|---|
| 0 | 0.1697 | 0.5514 | 0.0791 | 0.1998 |
| 1 | 0.1733 | 0.5629 | 0.0801 | 0.1837 |
| 5 | 0.1735 | 0.5647 | 0.0790 | 0.1828 |
| 10 | 0.1735 | 0.5660 | 0.0779 | 0.1826 |
| 15 | 0.1737 | 0.5672 | 0.0766 | 0.1825 |
| 20 | 0.1745 | 0.5712 | 0.0713 | 0.1829 |
| 25 | 0.1756 | 0.5766 | 0.0642 | 0.1835 |
| 30 | 0.1764 | 0.5810 | 0.0587 | 0.1839 |
| 35 | 0.1771 | 0.5852 | 0.0539 | 0.1838 |
| 40 | 0.1777 | 0.5896 | 0.0493 | 0.1834 |
| 45 | 0.1782 | 0.5945 | 0.0449 | 0.1824 |
| 50 | 0.1782 | 0.5999 | 0.0407 | 0.1813 |
| 55 | 0.1766 | 0.6063 | 0.0368 | 0.1803 |
| 60 | 0.1722 | 0.6155 | 0.0332 | 0.1791 |
| 65 | 0.1641 | 0.6279 | 0.0304 | 0.1775 |
| 70 | 0.1521 | 0.6438 | 0.0285 | 0.1755 |
| 75 | 0.1364 | 0.6630 | 0.0275 | 0.1731 |
| 80 | 0.1168 | 0.6866 | 0.0272 | 0.1694 |
| 85 | 0.0955 | 0.7084 | 0.0273 | 0.1687 |

**Table 10.6** Cumulative distribution functions, $F^*_{\alpha x}$, of age at death for different causes, each acting in the presence of all causes. U. S. White Males, 1970

| Exact age $x$ | Malignant neoplasms $F^*_{1x}$ | Circulatory system $F^*_{2x}$ | Accidents (external) $F^*_{3x}$ | Other causes $F^*_{4x}$ | Total deaths $F_x$ |
|---|---|---|---|---|---|
| 0 | 0.0000 | 0.0000 | 0.0000 | 0.0000 | 0.0000 |
| 1 | 0.0003 | 0.0003 | 0.0077 | 0.0997 | 0.0208 |
| 5 | 0.0022 | 0.0005 | 0.0256 | 0.1068 | 0.0240 |
| 10 | 0.0046 | 0.0006 | 0.0407 | 0.1100 | 0.0263 |
| 15 | 0.0062 | 0.0008 | 0.0591 | 0.1127 | 0.0287 |
| 20 | 0.0088 | 0.0011 | 0.1302 | 0.1171 | 0.0358 |
| 25 | 0.0123 | 0.0017 | 0.2249 | 0.1229 | 0.0454 |
| 30 | 0.0161 | 0.0026 | 0.2973 | 0.1288 | 0.0534 |
| 35 | 0.0215 | 0.0047 | 0.3608 | 0.1371 | 0.0621 |
| 40 | 0.0307 | 0.0101 | 0.4231 | 0.1502 | 0.0743 |
| 45 | 0.0483 | 0.0227 | 0.4857 | 0.1722 | 0.0955 |
| 50 | 0.0806 | 0.0470 | 0.5496 | 0.2051 | 0.1241 |
| 55 | 0.1373 | 0.0885 | 0.6142 | 0.2519 | 0.1710 |
| 60 | 0.2305 | 0.1535 | 0.6818 | 0.3199 | 0.2416 |
| 65 | 0.3600 | 0.2462 | 0.7453 | 0.4117 | 0.3380 |
| 70 | 0.5161 | 0.3695 | 0.8051 | 0.5255 | 0.4600 |
| 75 | 0.6766 | 0.5162 | 0.8602 | 0.6514 | 0.5976 |
| 80 | 0.8218 | 0.6776 | 0.9110 | 0.7804 | 0.7411 |
| 85 | 0.9234 | 0.8252 | 0.9530 | 0.8851 | 0.8639 |

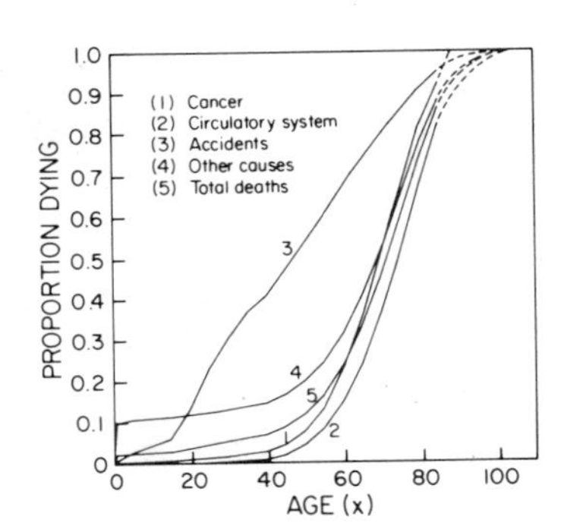


**Fig. 10.1** Cumulative distribution functions of age at death for different causes acting simultaneously in a population.

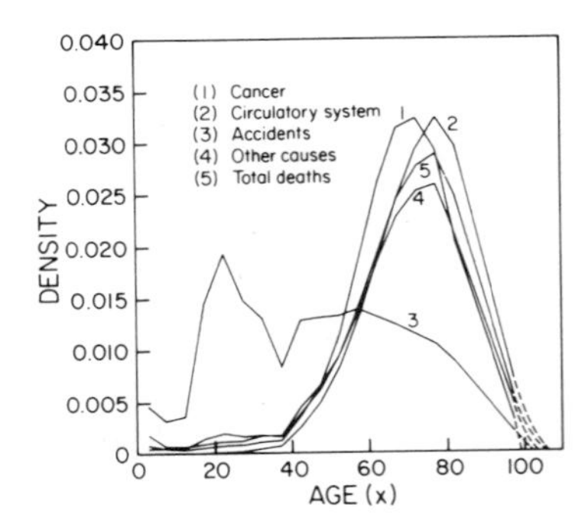


**Fig. 10.2** Curves of deaths (densities) for different causes acting simultaneously in a population.

density functions evaluated at midpoints, $x^* = x + n/2$, from the formula

$$f_\alpha^*(x^*) = \frac{al_{\alpha x} - al_{\alpha, x+n}}{n \cdot al_{\alpha 0}}. \tag{10.33}$$

The midpoint of the last open interval (85+), was arbitrarily chosen at $x^* = 97.5$. Above this point, the curves are extrapolated by hand.

The figures are self-explanatory. Notice the unusual shapes of the curves for accidental deaths. Also notice that the peak for deaths from cancer is between ages 65–75, whereas for deaths from circulatory system the modal age is approximately 80.

In summary, multiple decrement life tables provide information about the structure and patterns of mortality from various causes of death in the *population*—they provide a practical answer to question (1), stated in the Introduction to Chapter 9, about various "public" probabilities, as can be seen from the analogies between $al_{\alpha x}$ and $P_\alpha^*(x)$, and so on, emphasized throughout this chapter.

## REFERENCES

International Classification of Diseases Adapted for Use in the United States (ICDA) (1967). Eighth Revision, DHEW Publication No. 1693, Washington, D. C.

United States Life Tables by Causes of Death: 1969–1971. DHEW Publication No. (HRA) 75–1150 (1975), Rockville, MD.

Vital Statistics of the United States, 1970. Vol. II-Mortality. Part B (1974). DHEW Publication No. (HRA) 75–1101, Rockville, MD.

## EXERCISES

**10.1.** Use any population mortality records on multiple causes of death and construct a multiple decrement life table, following the rules described in Section 10.5. Use, for simplicity, ${}_nf_x = \frac{1}{2}$.

**10.2.** A stationary population (see Section 4.8.1) is a mixture of three different (ethnic) groups $\Pi_1$, $\Pi_2$, and $\Pi_3$. The number of births each year of these three types of individuals (which equals the corresponding number of deaths) is constant. Deaths are classified into three mutually exclusive causes, $C_1$, $C_2$, $C_3$. Denote the forces of mortality in this population for cause $C_\alpha$ on individuals in subpopulations $\Pi_1$, $\Pi_2$, $\Pi_3$ at age $x$ by $a\mu_{\alpha x}^{(1)}$, $a\mu_{\alpha x}^{(2)}$, $a\mu_{\alpha x}^{(3)}$, respectively. Assume that the ratios of the nine forces of mortality do not depend on $x$. Estimate the missing values in the following table.

| Age lbd | Proportion of population at this age | | | Proportion of deaths in this age group | | | Proportion of total population in this age group |
|---|---|---|---|---|---|---|---|
| | $\Pi_1$ | $\Pi_2$ | $\Pi_3$ | $C_1$ | $C_2$ | $C_3$ | |
| 20 | 0.50 | 0.30 | 0.20 | 0.80 | 0.10 | 0.10 | 0.015 |
| 50 | 0.53 | 0.27 | 0.20 | 0.75 | 0.08 | 0.17 | 0.013 |
| 65 | | | | 0.72 | 0.08 | 0.20 | 0.012 |
| 75 | | | | | | | 0.008 |

*Hints*: (*i*) First estimate the ratios $a\mu_x^{(1)}$: $a\mu_x^{(2)}$: $a\mu_x^{(3)}$ where $a\mu_x^{(j)} = a\mu_{1x}^{(j)} + a\mu_{2x}^{(j)} + a\mu_{3x}^{(j)}$. (*ii*) Use these values to fill out columns 2–4. (*iii*) From each of the first three rows, columns 2–7 provide two distinct *approximate* linear equations among the $a\mu_{\alpha x}^{(j)}$'s.

**10.3.** Using the data given in Table 10.1

(*a*) Construct a multiple decrement life table in which only one specific cause, denoted by $C_\alpha$, and all other causes, denoted by $C_{(-\alpha)}$ are distinguished. Use as a specific cause:

(*i*) malignant neoplasms,
(*ii*) circulatory system,
(*iii*) accidents.

(*b*) Calculate central rates, $am_{\alpha x}$, and $am_{(-\alpha)x}$ in each case.

(*c*) Represent graphically curves of deaths for each pair of causes $[C_\alpha, C_{(-\alpha)}]$.

**10.4.**

(*a*) What is the interpretation of CDF's given in Table 10.6?

(*b*) Suppose that you construct a CDF for a population of individuals of different ages who were diagnosed with cancer before they died. Why is it inappropriate (and wrong) to compare the CDF so obtained with that analogous to $F^*_{1x}$, given in Table 10.6 and obtained from the population mortality data of the corresponding general population?

*Hint*: Note that ${}_naq_{1x} < {}_naq_x < {}_nq_x^{(c)}$, where ${}_nq_x^{(c)}$, is the conditional probability of death in the cancer population.

CHAPTER 11

# Single Decrement Life Tables Associated with Multiple Decrement Life Tables: Their Interpretation and Meaning

## 11.1 ELIMINATION, PREVENTION, AND CONTROL OF A DISEASE

Toward the end of the last century, Farr (1875), known as the "first medical statistician", posed the question: *What would be the effect of life expectancy if a certain disease were eliminated as a cause of death?*

At that time, the focus of interest was on *infectious* diseases, such as smallpox, measles, typhus, etc. In this context, *elimination* of a cause has real natural meanings: removal of the sources of infection, or prevention of fatal development of the disease by vaccination.

It was also reasonable to assume that the times due to die from different infectious diseases are independent. This also implies that the force of mortality for a given cause is the same in the presence as in the absence of other causes [Section 9.5, (9.30)]. Using this assumption Farr (1875) described a quick method for constructing a life table that might result from "elimination" of specific cause. In fact, the problem was not new even then. Farr refers to a controversy on this topic between two French mathematicians, D. Bernoulli and d'Alembert, many years earlier. They discussed the expected consequences on mortality model if smallpox were to be eliminated as a cause of death. [See also, Karn (1933)].

At present, most infectious diseases are under rather effective control. *Chronic* diseases, associated with aging have now become the major causes of death. It is difficult to conceive, in the strict sense, of the elimination of a chronic disease as a cause of death. We may rather speak of the prevention or control of certain chronic conditions, with the corresponding increase in life expectancy.

Development of a disease is usually a complicated physiological process, involving metabolic errors (failures) of various components. Some of these components might be common to two or more diseases, others are different. As long as we do not identify the factors that take part in the disease process, we cannot decide what would be the effect of elimination of one cause of death on the mortality pattern from other causes. To simplify the problem, consider two causes of death. If the factors that cause failures of components common to both diseases are under control, then a decrease in the number of deaths from both diseases might be observed. On the other hand, if the factors specific for one disease are controlled, this may reduce the mortality from this disease, but have no effect on the mortality from the other disease. It is, then, difficult to speak about the effect of "elimination" of one cause in terms of the "gain" in survival. The part of the Life Tables by Causes of Death that is supposed to represent survival distributions after a specific cause is eliminated, the so called Single Decrement Life Tables (briefly, SDLT), might often be misleading. They have some historical aspects as we have mentioned, but often they do not represent what is claimed by their title. (Compare Tables 1–5, pp. 18–37 in the U. S. Life Tables 1969–1971.) A possible interpretation for them will be given in Section 11.3.

## 11.2 MORTALITY PATTERN FROM CAUSE $C_\alpha$ ALONE: 'PRIVATE' PROBABILITIES OF DEATH

We now return to question (2) introduced in Section 9.2, but will first pose it in the more general form: *What is the probability distribution of time at death from cause* $C_\alpha$ *alone*?

### 11.2.1 How might the SDF from cause $C_\alpha$ alone be interpreted?

We will discuss a few possibilities.

1. Ideally, one can imagine an isolated population, members of which can die from only one cause, $C_\alpha$. The corresponding SDF, $S_\alpha^+(x)$, say, will be one that represents the mortality pattern purely from cause $C_\alpha$ alone. Clearly, such a population does not exist, even under laboratory condi-

tions. In the present stage of knowledge, theoretical speculation about how such a population could be constructed is far from realistic.

2. Another possibility would be to suppose there is a population having the joint SDF, $S_{1\dots k}(x_1,\dots,x_k)$. Deaths from $C_\alpha$ alone will take place, if all other times due to die are removed or never exist by putting them equal to zero. Then the *marginal* SDF

$$S_\alpha(x_\alpha)=S_{1\dots k}(0,\dots,x_\alpha,\dots,0) \tag{11.1}$$

would be the one considered to correspond to time at death distribution from cause $C_\alpha$ alone [Gail (1975)]. This is usually called the net distribution.

3. Furthermore, we may consider the whole population as a *mixture* of homogeneous (isolated) subpopulations, each susceptible to different causes of death, and treat the marginal (net) survival function as a possible pattern of mortality for a given cause alone (cf. Section 9.9).

4. Another point of view is to assume that the conditions associated with causes other than $C_\alpha$ are, strictly speaking, not completely eliminated, but so well *controlled* that the times due to die from each other cause are increased, eventually approaching infinity. In this case, the *limiting* conditional SDF

$$S_{\alpha|\infty}(x|\infty)=\lim_{\substack{\{x_\beta\to\infty\}\\ \text{for all } \beta\neq\alpha}} \Pr\left\{(X_\alpha>x)\Big|\bigcap_{\beta\neq\alpha}(X_\beta=x_\beta)\right\}, \tag{11.2}$$

would be, perhaps, appropriate [see Elandt-Johnson (1976)]. If the times due to die are independent, (11.2) will coincide with $S_\alpha^+(x)$ and $S_\alpha(x)$.

We know, however, that the joint (and so the marginal and conditional) survival distributions cannot be uniquely identified, unless we assume independence (Section 9.5). We also know that the independence assumption does not hold for many diseases, so that we are going around in a circle.

To simplify the notation, and without loss of generality, we consider only two causes of death: a specific cause $C_\alpha$ and all other causes, except $C_\alpha$, denoted by $C_{(-\alpha)}$. Similarly, various lifetime functions will have subscripts $\alpha$ and $(-\alpha)$, respectively.

Clearly, by the same argument, we may introduce the same concepts for cause $C_{(-\alpha)}$, corresponding to elimination of $C_\alpha$. None of the distributions (1)–(4) can be identified in either case.

### 11.2.2 Estimable, Although Not Observable, Waiting Time Distributions

We must, then, be satisfied with something less. If we ask the question: "what is the chance that a newborn baby will die from heart attack, but not before age $x$?" we are not interested in a distribution of time at death from heart attack of babies living in an isolated population in which deaths from other causes are impossible. We are interested, indeed, in the individual (or private) waiting time distribution for which the force of mortality, $\mu_{\alpha x}$, is the same as the observable force of mortality, $a\mu_{\alpha x}$. We, then, assume (assumption A3)

$$\mu_{\alpha x} = a\mu_{\alpha x}. \tag{11.3}$$

In terms of probability functions, we are not concerned with the base, but with the core SDF (see Section 9.6).

Similarly for cause $C_{(-\alpha)}$, we are interested in the waiting time distribution with force of mortality

$$\mu_{(-\alpha)x} = a\mu_{(-\alpha)x}. \tag{11.4}$$

This is, of course, the old actuarial assumption. However, the SDF derived under this assumption does not give the answer to the old question: "What is the survival pattern when $C_\alpha$ is eliminated?" It only gives information on: "What is the probability that an individual will not die before age $x$ from cause different than $C_\alpha$ (i.e., from $C_{(-\alpha)}$), assuming that possibility of dying from cause $C_\alpha$ is *not* considered?"

## 11.3 ESTIMATION OF WAITING TIME DISTRIBUTION FOR CAUSE $C_\alpha$: SINGLE DECREMENT LIFE TABLE

In the parametric approach, the survival function $G_\alpha(x)$ defined in (9.35) represents the death waiting time distribution. However, this distribution is not directly observable (although it is estimable). It can be estimated by constructing a *single decrement life table* (SDLT) for cause $C_\alpha$. We will use the actuarial notation $\mu_{\alpha x}$, ${}_nq_{\alpha x}$, $l_{\alpha x}$, etc. (without prefix $a$) in this table. Our basic assumptions are

$$\mu_{\alpha x} = a\mu_{\alpha x}, \quad \text{and} \quad \mu_{(-\alpha)x} = a\mu_{(-\alpha)x}.$$

Our basic problem is to estimate ${}_nq_{\alpha x}$. There are several (approximate) methods of estimating ${}_nq_{\alpha x}$ in terms of the observable ${}_naq_{\alpha x}$, ${}_naq_{(-\alpha)x}$ and ${}_naq_x$, defined in Chapter 10. A few of these are presented below.

1. We have

$$a\mu_x = a\mu_{\alpha x} + a\mu_{(-\alpha)x} = \mu_{\alpha x} + \mu_{(-\alpha)x}. \tag{11.5}$$

It follows that

$$\int_0^n a\mu_{x+t}\,dt = \int_0^n \mu_{\alpha,x+t}\,dt + \int_0^n \mu_{(-\alpha),x+t}\,dt,$$

so that

$$\begin{aligned} {}_n ap_x &= \exp\left[-\int_0^n a\mu_{x+t}\,dt\right] \\ &= \exp\left[-\int_0^n \mu_{\alpha,x+t}\,dt\right]\exp\left[-\int_0^n \mu_{(-\alpha),x+t}\,dt\right] = {}_n p_{\alpha x}\cdot{}_n p_{(-\alpha)x}, \end{aligned} \tag{11.6}$$

or

$${}_n aq_x = 1-(1-{}_n q_{\alpha x})(1-{}_n q_{(-\alpha)x}). \tag{11.7}$$

We also have

$${}_n aq_{\alpha x} = \int_0^n a\mu_{\alpha,x+t}\cdot{}_t ap_x\,dt = \int_0^n (\mu_{\alpha,x+t}\cdot{}_t p_{\alpha,x})_t p_{(-\alpha),x}\,dt. \tag{11.8}$$

Assuming a *uniform* distribution of deaths from cause $C_\alpha$ over the interval $[x,x+n)$, we have

$$n\mu_{\alpha,x+t}\cdot{}_t p_{\alpha,x} \doteq {}_n q_{\alpha x}, \tag{11.9}$$

and so

$${}_n aq_{\alpha x} = \frac{1}{n}\,{}_n q_{\alpha x}\int_0^n {}_t p_{(-\alpha)x}\,dt. \tag{11.10}$$

Again, assuming *uniform* distribution of deaths from $C_{(-\alpha)}$, we have

$${}_t q_{(-\alpha)x} \doteq \frac{1}{n}t\cdot{}_n q_{(-\alpha)x}, \qquad 0 \leqslant t \leqslant n. \tag{11.11}$$

Hence (11.10) can be written in the form

$$\begin{aligned} {}_n aq_{\alpha x} &\doteq \frac{1}{n}\,{}_n q_{\alpha x}\int_0^n \left(1-\frac{t}{n}\,{}_n q_{(-\alpha)x}\right)dt \\ &= {}_n q_{\alpha x}\left(1-\tfrac{1}{2}\,{}_n q_{(-\alpha)x}\right). \end{aligned} \tag{11.12}$$

By a similar argument we obtain

$$ {}_n aq_{(-\alpha)x} \doteq {}_n q_{(-\alpha)x}\left(1-\tfrac{1}{2}{}_n q_{\alpha x}\right), \tag{11.13}$$

so that finally we have a set of two equations

$$\begin{aligned} {}_n aq_{\alpha x} &\doteq {}_n q_{\alpha}\left(1-\tfrac{1}{2}{}_n q_{(-\alpha)x}\right),\\ {}_n aq_{(-\alpha)x} &\doteq {}_n q_{(-\alpha)x}\left(1-\tfrac{1}{2}{}_n q_{\alpha x}\right). \end{aligned} \tag{11.14}$$

Note that [from (11.7)]

$$ {}_n aq_{\alpha x} + {}_n aq_{(-\alpha)x} = {}_n aq_x \doteq {}_n q_{\alpha x} + {}_n q_{(-\alpha)x} - \tfrac{1}{2}{}_n q_{\alpha x}\cdot {}_n q_{(-\alpha)x}. \tag{11.15}$$

Solving the two equations (11.14) for ${}_n q_{\alpha x}$, we obtain the quadratic equation

$$ {}_n q_{\alpha x}^2 - (2 - {}_n aq_{(-\alpha)x} + {}_n aq_{\alpha x})\,{}_n q_{\alpha x} + 2\,{}_n aq_{\alpha x} = 0, \tag{11.16}$$

from which

$$ {}_n q_{\alpha x} \doteq \tfrac{1}{2}\left[(2 - {}_n aq_{(-\alpha)x} + {}_n aq_{\alpha x}) - \sqrt{(2 - {}_n aq_{(-\alpha)x} + {}_n aq_{\alpha x})^2 - 8\,{}_n aq_{\alpha x}}\ \right]. \tag{11.17}$$

Interchanging $\alpha$ and $(-\alpha)$, we obtain a similar solution for ${}_n q_{(-\alpha)x}$.

2. If the ${}_n q_{\alpha x}$'s and ${}_n aq_{\alpha x}$'s are small, further approximation can be made. One of these, derived from (11.17), is

$$ {}_n q_{\alpha x} \doteq {}_n aq_{\alpha x}\,\frac{1-\tfrac{1}{2}{}_n aq_{(-\alpha)x}}{1-{}_n aq_{(-\alpha)x}}. \tag{11.18}$$

(The algebra is tedious, and we omit it.) Approximation (11.18) has been used in the construction of the U. S. 1959–1961 and 1969–1971 Life Tables.

3. For $C_\alpha$ alone, using the force of mortality $\mu_{\alpha x} \doteq a\mu_{\alpha x}$, we may construct a life table in the manner described in Chapter 4. We also have for central rates

$$ {}_n m_{\alpha x} \doteq {}_n am_{\alpha x} \doteq {}_n M_{\alpha x}. \tag{11.19}$$

Thus, from (4.46)

$$ {}_nq_{\alpha x} \doteq \frac{{}_nm_{\alpha x}}{\dfrac{1}{n}\left(1+\dfrac{n}{2}\,{}_nm_{\alpha x}\right)} = \frac{{}_nam_{\alpha x}}{\dfrac{1}{n}\left(1+\dfrac{n}{2}\,{}_nam_{\alpha x}\right)}. \tag{11.20} $$

[See also Neill (1977), Chapter 9.]

Compare (11.20) and (10.11), and notice the difference between the denominators of the two formulae.

Noticing that

$$ {}_nam_{\alpha x} = \frac{{}_nad_{\alpha x}}{{}_naL_x} \quad \text{and} \quad {}_naL_x \doteq n\left(al_x - \tfrac{1}{2}\,{}_nad_x\right), $$

we obtain

$$ \begin{aligned} {}_nq_{\alpha x} &= \frac{{}_nad_{\alpha x}}{\dfrac{1}{n}\left({}_naL_x + \dfrac{n}{2}\,{}_nad_{\alpha x}\right)} = \frac{{}_nad_{\alpha x}}{aL_x - \dfrac{1}{2}\left({}_nad_x - {}_nad_{\alpha(x)}\right)} \\ &= \frac{{}_nad_{\alpha x}}{aL_x - \frac{1}{2}\,{}_nad_{(-\alpha)x}} = \frac{{}_naq_{\alpha x}}{1-\frac{1}{2}\,{}_naq_{(-\alpha)x}}, \end{aligned} \tag{11.21} $$

where ${}_nad_{(-\alpha)x} = {}_nad_x - {}_nad_{\alpha x}$, and ${}_naq_{(-\alpha)x} = {}_naq_x - {}_naq_{\alpha x}$.

Note that in formula (11.21), deaths from other causes are treated as *withdrawals*. By a similar argument, we obtain

$$ {}_nq_{(-\alpha)x} \doteq \frac{{}_naq_{(-\alpha)x}}{1-\frac{1}{2}\,{}_naq_{\alpha x}}. \tag{11.22} $$

4. Substituting (11.22) into (11.21), we obtain

$$ {}_nq_{\alpha x} \doteq {}_naq_{\alpha x}\frac{1-\frac{1}{2}\,{}_naq_{\alpha x}}{1-\frac{1}{2}\,{}_naq_x}. \tag{11.23} $$

[see also Hooker and Longley-Cook (1957), p. 35]. Note that the right-hand side of (11.23) includes only the probabilities for the given cause $C_\alpha$ (in the numerator) and for all causes (in the denominator). In fact, this is a convenient formula since the denominator does not depend on which cause is under consideration. For other causes, $C_{(-\alpha)}$, we have, of course

$$ {}_nq_{(-\alpha)} = {}_naq_{(-\alpha)x}\frac{1-\frac{1}{2}{}_naq_{(-\alpha)x}}{1-\frac{1}{2}{}_naq_x}. \tag{11.24}$$

5. Another formula is obtained assuming that the ratio

$$\frac{a\mu_{\alpha,x+t}}{a\mu_{x+t}} = \frac{\mu_{\alpha,x+t}}{a\mu_{x+t}} = c_{\alpha x}, \tag{11.25}$$

where $0 \leqslant c_\alpha \leqslant 1$ is a constant, which does not depend on $t$ in the interval $(x, x+n)$, although it may be different for different intervals. In this case

$$\frac{{}_naq_{\alpha x}}{{}_naq_x} = \frac{\int_0^n a\mu_{\alpha,x+t}\cdot {}_tap_x\,dt}{\int_0^n a\mu_{x+t}\cdot {}_tp_x\,dt} = c_{\alpha x} = \frac{\mu_{\alpha,x+t}}{a\mu_{x+t}} \tag{11.26}$$

for $0 \leqslant t \leqslant n$.

Hence

$$\mu_{\alpha,x+t} = \frac{{}_naq_{\alpha x}}{{}_naq_x}a\mu_{x+t},$$

so that

$$\exp\left[-\int_0^n \mu_{\alpha,x+t}\,dt\right] = \left\{\exp\left[-\int_0^n a\mu_{x+t}\,dt\right]\right\}^{{}_naq_{\alpha x}/{}_naq_x},$$

or

$$1 - {}_nq_{\alpha x} = (1 - {}_naq_x)^{{}_naq_{\alpha x}/{}_naq_x},$$

or

$${}_nq_{\alpha x} \doteq 1 - (1 - {}_naq_x)^{{}_naq_{\alpha x}/{}_naq_x}. \tag{11.27}$$

[see, for example, Bailey and Haycocks (1946), p. 26, Greville (1948), Hooker and Longley-Cook (1957), p. 27, and Chiang (1961)].

We have now obtained five different approximations to ${}_nq_{\alpha x}$ given by (11.17), (11.18), (11.21), (11.23), and (11.27). Note that none of them depends on $n$.

**Table 11.1** Comparison of various approximations to $q_{\alpha x}$

| $aq_{\alpha x}$ | $aq_{(-\alpha)x}$ | (11.17) | (11.18) | (11.21) | (11.23) | (11.27) |
|---|---|---|---|---|---|---|
| 0.05 | 0.05 | 0.05132 | 0.05132 | 0.05128 | 0.05132 | 0.05132 |
| | 0.10 | 0.05271 | 0.05278 | 0.05263 | 0.05270 | 0.05273 |
| | 0.15 | 0.05418 | 0.05441 | 0.05405 | 0.05417 | 0.05426 |
| | 0.25 | 0.05738 | 0.05833 | 0.05714 | 0.05735 | 0.05771 |
| | 0.50 | 0.06745 | 0.07500 | 0.06667 | 0.06724 | 0.07002 |
| | 0.75 | 0.08211 | 0.12500 | 0.08000 | 0.08125 | 0.09570 |
| 0.10 | 0.05 | 0.10271 | 0.10263 | 0.10256 | 0.10270 | 0.10268 |
| | 0.10 | 0.10557 | 0.10556 | 0.10526 | 0.10556 | 0.10557 |
| | 0.15 | 0.10861 | 0.10882 | 0.10811 | 0.10857 | 0.10870 |
| | 0.25 | 0.11529 | 0.11667 | 0.11429 | 0.15515 | 0.11581 |
| | 0.50 | 0.13668 | 0.15000 | 0.13333 | 0.13571 | 0.14163 |
| | 0.75 | 0.16941 | 0.25000 | 0.16000 | 0.16522 | 0.20004 |
| 0.15 | 0.05 | 0.15418 | 0.15395 | 0.15385 | 0.15417 | 0.15410 |
| | 0.10 | 0.15861 | 0.15833 | 0.15790 | 0.15857 | 0.15853 |
| | 0.15 | 0.16334 | 0.16324 | 0.16216 | 0.16324 | 0.16334 |
| | 0.25 | 0.17379 | 0.17500 | 0.17143 | 0.17344 | 0.17433 |
| | 0.50 | 0.20805 | 0.22500 | 0.20000 | 0.20556 | 0.21515 |
| | 0.75 | 0.26411 | 0.37500 | 0.24000 | 0.25227 | 0.31871 |
| 0.25 | 0.05 | 0.25738 | 0.25658 | 0.25641 | 0.25735 | 0.25713 |
| | 0.10 | 0.26529 | 0.26389 | 0.26316 | 0.26515 | 0.26487 |
| | 0.15 | 0.27379 | 0.27206 | 0.27027 | 0.27344 | 0.27332 |
| | 0.25 | 0.29289 | 0.29167 | 0.28571 | 0.29167 | 0.29289 |
| | 0.50 | 0.35961 | 0.37500 | 0.33333 | 0.35000 | 0.37004 |
| | 0.75 | 0.50000 | 0.62500 | 0.40000 | 0.43750 | 0.38438 |
| 0.50 | 0.05 | 0.51745 | 0.51316 | 0.51282 | 0.51724 | 0.51612 |
| | 0.10 | 0.53668 | 0.52778 | 0.52632 | 0.53571 | 0.53400 |
| | 0.15 | 0.55805 | 0.54412 | 0.54054 | 0.55556 | 0.55405 |
| | 0.25 | 0.60961 | 0.58333 | 0.57143 | 0.60000 | 0.60315 |
| | 0.50 | 0.99976 | 0.75000 | 0.66667 | 0.75000 | 0.99976 |
| 0.75 | 0.05 | 0.78211 | 0.76974 | 0.76923 | 0.78125 | 0.77884 |
| | 0.10 | 0.81941 | 0.79167 | 0.78947 | 0.81522 | 0.81249 |
| | 0.15 | 0.86411 | 0.81618 | 0.81081 | 0.85227 | 0.85322 |
| | 0.25 | 1.00000 | 0.87500 | 0.85714 | 0.93750 | 1.00000 |

Table 11.1 gives values of ${}_nq_{\alpha x}$ (or equivalently of $q_{\alpha x}$) for selected values of ${}_naq_{\alpha x}$ ($aq_{\alpha x}$) and ${}_naq_{(-\alpha)x}$ ($aq_{(-\alpha)x}$). When the $aq_{\alpha x}$'s (and $aq_{\alpha x}$'s) are small, all five formulae give effectively equivalent results.

**Example 11.1** To illustrate the construction of a single decrement life table from a multiple decrement life table, we use the data given in Table 10.1 (U. S. White Males, 1970).

In calculating ${}_nq_{\alpha x}$, we used formula (11.23), using the values of ${}_naq_x$ and ${}_naq_{\alpha x}$ given in Table 10.2.

Once ${}_nq_{\alpha x}$ is calculated, we construct the life table as described in Chapter 4. We take the radix $l_{\alpha 0}=100{,}000$, and calculate progressively

$$l_{\alpha,x+n}=l_{\alpha x}(1-{}_nq_{\alpha x}), \tag{11.28}$$

and

$$ {}_nd_{\alpha,x}=l_{\alpha x}-l_{\alpha,x+n}=l_{\alpha x}\cdot {}_nq_{\alpha x}. \tag{11.29}$$

Table 11.2 gives the SDLTs for diseases of circulatory system ($C_2$) and for all causes except $C_2$ [i.e., $C_{(-2)}$]. Note that we have not calculated ${}_\infty q_{2,85}$ and ${}_\infty q_{(-2),85}$, because none of the approximate formulae apply over the open interval $(85,\infty)$. Usually, the probability for all other causes,

**Table 11.2** Single decrement life tables associated with circulatory system. U. S. White Males, 1970

| Age group | Circulatory system ($C_2$) | | | All other causes except circulatory system [$C_{(-2)}$] | | |
|---|---|---|---|---|---|---|
| $x$ to $x+n$ | ${}_nq_{2x}$ | $l_{2x}$ | ${}_nd_{2x}$ | ${}_nq_{(-2)x}$ | $l_{(-2)x}$ | ${}_nd_{(-2)x}$ |
| 0–1 | 0.00018 | 100,000 | 18 | 0.02058 | 100,000 | 2.058 |
| 1–5 | 0.00011 | 99,982 | 11 | 0.00323 | 97,942 | 316 |
| 5–10 | 0.00008 | 99,971 | 7 | 0.00229 | 97,626 | 224 |
| 10–15 | 0.00009 | 99,964 | 10 | 0.00233 | 97,402 | 227 |
| 15–20 | 0.00019 | 99,954 | 19 | 0.00714 | 97,175 | 694 |
| 20–25 | 0.00032 | 99,935 | 32 | 0.00958 | 96,481 | 925 |
| 25–30 | 0.00051 | 99,903 | 50 | 0.00792 | 95,556 | 756 |
| 30–35 | 0.00122 | 99,853 | 122 | 0.00802 | 94,800 | 760 |
| 35–40 | 0.00324 | 99,731 | 323 | 0.00973 | 94,040 | 915 |
| 40–45 | 0.00753 | 99,408 | 749 | 0.01337 | 93,125 | 1,245 |
| 45–50 | 0.01495 | 98,659 | 1,475 | 0.01903 | 91,880 | 1,749 |
| 50–55 | 0.02644 | 97,184 | 2,569 | 0.02786 | 90,131 | 2,511 |
| 55–60 | 0.04420 | 94,615 | 4,183 | 0.04286 | 87,620 | 3,755 |
| 60–65 | 0.06957 | 90,432 | 6,291 | 0.06191 | 83,865 | 5,193 |
| 65–70 | 0.10732 | 94,141 | 9,030 | 0.08613 | 78,672 | 6,776 |
| 70–75 | 0.15878 | 75,111 | 11,927 | 0.11412 | 71,896 | 8,205 |
| 75–80 | 0.23948 | 63,184 | 15,131 | 0.15351 | 63,691 | 9,777 |
| 80–85 | 0.34730 | 48,053 | 16,689 | 0.19323 | 53,914 | 10,418 |
| 85+ | | 31,364 | | | 43,496 | |

$_{\infty}q_{(-2),85}$ would be arbitrarily taken to be equal to 1, since everybody must die sometime. It is difficult, however, to decide what value should be assigned to $_{\infty}q_{2,85}$, without having data for very old ages. All we can say is that the chance of an individual dying from a disease of circulatory system $(C_2)$ *before* age 85 is $1-G_2(85)=(100{,}000-31{,}364)/100{,}000=0.68636 \doteq 69\%$. On the other hand, the proportion of deaths from $C_2$ in the population is $\pi_{2,0}=0.5514 \doteq 55\%$ (see Table 10.5).

## REFERENCES


Bailey, W. G. and Haycocks, H. W. (1946). *Some Theoretical Aspects of Multiple Decrement Life Tables*. Institute of Actuaries, London and Faculty of Actuaries, Edinburgh.

Chiang, C. L. (1961). On the probability of death from specific causes in the presence of competing risks. *Fourth Berkeley Symp.* **4**, 169–180.

Farr, W. (1874). Effect of the extinction of any single disease on the duration of life. *Suppl. 35th Ann. Rep. Registrar General*, **21**, 38.

Gail, M. (1975). A review and critique of some models used in competing risk analysis. *Biometrics* **31**, 209–222.

Greville, T. N. E. (1948). Mortality tables analyzed by cause of death. *Record: Amer. Inst. Actu.* **37**, 283–294.

Hooker, P. F. and Longley-Cook, L. H. (1957). *Life and Other Contingencies*, Vol. II, Chapters 19 and 20. Cambridge University Press, Cambridge.

Karn, M. N. (1933). A further study of methods of constructing life tables when certain causes of death are eliminated. *Biometrika* **25**, 91–101.

Neill, A. (1977). *Life Contingencies*. Heineman, London, Chapter 9.

United States Life Tables by Causes of Death: (1968). Vol. 1, No. 6, DHEW, Washington, D. C.

United States Life Tables by Causes of Death: (1975). DHEW, Rockville, MD.

Vital Statistics of the United States, 1970 Vol. II. Mortality, Part B. (1975). DHEW Publication No. (HRA) 75-1101, Rockville, MD.


## EXERCISES

**11.1.** (*a*) State the three assumptions (A1, A2, A3) commonly used in the analysis of competing causes which were discussed in detail in Chapter 9.

(*b*) Provided that the cause of death and time at death are recorded, which of the following statements are *true* or *false*.

The problem of nonidentifiability arises if:

(*i*) None of A1, A2, A3 holds;
(*ii*) A1 and A3 holds, but A2 does not hold;
(*iii*) A1 and A2 hold, but A3 does not hold;
(*iv*) A2 and A3 hold, but A1 does not hold;

In each case, give the reasons why your answer is yes or no.

**11.2.** Comment on the following argument: If a cause of death, $C_\alpha$, is eliminated, then the $ad_{\alpha y}$ deaths aged $y$ lbd so removed represent a saving of approximately $a\mathring{e}_{y+\frac{1}{2}}$ years of life each. The expected gain in lifetime from cause $C_\alpha$ is therefore

$$\frac{1}{l_0}\sum_{y=0}^{\infty} ad_{\alpha y}\, a\mathring{e}_{y+\frac{1}{2}}.$$

Would you expect this formula to underestimate or overestimate the increase in expectation of life consequent on elimination of $C_\alpha$?

**11.3.** Let $X>0, Y>0$ be two continuous random variables with PDF, $f_{XY}(x,y)>0$, and SDF, $S_{XY}(x,y)$.

Prove that if

$$\lim_{x\to\infty}\Pr\{Y>y|X=x\}=S_{Y|\infty}(y|\infty)$$

exists, then

$$\lim_{x\to\infty}\Pr\{Y>y|X>x\}=S_{Y|\infty}(y|\infty).$$

**11.4.** Derive formula (11.18) from formula (11.17). Assume that ${}_naq_{\alpha x}$ and ${}_naq_{(-\alpha x)}$ are small.

**11.5.** Use the data from the example given in Section 10.6 in constructing a single decrement life tables for accidents (cause $C_3$) analogous to that given in Table 11.2. We may assume that time due to die from an accident is independent of times due to die from other causes. Then the second part of SDLT (for $C_{(-3)}$) may represent a life table after accidents have been eliminated. Extend this life table by calculating the life expectancy, $\mathring{e}_{(-3)x}$, in the usual way as described in Section 4.9.4. Calculate also the life expectancy in the life table for all causes, $a\mathring{e}_x$, and the "gain" in life expectancy, $\mathring{e}_{(-3)x}-a\mathring{e}_x$.

**11.6.** Damiani ((1976) *J. Soc. Statist. Paris* **117**, 122–131) has proposed the model

$${}_naq_x={}_nq_{(-k)x}+A_k({}_naq_{k,x})^{b_k},$$

where (for fixed $n$) $A_k$, $b_k$ are constants depending on the cause of death $C_k$ but not on age.

Given estimated values of ${}_n aq_x$ and ${}_n aq_{k,x}$ for a set of ages $x_1, x_2, \ldots, x_m$, construct a method of estimating ${}_n q_{(-k)x}$ (for each of the $x$'s), $A_k$ and $b_k$.

**11.7.** Suggest tests of goodness-of-fit for Damiani's model (see Exercise 11.6):

(*i*) against the alternative that cause $C_k$ and the set of all other causes act independently,

(*ii*) against the alternative that $A_k$ (although not $b_k$) increases with age.

**11.8.** Applying the model of Exercise 11.6, Damiani (1976) used the following estimated values based on mortality data (covering ages 15–85) for France 1967–9. (M≡male; F≡female). Data for ages 45–85 were

| Estimated Values $\times 10^3$ | Cause of Death | Sex | $x$: 45 | 55 | 65 | 75 |
|---|---|---|---|---|---|---|
| ${}_{10}aq_x$ | all causes | M | 85 | 196 | 403 | 706 |
| | | F | 42 | 88 | 225 | 556 |
| ${}_{10}aq_{1x}$ | cancer | M | 21 | 69 | 123 | 196 |
| | | F | 15 | 30 | 58 | 111 |
| ${}_{10}aq_{2x}$ | heart ailments | M | 12 | 36 | 100 | 231 |
| | | F | 4 | 13 | 50 | 167 |
| ${}_{10}aq_{3x}$ | cerebrovascular diseases | M | 5 | 17 | 59 | 182 |
| | | F | 3 | 9 | 38 | 139 |
| ${}_{10}aq_{4x}$ | accidents, violent | M | 12 | 14 | 19 | 39 |
| | | F | 3 | 4 | 9 | 36 |

The following estimated values of probabilities of death if various causes were 'eliminated' were obtained.

Values of ${}_{10}q_{(-k)x} \times 10^3$

| $k$ | Sex | $x$: 45 | 55 | 65 | 75 |
|---|---|---|---|---|---|
| 1 | M | 53 | 96 | 189 | 353 |
| | F | 17 | 37 | 122 | 340 |
| 2 | M | 64 | 140 | 270 | 431 |
| | F | 35 | 67 | 155 | 357 |
| 3 | M | 77 | 167 | 306 | 406 |
| | F | 38 | 75 | 173 | 367 |
| 4 | M | 71 | 179 | 380 | 654 |
| | F | 38 | 83 | 212 | 484 |

The values of $A_k$ and $b_k$ used in these calculations are shown below

| | $A_k$ | | $b_k$ | |
|---|---|---|---|---|
| $k$ | male | female | male | female |
| 1 | 2.004 | 2.198 | 1.066 | 1.075 |
| 2 | 0.978 | 0.932 | 0.865 | 0.863 |
| 3 | 1.620 | 1.316 | 0.991 | 0.984 |
| 4 | 1.739 | 3.749 | 1.080 | 1.191 |

(*i*) Give an assessment of how well the model fits the part of the data shown.

(*ii*) Comment on the suggestions that:

(*a*) values of $b_k$ for males and females might be taken equal to each other for any specified $k$,

(*b*) this common value might be taken equal to 1 for all $k$.

**11.9.**

(*a*) Objections to the formula in Exercise 11.6 have been raised. If it is valid for one value of $n$, it will not be valid for other values of $n$. Formulate a reply to this objection, on the basis of *approximate* validity, with $A_k$ (but not $b_k$) depending on $n$.

(*b*) A similar objection can be raised against the model (proposed by Krall and Hickman [(1970) *Trans. Amer. Soc. Actu.* **64**, 163–179, discussion, 181–190)]

$$a_{(-k)}\mu_{ix} = a\mu_{ix}\left[1 + a\mu_{kx}\right], \qquad i \neq k,$$

where $a_{(-k)}$ means in presence of all other causes, $C_k$ having been eliminated. Formulate this objection. Why cannot it be answered in the same way as in (*a*)?

CHAPTER 12

# Estimation and Testing Hypotheses in Competing Risk Analysis

## 12.1 INTRODUCTION. EXPERIMENTAL DATA

In Chapters 10 and 11, we have been concerned with the construction of various survival functions to fit population mortality data. An important feature of methods described in these chapters was the estimation of the conditional probabilities of death, ${}_naq_x$, ${}_naq_{\alpha x}$, and $q_{\alpha x}$. Estimated (from population data) values were used as true values in construction of life tables; no random variation was assumed, and no tests of hypotheses were used.

In the present chapter we are concerned with *sample* data, and the *estimation* of various survival functions from different sampling schemes as well as testing hypotheses about these SDF's. Just as Chapters 10 and 11 are multivariate generalizations of Chapter 4, the present chapter can be treated as a generalization of Chapters 5 and 6, and also of Chapter 7 since we emphasize fitting parametric curves.

We will try to keep the notation as nearly as possible the same as in Chapters 5 or 6 to facilitate perception of the analogy between problems discussed there and here. The analysis presented here is mainly concerned with *cohort* or *follow-up* data, and we use $T$ to denote the time elapsed since occurrence of a certain initial event (e.g., beginning of observations). It need not always be age.

There are many methods by which the same results can be obtained, but space restriction does not allow the presentation of all of them in detail.

The emphasis here is on basic concepts and, in parametric cases, on construction of appropriate likelihoods.

## 12.2 GROUPED DATA. NONPARAMETRIC ESTIMATION

### 12.2.1 Complete data

Consider a cohort of $N$ individuals (where $N$ is rather large) with the data grouped into $M$ fixed intervals $[t_i, t_{i+1})$, $i=0,1,\ldots,M-1$. (The reader is advised to review Section 5.3.)

Here some additional information is obtained: the total number of deaths, $d_i$, in the interval $[t_i, t_{i+1})$ is separated into $k$ parts

$$d_i = d_{1i} + d_{2i} + \ldots + d_{ki}, \tag{12.1}$$

where $d_{\alpha i}$ is the number of deaths from cause $C_\alpha$, $\alpha = 1,2,\ldots,k$.

If $N_i$ denotes the number of survivors at $t_i$ and $N_{i+1}$ the number of survivors at $t_{i+1}$, we have

$$N_i = d_{1i} + d_{2i} \ldots + d_{ki} + N_{i+1} = d_i + N_{i+1} \tag{12.2}$$

(with $N_0 = N$).

Note that $d_{\alpha i}$ denotes here the *observed* number of deaths from cause $C_\alpha$ in the interval $[t_i, t_{i+1})$ in the *sample* (cohort) of $N$ individuals. These are *sample* analogues of life table deaths (which would be denoted by ${}_{h_i}ad_{\alpha t_i}$, where $h_i = t_{i+1} - t_i$ is the length of the interval). Clearly, $N_i$ is a sample analogue of $al_{t_i}$.

As in Chapter 10, $aq_i$ and $aq_{\alpha i}$ denote the conditional (on survival at $t_i$) probabilities of death from any cause and from cause $C_\alpha$, respectively.

Clearly,

$$aq_i = aq_{1i} + aq_{2i} + \ldots + aq_{ki}. \tag{12.3}$$

*Conditional on* $N_i$, the $d_{\alpha i}$'s and $N_{i+1}$, considered as random variables have a *multinomial* distribution, with probability function

$$\frac{N_i!}{d_{1i}!\ldots d_{ki}!\,N_{i+1}!}(aq_{1i})^{d_{1i}}\ldots(aq_{ki})^{d_{ki}}(ap_i)^{N_{i+1}}, \tag{12.4}$$

where

$$ap_i = 1 - aq_i.$$

Thus the maximum likelihood estimators are

$$\widehat{aq}_{\alpha i} = \frac{d_{\alpha i}}{N_i}, \qquad \alpha = 1, 2, \ldots, k, \tag{12.5}$$

and

$$\widehat{aq}_i = \frac{d_i}{N_i} \tag{12.6}$$

Substituting from (12.6) for $N_i$ in (12.5), we obtain

$$\widehat{aq}_{\alpha i} = \frac{d_{\alpha i}}{d_i} \widehat{aq}_i. \tag{12.7}$$

[Compare (10.4), (10.3) and (10.6), respectively.]

Further analogues of multiple decrement life table functions, expressed in terms of proportions rather than numbers, include the crude probability functions. We denote by $aP_i$ the overall survival function, and by $aP_{\alpha i}$ the "survival" function for cause $C_\alpha$ in the presence of all causes. These are analogues of $al_{t_i}/al_0$, and $al_{\alpha, t_i}/al_0$, respectively.

Clearly, the estimated $aP_i$ is

$$\widehat{aP}_i = \widehat{ap}_0 \cdot \widehat{ap}_1 \ldots \widehat{ap}_{i-1}. \tag{12.8}$$

Noticing that $\widehat{ap}_j = N_{j+1}/N_j$, (12.8) takes the form

$$\widehat{aP}_i = \frac{N_i}{N} \tag{12.9}$$

[cf. Section 5.2.1]. Furthermore, we also have

$$\begin{aligned} \widehat{aP}_{\alpha i} &= \widehat{aq}_{\alpha i} \cdot \widehat{aP}_i + \widehat{aq}_{\alpha, i+1} \cdot \widehat{aP}_{i+1} + \ldots + \widehat{aq}_{\alpha, M-1} \cdot \widehat{aP}_{M-1} \\ &= \sum_{j=i}^{M-1} \widehat{aq}_{\alpha j} \cdot \widehat{aP}_{\alpha j} \end{aligned} \tag{12.10}$$

[Compare (10.32a).]

Substituting for $\widehat{aq}_{\alpha i}$ from (12.5) and for $\widehat{aP}_i$ from (12.8), we obtain

$$\widehat{aP}_{\alpha i} = \frac{1}{N} \sum_{j=i}^{M-1} d_{\alpha j} \tag{12.11}$$

[cf. (10.32).]

Clearly, when the data are complete, the sample data (experience) is an analogue of a life table with the radix equal to the original sample size $N$.

Further calculations are straightforward. Thus

$$\frac{\widehat{aP}_{\alpha i}}{\widehat{aP}_i} = \hat{\pi}_{\alpha i} \tag{12.12}$$

represents the estimated conditional (on surviving) probability of death after age $x$ from cause $C_\alpha$ [cf. (10.18)].

We can also estimate the waiting time distributions using $\widehat{aq}_{\alpha i}$ and $\widehat{aq}_i$, as described in Chapter 11.

### 12.2.2 Follow up Data

With variable times at entry, the follow up data will include *withdrawals* (cf. Chapter 6). Let $w_i$ be the number of withdrawals in $[t_i, t_{i+1})$. In this case, we may use the approximations

$$\widehat{aq}_i \doteqdot \frac{d_i}{N_i - \frac{1}{2}w_i}, \tag{12.13}$$

and

$$\widehat{aq}_{\alpha i} \doteqdot \frac{d_{\alpha i}}{N_i - \frac{1}{2}w_i} = \frac{d_{\alpha i}}{d_i}\widehat{aq}_i, \tag{12.14}$$

where $N_i - \frac{1}{2}w_i = N_i'$, say is the effective number of those who passed through the exact time point, $t_i$.

The further analysis is similar to that in Section 12.2.1, but noticing that the simple formulae (12.9) and (12.11) no longer hold.

### 12.2.3 Truncated Samples

In many medical investigations the duration of the experiment is limited to a certain predetermined period of observation, from $t=0$, till $t=t_e$, say. The data are then *truncated*, since at time $t_e$, there might be some survivors. It is convenient (for counting and calculation purposes) to divide the whole period into $M$ equal observational units (often years or months).

The estimation of survival functions is as described in Sections 12.2.1 or 12.2.2, but care should be taken to make suitable allowance for the relationships among them.

Let $\widehat{aP}_M$ denote the (observed) proportion of survivors at time $t_e = M$ units. This means that, on the average, only a proportion $(1-\widehat{aP}_M)$ died during the period of the study. Clearly, we have

$$\widehat{aP}_{1i} + \widehat{aP}_{2i} + \ldots + \widehat{aP}_{ki} = \widehat{aP}_i - \widehat{aP}_M. \tag{12.15}$$

[Compare with (9.20) or (10.2).]

Further, the estimated conditional (on surviving) probability of dying from $C_\alpha$ in the presence of all causes is

$$\frac{aP_{\alpha i}}{\widehat{aP}_i - \widehat{aP}_M} = \pi'_{\alpha i} \tag{12.16}$$

[cf. (10.18)]. In particular,

$$\frac{aP_{\alpha 0}}{\widehat{aP}_0 - \widehat{aP}_M} = \hat{\pi}'_\alpha \tag{12.17}$$

estimates the proportion of deaths from $C_\alpha$ in a given study.

**Example 12.1** The data in the first part of Table 12.1 are adapted from Chiang (1961), and represent the mortality experience of 5982 women admitted to California hospitals between January 1, 1942 and December 31, 1954 with the diagnosis of cancer of *cervix uteri*. The major cause ($C_1$) of death was cancer of the *cervix uteri*, but there were also other causes, denoted here by $C_2$, rather than by $C_{(-1)}$. At the date of termination of the study there were 72 patients alive, so we have *truncated* data. Note that these 72 patients represent, in fact, the withdrawals at the end of the study. There were also withdrawals (planned or random) in each interval (column $w_i$). The second part of Table 12.1 gives the estimated [from (12.13)] values of $aq_i$, and the estimated overall survival function, $\widehat{aP}_i$.

Table 12.2 is a multiple decrement life table (with radix $l_0 = 1$) for causes $C_1$ and $C_2$ acting *in the presence* of each other. The $\widehat{aq}_{\alpha i}$ were calculated from (12.14), and the $\widehat{aP}_{\alpha i}$'s from (12.10). Since the data are truncated, the estimated (conditional on surviving) proportions of deaths after age $x$, from $C_1$ and $C_2$, were calculated from (12.17), with $\widehat{aP}_M = \widehat{aP}_{13} = 0.3437$ ($M = 13$ years).

It is clear from this table that $\hat{\pi}'_{1i}$ and $\hat{\pi}'_{2i}$ (as functions of $i$) are far from constant. We notice that in the first 5 years (and even more noticeably in the first 2 or 3 years), most deaths were from cancer of the *cervix uteri*, but after this period the proportions of deaths from $C_1$ and $C_2$ are nearly in the ratio 1 : 1.

**Table 12.1** Mortality experience of 5982 women diagnosed with cancer of the *cervix uteri*

| Years of follow-up $t_i$ to $t_{i+1}$ | Total number of living at $t_i$ ($N_i$) | Number of deaths in $[t_i, t_{i+1})$: All causes $d_i$ | *Cervix uteri* $d_{1i}$ | Other causes $d_{2i}$ | Withdrawn alive $w_i$ | $\widehat{aq_i}$ | $\widehat{aP_i}$ |
|---|---|---|---|---|---|---|---|
| 0–1 | 5982 | 1376 | 1175 | 201 | 576 | 0.2417 | 1.0000 |
| 1–2 | 4030 | 684 | 588 | 96 | 501 | 0.1810 | 0.7583 |
| 2–3 | 2845 | 269 | 221 | 48 | 459 | 0.1038 | 0.6211 |
| 3–4 | 2117 | 165 | 121 | 44 | 379 | 0.0856 | 0.5572 |
| 4–5 | 1573 | 91 | 63 | 28 | 306 | 0.0641 | 0.5095 |
| 5–6 | 1176 | 61 | 27 | 34 | 254 | 0.0582 | 0.4769 |
| 6–7 | 861 | 34 | 18 | 16 | 167 | 0.0437 | 0.4491 |
| 7–8 | 660 | 25 | 13 | 12 | 161 | 0.0431 | 0.4295 |
| 8–9 | 474 | 14 | 6 | 8 | 116 | 0.0337 | 0.4110 |
| 9–10 | 344 | 14 | 9 | 5 | 85 | 0.0464 | 0.3971 |
| 10–11 | 245 | 9 | 5 | 4 | 78 | 0.0437 | 0.3787 |
| 11–12 | 158 | 6 | 2 | 4 | 80 | 0.0508 | 0.3622 |
| 12–13 | 72 | 0 | 0 | 0 | 72 | 0.0000 | 0.3437 |

From Chiang (1961), *Biometrics* **17**, 58–78.

**Table 12.2** Competing risk analysis of data on cancer of the *cervix uteri* given in Table 12.1

| $t_i$ to $t_{i+1}$ | $a\hat{q}_{1i}$ | $a\hat{P}_{1i}$ | $a\hat{q}_{2i}$ | $a\hat{P}_{2i}$ | $a\hat{P}_i - a\hat{P}_{13}$ | $\hat{\pi}'_{1i}$ | $\hat{\pi}'_{2i}$ |
|---|---|---|---|---|---|---|---|
| 0–1 | 0.2064 | 0.4998 | 0.0353 | 0.1564 | 0.6563 | 0.7616 | 0.2384 |
| 1–2 | 0.1556 | 0.2935 | 0.0254 | 0.1211 | 0.4146 | 0.7078 | 0.2922 |
| 2–3 | 0.0845 | 0.1755 | 0.0184 | 0.1019 | 0.2774 | 0.6327 | 0.3673 |
| 3–4 | 0.0628 | 0.1230 | 0.0228 | 0.0905 | 0.2135 | 0.5762 | 0.4238 |
| 4–5 | 0.0444 | 0.0880 | 0.0197 | 0.0778 | 0.1658 | 0.5310 | 0.4690 |
| 5–6 | 0.0257 | 0.0654 | 0.0324 | 0.0677 | 0.1331 | 0.4914 | 0.5086 |
| 6–7 | 0.0232 | 0.0531 | 0.0206 | 0.0523 | 0.1054 | 0.5042 | 0.4958 |
| 7–8 | 0.0224 | 0.0427 | 0.0207 | 0.0430 | 0.0858 | 0.4984 | 0.5015 |
| 8–9 | 0.0144 | 0.0331 | 0.0192 | 0.0341 | 0.0672 | 0.4925 | 0.5075 |
| 9–10 | 0.0299 | 0.0272 | 0.0166 | 0.0262 | 0.0534 | 0.5091 | 0.4909 |
| 10–11 | 0.0243 | 0.0153 | 0.0194 | 0.0196 | 0.0350 | 0.4385 | 0.5615 |
| 11–12 | 0.0169 | 0.0061 | 0.0339 | 0.0123 | 0.0184 | 0.3333 | 0.6667 |
| 12–13 | 0 | 0 | 0 | 0 | 0 | 0 | 0 |

We also estimated the waiting time distribution functions, denoted here by $P_{1i}$ and $P_{2i}$ (these correspond to $l_{1t_i}$ and $l_{2t_i}$ in single-decrement life tables with radix $l_{1,0}=l_{2,0}=1$) (Table 12.3). The values of $\hat{q}_{1i}$, $\hat{q}_{2i}$ were calculated from formula (11.23) as described in Chapter 11. Note that in the population of women with diagnosed cancer, the personal (or private) probability of death from cancer of the *cervix uteri* ($C_1$) was still 50% five years after diagnosis, whereas for other causes ($C_2$) it was only 13% at that time.

**Table 12.3** Waiting time distribution (cancer of the *cervix uteri*)

| $t_i$ to $t_{i+1}$ | $\hat{q}_{1i}$ | $\hat{P}_{1i}$ | $\hat{q}_{2i}$ | $\hat{P}_{2i}$ |
|---|---|---|---|---|
| 0–1 | 0.2440 | 1.0000 | 0.0457 | 1.0000 |
| 1–2 | 0.1752 | 0.7560 | 0.0306 | 0.9543 |
| 2–3 | 0.0902 | 0.6235 | 0.0203 | 0.9251 |
| 3–4 | 0.0665 | 0.5673 | 0.0247 | 0.9063 |
| 4–5 | 0.0464 | 0.5296 | 0.0209 | 0.8839 |
| 5–6 | 0.0270 | 0.5050 | 0.0339 | 0.8655 |
| 6–7 | 0.0239 | 0.4914 | 0.0213 | 0.8362 |
| 7–8 | 0.0232 | 0.4796 | 0.0214 | 0.8184 |
| 8–9 | 0.0148 | 0.4685 | 0.0197 | 0.8009 |
| 9–10 | 0.0308 | 0.4616 | 0.0172 | 0.7851 |
| 10–11 | 0.0251 | 0.4473 | 0.0201 | 0.7715 |
| 11–12 | 0.0177 | 0.4361 | 0.0351 | 0.7560 |
| 12–13 | 0 | 0.4284 | 0 | 0.7295 |

## 12.3 GROUPED DATA. FITTING PARAMETRIC MODELS

When fitting parametric models, various techniques of the kind described in Chapter 7 can be used. One may fairly easily adapt them to multiple cause data. In this chapter, however, we confine ourselves to maximum likelihood estimation.

In view of equivalence and nonidentifiability in competing risks (see Section 9.6), the question of choice of the parametric model represents a serious problem. It can be handled in at least two different ways.

### 12.3.1 Choice of a Joint Survival Function

If we have some idea how the diseases, as causes of death, might be related, we can construct a parametric model of the joint survival function in which parameters have some meaning. As has been already mentioned in Chapter 11, it is rather difficult to construct a general model.

Another possibility is to choose, more or less arbitrarily, some well known family of multivariate survival distributions with the hope that it may fit our data. A drawback of such an approach is that we have difficulty in interpreting the parameters. We also have to be sure that the members of such a family can be uniquely identified, that is, the parameters will be estimable from available mortality data (cf. Section 9.6.2).

Anyway, suppose that for some reason or other we select such a joint base SDF, $S_{1\ldots k}(y_1,\ldots,y_k;\, \gamma)$, say, where $Y_1,\ldots, Y_k$ denote times due to die since occurrence of a certain initial event, and $\gamma=(\gamma_1,\gamma_2,\ldots,\gamma_s)$ is a vector of $s$ parameters.

The likelihood has to be constructed using crude probability functions, and available sample mortality data. If the data are complete, it is possible to construct the unconditional likelihood (see Exercise 12.5). Here, we only discuss the *conditional* (on $N_i$ survivors) likelihood contributions to the overall likelihood, since these have a wide application to different types of experimental data, including those obtained from several separate, although similar experiments.

Let $S_T(t)=S_T(t;\gamma)$ be the overall SDF defined in (9.6), with $T=\min(Y_1,\ldots, Y_k)$, and $P_\alpha^*(t)=P_\alpha^*(t;\gamma)$ be the crude survival function defined in (9.17).

We now introduce $q_{\alpha i}^*=q_{\alpha i}^*(\gamma)$, which is the conditional (parametric) probability of death in $[t_i,t_{i+1})$ from cause $C_\alpha$ given alive at $t_i$. This is an analogue of $aq_{\alpha i}$ in the nonparametric approach.

We have

$$q_{\alpha i}^*(\gamma)=\frac{1}{S_T(t_i;\gamma)}\int_{t_i}^{t_{i+1}} h_\alpha(t;\gamma)S_T(t;\gamma)\,dt$$

$$=\frac{1}{S_T(t_i;\gamma)}\left[P_\alpha^*(t_i;\gamma)-P_\alpha^*(t_{i+1};\gamma)\right]. \tag{12.18}$$

Similarly, an analogue of $aq_i$ is

$$q_i^*(\gamma)=\frac{1}{S_T(t_i;\gamma)}\int_{t_i}^{t_{i+1}} h_T(t;\gamma)S_T(t;\gamma)\,dt=\frac{S_T(t_i;\gamma)-S_T(t_{i+1};\gamma)}{S_T(t_i;\gamma)}, \tag{12.19}$$

and an analogue of $ap_i$ is

$$p_i^*(\gamma)=1-q_i^*(\gamma)=\frac{S_T(t_{i+1};\gamma)}{S_T(t_i;\gamma)}. \tag{12.20}$$

Thus, *conditional on* $N_i$ individuals available at $t_i$, the likelihood, $L_i = L_i(\gamma)$, is

$$L_i(\gamma) \propto [q_{1i}^*(\gamma)]^{d_{1i}} \dots [q_{ki}^*(\gamma)]^{d_{ki}} [p_i^*(\gamma)]^{N_i - d_i}, \tag{12.21}$$

and the overall likelihood is

$$L(\gamma) = \prod_{i=0}^{M-1} L_i(\gamma). \tag{12.22}$$

If we observe *withdrawals*, an approximate likelihood of the kind (12.21), conditional on the effective number of survivors, $N_i' = N_i - \frac{1}{2}w_i$, can be constructed.

Moreover, the cross-sectional data discussed in Chapter 10, can be approximately treated as $M$ ($M$-the number of age groups) *independent* samples; the intervals are $[x, x+n)$, observed deaths are ${}_nD_x$, and effective number of survivors is $N_x'$ given by (10.23).

To test the hypothesis that the parametric distribution fits the model, one can calculate in each time interval the expected number of deaths from the parametric model for each cause $C_\alpha$ and apply separate $X^2$ tests of goodness of fit. Graphical presentation also gives us some idea how well the model fits the data.

Suppose that the selected parametric model does fit the data. Clearly, we then can estimate the marginal SDF's of the base joint distribution. As has been mentioned in Section 11.2, our interest should be in the marginal SDFs, $G_\alpha(t)$, of the core function defined by (9.35).

The $X^2$ test of goodness of fit for $G_\alpha(t)$'s distributions can be constructed in the usual manner.

### 12.3.2 Fitting Crude Parametric Distribution to Mortality Data from Each Cause Separately

It is easier, perhaps, to fit a parametric distribution to grouped mortality data for each cause separately.

Let $S_\alpha^*(t) = S_\alpha^*(t; \gamma_\alpha)$ be the proper crude SDF for cause $C_\alpha$ defined in (9.22). Thus the improper crude probability function is

$$P_\alpha^*(t; \gamma_\alpha) = \pi_\alpha S_\alpha^*(t; \gamma_\alpha),$$

and the overall survival function is

$$S_T(t; \gamma, \pi) = \sum_{\alpha=1}^{k} \pi_\alpha S_\alpha^*(t; \gamma_\alpha), \tag{12.23}$$

where $\boldsymbol{\gamma}=(\gamma_1,\ldots,\gamma_s)$ is the vector of the parameters, and $\boldsymbol{\pi}=(\pi_i,\ldots,\pi_k)$.

The probability, $q_{\alpha i}^*(\gamma)$ is, from (12.18)

$$q_{\alpha i}^*(\gamma_\alpha,\boldsymbol{\pi})=\frac{\pi_\alpha\left[S_\alpha^*(t_i;\gamma_\alpha)-S_\alpha^*(t_{i+1};\gamma_\alpha)\right]}{\sum_{\beta=1}^{k}\pi_\beta S_\beta^*(t_i;\gamma_\beta)},\tag{12.24}$$

and the conditional likelihood is calculated from (12.21). Note that it is a function of $\pi_1,\pi_2,\ldots,\pi_k$, as well as $\gamma_1,\gamma_2,\ldots,\gamma_s$.

## 12.4 COHORT MORTALITY DATA WITH RECORDED TIMES AT DEATH OR CENSORING. NONPARAMETRIC ESTIMATION

### 12.4.1 Complete Data

1. Consider a cohort of $n$ individuals subject to death from $k$ causes and observed from time $t=0$, until the last member in this group dies. Let

$$t_1'<t_2'<\ldots<t_r'<\ldots<t_n'\tag{12.25}$$

denote distinct ordered times at death, so that $r$ is the number of deaths observed up to and including time $t_r'$. The empirical CDF and SDF for all causes, denoted by $F_T^o(t)$, and $S_T^o(t)$, respectively are

$$F_T^o(t)=\frac{r}{n}\quad\text{and}\quad S_T^o(t)=\frac{n-r}{n}\quad\text{for } t_r'\leqslant t<t_{r+1}'\tag{12.26}$$

[cf. (5.2a)]. If there are multiple deaths at some time points, formula (5.5) should be used.

2. Let

$$\delta_{\alpha j}=\begin{cases}1 & \text{if the cause of death for the } j\text{th individual to die, is } C_\alpha,\\ 0 & \text{otherwise.}\end{cases}\tag{12.27}$$

Then

$$\sum_{j=1}^{r}\delta_{\alpha j}=r_\alpha\tag{12.28}$$

is the number of deaths from cause $C_\alpha$ up to and including $t'_r$. In particular

$$\sum_{j=1}^{n} \delta_{\alpha j} = n_\alpha \tag{12.29}$$

is the total number of deaths from cause $C_\alpha$, and

$$\frac{n_\alpha}{n} = \pi_\alpha^o \tag{12.30}$$

is the observed proportion of deaths from cause $C_\alpha$.

The empirical crude distribution (simply, observed proportion of deaths up to and including $t'_r$) from cause $C_\alpha$ is

$$aQ_\alpha^o(t) = \frac{r_\alpha}{n} \qquad \text{for } t'_r \leqslant t < t'_{r+1}, \tag{12.31}$$

and

$$aP_\alpha^o(t) = \frac{n_\alpha - r_\alpha}{n} \qquad \text{for } t'_r \leqslant t < t_{r+1} \tag{12.32}$$

is the observed proportion of such deaths after time $t'_r$.

Perhaps, the simplest technique for evaluating $aQ_\alpha^o(t)$ is to order the times at death from cause $C_\alpha$

$$t'_{\alpha 1} < t'_{\alpha 2} < \ldots < t'_{\alpha, r_\alpha} < \ldots < t'_{\alpha, n_\alpha}. \tag{12.33}$$

Then

$$aQ_\alpha^o(t) = \frac{r_\alpha}{n} \qquad \text{for } t'_{\alpha, r_\alpha} \leqslant t < t'_{\alpha, r_\alpha + 1}. \tag{12.31a}$$

Clearly,

$$\frac{r_1}{n} + \frac{r_2}{n} + \ldots + \frac{r_k}{n} = \frac{r}{n},$$

and so

$$aQ_1^o(t) + aQ_2^o(t) + \ldots + aQ_k^o(t) = aQ^o(t) = F_T^o(t), \tag{12.34a}$$

$$aP_1^o(t) + aP_2^o(t) + \ldots + aP_k^o(t) = aP^o(t) = S_T^o(t). \tag{12.34b}$$

Also the (proper) empirical SDF for cause $C_\alpha$ is

$$S_\alpha^{*o}(t) = \frac{aP_\alpha^o(t)}{\pi_\alpha^o} = \frac{n_\alpha - r_\alpha}{n_\alpha}. \tag{12.35}$$

3. We also can construct the waiting time (giving private probabilities of death) survival functions, $G_\alpha(t)$. In this case, we may use the full array of times at death (12.25), but regard as effective times at death only those that correspond to deaths from cause $C_\alpha$; the other deaths are treated as if they were *withdrawals*. The empirical $G_\alpha^o(t)$ is

$$G_\alpha^o(t) = \prod_{j=1}^{r} \left( \frac{n-j}{n-j+1} \right)^{\delta_{\alpha j}} \qquad \text{for } t'_r \leqslant t < t'_{r+1}. \tag{12.36}$$

In practice, we may consider the ordered times at death from cause $C_\alpha$ as in (12.33). Suppose that $d_{(-\alpha)j}$ die from causes other than $C_\alpha$ in the interval $(t'_{\alpha,j-1}, t'_{\alpha j}]$. Let $R_{\alpha j}$ denote the number of individuals exposed to risk just before $t'_{\alpha j}$. Then

$$R_{\alpha 0} = n, \qquad R_{\alpha 1} = R_{\alpha 0} - d_{(-\alpha)1},$$

and generally,

$$R_{\alpha j} = R_{\alpha,j-1} - d_{(-\alpha)j} - 1 \tag{12.37}$$

for $j = 2, 3, \ldots, n_\alpha$. We have

$$G_\alpha^o(t) = \prod_{j=1}^{r_\alpha} \frac{R_{\alpha j} - 1}{R_{\alpha j}}. \tag{12.38}$$

(See product-limit method described in Section 6.9.)

### 12.4.2 Incomplete Data

When there are withdrawals between successive deaths from cause $C_\alpha$, the simple formulae in (1) and (2) of Section 12.4.1 no longer hold. We first apply the product-limit method to all data, and evaluate $aP^o(t)$ from formula (6.56).

1. If we desire to estimate the proportion of deaths that are due to cause $C_\alpha$, it is necessary to allow for the fact that the observed risk sets do

not reflect the numbers expected to survive in the population, because they are reduced by the withdrawals.

We can calculate an expected size of risk set at $t_j'$ as $E_j = n\cdot aP^o(t_j')$, say. We then weight the death at $t_j'$ by $E_j/R_j$, where $R_j$ is the risk set size at $t_j$ (see Section 6.7).

The $r_\alpha$ of Section 12.4.1 is replaced by

$$r_\alpha^* = \sum_{j=1}^{r} \delta_{\alpha j}\cdot \frac{E_j}{R_j} \tag{12.39}$$

in (12.31), (12.32), (12.34), and (12.36). [$\delta_{\alpha j}$ is defined in (12.27).]

2. In the estimation of $G_\alpha(t)$ the product-limit can be applied by simply including deaths from causes other than $C_\alpha$ with the true withdrawals. Formula (12.38) still applies, although the risk set, $R_{\alpha j}$, is now formed not merely by losses through death, but through withdrawals, also.

## 12.5 COHORT MORTALITY DATA WITH RECORDED TIMES AT DEATH OR CENSORING. PARAMETRIC ESTIMATION

The parametric model can be introduced either: (1) by the joint SDF (see Section 12.3.1) or (2) by fitting a model to crude probability functions (see Section 12.3.2). Here we present a method of construction of the likelihood function for case (1); the reader should be able to work out the mathematics appropriate to case (2).

Let $S_{1\ldots k}(y_1,\ldots,y_k;\gamma)$ be the joint SDF. As in Section 12.4.1 let $t_{\alpha j}'$ be the observed time at death from cause $C_\alpha$ for individual ($j$). Then the contribution to the likelihood for the $n_\alpha$ deaths from cause $C_\alpha$ is

$$L_\alpha(\gamma) \propto \prod_{j=1}^{n_\alpha} h_\alpha(t_{\alpha j}';\gamma) S_T(t_{\alpha j}';\gamma) \tag{12.40}$$

$$= \pi_\alpha^{n_\alpha} \prod_{j=1}^{n_\alpha} f_\alpha^*(t_{\alpha j}';\gamma), \tag{12.40a}$$

where $f_\alpha^*(t;\gamma)$ is the PDF of the (proper) crude distribution $S_\alpha^*(t;\gamma)$ (compare Sections 9.4.3 and 12.3.1).

Let $u_l$ be the time of withdrawal (censoring) alive of individual ($l$); let $N$ be the number of individuals ever entered the study; and let $n$ be the observed number of deaths during the study. Thus $N-n$ is the number withdrawn alive. Their contribution to the likelihood is

$$L_W(\gamma)=\prod_{l=1}^{N-n} S_T(u_l;\gamma) \tag{12.41}$$

The total likelihood is then

$$L=\left\{\prod_{\alpha=1}^{k} L_\alpha(\gamma)\right\} L_W(\gamma). \tag{12.42}$$

**Example 12.2** We use the (complete) mortality data for irradiated mice from Hoel (1972), given already in Table 7.1 in this book. We consider two specific causes of death: $C_1$ *Thymic Lymphoma*; and $C_2$ *Reticulum Cell Sarcoma*; all other causes are denoted by $C_3$.

We have $n=99$, $n_1=22$, $n_2=38$, and $n_3=39$. Of course, $n_1+n_2+n_3=n$. Hence $\pi_1^o=22/99=0.222$, $\pi_2^o=33/99=0.384$ and $\pi_3^o=39/99=0.394$ are the estimated proportions of deaths from $C_1$, $C_2$, and $C_3$, respectively.

***Nonparametric Estimation.*** We apply the results of Section 12.4.1 (1) and (2) to estimate $S_T^o(t)$ (from 12.26) and the $S_\alpha^o(t)$'s ($\alpha=1,2,3$) (from 12.36). Graphical representations of the results (step-functions) are given in Fig. 12.1 and 12.2a, b, c, respectively.

We also evaluated the $G_1^o(t)$ survival function, for deaths from thymic lymphoma alone. Formula (12.38) has been used for these calculations. The results are given in Table 12.4.

Note that the time at death of the last member in this group is $t_{1,22}=432$ days, and we have

$$G_1^o(432) \doteq G_1^o(\infty)=0.7429,$$

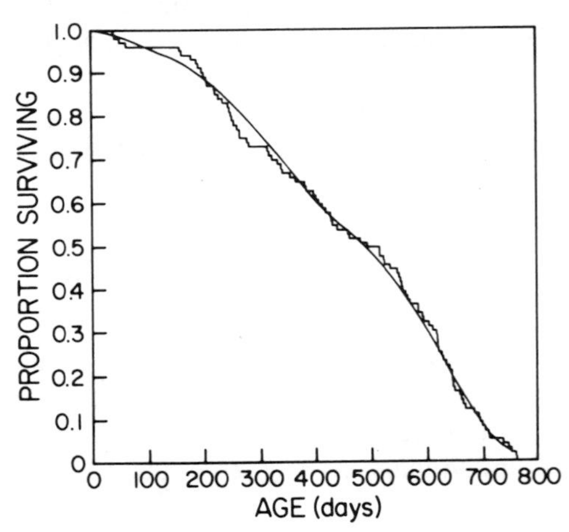


**Fig. 12.1** Estimated and fitted overall survival function (mice data from Table 7.1)

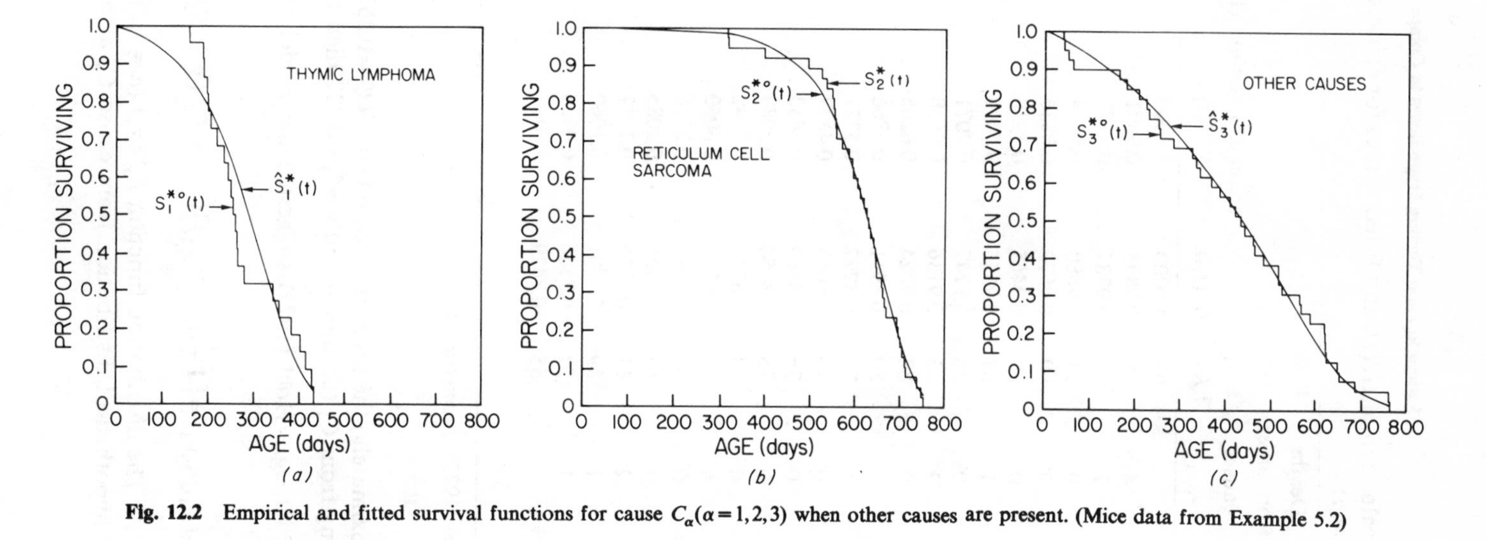


**Fig. 12.2** Empirical and fitted survival functions for cause $C_\alpha(\alpha=1,2,3)$ when other causes are present. (Mice data from Example 5.2)

**Table 12.4** Evaluation of $G_1^o(t)$, and fitted curves for cause $C_1$ (*thymic lymphoma*)

| $j$ | Time at death (in days) $t_{1j}$ | Deaths from other causes $d_{(-1)j}$ | Risk set at $t'_{1j}$ $R_{1j}$ | $G_1^o(t'_{1j})$ | (Nonparametric) $S_1^{+o}(t'_{1j})$ | (Parametric) $\hat{S}_1^+(t'_{1j})$ |
|---|---|---|---|---|---|---|
| 0 | | | 99 | 1.0000 | 1.0000 | 1.0000 |
| 1 | 159 | 4 | 95 | 0.9895 | 0.9591 | 0.8816 |
| 2 | 189 | 2 | 92 | 0.9787 | 0.9172 | 0.8299 |
| 3 | 191 | 0 | 91 | 0.9680 | 0.8754 | 0.8259 |
| 4 | 198 | 0 | 90 | 0.9572 | 0.8336 | 0.8114 |
| 5 | 200 | 0 | 89 | 0.9465 | 0.7918 | 0.8071 |
| 6 | 207 | 1 | 87 | 0.9356 | 0.7494 | 0.7915 |
| 7 | 220 | 0 | 86 | 0.9247 | 0.7071 | 0.7600 |
| 8 | 235 | 2 | 83 | 0.9136 | 0.6638 | 0.7195 |
| 9 | 245 | 0 | 82 | 0.9024 | 0.6205 | 0.6899 |
| 10 | 250 | 1 | 80 | 0.8911 | 0.5766 | 0.6743 |
| 11 | 256 | 1 | 78 | 0.8797 | 0.5322 | 0.6549 |
| 12 | 261 | 0 | 77 | 0.8683 | 0.4877 | 0.6382 |
| 13 | 265 | 0 | 76 | 0.8569 | 0.4433 | 0.6245 |
| 14 | 266 | 0 | 75 | 0.8454 | 0.3989 | 0.6210 |
| 15 | 280 | 0 | 74 | 0.8340 | 0.3545 | 0.5703 |
| 16 | 343 | 6 | 67 | 0.8216 | 0.3060 | 0.3127 |
| 17 | 353 | 0 | 66 | 0.8091 | 0.2576 | 0.2596 |
| 18 | 383 | 1 | 64 | 0.7965 | 0.2085 | 0.1599 |
| 19 | 403 | 2 | 61 | 0.7834 | 0.1577 | 0.1003 |
| 20 | 414 | 1 | 59 | 0.7701 | 0.1060 | 0.0740 |
| 21 | 428 | 1 | 57 | 0.7566 | 0.0535 | 0.0473 |
| 22 | 432 | 1 | 55 | 0.7429 | 0 | 0.0411 |
| | | 54 | | | | |

Data from Hoel (1972). *Biometrics* **28**, 475.

so that (approximately) the proportion $\hat{\phi}_1 = 1 - 0.7429 = 0.2571$ was exposed to risk of dying from $C_1$ (though actually $\pi^o{}_1 = 0.222$ died from $C_1$).

***Parametric Fitting.*** Hoel (1972) suggested the model of joint survival function

$$S_{123}(y_1, y_2, y_3) = [1 - \phi_1 F_1^+(y_1)][1 - \phi_2 F_2^+(y_2)] S_3^+(y_3) \quad (12.43)$$

(cf. Section 9.9). The distribution function $F_\alpha^+(y_\alpha)$ $(\alpha = 1, 2)$ is supposed to represent the hypothetical distribution from cause $C_\alpha$ alone among those who are liable to die from this cause; $\phi_\alpha$ is the proportion of individuals liable to die from $C_\alpha$.

For $S_\alpha^+(t)$, Hoel (1972) suggested a Gompertz distribution

$$S_\alpha^+(t)=\exp\left[\frac{R_\alpha}{a_\alpha}(1-e^{a_\alpha t})\right], \qquad \alpha=1,2,3. \tag{12.44}$$

Using model (12.43) and the observed times at death, $t'_{\alpha,j}$, for the $j$th individual dying from cause $C_\alpha$ *in the presence* of all other causes, the overall likelihood from (12.40) is proportional to

$$\prod_{j=1}^{n_1} \phi_1 f_1^+(t'_{1j}) S_3^+(t'_{1j})\left[1-\phi_2 F_2^+(t'_{1j})\right]$$
$$\times \prod_{j=1}^{n_2} \phi_2 f_2^+(t'_{2j}) S_3^+(t'_{2j})\left[1-\phi_1 F_1^+(t'_{2j})\right] \tag{12.45}$$
$$\times \prod_{j=1}^{n_3} f_3^+(t'_{3j})\left[1-\phi_1 F_1^+(t'_{3j})\right]\left[1-\phi_2 F_2^+(t'_{3j})\right]$$

We have obtained the following maximum likelihood estimates of the parameters:

$$\begin{aligned} \hat\phi_1&=0.257, & \hat\phi_2&=0.973,\\ \hat a_1&=1.119\cdot 10^{-2}, & \hat R_1&=2.863\cdot 10^{-4},\\ \hat a_2&=1.383\cdot 10^{-2}, & \hat R_2&=1.207\cdot 10^{-6},\\ \hat a_3&=4.010\cdot 10^{-3}, & \hat R_3&=2.184\cdot 10^{-4}, \end{aligned}$$

The overall SDF, $\hat S_T(t)$ was calculated from the formula

$$S_T(t)=\hat S_{123}(t,t,t), \tag{12.46}$$

and its graphical presentation (continuous) line is shown in Fig. 12.1.

The estimated crude SDF's were obtained (by numerical integration) from (9.17). For example,

$$\hat P_1^*(t)=\int_t^\infty \hat\phi_1 \hat f_1^+(y)\left[1-\hat\phi_2 \hat F_2^+(y)\right]\hat S_3(y)\,dy, \tag{12.47}$$

with $\hat\pi_1=\hat P_1^*(0)$, and then

$$\hat S_1^*(t)=\frac{\hat P_1^*(t)}{\hat\pi_1}. \tag{12.48}$$

In our Example, we obtained $\hat{\pi}_1=0.2244$, $\hat{\pi}_2=0.3789$, $\hat{\pi}_3=0.3966$. The continuous lines on each of the Figs. 12.2a, b, c, represent the parametric fitted $\hat{S}^*_\alpha(t)$'s.

In Table 12.4, we also give the estimates of $S_1^+(t)$. Since $Y_1$, $Y_2$, $Y_3$ are *independent* the nonparametric estimator of $S_1^+(t)$ is calculated as

$$S_1^{+o}(t)=\frac{G_1^o(t)-G_1^o(\infty)}{1-G_1^o(\infty)}\,, \tag{12.49}$$

where $G_1^o(\infty)=\phi_1^o=0.257$ and the fitted (parametric) estimator, $\hat{S}_1^+(t)$, from (12.44) substituting the ML-estimates of the parameters. These are given in the last two columns of Table 12.4, respectively. As we can see from the data, the fit is not good, but we do not apply any formal test.

More details on estimation in competing risk theory can be found in papers by Moeschberger and David (1971), Herman and Patell (1971), and in David and Moeschberger (1978) among others.

## REFERENCES

David, H. A. and Moeschberger, M. L. (1978). *The Theory of Competing Risks*, Griffin's Statistical Monographs and Courses, No. 39. Griffin: London.

Herman, R. J. and Patell, R. K. N. (1971). Maximum likelihood estimation for multi-risk model. *Technometrics* **13**, 385–396.

Hoel, D. G. (1972). A representation of mortality data by competing risks. *Biometrics* **28**, 475–488.

Moeschberger, M. L. and David, H. A. (1971). Life tests under competing causes of failure and the theory of competing risks. *Biometrics* **27**, 909–933.

## EXERCISES

**12.1.** Use the mice data given in Table 7.1. Consider each cause separately.

(*a*) Fit to each set of data the following distributions:

(*i*) Gompertz with different parameters for each cause $[S^*_\alpha(x;R_\alpha,a_\alpha)]$.

(*ii*) Exponential for $C_1$ and Gompertz (with different parameters) for $C_2$ and $C_3$.

(*b*) In each case construct the overall SDF (see Sections 12.3.2 and 9.4.3).

**12.2.** Use the same data as in Exercise 1.21.

(*a*) Estimate the hazard rates of empirical distributions, $S_{\alpha}^{*o}(x)$'s denoting them by $\lambda^{*o}_{\alpha}(x)$.
*Hint*: Use methods discussed in Section 7.7.

(*b*) Estimate the hazard rates of fitted distributions, $\lambda_{\alpha}^{*}(x)$, discussed in Example 12.1 (*i*). (Note that these are exponential hazards).

(*c*) Make a graphical comparison of the results obtained in (*a*) and (*b*).

**12.3.** Suggest a method of nonparametric estimation of crude hazard rates from complete data (e.g., cf. Section 9.4.3).

**12.4.** Estimate parametric crude hazard rates $h_{\alpha}(x)$'s from the models suggested in Example 12.2 and in Exercise 12.1, respectively. Comment on the results.

**12.5.** Suppose that we have complete mortality data grouped in fixed intervals $(t_i, t_{i+1}]$ as described in Section 12.2.1.

Let $S_{1\ldots k}(y_1,\ldots,y_k;\gamma_1,\ldots,\gamma_s)$ denote the joint SDF defined in Section 12.3.1. Construct the unconditional likelihood and show how to use it to estimate parameters $\gamma_1,\ldots,\gamma_s$.

# Part 4
# SOME MORE ADVANCED TOPICS

CHAPTER 13

# Concomitant Variables in Lifetime Distributions Models

## 13.1 CONCOMITANT VARIABLES

In Chapter 1, we stated that the primary interest in survival analysis is the study of lifetime distributions (life tables) in some chosen populations. These populations are defined in terms of specific factors (e.g., demographic factors such as race and sex, geographical location, assigned treatment, type of tool, etc.).

In clinical, and other experimental enquiries, however, measurements of further characters, beyond just time (or age) and mortality, are often obtained. Some, if not all, of these may be expected to have some association with failure rates. We recall (Section 1.3) that the variables corresponding to such characters are called *concomitant* ("going along with"), or *explanatory* variables, or briefly *covariates* or *covariables*.

Note that the association is usually not quantitatively causative, nor is such causation as exists, always in the same direction. If it is believed that a variable (e.g., number of cigarettes smoked per day) represents a factor that may, in itself, contribute to failure or disease, we speak of a *risk factor*. If causation is regarded as being in the other direction (e.g., elevated blood sugar of diabetic patients), we speak of a *symptom*. The distinction is often not clear, and in many cases a character may be both—a risk factor *and* a symptom. For example, a high level of cholesterol is usually considered as a risk factor in cardiovascular diseases; it might well be also a symptom of such a disease.

From the point of view of model construction, it is not always necessary to make this distinction, although it should be helpful in choosing among different forms of model.

## 13.2 THE ROLE OF CONCOMITANT VARIABLES IN PLANNING CLINICAL TRIALS

When the association of a concomitant variable with mortality is marked but, perhaps, not quantifiable in fairly simple terms, allowance can be made for it by the use of select tables, as described in Section 4.13. For example, Tallis et al. (1973) in constructing select life tables for patients diagnosed with breast cancer, used time elapsed since diagnosis as a primary variable, and age as a concomitant variable. As we have noted, select tables are, in fact, a set of separate life tables, one for each of a number of different levels of the relevant concomitant variable. From this point of view, separate tables for female and male lives could be regarded as "select" tables with respect to the concomitant variable "sex."

At the opposite extreme, reduction of effects of concomitant variables may be sought by trying to even out these effects. In planning experiments on the basis of which SDF's of two (or more) treatment groups are to be compared, one may assign individuals deliberately to chosen populations, using:

1. *Purposive selection*, that is, choosing individuals to have "representative" values for the concomitant variables, balanced among the populations.
2. *Randomization*, that is, allocating individuals to populations by some "objective" scheme (e.g., by the use of random numbers).

A special and very desirable case, of (1) is when each individual (or groups of individuals) from one population is paired with one individual (or group of individuals) from the other population(s), so that values of all measured concomitant variables nearly agree. This is called *individual* (*or group*) *matching*.

On the other hand, randomization does not require measurement of concomitant variables, although it is still desirable that this be done. Randomization has the merit of evening out, *in the long run*, the values of all concomitant variables (even those of which we are not aware) among individuals assigned to different populations. A particular randomized sample may, however, be very unrepresentative.

## 13.3 GENERAL PARAMETRIC MODEL OF HAZARD FUNCTION WITH OBSERVED CONCOMITANT VARIABLES

With either purposive selection or randomization, or even when neither has been used, it is sometimes possible to use measurements of concomitant variables to reduce the effects of such variables on survivorship comparisons. This can be done by incorporating such variables explicitly in the *parametric form of the hazard function*.

As usual, let $T$ denote time elapsed since starting point ($t=0$) of observation, and let $z_1, z_2, \dots, z_s$ denote $s$ concomitant (or explanatory) variables.

### 13.3.1 Types of Concomitant Variables

Any of the concomitant variables can be discrete or continuous. Numbers of disease attacks before treatment, numbers of hospitalizations for illness other than that under consideration, and so on, are examples of *discrete* variables. Included among discrete variables are the so-called *indicator variables*. For example, let

$$z_{1j} = \begin{cases} 0 & \text{if individual } (j) \text{ is a male,} \\ 1 & \text{if individual } (j) \text{ is a female,} \end{cases}$$

or

$$z_{2j} = \begin{cases} 0 & \text{if } (j) \text{ belongs to control group,} \\ 1 & \text{if } (j) \text{ belongs to treatment group.} \end{cases}$$

Indicator variables can be used to assign each individual to a separate group (subpopulation).

Among *continuous* variables, *age* plays an important role. It is almost always recorded at the date of entry. Other variables, such as systolic blood pressure, level of serum cholesterol, and so on, are usually not only *age dependent*, but also vary from time to time, somewhat irregularly, for the same individual. These variables are usually measured at the entry and also periodically during the course of the experiment.

It is important to note that if the model involves values of concomitant variables measured *after* treatment has commenced, *and these values are affected by the treatment itself*, there is need for special care in interpretation of results of analyses. For example, if mortality among treated and control groups with common *current* values of risk factor(s) is compared, it is not unlikely that the treated group will exhibit higher mortality, because it needed treatment to reduce its risk factor(s) to the value(s) the control group possessed without treatment. The observed higher mortality should clearly not be regarded as an indication that treatment has an adverse effect. In fact, *for treated individuals* the reduction in risk factor tends to reduce expected mortality. Valid comparisons would, in effect, compare treatment and control groups which have the same risk factor value(s) *before* commencement of treatment.

Furthermore, an observed high level of a concomitant variable among those suffering from a disease may not signify that the variable is either a risk factor or, in the usual sense, a symptom of the disease. It may be that a high level of the variable in question greatly improves chances of survival when the disease develops; consequently the average level *among survivors* with the disease is higher than in the general population. Clearly, interpretation as a risk factor would be quite inappropriate. For example, treatment resulting in reduction of the level of the variable, far from increasing resistance to onset of the disease, would tend to decrease chances of survival should the disease occur.

Sometimes exact values of continuous concomitant variables are not available; for example, we may have only records on age lbd, systolic blood pressure between certain limits, cholesterol level less than a certain value. In such cases an average value for the group, for example, 54.5 for age lbd 54, can often be used. Choice of such average values (sometimes referred to as coding) is often somewhat arbitrary. This must be so when the group is open-ended, for example, cholesterol above a certain value. Careful consideration should always be given to coding. Of course, when there are only two groups (e.g., less than $\xi$, and greater than $\xi$), they can be designated "0" and "1" unequivocally, as no relative magnitudes are involved.

### 13.3.2 General Model

Let $\mathbf{z}'=(z_1,z_2,\ldots,z_s)$ denote a $1\times s$ vector of concomitant variables. We denote the hazard rate by $\lambda(t;\mathbf{z})$. Of course, this can also depend on some parameters, $\boldsymbol{\beta}'=(\beta_1,\ldots,\beta_k)$, so that, in fact

$$\lambda(t;\mathbf{z})=\lambda(t;\mathbf{z};\boldsymbol{\beta}). \tag{13.1}$$

For brevity, we will use the notation $\lambda(t;\mathbf{z})$, unless some ambiguity might arise thereby. Note that we must have $\lambda(t;\mathbf{z})\geqslant 0$.

The cumulative hazard function is

$$\Lambda(t;\mathbf{z})=\int_0^t\lambda(u;\mathbf{z})\,du. \tag{13.2}$$

The survival function (SDF) and the density function (PDF) are

$$S(t;\mathbf{z})=\exp[-\Lambda(t;\mathbf{z})], \tag{13.3}$$

and

$$f(t;\mathbf{z})=\lambda(t;\mathbf{z})S(t;\mathbf{z}), \tag{13.4}$$

respectively.

Let $\mathbf{z}_j' = (z_{1j}, z_{2j}, \ldots, z_{sj})$ denote the vector of *observed* concomitant variables for individual $(j)$. The vector $(z_{1j}, \ldots, z_{sj})$ can, for example, represent a set of measurements taken at entry of individual $(j)$, or a set of average measurements over the course of study, or certain indicator variables as discussed in Section 13.3.1. But once they have been observed, they represent a set of *fixed* values. We first assume that the $z$'s do not depend on $t$.

For each individual $(j)$, we define his own hazard rate by $\lambda(t;\mathbf{z}_j)$ and survival function by $S(t;\mathbf{z}_j)$.

***Likelihood function.*** Consider a sample of $N$ individuals participating, at some time or other in a study.

Let $\tau_j$ be the time at which individual $(j)$ *entered* the study, and $t_j$ the time at which $(j)$ was *last observed*.

We denote by $\mathfrak{D}$ the set of $d$ individuals dying, and by $\overline{\mathfrak{D}}$ the set of $(N-d)$ individuals who were alive when last observed, in the whole study. Then the likelihood function for this sample is

$$\begin{aligned} L &= \prod_{(j)\in\mathfrak{D}} \left[ \frac{f(t_j;\mathbf{z}_j)}{S(\tau_j;\mathbf{z}_j)} \right] \times \prod_{(l)\in\overline{\mathfrak{D}}} \left[ \frac{S(t_l;\mathbf{z}_l)}{S(\tau_l;\mathbf{z}_l)} \right] \\ &= \prod_{(j)\in\mathfrak{D}} \left[ \lambda(t_j;\mathbf{z}_j) \right] \times \prod_{i=1}^{N} \frac{S(t_i;\mathbf{z}_i)}{S(\tau_i;\mathbf{z}_i)}. \end{aligned} \tag{13.5}$$

We may express (13.5) in a different form. Let

$$\delta_j = \begin{cases} 1 & \text{if } (j) \text{ died at } t_j, \\ 0 & \text{if } (j) \text{ was alive at } t_j. \end{cases} \tag{13.6}$$

Then the likelihood function (13.5) takes the form

$$L = \prod_{j=1}^{N} \left[ \lambda(t_j;\mathbf{z}_j) \right]^{\delta_j} \frac{S(t_j;\mathbf{z}_j)}{S(\tau_j;\mathbf{z}_j)}, \tag{13.7}$$

and

$$\log L = \sum_{j=1}^{N} \left[ \delta_j \log \lambda(t_j;\mathbf{z}_j) + \log S(t_j;\mathbf{z}_j) - \log S(\tau_j;\mathbf{z}_j) \right].$$

When there are no new entries, for example, when $N$ individuals are followed up from $t=0$, then $\tau_j = 0$ and so $S(\tau_j;\mathbf{z}_j) = 1$ for all $j$.

In such situations, which are common in clinical trials, the likelihood (13.7) takes the form

$$L=\prod_{j=1}^{N}\left[\lambda(t_j;\mathbf{z}_j)\right]^{\delta_j}S(t_j;\mathbf{z}_j). \tag{13.7a}$$

When the parametric form of the hazard function, $\lambda(t;\mathbf{z})$ is known, their ML-estimators can be obtained by maximizing (13.7a). Some specific forms of hazard functions are discussed in Sections 13.4 and 13.5.

**Example 13.1** A popular method of incorporating explanatory variables into the model, is *to express the parameters of the model as simple functions of these variables*. There is an example of this kind in Bailey et al. (1977) where data on survival of 748 cases of kidney transplant are analyzed using a hazard rate model in the form

$$\lambda(t)=\alpha e^{-\gamma t}+\delta, \tag{13.8}$$

where $t>0$ is time elapsed since graft, and $\alpha>0$, $\gamma>0$, and $\delta>0$ are parameters. The authors introduced two concomitant variables: $z_1=1$ if male, and $=2$ if female, and age $z_2$ (in years) of the recipient. The parameters $\alpha$, $\gamma$, and $\delta$ in (13.8) were replaced by

$$\begin{aligned}\alpha&=\exp(\alpha_0+\alpha_1z_1+\alpha_2z_2),\\ \gamma&=\exp(\gamma_0+\gamma_1z_1+\gamma_2z_2),\\ \delta&=\exp(\delta_0+\delta_1z_1+\delta_2z_2).\end{aligned} \tag{13.9}$$

Note that this formulation ensures that $\alpha$, $\gamma$, and $\delta$ cannot be negative. Since the only covariates used are *sex and age*, fitting the model is equivalent to constructing a select life table for each sex. Since $\gamma$ is positive, the selection is negative in the sense described in Section 4.13. The ultimate life table corresponds to $t\to\infty$, and has hazard rate

$$\delta=\exp(\delta_0+\delta_1z_1+\delta_2z_2).$$

This corresponds to a Gompertz law (with respect to age $z_2$), with the ratio of female to male hazard rates constant [$=\exp(\delta_1)$] at all ages. According to the model, the ultimate hazard rate is never reached (for finite $t$), but for practical purposes, the effect of selection can be ignored for sufficiently large $t$.

Values of the nine parameters were estimated by the method of maximum likelihood. Table 13.1 [adapted from Bailey et al. (1977)] gives the estimates obtained together with their estimated standard errors.

**Table 13.1** Parameter estimates and their standard errors for kidney transplant data

| Parameter | Estimate | Standard error | $t=\dfrac{\text{Estimate}}{\text{Standard error}}$ |
|---|---|---|---|
| $\alpha_0$ | −4.77 | 0.34 | −14.03** |
| $\gamma_0$ | −3.97 | 0.38 | −10.45** |
| $\delta_0$ | −8.00 | 0.47 | −17.02** |
| $\alpha_1$ | −0.19 | 0.21 | −0.90 |
| $\gamma_1$ | 0.20 | 0.18 | 1.11 |
| $\delta_1$ | −0.09 | 0.17 | −0.53 |
| $\alpha_2$ | 0.0018 | 0.0063 | 0.29 |
| $\gamma_2$ | −0.0125 | 0.0071 | −1.76 |
| $\delta_2$ | 0.0061 | 0.0093 | 0.66 |

As a crude set of tests of significance one might use a $t$-test for each individual parameter. If we consider two-sided alternatives for each parameter, only $\alpha_0$, $\gamma_0$, and $\delta_0$ seem to be significant on this basis. We should realize, however, that such crude tests on each individual parameter might be misleading; more appropriate (although still approximate) methods based on log likelihood ratio tests are discussed in Section 13.8.

### 13.3.3 Some Other Expressions for the General Model

1a. Suppose that there are $k$ groups (levels of a factor, treatments) in the experiment. In the general case, the hazard rate function for each group might be of *different form*. For the $g$th group we define [cf., (13.1)]

$$\lambda_g(t;\mathbf{z})=\lambda_g(t;\mathbf{z};\boldsymbol{\beta}_g), \qquad g=1,2,\ldots,k \tag{13.10}$$

[e.g., see (13.25)].

***Likelihood function.*** Let $(gj)$ denote the $j$th individual in the $g$th group, $t_{gj}$ be the time at which $(gj)$ is last observed, and

$$\mathbf{z}_{gj}=(z_{1gj},\ldots,z_{sgj})$$

be the $1\times s$ vector of covariates. We introduce

$$\delta_{gj}=\begin{cases}1 & \text{if } (gj) \text{ died at } t_{gj},\\ 0 & \text{if } (gj) \text{ was alive at } t_{gj}.\end{cases} \tag{13.11}$$

Let $N_g$ be the number of individuals followed up from $t=0$ in the $g$th subpopulation with $N_1+N_2+\cdots+N_k=N$. The likelihood function is

$$L=\prod_{g=1}^{k}\prod_{j=1}^{N_g}\left[\lambda_g(t_{gj};\mathbf{z}_{gj})\right]^{\delta_{gj}}S_g(t_{gj};\mathbf{z}_{gj}), \tag{13.12}$$

where, of course,

$$S_g(t_{gj};\mathbf{z}_{gj})=\exp\left[-\int_0^{t_{gj}}\lambda_g(t;\mathbf{z}_{gj})\,dt\right].$$

[See Kay (1977).]

1b. We may also distinguish more than one grouping factor. As a simple example, suppose that one grouping factor represents sex (males and females), and the other grouping factor is represented by $k$ different treatments. For the $(gh)$th group $(g=1,2;\ h=1,2,\ldots,k)$ we may then define the hazard rate function

$$\lambda_{gh}(t;\mathbf{z})=\lambda_{gh}(t;\mathbf{z};\boldsymbol{\beta}_{gh}) \tag{13.13}$$

for $g=1,2$; $h=1,2,\ldots,k$ [cf., Example 13.4, formula (13.29)].

Construction of the likelihood function in this case is straightforward.

2a. Often the form of the hazard rate function in different groups is the same; only the parameters may have different values. In such cases it might be convenient to use a single expression for the hazard rate by introducing indicator-type variables for the groups. To distinguish these from the "ordinary" covariables ($z$), we denote them by $y$'s. Then

$$\lambda(t;\mathbf{y};\mathbf{z})=\lambda(t;\mathbf{y};\mathbf{z};\boldsymbol{\beta}), \tag{13.14}$$

where $\mathbf{y}$ is a vector of appropriately chosen indicator variables, and is equivalent to model (13.10). If there are $k$ levels of a factor, it is necessary to introduce $(k-1)$ indicator variables to represent it in (13.10) by (13.14). For example, suppose that the hazard rate of type (13.10) is given by the formula

$$\lambda_g(t;z;\boldsymbol{\alpha},\boldsymbol{\beta})=\beta_{0g}+\beta_{1g}z, \qquad g=1,2. \tag{13.15a}$$

Let

$$y=\begin{cases}0 & \text{if } g=1,\\ 1 & \text{if } g=2.\end{cases}$$

Then

$$\lambda(t;\mathbf{y};\mathbf{z};\boldsymbol{\alpha},\boldsymbol{\beta})=(\alpha_1+\alpha_2 y)+(\beta_1+\beta_2 y)z \tag{13.15b}$$

is equivalent to the model (13.15a), with $\beta_{11}=\alpha_1$, $\beta_{21}=\beta_1$, $\beta_{12}=\alpha_1+\alpha_2$, and $\beta_{22}=\beta_1+\beta_2$.

2b. If there is more than one factor, indicator variables must be appropriately chosen to represent each factor. In (13.15a), we need $2-1=1$ indicator variable to represent sex ($g$), and $(k-1)$ indicator variables to represent treatments ($h$), that is $1+(k-1)=k$ indicator variables in all. (See Example 13.4.)

## 13.4 ADDITIVE MODELS OF HAZARD RATE FUNCTION

Two special types of hazard rate—additive and multiplicative—have attracted considerable attention. Their attraction seems to arise from their mathematical simplicity, and they are in common use. In this section, we discuss additive models. We define the general *additive* model as

$$\lambda(t;\mathbf{z})=\lambda(t)+\sum_{u=1}^{s} h_u(t)g_u(z_u), \tag{13.16}$$

where $\lambda(t)$ is the so called *underlying* hazard rate. Note that the $h_u(t)$'s are functions of $t$ only (in special cases they can be polynomial in $t$), while the functions $g_u(z_u)=g_u(z_u;\boldsymbol{\beta}_u)$, $u=1,2,\ldots,s$, do not depend on $t$.

Often, by appropriate definition of $z_u$'s (e.g., $z_u=\log x_u$, where $x_u$ is the original measurement), we can represent (13.16) as a *linear* function of $z_u$'s, that is

$$\lambda(t;\mathbf{z})=\lambda(t)+\sum_{u=1}^{s} h_u(t)z_u, \tag{13.17}$$

and special cases, when the hazard rate does not, in fact, vary with time are of the form

$$\lambda(t;\mathbf{z})=\beta_0+\sum_{u=1}^{s}\beta_u z_u=\sum_{u=0}^{s}\beta_u z_u=\boldsymbol{\beta}'\mathbf{z}, \tag{13.18}$$

where $z_0=1$ is a dummy variable. Note that the values of the $\beta$'s must be such that condition $\boldsymbol{\beta}'\mathbf{z}\geqslant 0$ is satisfied.

For a given set of $z_u$'s the linear function on the right-hand side of (13.18) yields a constant value; the model of SDF is then *exponential*. For this reason, survival models with hazard rate defined by (13.18) are sometimes referred to as *linear-exponential*.

For the $j$th individual, we have

$$\lambda(t;\mathbf{z}_j)=\sum_{u=0}^{s}\beta_u z_{uj}. \tag{13.19}$$

Of course, it is necessary that the right-hand side of (13.19) is positive. For a random sample of size $N$, the likelihood is

$$L=\prod_{j=1}^{N}\left\{\left(\sum_{u=0}^{s}\beta_u z_{uj}\right)^{\delta_j}\exp\left[-\left(\sum_{u=0}^{s}\beta_u z_{uj}\right)t_j\right]\right\}, \tag{13.20}$$

where $\delta_j$ is defined in (13.6), or

$$\log L=\sum_{j=1}^{N}\left[\delta_j\log\left(\sum_{u=0}^{s}\beta_u z_{uj}\right)-\left(\sum_{u=0}^{s}\beta_u z_{uj}\right)t_j\right]. \tag{13.20a}$$

This leads to $(s+1)$ maximum likelihood equations

$$\frac{\partial\log L}{\partial\beta_v}=\sum_{j=1}^{N}\frac{\delta_j z_{vj}}{\sum_{u=0}^{s}\beta_u z_{uj}}-\sum_{j=1}^{N}z_{uj}t_j=0, \qquad v=0,1,\ldots,s \tag{13.21}$$

[see also Byar et al. (1974)].

**Example 13.2** Byar et al. (1974) used model (13.18) in the analysis of data on mortality of patients with cancer of the prostate [Mellinger et al. (1967)].

They used 11 binary (0,1) variables, as shown in Table 13.2. Each estimated regression coefficient, $b_u$, was tested for significance ($H_0$: $\beta_u=0$ for $u=0,1,\ldots,11$), using a (partial) $t$ test ($t=b_u/s_{b_u}$, where $s_{b_u}$ is the estimated standard error of $b_u$). The results are summarized in Table 13.2.

For each value of $t$, the number of degrees of freedom is very large, so that a normal approximation to $t$ distribution can be used. For example, with the significance level $\alpha=0.05$, the critical value for a one-sided test is $t_{0.95}\doteq z_{0.95}=1.645$.

**Table 13.2** Estimates of regression coefficients. Stages III and IV of tumor

| $u$ | Variable ($z_u$) coded as 1 if: | $b_u$ | $s_{b_u}$ | $t=b_u/s_{b_u}$ |
|---|---|---|---|---|
| 0 | $z_0=1$ always | 7.88 | 0.64 | 12.31 |
| 1 | Pain due to cancer | −0.20 | 1.75 | −0.11 |
| 2 | Acid phosphatase 1.1–2.0 | 4.15 | 1.40 | 2.96 |
| 3 | Acid phosphatase 2.1–5.0 | 6.94 | 1.93 | 3.60 |
| 4 | Acid phosphatase $>5.0$ | 11.04 | 2.90 | 5.28 |
| 5 | Ureteral dilation | 4.68 | 2.31 | 2.03 |
| 6 | Metastases | 7.36 | 1.91 | 3.84 |
| 7 | Partially bedridden | 11.18 | 2.39 | 4.68 |
| 8 | Totally bedridden | 21.55 | 8.48 | 2.54 |
| 9 | Weight $<130$ lbs | 3.50 | 1.34 | 2.61 |
| 10 | Hemoglobin $<12$ g/100 ml | 5.44 | 1.49 | 3.65 |
| 11 | Age $>70$ | 3.02 | 0.84 | 3.55 |

From Byar et al. (1974). *J. Natl. Cancer Inst.* **52**, 321.

(*Note*: For example,

$$z_9=\begin{cases} 1 & \text{if weight is } <130 \text{ lb,} \\ 0 & \text{if weight is } >130 \text{ lb.} \end{cases}$$

Some variables are highly correlated. If, for example, $z_4=1$, then automatically $z_3=z_2=0$.)

A natural extension of linear-exponential models would be to use a polynomial function of the covariables. For example,

$$\lambda(t;\mathbf{z})=\alpha+\beta_{11}z_1^2+\beta_{21}z_2+\beta_{22}z_2^2 \tag{13.22}$$

represents an (additive) *quadratic-exponential* model in covariates $z_1$, $z_2$.

**Example 13.3** Feigl and Zelen (1965) suggested an exponential model, in which the *mean survival time is a linear function of a concomitant variable* $z$, that is

$$\mathcal{E}(T)=[\lambda(t;z)]^{-1}=\alpha+\beta z. \tag{13.23}$$

A possible extension to $s$ concomitant variables could be

$$\mathcal{E}(T)=[\lambda(t;\mathbf{z})]^{-1}=\sum_{u=0}^{s}\beta_u z_u, \tag{13.24}$$

or

$$\lambda(t;\mathbf{z})=\left[\sum_{u=0}^{s}\beta_u z_u\right]^{-1}. \tag{13.24a}$$

Model (13.23) was applied to mortality data of patients with acute leukemia, classified into $k=2$ groups AG-positive and AG-negative, depending on whether the Auer rods were absent or present, respectively, in bone marrow. The concomitant variable, $z$, was the white blood count at the time of diagnosis.

The authors used model (13.23) for each group separately, that is

$$\lambda_g(t;\mathbf{z})=(\alpha_g+\beta_g z)^{-1}, \qquad g=1,2. \tag{13.25}$$

## 13.5 MULTIPLICATIVE MODELS

We define the general *multiplicative* model

$$\lambda(t;\mathbf{z})=\lambda(t)g(\mathbf{z}), \tag{13.26}$$

where $\lambda(t)$ is the underlying hazard rate, and $g(\mathbf{z})=g(\mathbf{z};\boldsymbol{\beta})$.

### 13.5.1 Exponential-Type Hazard Functions

Cox (1972) discussed models in which $g(\mathbf{z};\boldsymbol{\beta})=\exp(\boldsymbol{\beta}'\mathbf{z})$, so that

$$\lambda(t;\mathbf{z})=\lambda(t)\exp(\boldsymbol{\beta}'\mathbf{z})=\lambda(t)\exp\left(\sum_{u=1}^{s}\beta_u z_u\right) \tag{13.27}$$

or

$$\log\lambda(t;\mathbf{z})=\log\lambda(t)+\sum_{u=1}^{s}\beta_u z_u. \tag{13.27a}$$

An advantage of model (13.27) is that no restriction need to be placed on the value of the expression $\boldsymbol{\beta}'\mathbf{z}$, since always $\exp(\boldsymbol{\beta}'\mathbf{z})>0$.

When the underlying hazard rate is constant $[\lambda(t)=\lambda]$, then

$$\lambda(t;\mathbf{z})=\lambda\exp\left(\sum_{u=1}^{s}\beta_u z_u\right) \tag{13.28}$$

or

$$\log\lambda(t;\mathbf{z})=\sum_{u=0}^{s}\beta_u z_u, \tag{13.28a}$$

where $\beta_0=\log\lambda$ and $z_0=1$.

This model is quite popular. [See, for example, Feigl and Zelen (1965), Glasser (1967), Prentice (1973).] It is sometimes called a *log-linear* (in $z_u$'s) model for hazard rate (or *log-linear exponential* model for survival function).

Another variant for the model would be to replace $\Sigma_u \beta_u z_u$ in (13.28) by a polynomial function (in $z_u$'s). We would then have log-quadratic, log-cubic, and so on, models of hazard rate function.

Further modification of model (13.28) is shown in the following example.

**Example 13.4** Prentice (1973) analyzed survival data of 137 patients with advanced lung cancer. Patients were randomly assigned to two groups of treatment with chemotherapeutic agents ("standard" and "test", $g=1,2$) and four types of tumor (squamous, small, adeno, and large, $h=1,2,3,4$, respectively).

The concomitant variables were:

$z_1$ = performance status (9 levels); $z_3$ = age at diagnosis (years);
$z_2$ = time since diagnosis (months); $z_4$ = prior therapy (yes or no).

Further, we denote by $\bar{z}_u$ the arithmetic mean of covariable $z_u$ ($u=1,\ldots,4$). The model used was

$$\lambda_{gh}(t;\mathbf{z}) = \lambda_{gh} \exp\left[\sum_{u=1}^{4} \beta_u (z_u - \bar{z}_u)\right]. \tag{13.29}$$

For the $j$th individual in the $(gh)$th treatment × tumor type combination, we then have

$$\lambda_{gh}(t;\mathbf{z}_{ghj}) = \lambda_{gh} \exp\left[\sum_{u=1}^{4} \beta_u (z_{ughj} - \bar{z}_u)\right]. \tag{13.30}$$

The estimates of $\beta$'s were $\hat{\beta}_1 = 0.029$, $\hat{\beta}_2 = 0.006$; $\hat{\beta}_3 = -0.003$; and $\hat{\beta}_4 = -0.012$. From further analysis, it appears that only $\hat{\beta}_1$ is significantly different from zero.

Note that $\lambda_{gh}$ could be expressed in terms of indicator variables for treatments ($y_1$), and for types of tumor ($y_2, y_3, y_4$), taking the values

$$y_1 = \begin{cases} 0 & \text{for "standard" treatment} \quad (g=1), \\ 1 & \text{for "test" treatment} \quad (g=2). \end{cases}$$

| Type of tumor | $y_2$ | $y_3$ | $y_4$ | |
|---|---|---|---|---|
| Squamous | 0 | 0 | 0 | $(h=1)$ |
| Small | 1 | 0 | 0 | $(h=2)$ |
| Adeno | 0 | 1 | 0 | $(h=3)$ |
| Large | 0 | 0 | 1 | $(h=4)$ |

Model

$$\lambda(t;\mathbf{y},\mathbf{z})=\lambda\exp\left[\sum_{v=1}^{4}\alpha_v y_v+\sum_{u=1}^{4}\beta_u(z_u-\bar{z}_u)\right] \tag{13.31}$$

is equivalent to (13.29) with

$$\left.\begin{array}{ll}\lambda_{11}=\lambda; & \lambda_{1h}=\lambda\exp(\alpha_h)\\ \lambda_{21}=\lambda\exp(\alpha_1); & \lambda_{2h}=\lambda\exp(\alpha_1+\alpha_h)\end{array}\right\} h=2,3,4. \tag{13.32}$$

### 13.5.2 Gompertz and Weibull Models with Covariates

We now give two further examples of multiplicative models.

*Gompertz Model.* The hazard rate for this model is obtained by replacing $R$ in (3.51) by $R\exp(\Sigma_{u=1}^{s}\beta_u z_u)$, giving

$$\lambda(t;z)=Re^{at}\exp\left(\sum_{u=1}^{s}\beta_u z_u\right)=e^{at}\exp\left(\sum_{u=0}^{s}\beta_u z_u\right), \tag{13.33}$$

with $\beta_0=\log R$ and $z_0=1$.

Taking logarithms of both sides, we obtain

$$\log\lambda(t;\mathbf{z})=at+\sum_{u=0}^{s}\beta_u z_u, \tag{13.34}$$

*Weibull Model.* The hazard rate in this case is obtained by replacing $\theta^{-c}$ in (3.55) (with $\xi=0$) by $\theta^{-c}\exp(\Sigma_{u=1}^{s}\beta_u z_u)$, giving

$$\lambda(t;z)=c\theta^{-c}t^{c-1}\exp\left(\sum_{u=1}^{s}\beta_u z_u\right). \tag{13.35}$$

Prentice (1973) fitted this model to lung cancer mortality data.

Of course the parameters $a$ in the Gompertz model, and $c$ in the Weibull model might also depend on covariates, but such models are not in common use.

## 13.6 ESTIMATION IN MULTIPLICATIVE MODELS

We consider the general multiplicative model (13.26). For individual $(j)$ this is

$$\lambda(t;\mathbf{z}_j)=\lambda(t)g(\mathbf{z}_j;\boldsymbol{\beta}), \tag{13.36}$$

where $g(\mathbf{z}_j;\boldsymbol{\beta})$ has a known form, while $\lambda(t)$ is unspecified. As before, we assume that the $z$'s do not depend on $t$; then for two individuals $(j)$ and $(l)$, the hazard rates are in the constant ratio, $g(\mathbf{z}_j;\boldsymbol{\beta})/g(\mathbf{z}_l;\boldsymbol{\beta})$, whatever the time $t$. A model of this kind is sometimes called a *proportional hazard rate model*. We now wish to obtain the ML-estimators of $\beta$'s when the parametric form of $g(\mathbf{z};\boldsymbol{\beta})$ is specified.

### 13.6.1 Construction of the Likelihood Function

To apply maximum likelihood estimation, it would be convenient if $\lambda(t)$ could be expressed in terms of some (not too numerous) parameters. One way of doing this is to suppose it to be constant over a number of intervals of time.

Dividing the range of variation of $t$ in $M$ *fixed* (consecutive) intervals $I_k \equiv (t_{k-1}, t_k]$, $k=1, 2,\dots,M$, we suppose

$$\lambda(t)=\lambda_k \qquad \text{for } t_{k-1}<t\leqslant t_k, \tag{13.37}$$

and so the hazard rate is

$$\lambda(t;\mathbf{z})=\lambda_k g(\mathbf{z};\boldsymbol{\beta}) \qquad \text{for } t_{k-1}<t\leqslant t_k \tag{13.38}$$

(cf. Section 6.7, which corresponds to $g(\mathbf{z};\boldsymbol{\beta})\equiv 1$).

Let $\mathbb{S}_k$ denote the set of all individuals who were in the study at any time in $I_k$; let $I_{kl}$ denote the part of $I_k$ for which individual $(l)$ was in the study, and $h_{kl}$ is the *length* of $I_{kl}$. The length of $I_k$ is $t_k - t_{k-1} = h_k$. (Note that if the notation of Chapters 5 and 6 were used $h_k$ would be $h_{k-1}$.) If $(l)$ is in the study for the whole of $I_k$, then $h_{kl}=h_k$. If information on exact time of withdrawal is not available, we might use the approximation $h_{kl} \doteqdot \frac{1}{2}(t_k - t_{k-1})$; if entry and withdrawal, or death, both occur in $I_k$, one might take $h_{kl} \doteqdot \frac{1}{3}(t_k - t_{k-1})$.

For the period exposed to risk in $I_k$, the cumulative hazard function for the individual $(l)$ is [using the model (13.38)]

$$\Lambda_k(\mathbf{z}_l)=\lambda_k \int_{I_{kl}} g(\mathbf{z}_l;\boldsymbol{\beta})\,dt=\lambda_k h_{kl} g(\mathbf{z}_l;\boldsymbol{\beta}). \tag{13.39}$$

Let

$$\delta_{kl}=\begin{cases}1 & \text{if } (l) \text{ dies in } I_k,\\ 0 & \text{otherwise.}\end{cases} \tag{13.40}$$

The contribution to the likelihood for observations over the interval $I_k$ is

$$L_k(\lambda_k;\boldsymbol{\beta})=\prod_{l\in\mathbb{S}_k}\left[\lambda_k g(\mathbf{z}_l;\boldsymbol{\beta})\right]^{\delta_{kl}}\exp\left[-\lambda_k h_{kl} g(\mathbf{z}_l;\boldsymbol{\beta})\right]. \tag{13.41}$$

The overall likelihood function is

$$L(\lambda_1,\dots,\lambda_M;\boldsymbol{\beta})=\prod_{k=1}^{M}L_k(\lambda_k;\boldsymbol{\beta}). \tag{13.42}$$

If the values of the parameters $\boldsymbol{\beta}$ be fixed, the ML-estimator of $\lambda_k$ is

$$\hat{\lambda}_k(\boldsymbol{\beta})=\frac{\sum_{l\in\mathbb{S}_k}\delta_{kl}}{\sum_{l\in\mathbb{S}_k}h_{kl}g(\mathbf{z}_l;\boldsymbol{\beta})}. \tag{13.43}$$

The numerator, $\Sigma_{l\in\mathbb{S}_k}\delta_{kl}$, is, of course, the total number of deaths in $I_k$. If there are no deaths observed in $I_k$, then $\hat{\lambda}_k=0$, whatever be $\boldsymbol{\beta}$. Note that if there are no concomitant variables [i.e., $g(\mathbf{z}_l;\boldsymbol{\beta})\equiv 1$], we obtain the formulas $\hat{\lambda}_k=$(number of deaths)/(exposed to risk) given in (6.27).

Substituting (13.43) into (13.41), we obtain the maximized value of the likelihood factor $L_k(\lambda_k;\boldsymbol{\beta})$ given $\boldsymbol{\beta}$,

$$\hat{L}_k(\boldsymbol{\beta})=\left[e^{-1}\frac{\sum_{l\in\mathbb{S}_k}\delta_{kl}}{\sum_{l\in\mathbb{S}_k}h_{kl}g(\mathbf{z}_l;\boldsymbol{\beta})}\right]^{\Sigma_l\delta_{kl}}\prod_{l\in\mathbb{S}_k^*}g(\mathbf{z}_l;\boldsymbol{\beta}), \tag{13.44}$$

where $\mathbb{S}_k^*$ is the set of individuals dying in $I_k$.

The maximum likelihood estimators of $\boldsymbol{\beta}$, $\hat{\boldsymbol{\beta}}$, maximize

$$L(\hat{\boldsymbol{\beta}})=\prod_{k=1}^{M}\hat{L}_k(\hat{\boldsymbol{\beta}}) \tag{13.45}$$

Usually, the estimates $\hat{\boldsymbol{\beta}}$ have to be obtained by numerical analysis, with the aid of an electronic computer.

The choice of intervals $I_k$ is somewhat arbitrary. Using finer and finer intervals makes it possible for the model to provide closer approximation to $\lambda(t)$, but because the volume of observation (amount of exposed to risk) in each interval becomes smaller, the *accuracy* of the estimates of $\lambda_k$'s [and so of $\lambda(t)$] is likely to decrease.

Assuming model (13.38) to be valid, and no two deaths to coincide exactly, we obtain in the limit [instead of (13.43)]

$$\hat{\lambda}_j = \hat{\lambda}_j(\boldsymbol{\beta}) = \frac{1}{h_j \sum_{l \in \mathcal{R}_j} g(\mathbf{z}_l; \boldsymbol{\beta})}, \tag{13.46}$$

where $h_j = t'_j - t'_{j-1}$ is now the time elapsing between the $(j-1)$th and $j$th deaths, and $\mathcal{R}_j$ is the risk set consisting of those individuals in the study alive just before $t'_j$, where $t'_j$ is the $j$th ordered time at death.

The ML-estimators of the $\beta$'s maximize

$$\left(\prod_{j=1}^{n} h_j\right) e^n \hat{L}(\boldsymbol{\beta}) = \prod_{j=1}^{n} \frac{g(\mathbf{z}_{i(j)}; \boldsymbol{\beta})}{\sum_{l \in \mathcal{R}_j} g(\mathbf{z}_l; \boldsymbol{\beta})}, \tag{13.47}$$

where $n$ is the total number of deaths, and $[i(j)]$ is the individual $(i)$ who is the $j$th to die.

This result can also be obtained by using, in place of (fixed) intervals $I_k$, the intervals $I'_j$'s, between two successive deaths at times $t'_{j-1}$ and $t'_j$, that is, $I'_j \equiv (t'_{j-1}, t'_j)$. Of course, the latter intervals are not fixed, but depend on the times of occurrence of deaths in the experience [Breslow (1972)]. It is also obtained if the intervals $I'_j$'s are used, and the withdrawals during the interval are ignored, so that $\mathcal{S}_k$ is replaced by $\mathcal{R}_k$.

For computational purposes, it might be useful to represent $\Sigma_{l \in \mathcal{R}_j}$ in (13.46) and (13.47) in the more convenient form. Let $\tau_l$ denote the time of departure (by failure or withdrawal) for the $l$th individual.

We define

$$\theta_{jl} = \begin{cases} 0 & \text{if } \tau_l < t'_j, \\ 1 & \text{if } \tau_l \geqslant t'_j. \end{cases} \tag{13.48}$$

Thus

$$\sum_{l \in \mathcal{R}_j} g(\mathbf{z}_l; \boldsymbol{\beta}) = \sum_{l=1}^{N} \theta_{jl} g(\mathbf{z}_l; \beta). \tag{13.49}$$

Introducing the specification $g(\mathbf{z}_l;\boldsymbol{\beta})=\exp(\boldsymbol{\beta}'\mathbf{z})$ (Cox's model), and using (13.49), formula (13.46) becomes

$$\lambda_j=\lambda_j(\boldsymbol{\beta})=\frac{1}{h_j\sum_{l=1}^{N}\theta_{jl}\exp(\boldsymbol{\beta}'\mathbf{z}_l)}, \tag{13.50}$$

and from (13.47)

$$\left(\prod_{j=1}^{n}h_j\right)e^n\hat{L}(\boldsymbol{\beta})=\prod_{j=1}^{n}\frac{\exp(\boldsymbol{\beta}'\mathbf{z}_{i(j)})}{\sum_{l=1}^{N}\theta_{jl}\exp(\boldsymbol{\beta}'\mathbf{z}_l)}. \tag{13.51}$$

Formula (13.51) was obtained by Cox (1972) for the model (13.27), which he introduced. He obtained it as a conditional likelihood making no assumptions about $\lambda(t)$, except $\lambda(t)>0$ (see Exercise 13.5). Cox (1975) called it a *partial* likelihood; it utilizes information only on those individuals who were present in the study just before each failure, and does not utilize times of withdrawals between failures.

In special cases, when there are *no concomitant variables* (i.e., $g(\mathbf{z}_l;\boldsymbol{\beta})\equiv 1$ for all $l$), (13.46) takes the form

$$\hat{\lambda}_j=\frac{1}{h_j\mathrm{R}_j}, \tag{13.52}$$

where $\mathrm{R}_j$ is the number of individuals in $\mathcal{R}_j$ so that

$$\hat{S}(t_i')=\exp\left(-\sum_{j=1}^{i}\frac{1}{R_j}\right) \tag{13.53}$$

[cf. (6.61)].

### 13.6.2 Multiple Failures

When more than one individual is observed to fail "at $t_j'$," this usually means that the recording unit (e.g., week) is too broad to distinguish the times of two failures. Of course, a natural suggestion would be to make the recording unit smaller. In practice, however, this might be difficult from a technical viewpoint (possibly including the high costs involved).

Suppose that we observe $\delta_j$ individuals: $[i(j;1)]$, $[i(j;2)],\ldots,[i(j;\delta_j)]$, which failed at $t_j'$. Using an approach like that in Section 13.6.1, we notice that the likelihood function will be of the same form as in (13.41), in which multiple deaths are included in the denominator. Using finer and finer division of the time scale, it is easy to show that we obtain in the limit

$$\hat{\lambda}_j = \hat{\lambda}_j(\boldsymbol{\beta}) = \frac{\delta_j}{h_j \sum_{l \in \mathcal{R}_j} g(\mathbf{z}_l; \boldsymbol{\beta})}, \tag{13.54}$$

and

$$\hat{L}(\boldsymbol{\beta}) \propto \prod_{j=1}^{n} \frac{\prod_{p=1}^{\delta_j} g(\mathbf{z}_{i(j;p)}; \boldsymbol{\beta})}{\left[ \sum_{l \in \mathcal{R}_j} g(\mathbf{z}_l; \boldsymbol{\beta}) \right]^{\delta_j}}. \tag{13.55}$$

[Compare (13.46) and (13.47), respectively.] The special case, when $g(\mathbf{z}_l; \boldsymbol{\beta}) = \exp(\boldsymbol{\beta}'\mathbf{z}_l)$, is straightforward.

**Example 13.5** To illustrate the construction of the likelihood, we use the (artificial) data given in Table 13.3. There were $N = 14$ individuals who participated in the trial. There were $n = 7$ deaths ($D$) and $N - n = 7$ withdrawals ($W$). Two treatments $A$ and $B$ are randomly assigned, and two concomitant (continuous) variables, $z_1$ and $z_2$, were observed. The trial was conducted over a period of 20 units (e.g., months).

**Table 13.3** Data

| $(i)$ | $t'_{(i)}$ | Time at entry | Mode of departure | Treatment | $z_{1i}$ | $z_{2i}$ |
|---|---|---|---|---|---|---|
| (1) | 10.3 | 4.5 | W | A | −0.4 | 1.2 |
| (2) | 11.1 | 0 | D | A | 0.0 | 0.8 |
| (3) | 12.7 | 0 | D | B | 0.4 | 0.9 |
| (4) | 13.0 | 11.8 | D | B | 0.1 | 0.9 |
| (5) | 14.2 | 0 | W | A | 0.1 | 0.7 |
| (6) | 15.0 | 0 | D | B | 0.5 | 0.7 |
| (7) | 15.2 | 0 | W | A | 0.4 | 1.0 |
| (8) | 17.3 | 12.5 | D | B | 0.5 | 0.9 |
| (9) | 18.0 | 5.0 | D | A | 0.4 | 0.4 |
| (10) | 19.1 | 0 | D | A | 0.3 | 0.5 |
| (11) | 20 | 0 | W | B | 0.2 | 0.9 |
| (12) | 20 | 0 | W | A | −0.2 | 1.4 |
| (13) | 20 | 16.5 | W | B | 0.5 | 0.8 |
| (14) | 20 | 0 | W | B | 0.7 | 0.8 |

For individual $(i)$, $t'_{(i)}$ is its time elapsed from the beginning of the treatment (e.g., operation) to the point of leaving the study. For a patient who also was *observed* (was actually in the study), $t'_{(i)}$ is its follow-up time. Note that there were "new entries" [patients: (1), (4), (8), (9) and (13)] for which the times of entry were different from zero.

The times at death, $t'_j$, and the length of the interval between two consecutive deaths, $h_j = t'_j - t'_{j-1}$, are listed in Table 13.4.

**Table 13.4** Intervals and risk sets

| $j$ | Individual dying $[i(j)]$ | $t'_j$ | $h_j$ | Risk set $\mathcal{R}_j$ | $R_j$ |
|---|---|---|---|---|---|
| 1 | (2) | 11.1 | 11.1 | (2), (3), (5), (6), (7), (9), (10), (11) (12), (14) | 10 |
| 2 | (3) | 12.7 | 1.6 | (3), (4), (5), (6) (7), (8), (9), (10) (11), (12), (14) | 11 |
| 3 | (4) | 13.0 | 0.3 | (4), (5), (6), (7), (8), (9), (10), (11) (12), (14) | 10 |
| 4 | (5) | 15.0 | 2.0 | (6), (7), (8), (9) (10), (11), (12), (14) | 8 |
| 5 | (8) | 17.3 | 2.3 | (8), (9), (10), (11) (12), (13), (14) | 7 |
| 6 | (9) | 18.0 | 0.7 | (9), (10), (11), (12), (13), (14) | 6 |
| 7 | (10) | 19.1 | 1.1 | (10), (11), (12), (13), (14) | 5 |

There are also listed the risk sets $\mathcal{R}_j$–the individuals who were alive and present in the study just before $t'_j$.

We will apply Cox's model in the form

$$\lambda(t;y;z_1,z_2)=\lambda(t)\exp(\alpha y+\beta_1 z_1+\beta_2 z_2),$$

where

$$y=\begin{cases}0 & \text{if treatment } A,\\ 1 & \text{if treatment } B,\end{cases}$$

and assume $\lambda(t)\doteqdot\lambda_j$ in $(t'_{j-1},t'_j]$.

For illustration we evaluate [from (13.41)] the contribution to the likelihood from the data available in $(t'_4, t'_5]$, that is, $L_5(\lambda_5; \alpha, \beta_1, \beta_2)$. Note that individual (7) is not in risk set $\mathcal{R}_5$ of Table 13.4.

**Table 13.5** Evaluation of the likelihood factor $L_5(\lambda_5; \alpha; \beta_1, \beta_2)$

| $(l)$ | Mode of departure | $h_{jl}$ | $y$ | $z_{1l}$ | $z_{2l}$ | Likelihood factor $L_{5l}(\lambda_5; \alpha; \beta_1, \beta_2)$ |
|---|---|---|---|---|---|---|
| (7) | W | 0.2 | 0 | 0.4 | 1.0 | $\exp[-0.2\lambda_5 \exp(0.4\beta_1 + 1.0\beta_2)]$ |
| (8) | D | 2.3 | 1 | 0.5 | 0.9 | $\lambda_5 \exp(\alpha + 0.5\beta_1 + 0.9\beta_2)$ $\times \exp[-2.3\lambda_5 \exp(\alpha + 0.5\beta_1 + 0.9\beta_2)]$ |
| (9) | S | 2.3 | 0 | 0.4 | 0.4 | $\exp[-2.3\lambda_5 \exp(0.4\beta_1 + 0.4\beta_2)]$ |
| (10) | S | 2.3 | 0 | 0.3 | 0.5 | $\exp[-2.3\lambda_5 \exp(0.3\beta_1 + 0.5\beta_2)]$ |
| (11) | S | 2.3 | 1 | 0.2 | 0.9 | $\exp[-2.3\lambda_5 \exp(\alpha + 0.2\beta_1 + 0.9\beta_2)]$ |
| (12) | S | 2.3 | 0 | −0.2 | 1.4 | $\exp[-2.3\lambda_5 \exp(-0.2\beta_1 + 1.4\beta_2)]$ |
| (13) | S | 0.8 | 1 | 0.5 | 0.8 | $\exp[-2.3\lambda_5 \exp(\alpha + 0.5\beta_1 + 0.8\beta_2)]$ |
| (14) | S | 2.3 | 1 | 0.7 | 0.8 | $\exp[-2.3\lambda_5 \exp(\alpha + 0.7\beta_1 + 0.8\beta_2)]$ |
| Total | | | | | | $L_5 = \prod_l L_{5l}$ |

(S = survival)

## 13.7 ASSESSMENT OF THE ADEQUACY OF A MODEL: TESTS OF GOODNESS OF FIT

In the preceding sections, we have presented several models of hazard rates with covariates, without discussing how to assess their appropriateness. Three basic questions arise:

1. Supposing that the concomitant variables have been correctly chosen (i.e., have some meaning in explanation of the course of the disease), how can we judge whether a correct *model* has been selected?
2. Which covariates (individually or jointly) have *important* effects on mortality?
3. Is there any treatment-covariate *interaction*; that is, are some explanatory variables more important for one treatment, while a different set of variables is more important for another treatment?

In this section we discuss some methods relevant in answering question (1), and assuming that the coefficients $\boldsymbol{\beta}$ in the model are the same whatever the treatment might be. Questions (2) and (3) will be discussed in Sections 13.8 and 13.9, respectively.

The tests, for which we are seeking, are goodness-of-fit-type tests. Various methods discussed in Chapter 7 and 8 can be adapted, bearing in mind that now each individual has its own survival function.

### 13.7.1 Cumulative Hazard Plottings

If $\lambda(t;\mathbf{z})$ is a hazard rate, then

$$\Lambda(t;\mathbf{z})=\int_0^t \lambda(u;\mathbf{z})\,du=-\log S(t;\mathbf{z})$$

is the cumulative hazard function [see (3.5)]. Define a random variable

$$W=\Lambda(T;\mathbf{z}). \tag{13.56}$$

We know (from Section 3.12) that $W$ has a standard exponential distribution (with parameter $\lambda=1$).

If we knew the population parameters, we could calculate for individual ($j$) the quantity

$$W_j=\Lambda(T_j;\mathbf{z}), \tag{13.56a}$$

for $j=1,2,\ldots,N$. The $W_j$'s would correspond to $N$ mutually independent variables, each having a standard exponential distribution. The set of variables, $W_1$, $W_2,\ldots$, $W_N$ was discussed by Cox and Snell (1968) [see also Kay (1977)], who called these quantities "residuals." If we use only *estimated* values of the parameters, we obtain a set of so-called crude residuals, $\hat{W}_1$, $\hat{W}_2,\ldots$, $\hat{W}_N$. It is presumed that these should behave *approximately* as a random sample from an exponential distribution with parameter $\lambda=1$. Goodness of fit tests are then tests of whether the sample *does* behave as if this were so.

Suppose that we have *complete data*, with deaths of $N$ individuals occurring at times $t_1'<t_2'<\cdots<t_N'$.

1. We may plot the set of *ordered estimated* residuals, $\hat{W}_1$, $\hat{W}_2,\ldots$; $\hat{W}_N$ $[=\hat{\Lambda}(t_1';\mathbf{z}_1)$, $\hat{\Lambda}(t_2';\mathbf{z}_2),\ldots$, $\hat{\Lambda}(t_N';\mathbf{z}_N)]$ against the nonparametric estimates of CHF $\Lambda^o(t_j')$, where

$$\Lambda^o(t_j')\doteqdot\sum_{i=1}^{j}\frac{1}{N-i+1} \tag{13.57}$$

[see (7.31)].

Plotting $\hat{\Lambda}(t_j';\mathbf{z}_j)$ against $\Lambda^o(t_j')$, we should obtain approximately a straight line through the origin with slope 1, if the model is appropriate.

2. We may also treat $\Lambda(t_j';\mathbf{z}_j)$ as the *observed* value of a CHF, and plot it (or a function of it) against some function of $t$, to which it might be

linearly related. For example, for Weibull distributions we have

$$\Lambda(t;\mathbf{z}) = \theta^{-1} t^c \sum_{u=1}^{s} \beta_u z_u, \tag{13.58}$$

hence

$$\log \Lambda(t,\mathbf{z}) = c \log t + \log \theta^{-1} \sum_{u=1}^{s} \beta_u z_u, \tag{13.59}$$

so that the set of points $[\log t_j'; \Lambda(t_j',\mathbf{z})]$ should approximately fit a straight line if the Weibull model with covariates is adequate.

When we observe withdrawals, their contribution to the construction of plots is (partially) neglected, as discussed in Section 7.7 (2).

### 13.7.2 Method of Half-Replicates

Another method of checking whether a given model is appropriate would be the use of the so-called *half-replicates*, [McCarthy (1966), Greenberg et al. (1974), Snee (1977)].

Suppose that the number of observations, $N=2n$, is even. We first divide the data *randomly* into $n$ pairs. From each pair, we select at random one unit, so that we obtain a *sample* of $n$ units. From this sample, we estimate the parameters of the model. We then check how well the resultant SDF fits the second half of $n$ units. Formal tests (e.g., $\chi^2$, Kolmogorov-Smirnov) may be used, regarding the first fitted SDF as if it were a hypothetical population SDF. These tests are not strictly valid as they do not allow for variability in the estimates of parameters in the first fitted SDF, but they are useful for checking purposes. The procedure can be repeated on the same data as many times as one wishes. Greenberg et al. (1974) used this technique in analyzing the data on cancer of prostate mentioned earlier in this chapter. They repeated the drawing of $n(=151)$ half-replicates 12 times from the same $2n(=302)$ random pairs, and calculated $\chi^2$ values as checks.

## 13.8 SELECTION OF CONCOMITANT VARIABLES

Apart from the question whether we have an appropriate *form* of model, we have also to decide whether we *need* all the concomitant variables included in the model. In the initial stages of an inquiry, a wide range of variables may be measured, because it is felt that each might have some influence on survival. It is very often found that some of the variables have, in fact, little practical effect, whereas it is possible to choose a fairly small subset of the others that are only slightly less effective than all combined. Choice of such a subset is important, both on the grounds of economy (since measurement can be costly) and also of scientific insight.

Problems of a similar kind arise in the selection of variables for inclusion in any multiple regression model. Procedures utilized for this purpose have been described in many textbooks on multiple regression. Among these, the so called *step-up* (or *forwards*) and *step-down* (or *backwards*) procedures are most common. In regression analysis, sums of squares "due to regression" are tested against residual sums of squares by F-tests, which are, in fact, the appropriate likelihood ratio tests. In survival analysis, also, likelihood ratio tests are used.

### 13.8.1 Likelihood Ratio Tests of Composite Hypotheses

Suppose that a (parametric) survival model includes $m$ parameters, $\theta_1,\ldots,\theta_m$ say, which have to be estimated. Denote the maximum likelihood estimators (MLE's) of these parameters by $\hat{\theta}_1,\ldots,\hat{\theta}_m$, and the maximized likelihood, $L(\hat{\theta}_1,\ldots,\hat{\theta}_m)$ briefly by $\hat{L}(m)$.

Suppose (without loss of generality) that the values of the first $r$ parameters $(\theta_1,\ldots,\theta_r)$ are specified by a hypothesis $H_0$: $\theta_1=\theta_{10}$, $\theta_2=\theta_{20},\ldots,\theta_r=\theta_{r0}$. We say there are $r$ *restrictions* on the parameters. (The particular form used here is obtained, if necessary, by reparameterization.) The maximized likelihood under these restrictions is

$$\max_{\theta_{r+1},\ldots,\theta_m} L(\theta_{10},\ldots,\theta_{r0};\theta_{r+1},\ldots,\theta_m)=\hat{L}(m-r)$$

(read: maximized likelihood with respect to $(m-r)$ parameters $\theta_{r+1},\ldots,\theta_m$).

Clearly,

$$\hat{L}(m-r)\leqslant\hat{L}(m)$$

If $H_0$ is valid, the statistic

$$X^2(r)=-2\log\left\{\frac{\hat{L}(m-r)}{\hat{L}(m)}\right\} \tag{13.60}$$

is approximately distributed as Chi square with $r$ degrees of freedom (d. f.). (Note that $r$ is the number of restrictions on the parameters.)

### 13.8.2 Step-Down Procedure

Suppose that there are $s$ covariables $z_1, z_2, \ldots, z_s$ and we wish to select for inclusion in a model only those covariables that are really needed, assuming that the *form* of the model has been chosen correctly. We use the notation $\alpha$, $\beta_1, \ldots, \beta_s$ for parameters (so that in the notation of Section 13.8.1, $m=s+1$) such that $\beta_i=0$ if and only if $z_i$ has no real effect $(i=1,2,\ldots,s)$; $\alpha$ may, in fact, be a vector containing several parameters. Some of the $z_i$'s may be functions of others (for example, we might have $z_2=z_1^2$) so that the number of directly measured covariables may be less than $s$. This possibility does not affect our subsequent discussion.

A *step-down* procedure, based on likelihood ratios, can be constructed in the following way:

1. Obtain the MLE's $\hat{\alpha}$, $\hat{\beta}_1, \ldots, \hat{\beta}_s$ when all covariables are included and denote the maximized likelihood by

$$L(\hat{\alpha}, \hat{\beta}_1, \ldots, \hat{\beta}_s) = \hat{L}(s+1)$$

2. Remove variable $z_1$; equivalently put $\beta_1=0$, and calculate a new maximized likelihood

$$\hat{L}_1(s) = \max_{\alpha, \beta_2, \ldots, \beta_s} L(\alpha, 0, \beta_2, \ldots, \beta_s)$$

3. Repeat for each $z_u$ $(u=2,\ldots,s)$ in turn.

4. We thus obtain $s$ restricted maximized likelihoods $\hat{L}_1(s)$, $\hat{L}_2(s), \ldots, \hat{L}_s(s)$. Select the maximum among these quantities $\hat{L}_{[-1]}(s)$, say. Then the corresponding covariable $z_{[-1]}$ is our first candidate for exclusion from the model.

5. If removal of $z_{[-1]}$ makes no (or little) difference to the model (i.e., if $\beta_{[-1]}=0$), then we would expect the ratio $\hat{L}_{[-1]}(s)/\hat{L}(s+1)$ to be near to 1. From Section 13.8.1, if $\beta_{[-1]}=0$ then

$$X^2_{[-1]}(1) = -2\log\frac{\hat{L}_{[-1]}(s)}{\hat{L}(s+1)} \tag{13.61}$$

would be distributed approximately as Chi square with 1 $[=(s+1)-s]$ degree of freedom. We can apply a formal test of significance and decide to remove $z_{[-1]}$ if the value of $X^2_{[-1]}(1)$ is *not* significant at some (approximate) chosen level $100\varepsilon\%$, that is, if $X^2_{[-1]}(1) < \chi^2_{1,1-\varepsilon}$. Since $z_{[-1]}$ has been chosen to make $X^2_{[-1]}$ as great as possible (for the data available) the values of the latter will tend to be larger than would otherwise be the case, and allowance should be made for this. However, it is difficult to do this at all precisely and in practice the only allowance made is to use a rather smaller value of $\varepsilon$ (and so a larger value of $\chi^2_{1,1-\varepsilon}$) than would usually be employed. The test is then a more or less arbitrary rule, formulated against a broad theoretical background, rather than a true test of significance.

6. If it is decided to remove $z_{[-1]}$, repeat the procedure starting with the remaining $(s-1)$ covariables. The restricted maximized likelihood $\hat{L}_{[-1]}(s)$ now plays the role previously played by $\hat{L}(s+1)$, and $z_{[-2]}$ denotes that one of the remaining $(s-2)$ covariables for which the restricted maximum of $L(\alpha,\beta_1,\ldots,\beta_s)$ with $\beta_{[-1]}=0$ and $\beta_{[-2]}=0$ is maximum. Call this maximum $\hat{L}_{[-1],[-2]}(s-1)$.

The covariable $z_{[-2]}$ is omitted if

$$-2\log\frac{L_{[-1],[-2]}(s-1)}{L_{[-1]}(s)}<\chi^2_{1,1-\varepsilon}.$$

7. Continue the procedure in a similar way until, for the first time, we reach a decision not to omit a variable $z_{[-(s-p+1)]}$, say. The procedure then terminates and we retain the $p$ variates not omitted in the model. The procedure may be modified by removing the covariables two (or more) at a time, instead of singly. Of course, the modified procedure does not always give the same results as if covariables are removed one at a time.

### 13.8.3 Step-Up Procedure

This procedure is, briefly, as follows:

1. Taking each covariable $z_u$ $(u=1,\ldots,s)$ in turn, obtain the restricted maximized likelihoods,

$$\max_{\alpha,\beta_u} L(\alpha,0,\ldots,0,\beta_u,0,\ldots,0)=\hat{L}_u(2).$$

2. If $\max_u \hat{L}_u(2)=\hat{L}_{[1]}(2)$ then $z_{[1]}$ is considered for inclusion in the model. This variable is included if

$$-2\log(\hat{L}(1)/L_{[1]}(2))>\chi^2_{1,1-\varepsilon}$$

with

$$\hat{L}(1)=\max_{\alpha} L(\alpha,0,\ldots,0),$$

that is, if $\hat{L}_{[1]}(2)$ is (formally) significant.

3. If $z_{[1]}$ is included, proceed to calculate the restricted maximized likelihoods $\hat{L}_{[1],v}(3)$ obtained by including each $z_v$ ($v=1,\ldots,s$, excluding $v=u$) in addition to $z_{[1]}$. Select $z_{[2]}$ if $\hat{L}_{[1],v}(3)$ is maximized for $v=[2]$ and include it in the model if

$$-2\log\frac{\hat{L}_{[1]}(2)}{\hat{L}_{[1],[2]}(3)} \geqslant \chi^2_{1,1-\varepsilon}.$$

4. Continue similarly until a nonsignificant result is obtained. The procedure terminates with the covariables so far included.

Thus if

$$-2\log(\hat{L}_{[1],\ldots,[p]}(p+1)/\hat{L}_{[1],\ldots,[p+1]}(p+2))<\chi^2_{1,1-\varepsilon}$$

is the first nonsignificant result the covariables $z_{[1]},\ldots,z_{[p]}$ (and no others) are included in the model.

Note that

$$z_{[1]}, z_{[2]},\ldots,z_{[p]}$$

and

$$z_{[-1]}, z_{[-2]},\ldots,z_{[-(s-p)]}$$

are not necessarily complementary, although they are often nearly so.

Modified procedures, by which two (or more) covariables can be included at each stage, can be constructed by methods similar to those described near the end of Section 13.8.2.

**Example 13.6** Table 13.6 [taken from Bailey et al. (1977)] is based on the kidney transplant data discussed in Example 13.1. To test, for example, whether age ($z_2$) is a significant covariable, the statistic

$$X^2=-2[\log(\text{“None” with 3 parameters})-\log(\text{“Age” with 6 parameters})]$$

$$=-2[-3086.409+3081.930]=8.959 \text{ (with 3 d.f.)}$$

**Table 13.6** Likelihood ratio tests in assessing significance of sex and age in kidney transplant data

| Variables | Log likelihood | $X^2$ | d.f. | Significance level |
|---|---|---|---|---|
| None | −3086.409 | | | |
| Sex ($z_1$) | −3083.021 | 6.777 | 3 | 0.079 |
| Age ($z_2$) | −3081.930 | 8.959 | 3 | 0.030 |
| Age and sex | −3078.887 | 15.045 | 6 | 0.020 |
| Age adjusted for sex | | 8.268 | 3 | 0.041 |
| Sex adjusted for age | | 6.086 | 3 | 0.108 |

is calculated. It is significant at 5% level (the observed significance level is $\hat{\alpha}=0.030$). Applying a step-up procedure, we would choose age ($z_2$) rather than sex ($z_1$) as first variable because the likelihood is greater for age. The last line of Table 13.6 shows that if age is already used, inclusion of sex as an additional variable does not contribute significant improvement.

$$X^2 = -2(-3081.930+3078.887)=6.086.$$

## 13.9 TREATMENT-COVARIATE INTERACTION

Variation in the relative effects of different treatments with variation in values of concomitant variables is an *interaction* between treatments and covariates.

If such an interaction exists, then at least some concomitant variables which are important with one treatment, are less important (or even have no real importance) in some other treatment.

To assess whether interaction exists, a separate model is fitted to each treatment group.

1. If a method of selection of covariables, as described in Section 13.8 is applied to each treatment, different sets of variables may be obtained for different treatments. This would indicate that there is an interaction between treatments and covariates.
2. Formal tests can also be applied, using likelihood ratios. Such tests have been discussed by Byar and Corle (1977).

Suppose there are $k$ treatments. Let $\alpha_1,\ldots,\alpha_k$ be parameters defining the main (average) effects of these treatments and let $\boldsymbol{\beta}'_g=(\beta_{g1},\ldots,\beta_{gs})$ $(g=1,\ldots,k)$ be a vector of parameters defining the effect of the concomitant variables $z_1,\ldots,z_s$ respectively with the $g$th treatment. Also, let $\boldsymbol{\beta}'=(\beta_1,\ldots,\beta_s)$ represent the common values of $\boldsymbol{\beta}'_g$ under the hypothesis

$$H_\beta\colon \boldsymbol{\beta}_1=\boldsymbol{\beta}_2=\cdots=\boldsymbol{\beta}_g$$

(i.e., that there is no interaction between treatments and concomitant variables).

We further write

$$\hat{L}_{\alpha\beta}(s+1)$$

for the restricted maximized likelihood under the hypothesis that the *same* model applies for each treatment, that is, under the hypothesis

$$H_{\alpha\beta}\colon (\alpha_1=\alpha_1=\cdots=\alpha_k)\cap(\boldsymbol{\beta}_1=\boldsymbol{\beta}_2=\cdots=\boldsymbol{\beta}_g). \tag{13.62}$$

[The $(s+1)$ indicates that $(s+1)$ parameters $(\alpha, \beta_1, \ldots, \beta_s)$ are estimated in maximizing the likelihood.]

With similar notation

$$\hat{L}_{\alpha}(ks+1)$$

denotes the restricted maximized likelihood under the hypothesis

$$H_{\alpha}: (\alpha_1 = \alpha_2 = \cdots = \alpha_k),$$

that is, there are, on the average, no differences among treatments (but covariables may play different roles for different treatments); further,

$$\hat{L}_{\beta}(k+s)$$

denotes the restricted maximized likelihood under $H_{\beta}$ (see above); and finally, when no restrictions are imposed, we denote the maximized likelihood by

$$\hat{L}(k(s+1)) \left( = \max_{\substack{\alpha_1, \ldots, \alpha_k \\ \boldsymbol{\beta}_1, \ldots, \boldsymbol{\beta}_k}} L(\alpha_1, \ldots, \alpha_k; \boldsymbol{\beta}_1, \ldots, \boldsymbol{\beta}_k) \right). \tag{13.63}$$

Likelihood ratio tests can be used to obtain answers to the following questions:

1. *Is it necessary to use different models for different treatments?* The null hypothesis is $H_{\alpha\beta}$. The test statistic is

$$X^2[(k-1)(s+1)] = -2\log\left\{ \frac{\hat{L}_{\alpha\beta}(s+1)}{\hat{L}(k(s+1))} \right\}. \tag{13.64}$$

If $H_{\alpha\beta}$ is valid this statistic should be approximately distributed as Chi square with $(k-1)(s+1)$ $[=k(s+1)-(s+1)]$ d.f.

2. *Do the contributions of covariables depend on the treatment?* (In other words, is there treatments$\times$covariable interaction?) The null hypothesis is $H_{\beta}$, and the test statistic is

$$X^2[s(k-1)] = -2\log\left( \frac{\hat{L}_{\beta}(k+s)}{\hat{L}(k(s+1))} \right). \tag{13.65}$$

If $H_{\beta}$ is valid $X^2[s(k-1)]$ should be approximately distributed as Chi square with $s(k-1)$ $[=k(s+1)-(k+s)]$d.f.

3. *Are there differences in the average (main) effects among the treatments?* The null hypothesis is $H_\alpha$. The test statistic is

$$X^2(k-1) = -2\log\left[\frac{\hat{L}_\alpha(ks+1)}{\hat{L}(k(s+1))}\right]. \tag{13.66}$$

If $H_\alpha$ is valid this statistic should be approximately distributed as Chi square with $k-1[=ks+1-k(s+1)]$ d.f.

Other possibilities arise if the field of alternative hypotheses is restricted. For example:

4. *Assuming that there are no main treatment effects, is there any treatment × covariable interaction?* (cf. 3). The null hypothesis is now $H_{\alpha\beta}$, and the alternatives are restricted to $H_\alpha$, so that the test statistic is

$$X^2[s(k-1)] = -2\log\left(\frac{\hat{L}_{\alpha\beta}(s+1)}{\hat{L}_\alpha(ks+1)}\right). \tag{13.67}$$

If $H_{\alpha\beta}$ is valid, this statistic should be approximately distributed as Chi square with $s(k-1)$ $[=ks+1-(s+1)]$ d.f.

Similarly, one can consider the question:

5. *Assuming that the β's are the same for all treatments are there any main treatment effects?* Construction of the appropriate test is left to the reader.

## 13.10 LOGISTIC LINEAR MODELS

Another popular type of model with concomitant variables used for some time by epidemiologists is the so-called logistic linear model. Its background is as follows.

Let $\mathbf{z}'=(z_1,\ldots,z_s)$ be $s$ concomitant variables. Consider a population composed of individuals, some with disease $A$ and the remainder free of disease $A$. Let $Q_{(j)}=Q_{(j)}(\mathbf{z})$ be the probability that an individual $(j)$ selected at random from the population is affected by disease $A$; $P_{(j)}=1-Q_{(j)}$ is the probability that he is free of it. The *postulated logistic linear model* is

$$\log\frac{Q_{(j)}}{P_{(j)}} = \beta_0 + \sum_{u=1}^{s} \beta_u z_{uj}.$$

From (13.68) we obtain

$$Q_{(j)} = \exp\left(\beta_0 + \sum_{u=1}^{s}\beta_u z_{uj}\right)\left[1+\exp\left(\beta_0+\sum_{u=1}^{s}\beta_u z_{uj}\right)\right]^{-1}, \tag{13.69a}$$

and

$$P_{(j)}=\left[1+\exp\left(\beta_0+\sum_{u=1}^{s}\beta_u z_{uj}\right)\right]^{-1}. \tag{13.69b}$$

Notice that model (13.68) does not include time; the $\beta$'s can be estimated from cross-sectional type data.

Rather unjustifiably, model (13.68) is sometimes used in the analysis of time dependent response (prospective) experiments. It is then assumed that (13.68) approximately holds for a *specified* ("*unit*") *period of time*. If this assumption were true and the data on response (incidence of disease or death) were collected on a cohort of individuals, all starting at $t=0$, then one could use the likelihood

$$L(\beta)=\prod_{j=1}^{n} Q_{(j)}^{\delta_j}P_{(j)}^{1-\delta_j}=\prod_{j=1}^{n}\left(\frac{Q_{(j)}}{P_{(j)}}\right)^{\delta_j}P_{(j)} \tag{13.70}$$

where $n$ is the sample size, and

$$\delta_j=\begin{cases}1 \text{ if } (j) \text{ responds}\\ 0 \text{ if } (j) \text{ does not respond.}\end{cases}$$

The *expected* number of individuals with the event (e.g., deaths) is calculated from the formula

$$E=\sum_{j=1}^{n} Q_{(j)}.$$

If there are varying times at entry, it is necessary to introduce further assumptions about the distribution of time at which response occurs (e.g., time at death). Also, model (13.68) is not time invariant, and cannot be used for any other length of period. To meet these difficulties, Myers et al. (1973) modified (13.68) by introducing time $t$ explicitly, together with an additional parameter $\tau$, in the formula

$${}_tP_{(j)}=\left[1+\exp\left(\beta_0+\sum_{u=1}^{s}\beta_u z_{uj}\right)\right]^{-t/\tau}. \tag{13.71}$$

which, of course, corresponds to an exponential distribution of time to death, with hazard rate

$$\lambda_{(j)}=\frac{1}{\tau}\log\left[1+\exp\left(\beta_0+\sum_{u=1}^{s}\beta_u z_{uj}\right)\right].$$

If $t_j$ denotes the time at which individual $(j)$ was last seen (e.g. time of death or withdrawal), then the likelihood is

$$L(\boldsymbol{\beta}) = \prod_{j=1}^{n} \lambda_{(j)}^{\delta_j} \cdot {}_{t_j}P_{(j)}. \tag{13.72}$$

**Example 13.7** Dyer (1975) studied the mortality of 1233 white male employees of the Chicago Peoples Gas, Light, and Coke company, aged 40–59 at entry, who were free of a definite heart disease at the beginning of the study, 1958. The period of study was 14 years, with 246 deaths observed. The risk factors considered were age in years ($z_1$), serum cholesterol in mg/ml ($z_2$), systolic blood pressure in mm Hg ($z_3$), and cigarette smoking in number per day ($z_4$).

The author applied the logistic linear model

$$\log \frac{Q_{(j)}}{1-Q_{(j)}} = \beta_0 + \sum_{u=1}^{s} \beta_u z_{uj},$$

and a parametric exponential-Weibull model of the form

$$S(t|\mathbf{z}) = \exp\left[-\left(\beta_0 + \sum_{u=1}^{s} \beta_u z_u\right) t^c\right],$$

with $Q_{(j)} = 1 - S(14|\mathbf{z}_j)$ [compare model (13.58)].

The estimates of $\beta$'s and the value of $t$-statistic for each $\beta$ separately, for each of the two models are given in Table 13.7.

Using these estimates, the individual probabilities, $Q_{(j)}$, were calculated and ordered from the smallest to the largest. Then they were grouped into deciles, and the number of observed and expected deaths calculated in each decile, for each model. The results are exhibited in Table 13.8. Both logistic and Weibull models give close results.

**Table 13.7** Fitting logistic-linear model and exponential-Weibull model to Gas Company mortality data

| | Logistic linear model | | Exponential-Weibull model | |
|---|---|---|---|---|
| Variable | $\hat{\beta}_u$ | $t$-value | $\hat{\beta}_u$ | $t$-value |
| Constant | −6.50298 | — | −10.31166 | — |
| Age ($z_1$) | 0.10334 | 6.98 | 0.08620 | 6.87 |
| Cholesterol ($z_2$) | 0.00257 | 1.53 | 0.00217 | 1.56 |
| Systolic pressure ($z_3$) | 0.02108 | 5.77 | 0.01842 | 6.59 |
| Smoking ($z_4$) | 0.03403 | 5.38 | 0.02681 | 5.34 |

**Table 13.8** Observed and expected deaths

| Decile | Logistic-linear model | | Exponential-Weibull model ($c=1.6743$) | |
|---|---|---|---|---|
| | Observed deaths | Expected deaths | Observed deaths | Expected deaths |
| 1 | 5 | 5.4 | 5 | 6.3 |
| 2 | 5 | 8.8 | 4 | 9.6 |
| 3 | 13 | 11.7 | 13 | 12.3 |
| 4 | 16 | 15.0 | 17 | 15.3 |
| 5 | 18 | 18.2 | 16 | 18.2 |
| 6 | 23 | 22.2 | 22 | 21.9 |
| 7 | 34 | 27.1 | 35 | 26.4 |
| 8 | 31 | 33.1 | 31 | 31.9 |
| 9 | 39 | 41.5 | 43 | 40.1 |
| 10 | 62 | 63.0 | 60 | 64.0 |

The next four sections (13.11–13.14) discuss some relatively new problems in the analysis of survival models with concomitant variables. The theory is not yet completely developed, so we only outline these topics, in order that the reader should be aware of them and of their potential value.

## 13.11 TIME DEPENDENT CONCOMITANT VARIABLES

So far we have discussed models of hazard rates, in which the covariables ($z$'s) were independent of $t$. Therefore, it was irrelevant to the model at which time point their values were measured. Indicator type variables (such as sex, race), height of an adult person, etc, are examples of such covariables.

However, many covariables vary with $t$. Some are subject to individual and seasonal variation (e.g., blood counts); others may be expressed by functional (usually linear) relationship to age, but are also subject to stochastic variation (e.g., blood pressure, level of serum cholesterol, etc.).

Formally, this problem is handled by introducing $\mathbf{z}(t)$ instead of $\mathbf{z}$ in the model, that is, using $\lambda[t,\mathbf{z}(t)]$ *without* specifying the functional relationship between $z$ and $t$. In the multiplicative model, for example, $\lambda[t;\mathbf{z}(t)]=\lambda(t)g[\mathbf{z}(t)]$. This does not give a simple formula for the cumulative hazard function $\Lambda[t;\mathbf{z}(t)]$, since $g[\mathbf{z}(t)]$ is now a function of $t$. Nevertheless, following the methodology described in Section 13.6, it turns out, that the formulae (13.46) and (13.47) for $\hat{\lambda}_j(\boldsymbol{\beta})$ and $L(\hat{\boldsymbol{\beta}})$, are the same, except that

$z_{ul}$'s which depend on $t$ are now replaced by $z_{ul}(t'_j)$'s, and $z_{i(j)}$'s by $z_{i(j)}(t'_j)$'s.

This means, that when someone dies, we should, in principle, record the values of the $z_u$'s on all individuals (including the one who died). In practice, this is rather difficult, or even impossible. To avoid this difficulty, some authors introduce discrete type variables such as: blood pressure below and above certain values.

The resulting discrete (indicator) variables can be measured at the entry of each individual, under the assumption that the individual who scores "0" at entry, is unlikely to score "1" at time $t'_j$ (although this might happen). Another way of facing this difficulty, is to use the measurements recorded at the time *nearest to* $t'_j$. This is especially suitable for variables, such as blood pressure or temperature, which we measured periodically on a routine basis.

Another approach, which is only occasionally possible, would be to specify a *functional relationship between* $z$ and $t$. For simplicity, we present this situation, assuming a single covariable.

Let $z_0$ denote the value of the covariable $z$ measured at time $t=0$, and $z_t$ its value at time $t$. If it happens that there is a functional relationship between $z_t$ and $z_0$

$$z_t=\psi(t,z_0;\gamma), \tag{13.73}$$

where $\gamma$ is a vector of suitable parameters, then

$$\lambda(t;z_t)=\lambda[t,\psi(t,z_0;\gamma)]=\lambda^*(t;z_0;\gamma), \tag{13.74}$$

-this means that $\lambda(t;z_t)$ can be expressed in terms of the initial observations $z_0$'s.

**Example 13.8** A simple example is when $z_0$ is age. Thus we have $z_t=z_0+t$ which is a straightforward deterministic (exact) relation. Suppose $\lambda(t;z_t)=\lambda\exp(\beta z_t)$. Thus $\lambda^*(t,z_0;\beta)=\exp[\beta(z_0+t)]=\lambda e^{\beta t}e^{\beta z_0}$ which is a Gompertz model [see (13.33)].

Usually, however, there is also a stochastic element involved in the relation between $z_t$ and $z_0$. For example, assuming that blood pressure is a function of age, we have, on average, $z_t=z_0+bt$, where $b$ is often estimated from the sample (regression coefficient). In consequence, $z_0$ can only be estimated with some sampling error, which has to be taken into account in our calculations.

## 13.12 CONCOMITANT VARIABLES REGARDED AS RANDOM VARIABLES

Concomitant variables, **z**, observed on individuals can be highly selective (in a risk factor controlled experiment) or they can represent a random sample of **z** values from a population. In either case, we might observe the empirical distribution of $\mathbf{z}_0$'s, that is, of **z**'s recorded at $t=0$. Again, for simplicity, we confine ourselves to a single variable, $z_0$, bearing in mind that the extension to $s$ variables is straightforward. We assume that $z_0$ is continuous; for discrete variables summation would be used in place of integration.

Let $S_T(t;z_0)$ be the SDF for a given value of $z_0$, and $f_{Z_0}(z_0)$ be the PDF of $Z_0$. Then

$$S_T(t)=E_{Z_0}[S_T(t;Z_0)]=\int_{-\infty}^{\infty} S_T(t;z_0)f_{Z_0}(z_0)\,dz_0. \quad (13.75)$$

This is a *compound* SDF with compounding PDF $f_{Z_0}(z_0)$. (Compare Section 3.13.) In fact, this is the average (expected) survival function over all possible values of $z_0$'s.

**Example 13.9** Consider a simple additive model of hazard rate

$$\lambda(t;z_0)=\alpha t+\beta z_0, \qquad t>0, \qquad \alpha>0, \qquad \beta>0, \quad (13.76)$$

[see (13.17)]. Thus

$$S_T(t;z_0)=\exp\left[-\left(\tfrac{1}{2}\alpha t^2+\beta z_0 t\right)\right]. \quad (13.77)$$

Suppose that $Z_0$ has a normal distribution with PDF $f_{Z_0}(z_0)=(\sqrt{2\pi}\,\sigma_0)^{-1}\exp\{-\frac{1}{2}[(z_0-\zeta_0)/\sigma_0]^2\}$.

Then the compound SDF is

$$S_T(t)=E_{Z_0}[S_T(t;z_0)]=\exp\left(-\tfrac{1}{2}\alpha t^2\right)\int_{-\infty}^{\infty} e^{-\beta t z_0}f_{Z_0}(z_0)\,dz_0. \quad (13.78)$$

The integral on the right-hand side of (13.78) is equal to

$$\exp\left[-\zeta_0\beta t+\tfrac{1}{2}\sigma_0^2(-\beta t)^2\right].$$

(It can be recognized as moment generating function of normal distribution with the dummy variable $-\beta t$, that is $M_{Z_0}(-\beta t)$.)

Substituting it into (13.78), we obtain

$$S_T(t)=\exp\left\{-\left[\tfrac{1}{2}(\alpha-\beta^2\sigma_0^2)t^2+\beta\zeta_0 t\right]\right\} \tag{13.79}$$

$$=\exp\left[-\left(\tfrac{1}{2}At^2+Bt\right)\right]. \tag{13.79a}$$

Note that if the assumptions (that the hazard rate is of the additive form and $Z_0$ is normally distributed) are correct, $S_T(t)$ and $S_T(t;z_0)$ are of the same form.

Suppose that we can obtain unbiased estimates $\hat{\zeta}_0$ and $\hat{\sigma}_0^2$ of $\zeta_0$ and $\sigma_0^2$, respectively. Furthermore, from the mortality data, we may be able to estimate $\hat{A}$ and $\hat{B}$ for the $S_T(t)$. Then from

$$\left.\begin{aligned}\hat{\alpha}-\hat{\beta}^2\hat{\sigma}_0^2&=\hat{A},\\ \hat{\beta}\hat{\zeta}_0&=\hat{B},\end{aligned}\right\} \tag{13.80}$$

we can estimate $\alpha$ and $\beta$ in the model (13.76).

In this example, we have shown that if $Z_0$ is normally distributed, then the model of hazard rate given by (13.76) results in the average SDF given by (13.79). A question arises: can we establish that (13.79) always leads to (13.76)? If so, the additive model (13.76) could be derived from the compound distribution $S_T(t)$. This inverse problem is usually not simple; it involves integral equations and is not considered here.

## 13.13 POSTERIOR DISTRIBUTION OF CONCOMITANT VARIABLES

As before, we consider a single (continuous) variate $Z_0$. At $t=0$, $Z_0$ has PDF $f_{Z_0}(z_0)$, which we can think of as a *prior* distribution. If this variable has some effect on mortality (i.e., represents a risk factor), then its *posterior* distribution among the survivors to time $t$ will be different from that at $t=0$, even when the value of $Z_0$ does not depend on $t$.

The *posterior* PDF of $Z_0$ among the survivors to time $t$ is

$$f_{Z_0|t}(z_0|t)=\frac{S_T(t;z_0)f_{Z_0}(z_0)}{E_{Z_0}[S_T(t;z_0)]}. \tag{13.81}$$

The denominator of the right-hand side of (13.81) has already been discussed in the preceding section. Extension to the case when $\mathbf{Z}_0$ is a vector, is straightforward.

There are four different functions of interest in (13.81). Knowing (or estimating) three of them, the fourth can be found. For example, in certain experiments, we can estimate $f_{Z_0}(z_0)$, $f_{Z_0|t}(z_0|t)$ and $E_{Z_0}[S(t;z_0)]=S_T(t)$ from measurements on entrants, from survivors to time $t$, and from mortality data, respectively. Then we can obtain $S_T(t;z_0)$, and so the hazard function, $\lambda(t;z_0)$.

Further details, including cases when the $z$'s are time dependent, are given in a paper by Elandt-Johnson (1980).

**Example 13.10** In this example, we shall prove that when $Z_0$ is $N(\zeta_0,\sigma_0^2)$ and the hazard rate function is of additive form (13.76), then the posterior distribution is also normal.

Using the results of Example 13.9 and formula (13.81), we obtain the *posterior* PDF among the survivors to time $t$,

$$f_{Z_0|t}(z_0|t)=\frac{e^{-(1/2)\alpha t^2}\cdot e^{-\beta z_0 t}\times\left(\sqrt{2\pi}\,\sigma_0\right)^{-1}e^{-(1/2)[(z_0-\zeta_0)/\sigma_0]^2}}{e^{-(1/2)\alpha t^2}e^{\beta^2\sigma_0^2 t^2}e^{-\beta\zeta_0 t}}$$

$$=\frac{1}{\sqrt{2\pi}\,\sigma_0}\exp\left\{-\left[\frac{1}{2\sigma_0^2}(z_0-\zeta_0)^2+\beta(z_0-\zeta_0)+\frac{1}{2}\sigma_0^2\beta^2t^2\right]\right\}$$

$$=\frac{1}{\sqrt{2\pi}\,\sigma_0}\exp\left\{-\frac{1}{2\sigma_0^2}\left[z_0-(\zeta_0-\sigma_0^2\beta)\right]^2\right\}, \qquad (13.82)$$

which is the PDF of the normal distribution with mean $(\zeta_0-\sigma_0^2\beta)$ and variance $\sigma_0^2$.

## 13.14 CONCOMITANT VARIABLES IN COMPETING RISK MODELS

Concomitant variables may be incorporated into competing risk models of the kind discussed in Chapters 9–12.

First, since we now are taking into account different modes of exit (causes of death), a particular covariable $z_u$, which is a high risk in regard to disease $A$, may present no risk in regard to disease $B$. Therefore, for each cause of death a different model (or at least a different set of parameters) needs to be considered.

Second, because of nonidentifiability of the joint survival function (see Section 9.6), we should consider models of crude rather than net hazard functions [Cox (1972)], and construct the likelihood function using the observable (crude) probability functions of death *in the presence* of other causes.

Problems of this nature can be attacked by methods of Chapters 9–12 combined with those discussed in the present chapter. Here we will confine ourselves again to simple situations with only two causes of death, $C_1$ and $C_2$, a single covariable $z$ independent of $t$, and the *multiplicative model* (13.27).

Let

$$h_\alpha(t;z;\beta_\alpha)=\lambda_\alpha(t)e^{\beta_\alpha z} \tag{13.83}$$

be the crude hazard rate for cause $C_\alpha$ $(\alpha=1,2)$, and

$$h(t;z,\beta_1,\beta_2)=h_1(t;z,\beta_1)+h_2(t;z,\beta_2) \tag{13.84}$$

be the overall hazard rate [cf. (9.11)].

Further, let $n_1$ and $n_2$ be the number of observed deaths from $C_1$ and $C_2$, respectively, with the total deaths $n=n_1+n_2$ among $N$ individuals involved in the study. Denote by

$$t'_{\alpha 1}<t'_{\alpha 2}<\cdots<t'_{\alpha j}<\cdots<t'_{\alpha n_\alpha} \tag{13.85}$$

the ordered times at death from $C_\alpha$ $(\alpha=1,2)$.

Since $\lambda_\alpha(t)$ is unspecified, we approximate it by a sequence of constants, $\lambda_{\alpha,j}$ for $t'_{\alpha,j-1}\leqslant t<t'_{\alpha j}$ $(j=1,2,\ldots,n_\alpha)$. We assume that there are no new entries between $t'_{\alpha,j-1}$ and $t'_{\alpha j}$, and denote by $\mathfrak{R}(t'_{\alpha j})$ the risk set at $t'_{\alpha j}$, that is those individuals who were alive just before $t'_{\alpha j}$. The length of the interval between the $(j-1)$th and $j$th death from $C_\alpha$ will be denoted in this context by $a_{\alpha j}$, that is $t'_{\alpha j}-t'_{\alpha,j-1}=a_{\alpha j}$.

Following the arguments of Section 13.6, the contribution to the overall likelihood at $t'_{\alpha j}$, is

$$L_{\alpha j}(\lambda_{\alpha j},\beta_\alpha)=h_\alpha(t'_{\alpha j};z_{i(j)},\beta_\alpha)\prod_{l\in\mathfrak{R}(t'_{\alpha j})}\exp\left[-\int_{t'_{\alpha,j-1}}^{t'_{\alpha j}}h(t;z_l,\beta_1,\beta_2)\,dt\right] \tag{13.86}$$

$$=\lambda_{\alpha j}\exp(\beta_\alpha z_{i(j)})\prod_{l\in\mathfrak{R}(t'_{\alpha j})}\exp\left[-a_{\alpha j}(\lambda_{\alpha j}e^{\beta_\alpha z_l}+0.e^{\beta_{3-\alpha}z_l})\right]. \tag{13.86a}$$

Note that it does not depend on $\beta_{3-\alpha}$. Then

$$L_\alpha(\beta_\alpha)=\prod_{j=1}^{n_\alpha}L_\alpha(\lambda_{\alpha j},\beta_\alpha). \tag{13.87}$$

By procedures similar to those in Section 13.6, we obtain

$$\hat{\lambda}_{\alpha j}(\beta_\alpha) = \left[ a_{\alpha j} \sum_{l \in \mathcal{R}(t'_{\alpha j})} \exp(\beta_\alpha z_l) \right]^{-1} \tag{13.88}$$

[cf. (13.46)]. The likelihood $L_\alpha(\beta_\alpha)$ is proportional to

$$\prod_{j=1}^{n_\alpha} \frac{\exp(\beta_\alpha z_{i(j)})}{\sum_{l \in \mathcal{R}(t'_{\alpha j})} \exp(\beta_\alpha z_l)}, \qquad \alpha = 1,2. \tag{13.89}$$

[cf. (13.47)]. Extension to cases in which $z$ depends on $t$, and/or $z$'s are functions of $t$ are straightforward. [For more details, see Prentice et al. (1978).]

## REFERENCES


Aitchison, J. (1970). Statistical problems of treatment allocation. *J. R. Statist. Soc. Ser. A*, **133**, 206–238.

Anderson, T. A. (1972). Separate sample logistic discrimination. *Biometrika* **59**, 19–35.

Bailey, R. C., Homer, L. D., and Summe, J. P. (1977). A proposal for the analysis of kidney graft survival. *Transplantation* **24**, 309–315.

Breslow, N. E. (1974). Covariance analysis of censored survival data. *Biometrics*, **30**, 89–99.

Byar, D. P. and Corle, D. K. (1977). Selecting optimal treatment in clinical trials using covariate information. *J. Chron. Dis.* **30**, 445–449.

Byar, D. P., Huse, R., Bailar, J. C., and the Veterans Administration Urological Research Group (1974). An exponential model relating censored survival data and concomitant information for prostatic cancer patients. *J. Natl. Cancer Inst.* **52**, 321–326.

Cornfield, J. (1962). Joint dependence of risk of coronary heart disease on serum cholesterol and systolic blood pressure: a discriminant function approach. *Fed. Proc.* **21**, 4, Supp. 2, 58–61.

Cox, D. R. (1958). The regression analysis of binary sequences. *J. Roy. Statist. Soc., Ser. B.*, **20**, 215–232.

Cox, D. R. (1966). Some procedures associated with the logistic qualitative response curve. *Research Papers in Statistics: Essays in Honour of J. Neyman's 70th Birthday* (Ed., F. N. David), pp. 55–71. Wiley, London; University of California Press, Berkeley.

Cox, D. R. (1972). Regression models and life tables. *J. Roy. Statist. Soc. Ser. B* **33**, 187–202 (Discussion, 202–220.)

Cox, D. R. and Snell, E. J. (1968). A general definition of residuals. *J. Roy. Statist. Soc. Ser. B.* **34**, 826–838.

Day, N. E. and Kerridge, D. F. (1967). A general maximum likelihood discriminant. *Biometrics*, 313–323.

Dyer, A. R. (1975). An analysis of relationship of systolic blood pressures, serum cholesterol and smoking to 14-year mortality in the Chicago People's Gas Company study, I. *J. Chron. Dis.* **28**, 565–570.

Elandt-Johnson, R. C. (1971). *Probability Models and Statistical Methods in Genetics,* Chapter 11. Wiley, New York.

Elandt-Johnson, R. C. (1976). A class of distributions generated from distributions of exponential type. *Nav. Res. Log. Quart.* **23**, 131–138.

Elandt-Johnson, R. C. (1980). Some prior and posterior distributions in survival analysis: A critical insight on relationships derived from cross-sectional data. *J. R. Statist. Soc., Ser. B*, **42**, 96–106.

Epstein, F. H. (1968). Multiple risk factors and the prediction of coronary heart disease. *Bull. N. Y. Acad. Med.* **44**, 916–935.

Feigl, P. and Zelen, M. (1965). Estimation of exponential survival probabilities with concomitant information. *Biometrics* **21**, 826–838.

Glasser, M. (1967). Exponential survival with covariance. *J. Amer. Statist. Assoc.* **62**, 561–568.

Greenberg, R. A., Bayard, S., and Byar, D. P. (1974). Selecting concomitant variables using a likelihood ratio step-down procedure and a method of testing goodness of fit in a exponential survival model. *Biometrics* **30**, 601–608.

Gross, A. J. and Clark, V. A. (1975). *Survival Distributions and Reliability Applications in the Biomedical Sciences*. Wiley, New York.

Holford, T. R. (1976). Life tables with concomitant information. *Biometrics* **32**, 587–597.

Holt, J. D. and Prentice, R. L. (1974). Survival analysis in twin studies and matched pair experiments. *Biometrika* **61**, 17–30.

Johnson, N. L. and Elandt-Johnson, R. C. (1978). Estimation in general multiplicative model for survival, Institute of Statistics Mimeo Series No. 1187, University of North Carolina at Chapel Hill.

Kalbfleisch, J. and Prentice, R. L. (1973). Marginal likelihoods based on Cox's regression and life model. *Biometrika* **60**, 267–278.

Kay, R. (1977). Proportional hazard regression models and the analysis of censored survival data. *Appl. Statist.* **26**, (JRSS, Ser. C) 227–237.

Koch, G. G., Johnson, W. D., and Tolley, H. D. (1972). A linear models approach to the analysis of survival and extent of disease in multidimensional contingency tables. *J. Amer. Statist. Assoc.* **67**, 783–796.

Krall, J. M., Uthoff, V. A., and Harley, J. B. (1975). A step-up procedure for selecting variables associated with survival. *Biometrics* **31**, 49–57.

Lawless, J. F. and Singhal, K. (1978). Efficient screening of nonnormal regression models, *Biometrics*, **34**, 318–327.

McCarthy, P. J. (1966). Replication, an approach to the analysis of data from complex surveys. *Natl. Center Health Stat.*, Ser. 2, No. 14.

Mellinger, G. T., Bailar, J. C., Arduino, L. J., Veterans Administration Cooperative Urological Research Group (Writing Committee) (1967). Treatment and survival of patients with cancer of the prostate. *Surgery, Gynecology and Obstetrics* **124**, 1011–1017.

Myers, M. H., Hankey, B. E., and Mantel, N. (1973). A logistic-exponential model for use with response-time involving regressor variables. *Biometrics* **29**, 257–269.

Prentice, R. L. (1973). Exponential survivals with censoring and explanatory variables. *Biometrika* **60**, 279–288.

Prentice, R. L. (1976). Use of the logistic model in retrospective studies. *Biometrics* **32**, 599–606.

Prentice, R. L., Kalbfleisch, J. D., Peterson, A. V., Fluornoy, N., Farewell, V. T., and Breslow, N. E. (1978). The analysis of failure time in the presence of competing risks. *Biometrics* **34**, 541–554.

Snee, R. D. (1977). Validation of regression models: Methods and examples. *Technometrics* **19**, 415–428.

Truett, J., Cornfield, J., and Kannel, W. (1967). A multivariate analysis of the risk of coronary heart disease in Framingham. *J. Chron. Dis.* **20**, 511–524.

Tallis, G. M., Sarfaty, G., and Leppard, P. (1973). The use of a probability model for the construction of age specific life tables for women with breast cancer, Endocrine Research Unit, Cancer Institute, Melbourne, Australia.

Walker, S. H. and Duncan, D. B. (1967). Estimation of the probability of an event as a function of several independent variables. *Biometrika* **54**, 169–179.

Zippin, C. and Armitage, P. (1966). Use of concomitant variables and incomplete survival information in the estimation of an exponential survival parameter. *Biometrics* **22**, 665–672.

## EXERCISES

**13.1.** Consider a hazard rate of form

$$\lambda(t;\mathbf{z})=\alpha+\beta_1 z_1+\beta_2 z_2+\cdots+\beta_s z_s.$$

(*a*) Suppose that $N$ individuals are observed from time $t=0$ until death, and let $t_1'<t_2'<\cdots<t_N'$ be the recorded times at death. Construct the likelihood function and derive the maximum likelihood equations for estimating $\alpha$ and $\boldsymbol{\beta}$.

(*b*) Suppose that only $D(<N)$ deaths occur and $(N-D)$ individuals are withdrawn, while still alive, at recorded times. Construct the likelihood function for estimating $\alpha$ and $\boldsymbol{\beta}$.

**13.2.** The data in Table E13.1 represent the survivorship of patients with acute leukemia in two groups—AG-negative and AG-positive. (cf. Example 13.3.) The original data [from Feigl and Zelen (1965)] represent times at death (complete data) and white cell blood count (WBC) in thousands ($z$). Table E13.1 represents these data in a modified form to allow for withdrawals. Fit the following models to these data by the method of maximum likelihood.

(*a*) Linear-exponential with

$$\lambda(t;z)=\alpha+\beta z.$$

**Table E13.1** Survival data and white cell blood counts for patients with acute leukemia

| | Group I (AG-positive) | | | Group II (AG-negative) | | |
|---|---|---|---|---|---|---|
| | Time in weeks $t_j$ | Blood count $z_j$ | Mode of departure | Time in weeks $t_j$ | Blood count $z_j$ | Mode of departure |
| 1 | 65 | 2.3 | D | 56 | 4.4 | D |
| 2 | 156 | 0.75 | D | 65 | 3.0 | D |
| 3 | 100 | 4.3 | D | 17 | 4.0 | W |
| 4 | 134 | 2.6 | D | 7 | 1.5 | D |
| 5 | 16 | 6.0 | W | 16 | 9.0 | D |
| 6 | 108 | 10.5 | D | 22 | 5.3 | D |
| 7 | 121 | 10.0 | W | 3 | 10.0 | D |
| 8 | 4 | 17.0 | D | 4 | 19.0 | D |
| 9 | 39 | 5.4 | D | 2 | 27.0 | D |
| 10 | 143 | 7.0 | D | 3 | 28.0 | D |
| 11 | 56 | 9.4 | W | 8 | 31.0 | W |
| 12 | 26 | 32.0 | D | 4 | 26.0 | D |
| 13 | 22 | 35.0 | W | 3 | 21.0 | D |
| 14 | 1 | 100.0 | D | 30 | 79.0 | D |
| 15 | 1 | 100.0 | D | 4 | 100.0 | D |
| 16 | 5 | 52.0 | D | 43 | 100.0 | W |
| 17 | 65 | 100.0 | D | | | |

(*b*) Multiplicative-exponential with

$$\lambda(t;z)=\alpha e^{\beta z}.$$

(*c*) Piecewise exponential with

$$\lambda(t;z)=\begin{cases}\lambda_1 e^{\beta z} & \text{for } t<\tau,\\ \lambda_2 e^{\beta z} & \text{for } t\geqslant\tau,\end{cases}$$

where $\tau$ is fixed.

For each model construct the likelihood function for estimating the corresponding parameters.

**13.3.** The data in Table E13.2 are taken from Prentice [(1973). *Biometrika* **60**, 279], and represent survival data and concomitant variables for

**Table E13.2.** Patients with lung cancer

| Standard | | | | | | Test | | | | | |
|---|---|---|---|---|---|---|---|---|---|---|---|
| $(i)$ | $t_{(i)}$ | $z_1$ | $z_2$ | $z_3$ | $z_4$ | $(i)$ | $t_{(i)}$ | $z_1$ | $z_2$ | $z_3$ | $z_4$ |
| 1 | 72 | 60 | 7 | 69 | 0 | 1 | 999 | 90 | 12 | 54 | 10 |
| 2 | 411 | 70 | 5 | 64 | 10 | 2 | 112 | 80 | 6 | 60 | 0 |
| 3 | 228 | 60 | 3 | 38 | 0 | 3 | 87* | 80 | 3 | 48 | 0 |
| 4 | 126 | 60 | 9 | 63 | 10 | 4 | 231* | 50 | 8 | 52 | 10 |
| 5 | 118 | 70 | 11 | 65 | 10 | 5 | 242 | 50 | 1 | 70 | 0 |
| 6 | 10 | 20 | 5 | 49 | 0 | 6 | 991 | 70 | 7 | 50 | 10 |
| 7 | 82 | 40 | 10 | 69 | 10 | 7 | 111 | 70 | 3 | 62 | 0 |
| 8 | 110 | 80 | 29 | 68 | 0 | 8 | 1 | 20 | 21 | 65 | 10 |
| 9 | 314 | 50 | 18 | 43 | 0 | 9 | 587 | 60 | 3 | 58 | 0 |
| 10 | 100* | 70 | 6 | 70 | 0 | 10 | 389 | 90 | 2 | 62 | 0 |
| 11 | 42 | 60 | 4 | 81 | 0 | 11 | 33 | 30 | 6 | 64 | 0 |
| 12 | 8 | 40 | 58 | 63 | 10 | 12 | 25 | 20 | 36 | 63 | 0 |
| 13 | 141 | 30 | 4 | 63 | 0 | 13 | 357 | 70 | 13 | 58 | 0 |
| 14 | 25* | 80 | 9 | 52 | 10 | 14 | 457 | 90 | 2 | 64 | 0 |
| 15 | 11 | 70 | 11 | 48 | 10 | 15 | 201 | 80 | 28 | 52 | 10 |
| | | | | | | 16 | 1 | 50 | 7 | 35 | 0 |
| | | | | | | 17 | 30 | 70 | 11 | 63 | 0 |
| | | | | | | 18 | 44 | 60 | 13 | 70 | 10 |
| | | | | | | 19 | 283 | 90 | 2 | 51 | 0 |
| | | | | | | 20 | 15 | 50 | 13 | 40 | 10 |

*Withdrawn alive.

patients diagnosed with lung cancer and treated by a "standard" method (group 1), and a "test" method (group 2). The concomitant variables are:

$z_1$ = performance status (9 levels).
$z_2$ = time since diagnosis (months).
$z_3$ = age at diagnosis (years).
$z_4$ = prior therapy (yes = 10, no = 0).

Consider the following hazard function models:

(*a*) Multiplicative exponential model.
(*b*) Weibull model.
(*c*) Cox's model, assuming a constant hazard rate between successive deaths:

(*i*) In each case find the maximum likelihood estimates of the corresponding parameters, and fit the corresponding SDF.
(*ii*) In each case, discuss the appropriateness of the model; using cumulative hazard plotting.

(*iii*) In each case, examine the order of importance of the four variables using: (1) a step-up, (2) a step-down procedure.

(*iv*) Fit each of the models (*a*), (*b*), (*c*) to each treatment, and construct, in each case, a test for interaction treatments × covariates.

**13.4.** The Gompertz law $\lambda_X(x) = Re^{ax}$ may be modified by making $R$ and/or $a$ dependent on the values of concomitant variables, $z_1, z_2, \ldots, z_s$. It is proposed to use

$$R = \exp\left(\beta_0 + \sum_{u=1}^{s} \beta_u z_u\right) \qquad \text{and/or} \qquad a = \sum_{u=1}^{s} \alpha_u z_u.$$

(*i*) If $z_1 = 0$ or 1 according as an individual is from a population $A$ or $B$, respectively, under what conditions on the $\beta$'s, $\alpha$'s, and $z$'s is the model a proportional hazard rate model?

(*ii*) Suppose that $n_1$ deaths in population $A$ and $n_2$ deaths in population $B$ were observed at distinct times. Construct the appropriate likelihood function.

**13.5.** Let $r$ be the number of individuals present in the risk set just before time $t = \tau$ ($\tau$ is fixed). Let $\lambda_1(t), \lambda_2(t), \ldots, \lambda_r(t)$ be independent hazard rates for these individuals.

(*i*) Find the probability that *only* ($j$) fails in the interval $(\tau, \tau + (\delta\tau))$.

(*ii*) Show that

$$\lim_{(\delta\tau)\to 0} \Pr\left\{ \text{Individual } (j) \text{ fails in } (\tau, \tau + (\delta\tau)) \;\middle|\; \begin{array}{l} \text{Only one failure occurs in} \\ (\tau, \tau + (\delta\tau)) \text{ among } r \text{ individuals} \\ \text{alive just before } \tau \end{array} \right\} = \frac{\lambda_j(\tau)}{\sum_{l=1}^{r} \lambda_l(\tau)}.$$

(See Exercise 15.4.)

**13.6.** Suppose that in the $g$th population ($g = 1, 2, \ldots, k$), the hazard rate at time $t$ is

$$\lambda_g(t; \mathbf{z}) = \lambda(t)\left(\gamma_g + \sum_{u=1}^{s} \beta_u z_u\right),$$

where $z_1, z_2, \ldots, z_s$ are concomitant variables and the $\gamma$'s and $\beta$'s are fixed parameters.

(*i*) Is this a proportional hazard rate model?

(*ii*) For data given in Exercise 13.2 construct the likelihood function, assuming that the unknown $\lambda(t)$ between two consecutive deaths is approximately constant.

(*iii*) How would you try to decide whether the above model, or model of type

$$\lambda(t;\mathbf{z}) = \lambda(t)\exp\left(\gamma_g + \sum_{u=1}^{s} \beta_u z_u\right)$$

is more appropriate?

**13.7.** The lognormal SDF, according to which $\log X$ is distributed normally with expected value $\xi$ and variance $\sigma^2$, may be modified, by making $\xi$ and/or $\sigma^2$ dependent on the values of concomitant variables, $z_1, \ldots, z_s$. It is proposed to use

$$\xi = \xi_0 + \sum_{u=1}^{s} \xi_u z_u \qquad \text{and} \qquad \sigma^2 = \sigma_0^2\left(1 + \sum_{u=1}^{s} \gamma_u^2 z_u^2\right).$$

(*i*) Give an expression for the HRF.

(*ii*) Construct a likelihood function for $N = 10$ individuals who died ($D$) or withdrew ($W$) as shown below:

$$\begin{matrix} t'_1 & < & t'_2 & < & t'_3 & < & t'_4 & < & t'_5 & < & t'_6 & < & t'_7 & < & t'_8 & < & t'_9 & < & t'_{10}. \\ D & & D & & W & & D & & W & & W & & D & & D & & D & & D \end{matrix}$$

**13.8.**

(*i*) Show that if $k$ individuals, $B_1, \ldots, B_k$, have hazard rates proportional to $\theta_1, \theta_2, \ldots, \theta_k$ at all ages, the probability that $B_j$ has the youngest age at failure is

$$\frac{\theta_j}{\sum_{l=1}^{k} \theta_l}.$$

(*ii*) Show that this result also holds, conditional on the youngest age at failure exceeding any specified value $T$, say.

**13.9.**

(*i*) Show that under the conditions stated in (*i*) or (*ii*) of Exercise 13.8, the probability that $B_j$ has the *second* youngest age at failure is

$$\frac{\theta_j}{\sum_{i=1}^{k}\theta_i}\sum_{h\neq j}\frac{\theta_h}{\sum_{i=1}^{k}\theta_i-\theta_h}.$$

(*ii*) Find the probability that $B_j$ has the *third* youngest age at failure.

**13.10.** Consider a Gompertz distribution with hazard rate

$$\lambda(t)=(B+Rz)e^{at},$$

$B>0$, $R>0$, $a>0$, $t>0$, where $z$ is a continuous concomitant variable:

(*i*) following gamma distribution, with PDF

$$f_Z(z)=\frac{\beta^{-\alpha}}{\Gamma(\alpha)}z^{\alpha-1}e^{-y/\beta},\qquad y>0,\qquad \alpha>0,\qquad \beta>0;$$

(*ii*) following normal distribution $N(\mu,\sigma^2)$.

In each case, find the compound distribution of $T$.

**13.11.** Consider a joint distribution of two random variables, $T_1, T_2$ in the form

$$S_{12}(t_1,t_2|z)=S_1(t_1|z)\cdot S_2(t_2|z),$$

where

$$S_i(t|z)=\exp[-(\lambda_i+\gamma_i z)t],\qquad i=1,2,$$

and $z$ has gamma distribution with PDF as defined in Exercise 13.10.

(*i*) Find the joint SDF, $S_{12}(t_1,t_2)$.

(*ii*) Find the SDF of

$$T=\min(T_1,T_2).$$

(*iii*) Let $I$ be an indicator variable taking the value $i$ when $T=T_i$. We define

$$Y_i=\begin{cases}T & \text{if } I=i,\\ \text{otherwise undefined.}\end{cases}$$

Find the probability

$$\Pr\{Y_i>t\}=\Pr\{(T>t)\cap(I=i)\}.$$

**13.12.** For the data on *systemic lupus erythematosus* given in Table E2.1, evaluate a survival distribution function of time elapsed since diagnosis ($t$), when introducing two concomitant variables: age (in years) ($z_1$), and length of time (in years) between onset and diagnosis ($z_2$).

(*a*) Try various models discussed in this chapter, which might be suitable. To start with, perhaps, try the multiplicative model $\lambda(t;\mathbf{z}) = \lambda(t)\exp(\boldsymbol{\beta}'\mathbf{z})$, assuming constant $\lambda(t)$ between any two consecutive deaths.

(*b*) Assume $z_{1j} = z_{2j} = 0$ for all $j$. Construct the SDF in this case. (See also Section 6.8.)

**13.13.** Consider a model in which times $X_1, X_2, \ldots, X_k$ due to fail from causes $C_1, C_2, \ldots, C_k$ respectively are (*i*) independent and (*ii*) each exponentially distributed with constant hazard rate

$$\lambda(\mathbf{z}) = \exp(\text{linear function of covariates } z_1, \ldots, z_r) \tag{A}$$

Suppose that there are $M$ types of individual in a study, and each type may have different coefficients in the linear function in (A). Writing this function, for the $j$th type, as

$$\alpha_j + \sum_{i=1}^{r} \beta_i^{(j)} z_i$$

what interpretation can be given the hypotheses

(*a*) $\beta_i^{(j)} = \beta_i$ for all $j$ and all $i$?

(*b*) $\beta_i^{(j)} = \beta_i$ for all $j$ and selected $i$?

(*c*) $\alpha_j = \alpha$ for all $j$?

Assuming the $X$'s are mutually independent functions based on incomplete mortality data, in which time and cause of death, or time of last observation are recorded for each individual.

Discuss the introduction of a further variable $W$, representing time due to withdraw, into the model [S. Lagakos (1978). *Appl. Statist.* **27**, 235–241.]

**13.14.** Cross-sectional studies have established association between a disease $A$ and high levels of a variable $Z$. What additional statistical data might be used to distinguish whether $Z$ is a risk factor for $A$, or is a symptom of $A$, or increases chances of survival for individuals affected by $A$?

Explain how the additional data would be used.

CHAPTER 14

# Age of Onset Distributions

## 14.1 INTRODUCTION

In this chapter, we consider situations in which the event of interest is not death, but *first occurrence* (*onset*) of a chronic disease, or other age dependent phenomenon. In practice, *date of diagnosis* is often regarded as approximately equivalent to date of onset. Although estimation of the latent period between actual onset and diagnosis is not impossible [e.g., Tallis et al. (1973)], methods of estimation are rather laborious and require large sets of data. In further analysis, age at onset and age at diagnosis will often be used equivalently.

Different diseases have different ranges for age of onset. For example, rheumatic fever is a disease of childhood with peak age of onset at approximately 11–13 years; schizophrenia occurs mostly in adolescents and younger adults, whereas psoriasis can occur at any age, but with peak at older ages.

The main topics of this chapter are methods of estimation of age of onset distributions from different types of data, including cross-sectional and retrospective follow up experiments, in which either incidence or prevalence, or both types of data are available.

Two basic questions can be asked:

1. *What is the expected proportion of new cases occurring before age $x$ in a population*?

2. *For a newborn individual, what is the probability that he will experience onset before age $x$, supposing he survives so long* ?

As we shall see in the following sections, these problems have some analogues in the theory of competing risks, although the methods of estimating the probabilities are not quite the same, because: (1) after onset of the disease, an individual can still live for some time, but eventually must die, and (2) for some diseases, there might be differential mortality between affected and unaffected individuals.

## 14.2 MODELS OF ONSET DISTRIBUTIONS

### 14.2.1 Incidence Onset Distribution

We will use a somewhat different notation to distinguish the problems from those in competing causes. Let $U$ $(>0)$ denote the hypothetical time due to onset or waiting time for onset of a given disease, $A$, and $T$ $(>0)$ be the time due to die.

To build up a theoretical model, it is supposed that an individual is endowed at birth with values of $U$ and $T$. If he is not liable to get disease $A$, then $U=\infty$.

Generally, $U$ and $T$ might not be independent. We introduce the joint survival function

$$S_{U,T}(u,t)=\Pr\{U>u, T>t\}, \tag{14.1}$$

and the random variable

$$Y=\min(U,T). \tag{14.2}$$

Note that $Y$ represents the (observed) time of onset if $U<T$; it represents time of death if $U>T$. In the latter case, of course, onset is not observed. Thus

$$S_Y(x)=\Pr\{Y>x\}=S_{U,T}(x,x) \tag{14.3}$$

is the probability that an individual is alive and still free of disease, up to age $x$; it might be called a health function so far as disease $A$ is concerned.

Let $h_U(x)$ be the crude onset hazard rate,

$$h_U(x)=-\frac{1}{S_Y(x)}\left.\frac{\partial S_{U,T}(u,t)}{\partial u}\right|_{u=t=x}. \tag{14.4}$$

Then

$$Q_U^*(x)=\Pr\{(U\leqslant x)\cap(U\leqslant T)\}=\int_0^x h_U(y)S_Y(y)\,dy \tag{14.5}$$

gives the expected proportion of new cases before age $x$ in the population, and

$$F_U^*(x) = \frac{Q_U^*(x)}{Q_U^*(\infty)} \tag{14.6}$$

is the (proper) distribution of ages at onset among those who *actually* experience onset (incidence) of the disease. We will call it *incidence onset distribution*.

Clearly, incidence and death are competing events, and (14.6) is the distribution of time at onset among those who had the chance to experience onset (did not die) before age $x$ (cf. Section 9.4.3).

### 14.2.2 Waiting Time Onset Distribution

Question (2) posed in Section 14.1 concerning an individual's chance of experiencing onset before age $x$ and assuming, of course, that he is alive and liable to experience onset, is answered by the *waiting time onset distribution*. It is analogous to the waiting time death distribution discussed in Chapter 11. We assume that the instantaneous incidence rate is equal to $h_U(x)$ defined in (14.4).

Then the probability that a newborn individual is free of disease up to age $x$ is

$$\Pr\{U > x\} = G_U(x) = \exp\left[-\int_0^x h_U(y)\,dy\right]. \tag{14.7}$$

The quantity $\phi = 1 - G_U(\infty)$ gives the expected proportion of those who are liable to experience onset, and

$$F_U^+(x) = \frac{1 - G_U(x)}{1 - G_U(\infty)} = \frac{1 - G_U(x)}{\phi} \tag{14.8}$$

represents the (proper) waiting time onset distribution.

### 14.2.3 Life Tables and Onset Distributions

1. Imagine a cohort of $l_0$ newborn individuals whose mortality experience is represented by an appropriate life table. Let $a_x$ be the expected number of individuals, out of $l_0$ newborn, who will experience onset between age $x$ and $x+1$. The expected incidence (central) rate over this age interval is

$$i_x = \frac{a_x}{L_x}. \tag{14.9}$$

The expected proportion of those (among $l_0$ newborn) who will experience onset before age $x$ is

$$\frac{1}{l_0}\sum_{t=0}^{x-1} a_t = \frac{1}{l_0}\sum_{t=0}^{x-1} i_t L_t, \tag{14.10}$$

and the incidence onset distribution defined in (14.6) is then

$$F_U^*(x) = \frac{\sum_{t=0}^{x-1} a_t}{\sum_{t=0}^{\infty} a_t} = \frac{\sum_{t=0}^{x-1} i_t L_t}{\sum_{t=0}^{\infty} i_t L_t}. \tag{14.11}$$

2. Among a cohort originally containing $l_0$ newborn individuals, let $c_x$ be the number of those present age $x$ lbd who have at some time in the past experienced onset of disease $A$. Then

$$\Pi_x = \frac{c_x}{L_x} \tag{14.12}$$

is the *lifetime prevalence* for age $x$ lbd.

Let $q_{Ux}$ denote the (conditional) probability of onset between age $x$ and $x+1$, given free of the disease at age $x$. Then

$$q_{Ux} \doteq \frac{a_x}{L_x(1-\Pi_x)+\frac{1}{2}a_x} = \frac{i_x}{(1-\Pi_x)+\frac{1}{2}i_x}. \tag{14.13}$$

Clearly, the probability of being free of disease at age $x$ is

$$G_U(x) = \prod_{t=0}^{x-1}(1-q_{Ut}). \tag{14.14}$$

and the (proper) waiting time onset distribution is obtained from (14.8) with $G_U(x)$ given in (14.14).

## 14.3 ESTIMATION OF INCIDENCE ONSET DISTRIBUTION FROM CROSS-SECTIONAL INCIDENCE DATA

Let $A_x$ denote the (average) number of new cases over a calendar year in age group $x$ to $x+1$, and $K_x$—the midperiod population in this age group.

Thus

$$I_x = \frac{A_x}{K_x} \tag{14.15}$$

is the observed incidence rate [see formula (2.35)].

1. One may, very roughly, use the *observed* numbers of new cases and construct simply an *observed* incidence onset distribution [we denote it here by $\mathring{F}_U^*(x)$], that is

$$\mathring{F}_U^*(x) = \frac{\sum_{t=0}^{x-1} A_t}{\sum_{t=0}^{\infty} A_t}. \tag{14.16}$$

This, of course, is not correct, since it does not take into account the age structure of the population. (The nature of the problem is similar to that of constructing a life table from deaths only, without taking into account the survivors.)

2. We may, however, estimate the expected number of new cases, $a_x$, say, by adjusting $I_x$'s, according to an appropriate life table, that is

$$\hat{a}_x = I_x \cdot L_x = A_x \cdot \frac{L_x}{K_x}. \tag{14.17}$$

Substituting (14.17) into (14.11), we obtain

$$\hat{F}_U^*(x) = \frac{\sum_{t=0}^{x-1} I_t L_t}{\sum_{t=0}^{\infty} I_t L_t}. \tag{14.18}$$

(Note that $I_t \cdot 10^k$, instead of $I_t$, can be used in (14.18) since the factors $10^k$ cancel out.)

Incidence data, similar to mortality data, are usually recorded for $n$ year (customarily with $n=5$) age groups. We then define ${}_nI_x = {}_nA_x / {}_nK_x$, using an obvious notation (see Chapter 4). The estimated incidence onset distribution is then

$$\hat{F}_U^*(x) = \frac{\sum_{t=1}^{v-1} {}_nI_{nt} \cdot {}_nL_{nt}}{\sum_{t=0}^{\infty} {}_nI_{nt} \cdot {}_nL_{nt}} \tag{14.19}$$

for $x = nv$ ($v$ is an integer).

It is also worthwhile noticing that the method resembles direct standardization (Section 2.9), but with the standard population being the life table appropriate for a given population, not common to both populations.

**Example 14.1** For illustration, we use incidence data for cancer of digestive organs for U.S. White Males. The incidence data in Table 14.1 are from *Third National Cancer Survey*: *Incidence Data*, National Cancer Institute (1975). (See Example 2.9.) The numbers of cases in this survey are given for quinquennial age groups over a period of three years (1969–71); in fact, the second column in Table 14.1 has the heading $3 \cdot {}_5A_x$. The $L_x$ values are taken from the U.S. Life Tables, White Males, 1969–71. The ${}_5L_x$ (in the fourth column) are calculated from the formula

$${}_5L_x = L_x + L_{x+1} + \cdots + L_{x+4}.$$

The fifth column exhibits the observed, and the sixth column, the 'expected' cumulative proportions of males with cancer of digestive system up

**Table 14.1** Cancer of digestive organs in U.S. White Males, 1969–71. Incidence onset distributions

| Age group $x$ to $x+5$ | $3 \cdot {}_5A_x$ | ${}_5I_x \cdot 10^5$ | ${}_5L_x$ | $\mathring{F}_U^*(x)$ | $\hat{F}_U^*(x)$ |
|---|---|---|---|---|---|
| 0–5 | 25 | 1.1 | 489492 | 0.00000 | 0.00000 |
| 5–10 | 14 | 0.5 | 487745 | 0.00118 | 0.00070 |
| 10–15 | 7 | 0.2 | 486735 | 0.00184 | 0.00101 |
| 15–20 | 28 | 1.1 | 484399 | 0.00217 | 0.00114 |
| 20–25 | 36 | 1.7 | 480020 | 0.00349 | 0.00183 |
| 25–30 | 64 | 3.3 | 475553 | 0.00519 | 0.00289 |
| 30–35 | 101 | 6.3 | 471472 | 0.00820 | 0.00492 |
| 35–40 | 176 | 11.6 | 466395 | 0.01296 | 0.00876 |
| 40–45 | 419 | 25.3 | 458745 | 0.02126 | 0.01577 |
| 45–50 | 961 | 57.3 | 446572 | 0.04101 | 0.03079 |
| 50–55 | 1517 | 100.5 | 427547 | 0.08631 | 0.06392 |
| 55–60 | 2210 | 170.1 | 398442 | 0.15782 | 0.11955 |
| 60–65 | 2855 | 270.1 | 356831 | 0.26200 | 0.20729 |
| 65–70 | 3192 | 407.6 | 302111 | 0.39658 | 0.33206 |
| 70–75 | 3321 | 564.0 | 236680 | 0.54704 | 0.49148 |
| 75–80 | 3070 | 738.7 | 165221 | 0.70359 | 0.66430 |
| 80–85 | 2057 | 876.6 | 97233 | 0.84831 | 0.82230 |
| 85+ | 1161 | 827.5 | 62635 | 0.94527 | 0.93290 |
| | 21214 | | | | |

to age $x$ among White Males of all ages with cancer of digestive organs [formulae (14.16) and (14.19), respectively].

## 14.4 ESTIMATION OF INCIDENCE ONSET DISTRIBUTION FROM PREVALENCE DATA

Incidence onset distribution can also be estimated from prevalence data. The problem is that of estimating the incidence ($i_x$) from prevalence ($\Pi_x$). In this case, it is necessary to distinguish whether there is or not differential mortality between affected and unaffected. Methods presented in this section are really only applicable when the disease under consideration is not uncommon. When data are sparse, it is quite possible to get negative estimates for some estimates for some $i_x$.

It is convenient to derive the formulae, still using the concept of a cohort of $l_0$ newborn individuals. When only (cross-sectional) data from current population are available, the prevalence, $\Pi_x$, defined in (14.12) will be replaced by the observed prevalence $\hat{\Pi}_x = C_x / K_x$, where $C_x$ is the *observed* number of affected, present age $x$ lbd.

### 14.4.1 Estimation of Age Specific Incidence from Prevalence Data: No Differential Mortality

From now on, we assume that ascertainment of affected is complete. If we can further assume that there is no differential mortality between affected and unaffected then it is possible to estimate the incidence rates, $i_x$.

We recall that among $l_0$ newborns the expected number who experience onset of the disease between age $t$ and $t+1$ is $a_t = i_t L_t$ [from (14.9)]. Among these $i_t L_t$, the expected number surviving to be included in the $L_x$ aged $x$ lbd. $(x > t)$ is

$$i_t L_t \cdot \frac{L_x}{L_t} = L_x i_t, \tag{14.20}$$

where $L_x / L_t$ is the probability of surviving from age $t$ lbd to age $x$ lbd.

From (14.20), the expected number among $L_x$ aged $x$ lbd who have experienced onset at some time *and* survived is approximately

$$\sum_{t=0}^{x-1} i_t L_x + \tfrac{1}{2} i_x L_x = \sum_{t=0}^{x} i_t L_x - \tfrac{1}{2} i_x L_x. \tag{14.21}$$

[The last term in (14.21) is, of course, based on the assumption of uniform distribution of events over the current year of life.]

Thus the expected *lifetime prevalence* among those aged $x$ lbd is

$$\Pi_x = \frac{1}{L_x} \cdot L_x \left( \sum_{t=0}^{x-1} i_t + \tfrac{1}{2} i_x \right) = \sum_{t=0}^{x-1} i_t + \tfrac{1}{2} i_x \tag{14.22}$$

$$= \sum_{t=0}^{x} i_t - \tfrac{1}{2} i_x. \tag{14.22a}$$

Note first that this result does not depend on the actual values of $L_x$'s, although it does, of course, depend on the assumptions already mentioned. Second,

$$\Pi'_x = \sum_{t=0}^{x-1} i_t \tag{14.23}$$

represents the proportion affected among survivors when they reach age $x$, it is lifetime prevalence at *exact* age $x$. The observed prevalence, $\hat{\Pi}_x$, can be used as an estimator of $\Pi_x$. The equations (14.22) can then be solved to yield estimates $\hat{i}_x$ of the incidence rates. In fact, we have from (14.22)

$$\hat{\Pi}_{x+1} - \hat{\Pi}_x = \tfrac{1}{2}(\hat{i}_{x+1} + \hat{i}_x)$$

or

$$\hat{i}_{x+1} = 2(\hat{\Pi}_{x+1} - \hat{\Pi}_x) - \hat{i}_x, \tag{14.24}$$

with

$$\hat{\Pi}_0 = \tfrac{1}{2} \hat{i}_0 \qquad \text{so that} \qquad \hat{i}_0 = 2\hat{\Pi}_0. \tag{14.25}$$

Once the incidence rates are estimated, we can apply the method described in Section 14.3 to obtain the estimated incidence onset distribution.

If the data are grouped in age intervals $[x, x+n)$, $\hat{\Pi}_x$ approximates the proportion of affected among persons aged $x$ to $x+n$, and $\hat{i}_x$ corresponds to rates per person per *n years* (not per year). Of course, in evaluation of the age of onset distribution this is immaterial, since we are interested only in ratios and not in absolute values.

### 14.4.2 Affected Individuals Are Subject to Differential Mortality

If, in fact, mortality among affected is higher than among the unaffected, $\hat{\Pi}_x$ is likely to be an underestimate of $\Pi_x$ since fewer affected individuals than expected will survive to be included among those aged $x$ lbd. We now describe a way of allowing for this possibility. We will use primes to denote life table values for affected individuals. In fact, since mortality usually depends not only on age, but also on the duration of illness, *select life tables* (see Section 4.13) should be used.

The expected number of survivors till age $x$ lbd among those who experienced onset when aged $t$ lbd [i.e., with duration $(x-t)$ at $x$ lbd] is

$$i_t L_t \cdot \frac{L'_{[t]+x-t}}{L'_{[t]}}, \qquad x>t, \tag{14.26}$$

where $L'_{[t]+x-t}$ is the select life table value for affected with duration $x-t$, and $L'_{[t]+x-t}/L'_{[t]}$ is the probability of surviving $(x-t)$ years after onset at age $t$.

Hence, in place of (14.21) we have approximately (neglecting differential mortality since attainment of age $x$)

$$\Pi_x \doteqdot \frac{1}{L_x} \sum_{t=0}^{x-1} \frac{L'_{[t]+x-t}}{L'_{[t]}} \cdot i_t L_t + \tfrac{1}{2} i_x L_x. \tag{14.27}$$

Dividing both sides of (14.27) by $L'_{[x]}$, we obtain

$$\frac{L_x}{L'_{[x]}} \Pi_x = \sum_{t=0}^{x-1} \frac{L'_{[t]+x-t}}{L'_{[x]}} \left( \frac{L_t}{L'_{[t]}} i_t \right) + \tfrac{1}{2} \frac{L_x}{L'_{[x]}} i_x. \tag{14.28}$$

Writing

$$\frac{L_x}{L'_{[x]}} \Pi_x = \Pi_x^* \qquad \text{and} \qquad \frac{L_t}{L'_{[t]}} i_t = i_t^*, \tag{14.29}$$

we obtain

$$\Pi_x^* = \sum_{t=0}^{x-1} \frac{L'_{[t]+x-t}}{L'_{[x]}} i_t^* + \tfrac{1}{2} i_x^*. \tag{14.30}$$

When $x-t$ is greater than the select period, $d$, $L'_{[t]+x-t} = L'_x$, and so does not depend on $t$. So (14.30) can be written as

$$\Pi_x^* = \frac{L'_x}{L'_{[x]}} \sum_{t=0}^{x-d} i_t^* + \sum_{t=x-(d-1)}^{x-1} \psi_{x;x-t} i_t^* + \tfrac{1}{2} i_x^*, \tag{14.31}$$

where

$$\psi_{x;x-t} = \frac{L'_{[t]+x-t}}{L'_{[x]}}.$$

For a first approximation we may often be able to take $\psi_{x;x-t}$ as a function of the time since onset $(x-t)$ only.

If there is *no selection* (i.e., no effect of duration since onset) then $L'_x / L'_{[x]} = \psi_{x;x-t} = 1$, and (14.31) becomes

$$\Pi_x^* = \sum_{t=0}^{x-1} i_t^* + \tfrac{1}{2} i_x^*. \tag{14.32}$$

Note that (14.32) is of precisely the same form as (14.22). Of course, even for this approximation, it is necessary to know the life table (with $L'_x$) for the affected individuals, which is, unfortunately, not often available. When there is no selection, we can obtain estimates $\hat{i}_x^*$ of the $i_x^*$'s by calculating $\hat{\Pi}_x^* = \hat{\Pi}_x L_x / L'_x$ and using (14.31). Finally, we can calculate

$$\hat{i}_x = \hat{i}_x^* \frac{L'_x}{L_x}. \tag{14.33}$$

To construct the estimated age of onset distribution $\hat{G}_U(x)$, we then use the method described in Section 14.3. However, we notice that the terms $\hat{i}_t L_t$ [used in (14.10) and (14.11)] can now be represented as

$$\hat{i}_t L_t = \hat{i}_x^* \cdot \frac{L'_t}{L_t} \cdot L_t = \hat{i}_t^* L'_t. \tag{14.34}$$

Therefore, we do not, in fact, need to calculate the $\hat{i}_t$'s separately; the method presented in Section 14.3 can be applied directly to the $\hat{i}_t^*$'s combined with the life table for affected, that is, using $L'_t$.

Generalization for data grouped in age intervals $[x, x+n)$ is straightforward.

**Example 14.2** We apply the methods discussed in this Section to some data on breathlessness among British coal miners reported by Ashford et al. (1970) and shown in Table 14.2a, (columns 2, 3, and 4). The Table also gives values of $5 \cdot {}_5\hat{i}_x$, calculated from an analogue of (14.25), and (14.24) under the assumption of no differential mortality between affected and unaffected.

As we can see from the table, $5 \cdot {}_5\hat{i}_{55} = -0.0117$ is negative. This can arise from at least two causes (in addition to random variation): incomplete ascertainment or differential mortality. Since we do not have life tables appropriate for these coal miners (either affected *or* unaffected), we cannot estimate age of onset distribution.

However, purely for *illustrative purposes*, two sets of calculations were carried out, each under different assumptions.

**Table 14.2a.** Prevalence of breathlessness

| (1) Age group $x$ to $x+5$ | (2) Number observed ${}_5K_x$ | (3) Number with breath- lessness $C_x$ | (4) Prevalence $\hat{\Pi}_x$ | (5) Estimated incidence $5 \cdot {}_5\hat{i}_x$ | (6) ${}_5L_x$ | (7) $5\Sigma_5\hat{i}_x \cdot {}_5L_x$ | (8) $\hat{F}_X(x)$ |
|---|---|---|---|---|---|---|---|
| 20–25 | 1952 | 16 | 0.0082 | 0.0164 | 481185 | — | — |
| 25–30 | 1791 | 32 | 0.0179 | 0.0029 | 478683 | 7,891.4 | 0.0353 |
| 30–35 | 2113 | 73 | 0.0346 | 0.0306 | 476150 | 9327.5 | 0.0417 |
| 35–40 | 2783 | 169 | 0.0607 | 0.0218 | 472641 | 23802.4 | 0.1064 |
| 40–45 | 2274 | 223 | 0.0981 | 0.0529 | 467066 | 34106.0 | 0.1525 |
| 45–50 | 2393 | 357 | 0.1492 | 0.0493 | 457729 | 58860.5 | 0.2632 |
| 50–55 | 2090 | 521 | 0.2493 | 0.1509 | 441895 | 81380.8 | 0.3639 |
| 55–60 | 1750 | 558 | 0.3189 | −0.0117 | 415262 | 148106.9 | 0.6623 |
| 60–65 | 1136 | 478 | 0.4208 | 0.2755 | 372908 | 143206.8 | 0.6404 |
| | | | | | | 223605.8 | 1.00000 |

Data from: Ashford and Sowden (1970), *Biometrics*, **26**, 535.

**Table 14.2b.** Estimation of incidence onset distribution assuming higher mortality among affected

| (1) Age group $x$ to $x+5$ | (2) ${}_5L_x$ | (3) ${}_5L'_x$ | (4) $\hat{\Pi}^*_x$ | (5) $5_5\hat{i}^*_x$ | (6) $5_5i^*_x \cdot {}_5L'_x$ | (7) $\sum$ (6) | (8) $\hat{F}^*_X(x)$ |
|---|---|---|---|---|---|---|---|
| 20–25 | 481185 | 343937 | 0.01147 | 0.02294 | 7889.9 | — | — |
| 25–30 | 478683 | 333343 | 0.02570 | 0.00552 | 1840.1 | 7889.9 | 0.0298 |
| 30–35 | 476150 | 320446 | 0.05141 | 0.04590 | 14708.5 | 9730.0 | 0.0367 |
| 35–40 | 472641 | 304305 | 0.09498 | 0.04124 | 12549.5 | 24438.4 | 0.0923 |
| 40–45 | 467066 | 284325 | 0.16115 | 0.09110 | 25902.0 | 36988.0 | 0.1396 |
| 45–50 | 457729 | 260806 | 0.26185 | 0.11030 | 28766.9 | 62890.0 | 0.2374 |
| 50–55 | 441895 | 233060 | 0.47268 | 0.31136 | 72565.6 | 91656.9 | 0.3460 |
| 55–60 | 415262 | 200561 | 0.66028 | 0.06384 | 12803.8 | 164222.4 | 0.6200 |
| 60–65 | 372908 | 163241 | 0.96127 | 0.53814 | 87846.5 | 177026.3 | 0.6683 |
| | | | | | | 264872.3 | 1.0000 |

*Values in columns (2) and (3) are taken from Preston et al. (1972).

1. First, we assume complete ascertainment and no differential mortality, using ${}_5L_x$ values from the life table for England and Wales, Males, 1964. We also make the assumption that ${}_5\hat{i}_x = 0$ for $x < 20$, that is, no cases of breathlessness occur before age 20. Furthermore, the age of onset distribution is truncated from above at age 65.

The results are given in columns (6), (7), and (8) of Table 14.2a. Clearly the two values $F_X(55) = 0.6633$, $F_X(60) = 0.6404$ are inconsistent. This is because $5 \cdot {}_5\hat{i}_{55} = -0.0117$ is negative.

2. In the second set, we assumed differential mortality, but no selection. The England and Wales, Males, 1891 life table was used as a source of ${}_5L'_x$ values. The calculations are presented in Table 14.2b, using formulae given in Section 14.4.2.

## 14.5 ESTIMATION OF WAITING TIME ONSET DISTRIBUTION FROM POPULATION DATA

As was pointed out in Section 14.2.2, the waiting time onset distribution can be evaluated from life table functions, together with the expected number of *new cases* ($a_x$) and *prevalence* ($\Pi_x$) first calculating the (conditional) probability of onset, $q_{Ux}$, given in (14.13).

This probability can be estimated from the *observed* number of new cases ($A_x$), and *observed* prevalence ($\hat{\Pi}_x$), using the formula

$$\hat{q}_{Ux} = \frac{A_x}{K_x(1-\hat{\Pi}_x) + \frac{1}{2}A_x} = \frac{I_x}{(1-\hat{\Pi}_x) + \frac{1}{2}I_x}. \tag{14.35}$$

If the data are grouped for ages $x$ to $x+n$, then (cf. Section 4.9.4),

$${}_n\hat{q}_{Ux} = \frac{{}_nA_x}{\frac{1}{n}\left[{}_nK_x(1-\hat{\Pi}_x) + \frac{n}{2}\,{}_nA_x\right]} = \frac{n \cdot {}_nI_x}{(1-\hat{\Pi}_x) + \frac{n}{2}\,{}_nI_x}, \tag{14.36}$$

where $\hat{\Pi}_x$ is the prevalence among persons aged $x$ to $x+n$.

Estimation of $G_U(x)$ and $F_U^+(x)$, is straightforward (see Section 14.2.2).

## 14.6 ESTIMATION OF WAITING TIME ONSET DISTRIBUTION FROM RETROSPECTIVE DATA

Suppose that on the date of a survey on prevalence of a certain disease we record not only the number of affected aged $x$ lbd (prevalence data), but we also obtain information from each affected individual on the age of the first episode (diagnosis), $y$, say. Therefore, we have the prevalence at age $x$ lbd, and also some information on *age of incidence among survivors*. The data can be displayed in a two-way table as shown in Table 14.3.

**Example 14.3** The data in Table 14.3 are from Falconer et al. (1971) and give information on *diabetes mellitus* in Edinburgh, Scotland (January 1, 1968). The data are grouped in 10-year intervals.

The last column in Table 14.3 gives the estimated prevalence function. If there were no differential mortality we could estimate age of onset distribution using the $\hat{\Pi}_x$'s as described in Section 14.4.2. However, in this case, we do not obtain useful results since some incidence rates are negative (the reader may check this as an exercise).

The data in Table 14.3 indicate that the mortality among those affected is probably higher than in the general population; younger ages of onset are underrepresented among those persons now at more advanced ages.

The last two rows in Table 14.3 give the frequency and proportionate (percentage) distributions of *ages of onset among the affected survivors*. The latter, also, is sometimes called an age of onset distribution. Clearly, this does not give the distribution of all incidences, since those who were affected and had died subsequently, are now not taken into account. We now show that, under certain assumptions, it is possible to estimate the waiting time onset distribution from data of this kind.

### 14.6.1 No Differential Mortality Between Affected and Unaffected

1. The basic assumption for the method presented in this section is that there is no *differential mortality* between affected and unaffected.
2. We also assume that the *incidence rate* for each age group *does not depend on chronological time* (for example a person aged 20 in 1900 has the same chance of getting the disease before age 21 as a person age 20 in 1970).

It is no longer necessary to assume that everybody in the population is liable to contract the disease $A$.

Let $K_x$ denote the number of individuals aged $x$ lbd in the sample (survey) on the date of examination. Let $g_{xy}$ denote the number of individuals now aged $x$ lbd who had onset of the disease at age $y$ lbd ($y \leqslant x$). Then

$$C_x = \sum_{j=0}^{x} g_{xj} \tag{14.37}$$

**Table 14.3** Diabetic patients in Edinburgh, (Jan. 1, 1971).

| Current age | Age at onset ($y$ lbd) | | | | | | | | | Total | Population in thousands | Prevalence |
|---|---|---|---|---|---|---|---|---|---|---|---|---|
| $x$ lbd | 0–9 | 10–19 | 20–29 | 30–39 | 40–49 | 50–59 | 60–69 | 70–79 | 80+ | ${}_nC_x$ | ${}_nK_x$ | $\hat{\Pi}_x \cdot 10^3$ |
| 0–9 | 4 | | | | | | | | | 4 | 39.05 | .1024 |
| 10–19 | 16 | 17 | | | | | | | | 33 | 35.93 | .9184 |
| 20–29 | 5 | 33 | 21 | | | | | | | 59 | 28.27 | 2.0870 |
| 30–39 | 5 | 18 | 35 | 23 | | | | | | 81 | 26.44 | 3.0635 |
| 40–49 | | 10 | 32 | 59 | 68 | | | | | 169 | 26.37 | 6.4088 |
| 50–59 | | 3 | 7 | 23 | 91 | 159 | | | | 283 | 28.43 | 9.9954 |
| 60–69 | | 1 | 5 | 15 | 32 | 137 | 183 | | | 373 | 21.33 | 17.4871 |
| 70–79 | | | 1 | 2 | 2 | 28 | 91 | 83 | | 207 | 9.91 | 20.8880 |
| 80+ | | | | | 2 | 3 | 6 | 21 | 14 | 46 | 3.10 | 15.0820 |
| Total (${}_nb_x$) | 30 | 82 | 101 | 122 | 195 | 327 | 280 | 104 | 14 | 1255 | 218.83 | |
| Percent | 2.4 | 6.5 | 8.1 | 9.7 | 15.5 | 26.1 | 22.3 | 8.3 | 1.1 | | | |

From Falconer *et al.* (1971), p. 351.

is the number of living persons aged $x$ lbd who have experienced onset of the disease and $(K_x - C_x)$ is the number of those persons, who are still free of the disease.

Further, let $B_y$ denote the number of persons (whatever their present age), who have experienced onset between age $y$ and $y+1$. Clearly,

$$B_y = \sum_{i=y}^{\omega'} g_{iy} = \sum_{i=y}^{\infty} g_{iy}, \tag{14.38}$$

where $\omega'$ is the last recorded age. The values of $B_y$ are obtained by counting all the individuals in the sample (survey) who reported onset at age $y$ lbd.

Let $N_x$ denote the number of persons now living who passed through age $x$ exactly, never having previously experienced onset of the disease. We have

$$N_0 = \sum_{x=0}^{\omega'} K_x = \sum_{x=0}^{\infty} K_x. \tag{14.39}$$

[If the first possible manifestation of the disease can take place at age $\alpha$ $(>0)$, then we can replace $N_0$ by $N_\alpha = \Sigma_{x=\alpha}^{\omega'} K_x$.]

Furthermore,

$$N_{x+1} = N_x - (K_x - C_x) - B_x. \tag{14.40}$$

We can formally construct a survival function, which is the complement of the waiting-time onset distribution described in Section 6.2. Now, $N_x$ plays the role of survivors at exact age $x$ (in fact, these are the survivors free of the disease at age $x$); $(K_x - C_x)$ - plays the role of withdrawals alive between age $x$ and $x+1$; and finally, $B_x$ (the number of affected between ages $x$ and $x+1$) plays the role of deaths between ages $x$ and $x+1$. The number of "*initial exposed to risk*" (those who are alive and free of the disease at exact age $x$) is then, from (6.14)

$$N'_x \doteqdot N_x - \tfrac{1}{2}(K_x - C_x), \tag{14.41}$$

and the conditional probability of being affected between age $x$ and $x+1$ is

$$\hat{q}_{Ux} = \frac{B_x}{N'_x} = \frac{B_x}{N_x - \frac{1}{2}(K_x - C_x)}. \tag{14.42}$$

The proportion of affected on reaching exact age $x$ is estimated from (14.14)

$$1-\hat{G}_U(x)=1-\prod_{t=0}^{x-1}(1-\hat{q}_{Ut})$$

and

$$\hat{F}_U^+(x)=\frac{1-\hat{G}_U(x)}{1-\hat{G}_U(\infty)}, \tag{14.43}$$

[see (14.8)] estimates the onset waiting time distribution. The quantity $[1-\hat{G}_U(\infty)]$ represents the fraction of the population vulnerable to the disease $A$, that is, liable to onset of $A$.

Modification for data grouped into age intervals $[x, x+n)$ is straightforward. Derivation of (14.43) is given by Elandt-Johnson (1973), although the fact that $K_x - C_x$ can be treated as *withdrawals* was noticed by Bernstein et al. (1977), when using (independently) this method on gallbladder data. We will discuss their data in Example 14.4.

The assumptions (1) and (2) hold but rarely. The data in Table 14.3 indicate that the assumption of no differential mortality for diabetic patients probably does not hold. Therefore the method discussed in this section would not really be applicable even if fuller data were available. For illustrative purpose, however, we present the calculations in Table 14.4 *as if* there were no *differential mortality*. Only a very small proportion

**Table 14.4** Waiting time onset distribution. Data from Table 14.3

| Age group $x$ to $x+10$ | ${}_{10}K_x$ | ${}_{10}C_x$ | ${}_{10}B_x$ | $N_x$ | $N'_x$ | ${}_{10}\hat{q}_{Ux}$ | $1-\hat{G}_U(x)$ | $\hat{F}_U^+(x)$ |
|---|---|---|---|---|---|---|---|---|
| 0–10 | 39050 | 4 | 30 | 218830 | 218830 | 0.00014 | 0.00000 | 0 |
| 10–20 | 35930 | 33 | 82 | 179780 | 179754 | 0.00046 | 0.00014 | 0.00469 |
| 20–30 | 28270 | 59 | 101 | 143850 | 143801 | 0.00070 | 0.00059 | 0.01976 |
| 30–40 | 26440 | 81 | 122 | 115580 | 115538 | 0.00106 | 0.00130 | 0.04354 |
| 40–50 | 26370 | 169 | 195 | 89140 | 89099 | 0.00219 | 0.00235 | 0.07870 |
| 50–60 | 28430 | 283 | 327 | 62770 | 62744 | 0.00521 | 0.00453 | 0.15171 |
| 60–70 | 21330 | 373 | 280 | 34340 | 34296 | 0.00816 | 0.00978 | 0.32552 |
| 70–80 | 9910 | 207 | 104 | 13010 | 13103 | 0.00794 | 0.01781 | 0.59645 |
| 80+ | 3100 | 46 | 14 | 3100 | 3203 | 0.00437 | 0.02560 | 0.85733 |
| | | | | | | | 0.02986 | 1.00000 |
| Total | 218830 | 1255 | 1255 | | | | | |

(approximately 3%) is expected to be affected at all. The disease seems to have a serious effect on older people (after age 40).

**Example 14.4** We use a set of data on prevalence and incidence of gallbladder disease among women enrolled in a weight-reducing organization (TOPS Club, Inc.) discussed by Bernstein et al. (1977). At the time the data were collected, there were 62,739 living members. For each member, information was available on: (1) present age, and (2) whether the member had ever suffered from gallbladder disease. There were 9855 such women, but only for 6328 of them was the age of onset of the disease available. This raised some difficulties in applying the method presented in this section. The resolution of these difficulties will now be described.

The data were grouped into intervals of width $n=2$. Since each survivor of present age at least $x$ clearly had passed through age $x$ exactly, it was easy to find the number entering any specified age group $x$ to $(x+2)$. In particular, the number entering the age group 0–2 was 62,739. Passing from age group $x$ to $(x+2)$ to the next higher (i.e., $x+2$ to $x+4$), we need to calculate $N_{x+2}$. Since 3527 women did not give their age of onset, ${}_2B_x$ in (14.38), is not known exactly. To estimate the unknown ages of onset, Bernstein et al. (1977) assumed that the distribution of age of onset among women who did not give age of onset was the same as among women of the same present age, who did report age of onset. Thus, for women of present age $x$ to $x+2$, the *estimated number* of onsets between $y$ to $y+2$ is

$$ {}_2B_x^* = \frac{\begin{pmatrix}\text{number reporting}\\ \text{onset sometime}\end{pmatrix}}{\begin{pmatrix}\text{number reporting age}\\ x \text{ to } (x+2) \text{ of onset}\end{pmatrix}} \times \begin{pmatrix}\text{number reporting onset}\\ \text{between } x \text{ and } x+2\end{pmatrix}, \tag{14.44} $$

where "number" in each case means "number of women of present age $x$ to $x+2$."

Table 14.5 [based on Table 2 of Bernstein et al. (1977)] shows values of ${}_2B_x^*$ in column (3). Column (4) gives the number of "withdrawals" without the disease, $({}_2K_x - {}_2C_x)$.

The estimated number of women who have passed through age $x$ exactly, never having had the disease previously, denoted here by $N_x^*$, is calculated from (14.40) substituting ${}_2B_x^*$ for ${}_2B_x$. The *initial* exposed to risk

**Table 14.5** Estimated age of onset distribution for gallbladder disease in a population of 62,739 TOPS women

| Age group $x$ to $x+2$ | $N_x^*$ | ${}_2B_x^*$ | ${}_2K_x - {}_2C_x$ | $N_x^{*\prime}$ | ${}_n\hat{q}_U$ | $\hat{G}_U(x)$ | $\hat{F}_U^+(x)$ |
|---|---|---|---|---|---|---|---|
| 0–2 | 62739.0 | 1.6 | 0 | 62390.0 | 0.0000 | 1.0000 | 0.0000 |
| 2–4 | 62737.4 | 1.5 | 0 | 62737.4 | 0.0000 | 1.0000 | 0.0000 |
| ⋮ | ⋮ | 36.1 } | ⋮ | ⋮ | ⋮ | ⋮ | ⋮ |
| 16–18 | 62699.8 | 56.5 | 0 | 62699.8 | 0.0009 | 0.9995 | 0.0010 |
| 18–20 | 62643.3 | 208.9 | 0 | 62643.3 | 0.0033 | 0.9986 | 0.0029 |
| 20–22 | 62434.4 | 440.1 | 937 | 61965.9 | 0.0071 | 0.9953 | 0.0098 |
| 22–24 | 61057.3 | 588.0 | 1665 | 60224.8 | 0.0098 | 0.9882 | 0.0245 |
| 24–26 | 58804.3 | 712.2 | 2202 | 57703.3 | 0.0123 | 0.9785 | 0.0446 |
| 26–28 | 55820.1 | 719.5 | 3187 | 54296.6 | 0.0133 | 0.9665 | 0.0695 |
| 28–30 | 51983.6 | 716.4 | 3438 | 50264.6 | 0.0143 | 0.9536 | 0.0963 |
| 30–32 | 47829.2 | 705.0 | 3776 | 46091.2 | 0.0153 | 0.9400 | 0.1245 |
| 32–34 | 43648.2 | 672.9 | 3509 | 41893.7 | 0.0161 | 0.9256 | 0.1544 |
| 34–36 | 39466.3 | 634.3 | 3357 | 37787.8 | 0.0168 | 0.9107 | 0.1853 |
| 36–38 | 35475.0 | 577.8 | 3144 | 33903.0 | 0.0170 | 0.8954 | 0.2170 |
| 38–40 | 31753.2 | 574.5 | 3242 | 30132.2 | 0.0191 | 0.8802 | 0.2485 |
| 40–42 | 27936.7 | 516.4 | 3147 | 26363.2 | 0.0196 | 0.8634 | 0.2834 |
| 42–44 | 24273.3 | 460.2 | 2865 | 22840.8 | 0.0201 | 0.8465 | 0.3185 |
| 44–46 | 20948.1 | 430.6 | 2820 | 19538.1 | 0.0220 | 0.8295 | 0.3537 |
| 46–48 | 17697.5 | 340.3 | 2609 | 16393.0 | 0.0208 | 0.8113 | 0.3915 |
| 48–50 | 14748.2 | 309.6 | 2424 | 13536.2 | 0.0229 | 0.7944 | 0.4266 |
| 50–52 | 12014.6 | 281.1 | 1976 | 11026.6 | 0.0255 | 0.7762 | 0.4643 |
| 52–54 | 9757.5 | 259.0 | 1920 | 8797.5 | 0.0294 | 0.7564 | 0.5054 |
| 54–56 | 7578.5 | 161.9 | 1596 | 6780.5 | 0.0239 | 0.7342 | 0.5515 |
| 56–58 | 5820.6 | 131.5 | 1368 | 5136.6 | 0.0256 | 0.7167 | 0.5878 |
| 58–60 | 4321.1 | 102.3 | 1076 | 3783.1 | 0.0270 | 0.6984 | 0.6257 |
| 60–62 | 3142.8 | 74.3 | 853 | 2716.3 | 0.0274 | 0.6795 | 0.6649 |
| 62– | | | | | | 0.6609 | 0.7035 |
| ⋮ | | | | | | ⋮ | ⋮ |
| 84+ | | | | | | 0.5180 | 1.0000 |

From Bernstein et al. (1977). *J. Chron. Dis.* **30**, 529–541.

values [from (14.41)], $N_x^{*\prime} = N_x^* - \frac{1}{2}({}_2K_x - {}_2C_x)$, are given in column (5). Column (6) gives ${}_2\hat{q}_{Ux}$ and column (7), the estimated waiting time onset 'survival' function, $\hat{G}_U(x)$.

There were only a few onsets before age 16 and after age 60. The last available $\hat{G}_U(x)$ in the paper by Bernstein et al. (1977) is $\hat{G}_U(84) = 0.5180$.

This is the proportion of women who never had the disease before reaching age $\omega'=84$. The waiting time onset distribution function (over the age range 0–84) is then estimated from the formula (14.8)

$$\hat{F}_U^+(x)=\frac{1-\hat{G}_U(x)}{1-\hat{G}_U(84)}=\frac{1-\hat{G}_U(x)}{0.4820}.$$

The values of $\hat{F}_U^+(x)$ are given in column (8).

In this Example, selection of women for inclusion in the study can reasonably be regarded as being independent of occurrence at gallbladder disease. If inclusion depends on occurrence of the disease in question (e.g., family studies of genetically controlled diseases), then the analysis becomes more complicated.

### 14.6.2 Effects of Differential Mortality

If the mortality among affected persons (i.e., those who have already experienced onset) is different (using symbols $L'_x$) from that in the general population ($L_x$), some adjustment for the differential model is needed. To simplify the following discussion we suppose that there is no need to allow for selection.

We recall from Section 14.3 that among $l_0$ newborn, the expected number of new cases aged $y$ lbd is $i_y L_y$, of whom $i_y L_y \cdot L'_x / L'_y$ are expected to survive to age $x$ lbd at time of enumeration. The expected number of persons (affected or not) from this same group aged $x$ lbd is $L_x$, so that the "expected" proportion affected is

$$\frac{1}{L_x} i_y L_y \frac{L'_x}{L'_y} = i_y \frac{L_y}{L'_y} \cdot \frac{L'_x}{L_x}. \tag{14.45}$$

If we enumerate $K_x$ aged $x$ lbd, of whom $g_{xy}$ experienced onset when $y$ lbd, a natural estimator, $\hat{i}_{y|x}$, say, of $i_y$ is obtained by equating

$$\hat{i}_{y|x} \cdot \frac{L_y L'_x}{L'_y L_x} = \frac{g_{xy}}{K_x},$$

giving

$$\hat{i}_{y|x} = \frac{g_{xy}}{K_x} \cdot \frac{L'_y}{L_y} \cdot \frac{L_x}{L'_x}. \tag{14.46}$$

There will be a sequence $\hat{i}_{y|y}, \hat{i}_{y|y+1}, \ldots$ of such estimators. To obtain a combined estimator we try to estimate the total numbers of exposed to risk aged $y$ lbd and of new cases aged $y$ lbd (whatever the present age).

We estimate total (affected *and* unaffected) exposed to risk as

$$L_y \sum_{x=y}^{\infty} \frac{n_x}{L_x} = W_{xy}, \tag{14.47}$$

and total new cases as

$$L'_y \sum_{x=y}^{\infty} \frac{g_{xy}}{L'_x} = V_{xy}, \tag{14.48}$$

leading to

$$\hat{i}_y = \frac{V_{xy}}{W_{xy}} = \left[ L'_y \sum_{x=y}^{\infty} \frac{g_{xy}}{L'_x} \right] \left[ L_y \sum_{x=y}^{\infty} \frac{n_x}{L_x} \right]^{-1}. \tag{14.49}$$

Just as in Section 14.4.2, it is not possible to apply this adjustment unless at least the ratios

$$\frac{L'_y}{L_y} \cdot \frac{L_x}{L'_x}$$

are known.

Since the separate estimates $\hat{i}_{y|x}$ should each estimate $i_y$, they should be reasonably consistent with each other (taking into account the magnitude of the $K_x$'s).

## REFERENCES

Ashford, J. R. and Sowden, R. R. (1970). Multivariate probit analysis. *Biometrics* **26**, 535–546.

Bernstein, R. A., Giefer, E. E., Vieira, J. J., Werner, L. H., and Rimm, A. A., (1977). Gallbladder disease—I. Utilization of the life table method in obtaining clinically useful information. *J. Chron. Dis.* **30**, 529–541.

Elandt-Johnson, R. C. (1973). "Age-at-onset" distributions in chronic diseases. A life table approach to analysis of family data. *J. Chron. Dis.* **26**, 529–545.

Elandt-Johnson, R. C. (1980). Onset distributions of age dependent zero-one events. Estimation from population data *Scand. Actuar. J.*

Falconer, D. S., Duncan, L. J. P. and Smith, Ch. (1971). A statistical and genetical study of diabetes. *Ann. Hum. Genet. London* **34**, 347.

Freeman, D. H., Freeman, J. L., and Koch, G. G. (1974). A modified $\chi^2$ approach for fitting Weibull models to synthetic life tables. *Institute of Statistics Mimeo Series No. 958*, Department of Biostatistics, University of North Carolina, Chapel Hill, N.C.

Preston, S. H., Keyfitz, N., and Schoen, R. (1972). *Causes of Death: Life Tables for National Populations*. Seminar Press, New York.

Tallis, G. M., Sarfaty G., and Leppard, P. (1973). The use of a probability model for the construction of age specific life tables for women with breast cancer, University of Adelaide and Cancer Institute, Melbourne, Australia.

Third National Survey: Incidence Data (1975). Vol. I, Number 1, DHEW Publication No. (HRA) 74-1150. U.S. Government Printing Office, Washington, D.C.

## EXERCISES

**14.1.** Using the data in columns (1)–(3) of Table 14.2a, but with the U.S. White Males 1969-71 life table for ${}_5L_x$, estimate

(*a*) the incidence onset distribution;

(*b*) the waiting time onset distribution, over the range of ages for which this is feasible.

**14.2.** Explain why, although the waiting time onset distribution does not depend on mortality, it might change if mortality from a specified disease (different from that to which the onset refers) were reduced or eliminated. Your arguments should point out that the probability of onset within a given time might vary with the constitution of the population exposed to risk. Algebraic expressions representing the various probabilities should be used. Give assumptions which could ensure that the waiting time onset distributions would not change in the circumstances described.

**14.3.** Modify the method described in Section 14.6.2 to apply it to situations where data are recorded in 10 year age groups, as in Table 14.3. Apply your method to the data in Table 14.3 supposing that

$$\frac{{}_{10}L_y \cdot {}_{10}L'_x}{{}_{10}L'_y \cdot {}_{10}L_x} = e^{\alpha(y-x)} \qquad \text{with } \alpha = 0.0025.$$

Show that this assumption corresponds *roughly* to the addition of a constant to the force of mortality for affected as compared with unaffected persons.

**14.4.** Suppose that you are given data on the dates of onset for two diseases, $D_1$, $D_2$, among all individuals in a current population. You are asked to consider the conditional waiting time onset distribution for $D_2$, given that onset $D_1$ has occurred. Describe how a select life table form might be suitable for this distribution, and give directions for the calculations needed to construct such a table, assuming the data available are sufficient for the purpose.

**14.5.** Show that if the joint PDF of time due to die ($T$) and time due for onset ($U$) is $g(t,u)$ then the joint distribution of the observable random variables $T$, and $U$ (provided $U<T$) remains unchanged if $g(t,u)$ is replaced by

$$g^*(t,u)=\begin{cases} g(t,u) & \text{if } u \leqslant t, \\ \phi g(t,t+\phi(u-t)) & \text{if } u>t, \end{cases}$$

where $\phi>0$, but not, in general, if it is replaced by

$$g^{**}(t,u)=\begin{cases} g(t,u) & \text{if } u \geqslant t, \\ \phi g(u+\phi(t-u),u) & \text{if } u<t. \end{cases}$$

CHAPTER 15

# Models of Aging and Chronic Diseases

## 15.1 INTRODUCTION

So far our main concern has been mortality (or survivorship), failure (or survival) distributions, and their estimation. From time to time we have tried to fit parametric models to such data—Gompertz, Weibull, exponential are those most commonly used. They were often used for mathematical convenience, without attempting to establish biological bases for the models themselves.

Causes of death, though recognized as consequences of certain diseases, were not analyzed from that viewpoint. Simply, potential "times due to die" were arbitrarily introduced in the theory of competing causes and the observed time at death was regarded as the minimum of these potential times. Clearly, it would be more appropriate to discuss the problem in terms of disease processes and/or failures of certain parts or components. Unfortunately, such models are highly speculative in the present state of knowledge of the biological processes involved.

Nevertheless, we present in this chapter a few models of disease development in terms of multistage failures of some fairly simple systems. Sometimes, one might think of normal and damaged cells as corresponding to healthy and diseased biological entities, instead of regarding individuals as observational units.

It would be oversanguine to infer from the title of this chapter that the topics will really be adequately covered. This would need a complete book rather than a single chapter. We only outline some problems that seem to present features of general interest:

1. Stochastic modeling of disease processes in normal cells, with special reference to cancer (Sections 15.2–15.4).
2. Multistage models of illness-death processes introducing parallel and series type systems, and combinations of these (Sections 15.5–15.7).

## 15.2 AGING AND CHRONIC DISEASES

### 15.2.1 Biological Aging

Aging processes of living organisms are not fully understood as yet. It is, therefore, not possible to give *the* definition of aging, though gerontologists attempt to describe it at various organizational levels of an organism. We may, for example, speak about aging of organs, tissues, and cells. It is usually postulated that the aging process is a result of *changes* in cell functions. The basic role of a cell is to maintain itself and take part in maintaining the whole organism in equilibrium (homeostasis). Accumulation of errors in transmission of genetic information and reduction in mitotic cell division, which eventually leads to cell death, seem to be basic features of aging processes. The organism becomes less fit and eventually dies. For more details see Finch and Hayflick (1977).

### 15.2.2 Hazard Rate: A Measure of Aging

It is not our aim to review the vast literature on aging. But since aging leads eventually to death, mortality models reflect—to a certain extent—the aging process, though they are not identical with aging models. In particular, the *hazard rate function*, which is a relative rate of change (in mass probability) at a point of time (see Section 2.3.2) is appropriate in describing aging processes.

We recall that if the hazard rate is constant, $\lambda$, say, then the survival function is exponential, $\Pr\{X>x\}=e^{-\lambda x}$, $x>0$, and the conditional distribution $\Pr\{X>x+t|X>x\}=e^{-\lambda t}$, $t>0$ is also exponential. Therefore, a system that starts life at time 0, and has life expectancy $\mathcal{E}(X)=1/\lambda$, has the same life expectancy at time $x$ ($>0$) given it has survived till $x$. In other words: when the hazard rate of a survival distribution is constant, there is no aging. Aging processes can be conveniently measured in terms of increasing hazard rates.

Therefore, mathematical models of hazard rate functions, which express some changes in biological age dependent processes, might be appropriate for models of aging.

### 15.2.3 Models of Chronic Diseases

Chronic diseases, such as cardiovascular diseases, cancer, diabetes, and so on, are associated with, but not identical to, aging. Old (aged) individuals do not always suffer from a definite illness, though their health is usually impaired. On the other hand, very young persons may develop severe illnesses (e.g., cancer, heart disease) and die soon afterward. Some authors believe that such individuals may already have an *inborn fault* that eventually causes death.

Biological theories of chronic diseases are essentially based on assumptions similar to those of theories of aging: *changes in activity* of normal cells or changes in the *structure* of normal cells are the subjects of these basic assumptions.

In the next section, we present some models developed for carcinogenesis, which lead to age of onset distributions of well known forms—Gompertz and Weibull. As we already know, these distributions are commonly used in fitting mortality data (Chapter 7).

## 15.3 SOME MODELS OF CARCINOGENESIS

There are many biological hypotheses on initiation and development of various types of cancer. Essentially, these can be divided into two broad classes: those postulating that cancer cells are *normal cells* that have lost control of mitotic activity and divide indefinitely; and those assuming that cancer cells are *abnormal cells*, caused by mutation in one or more cell genes (somatic mutation), that also undergo infinite division.

The fact that a body (somatic) cell becomes a cancer cell requires that there must be a mechanism which *initiates* the process; it is usually assumed that a carcinogenic agent, radiation, or virus starts this process.

*One-hit* models assume that it is sufficient to change one type of cells or a single site in a cell, while *multi-hit* models require more than one distinct change to occur, to initiate the disease. Examples of such models will now be presented.

### 15.3.1 A One-Hit Model of Carcinogenesis: Gompertz Distribution

Arley (1961) and Iversen and Arley (1950, 1952, and some other papers) developed a model of cancer in which a certain number of the same type of somatic cells undergo "irreversible change in the genetic apparatus" due to a carcinogenic agent. The time to appearance of a tumor is supposed to be the sum of an "excitation time," $T$ say, at which there is a reaction between a cell and the agent, and a "growth time," $T'$ say, during which the affected cell gives rise to a group of cells of sufficient size to be observed. Here we are concerned with the distribution of the excitation time, $T$.

The model assumes that during the course of the interaction between the carcinogenic agent and cell constituents, the following events can occur: the defense mechanisms eliminate part of the agent (at rate $\alpha_e$), or some cells adapt the agent (at rate $\alpha_a$) but some others undergo multiplication (at rate $\alpha_g$). To initiate the disease, we must have $\alpha_g < \alpha_e + \alpha_a$. According to the standard theory of chemical reactions, the rate of cancer growth is

$$\lambda(t) = \lambda_0 e^{\alpha t}, \qquad t > 0, \qquad \lambda_0 > 0, \alpha > 0, \tag{15.1}$$

where $\alpha = \alpha_g - \alpha_e - \alpha_a$, and $\lambda_0$ is a constant depending on concentrations of various cell constituents, toxicity of the agent, and so on.

The probability of initiation of cancer is then

$$\Pr\{T \leqslant t\} = 1 - \exp\left[\frac{\lambda_0}{\alpha}(1 - e^{\alpha t})\right], \qquad t > 0. \tag{15.2}$$

This is the well known *Gompertz distribution* (see Section 3.10). It is commonly used as a model of mortality.

It will be appreciated that the arguments leading to (15.1) involve a number of assumptions that might be difficult to verify experimentally. Nevertheless, knowledge of the ways this family of distributions might arise—however speculative—is useful in survival analysis.

### 15.3.2 Multi-Hit Models of Carcinogenesis: Parallel Systems and Weibull Distribution

A number of models of carcinogenesis—or more generally of development of a chronic disease—were constructed, assuming that several different changes in body cells, so called *somatic mutations* of genes located in these cells, are necessary to initiate a disease. Suppose that, in order to initiate the disease, $r$ distinct (i.e., each associated with a different genetic locus) mutations generate a "forbidden clone" (*clone* is the asexual progeny of a single cell) whose growth leads to initiation of a disease [see e.g., Burch (1966)]. In other words, a cell fails (i.e., the disease is initiated) if $r$ components fail (i.e., $r$ different mutations in a cell occur). This is a *parallel system* as discussed in Section 3.11.2.

Let $\omega_j$ be the mutation rate at $j$th locus per cell, per unit time (e.g., per year). We assume that

$$\Pr\left\{\begin{array}{l}\text{mutation at } j\text{th locus occurs}\\ \text{in a cell before time } t\end{array}\right\}=\omega_j t, \tag{15.3}$$

provided that $\omega_j$ is so small that $\omega_j t$ is also small. Then, for each cell,

$$\Pr\left\{\begin{array}{l}\text{required } r \text{ mutations occur}\\ \text{in a cell before time } t\end{array}\right\}=\left(\prod_{j=1}^{r}\omega_j\right)t^r. \tag{15.4}$$

Suppose that the number of cells at risk in an individual is $L$. Then the average number of clones developed from mutated cells up to time $t$ is

$$\xi = L\cdot\left(\prod_{j=1}^{r}\omega_j\right)t^r = Bt^r. \tag{15.5}$$

It is further assumed that the number of forbidden clones at time $t$ has approximately a Poisson distribution with mean $\xi$ given by (15.5). This implies that the CDF of the waiting time to the first occurrence of the disease (age of onset distribution of the disease) is approximately

$$F_T(t)=1-\exp(-Bt^r), \qquad t>0, \qquad B>0. \tag{15.6}$$

This is, of course, a *Weibull distribution* (see Section 3.10). Its hazard rate is

$$\lambda_T(t)=rBt^{r-1}. \tag{15.7}$$

Weibull distributions as models for initiation of cancer have been derived by Nordling (1953), Armitage and Doll (1961), Burch (1966), and many others. Our presentation using the concept of forbidden clones is similar to that given by Burch (1966).

Distribution (15.6) was obtained by Fisher and Tippett (1928) as a limiting distribution (Type 2) of extreme values. Weibull (1951) obtained it as a model of failure of a mechanical device composed of several parts, and assuming changes in the strength of material of various parts. This distribution is commonly used in the theory of reliability.

## 15.4 SOME "MOSAIC" MODELS OF A CHRONIC DISEASE

Burch (1966) also considered some more complicated models. He argued that, although the cells at risk seem to be histologically homogeneous, they may, in fact, consist of a mosaic of different cells, each cell in the mosaic having a characteristic specificity. This implies that each type of cell may develop its own set of forbidden clones.

Suppose that a simple somatic mutation is necessary to develop a forbidden clone, and that $s$ distinct clones are required to initiate a disease.

Let $L_i$ denote the number of cells at risk of developing the $i$th specific type of clone. If $\alpha$ is the average rate of mutation per cell per time unit, then the expected number of clones of $i$th type developed up to time $t$ is

$$\xi_i = L_i \alpha t = \lambda_i t, \qquad i = 1, 2, \ldots, s.$$

Assume again a Poisson process for formation of forbidden clones. Let $T_i$ be the waiting time for initiation of the $i$th type of clones. Then

$$F_{T_i}(t) = 1 - \exp(-\lambda_i t) \tag{15.8}$$

represents the CDF of time of initiation of $i$th type of clones. Let

$$T = \max(T_1, T_2, \ldots, T_s). \tag{15.9}$$

Initiation of a disease takes place later than time $t$ if $T$ exceeds $t$.

Assuming that $T_1, T_2, \ldots, T_s$ are mutually independent, the function

$$F_T(t) = \Pr\{T \leqslant t\} = \prod_{i=1}^{s} (1 - e^{-\lambda_i t}) \tag{15.10}$$

represents the CDF of the waiting time for initiation of a disease with mosaic pattern.

In particular, when

$$\lambda_1 = \lambda_2 = \cdots = \lambda_s = \lambda,$$

formula (15.10) takes the form

$$F_T(t) = (1 - e^{-\lambda t})^s. \tag{15.11}$$

Burch (1966) suggests that, perhaps, *lupus erythematosus* might obey the mosaic model (15.11), with $s = 3$, [see also Burch and Rowell (1965)].

Assuming that $r$ mutations in each somatic cell *and* $s$ types of distinct forbidden clones are required for initiation of a disease, the distribution (15.11) can be generalized to the form

$$F_T(t) = [1 - \exp(-\lambda t^r)]^s. \tag{15.12}$$

## 15.5 "FATAL SHOCK" MODELS OF FAILURE

### 15.5.1 A Two Component Series System

Marshall and Olkin (1967) considered the following simple model of failure of a mechanical device.

Suppose that a device consists of two components, $\Gamma_1$ and $\Gamma_2$, say. Suppose that the device fails if either of the two components fails. This is a *series system* (see Section 3.11.1). The authors, however, assume that failure depends on external "shocks," which are always fatal for the device, and from these shocks either (1) $\Gamma_1$ alone fails; or (2) $\Gamma_2$ alone fails; or (3) $\Gamma_1$ and $\Gamma_2$ fail simultaneously. This may imply that there are three different kinds of external shocks, each causing failure of the device.

It was further assumed that each failure is governed by a Poisson process with intensity (hazard) rates $\lambda_{11}$, $\lambda_{22}$, and $\lambda_{12}$, respectively, and that the processes are independent.

Let $T_1$, $T_2$ be the "waiting times to failure" of $\Gamma_1$ and $\Gamma_2$, respectively. Then the joint SDF is

$$S_{T_1,T_2}(t_1,t_2)=\Pr\{T_1>t_1,T_2>t_2\}$$
$$=\exp\{-[\lambda_{11}t_1+\lambda_{12}t_2+\lambda_{12}\max(t_1,t_2)]\}. \tag{15.13}$$

Clearly, we have a problem of competing risks, in which not only $T_1$ and $T_2$ are *stochastically dependent* but also the events that the failure of a device is caused by failure of $\Gamma_1$ or $\Gamma_2$ are not mutually exclusive (assumption Al, Section 9.1). Let

$$T=\min(T_1,T_2). \tag{15.14}$$

Then the overall survival function is

$$S_T(t)=\Pr\{\min(T_1,T_2)>t\}=S_{T_1,T_2}(t,t)$$
$$=\exp[-(\lambda_{11}+\lambda_{22}+\lambda_{12})t]=\exp(-\lambda t), \tag{15.15}$$

where

$$\lambda=\lambda_{11}+\lambda_{22}+\lambda_{12}. \tag{15.16}$$

The crude hazard rate of failure caused by failure of component $\Gamma_1$ is

$$h_1(t)=-\left.\frac{\partial S_{T_1,T_2}(t_1,t_2)}{\partial t_1}\right|_{t_1=t_2=t}=\lambda_{11}+\lambda_{12} \tag{15.17}$$

(see Section 9.4.2). Thus the probability that the device fails before time $t$, due to failure of $\Gamma_1$, is

$$Q_1^*(t) = \int_0^t h_1(u) S_T(u)\, du = \int_0^t (\lambda_{11} + \lambda_{12}) e^{-\lambda u}\, du$$

$$= \frac{\lambda_{11} + \lambda_{12}}{\lambda}(1 - e^{-\lambda t}), \qquad (15.18)$$

and the proportion of failures of $\Gamma_1$ is

$$\pi_1 = \frac{\lambda_{11} + \lambda_{12}}{\lambda} \qquad (15.19)$$

(compare Section 9.4.3). In a similar way, we obtain the "crude" probability of failure before time $t$ due to failure of $\Gamma_2$,

$$Q_2^*(t) = \frac{\lambda_{22} + \lambda_{12}}{\lambda}(1 - e^{-\lambda t}), \qquad (15.20)$$

and proportion of failures of $\Gamma_2$ is

$$\pi_2 = \frac{\lambda_{22} + \lambda_{12}}{\lambda}. \qquad (15.21)$$

Note that

$$h_T(t) = \lambda_{11} + \lambda_{22} + \lambda_{12} = \lambda, \qquad h_1(t) = \lambda_{11} + \lambda_{12}, \qquad \text{and} \qquad h_2(t) = \lambda_{22} + \lambda_{12},$$

so that

$$h_T(t) \leqslant h_1(t) + h_2(t), \qquad (15.22)$$

and

$$1 - S_T(t) = F_T(t) \leqslant Q_1^*(t) + Q_2^*(t). \qquad (15.23)$$

Further, we notice that the (conditional) "survival" functions for $\Gamma_1$ and $\Gamma_2$, respectively, in the presence of each other are

$$S_1^*(t) = \frac{Q_1^*(\infty) - Q_1^*(t)}{Q_1^*(\infty)} = e^{-\lambda t}, \qquad (15.24a)$$

and

$$S_2^*(t) = \frac{Q_2^*(\infty) - Q_2^*(t)}{Q_2^*(\infty)} = e^{-\lambda t}. \qquad (15.24b)$$

These are identical with $S_T(t)$. This is because

$$\frac{h_1(t)}{h_T(t)}=\frac{\lambda_{11}+\lambda_{12}}{\lambda}=c_1 \qquad \text{and} \qquad \frac{h_2(t)}{h_T(t)}=\frac{\lambda_{22}+\lambda_{12}}{\lambda}=c_2$$

(though $c_1+c_2>1$), that is, the crude hazard-rates are *proportional*.

In Section 9.7, we proved that this result holds, whenever the events "failure is from cause $C_\alpha$, $\alpha=1,2,\ldots,k$," are mutually exclusive events (assumption A1), and hazard rates are *proportional*. It is clear from this example, and can be shown more generally (if neither A1 nor A2 nor A3 holds) that the proportionality assumption of crude hazard rates is that which is relevant for the result $S_\alpha^*(t)\equiv S_T(t)$ to hold. Of course, using similar kinds of argument, one can derive the results of this paragraph for marginal waiting time distributions $F_{T_1}(t)$, $F_{T_2}(t)$, which are not necessarily exponential (see Exercise 15.3).

Since the external "shock" is, in fact, the immediate cause of failure, one can consider three rather than two distinct causes of failure:

$C_{11}$: failure of component $\Gamma_1$ *alone*
$C_{22}$: failure of component $\Gamma_2$ *alone*
$C_{12}$: failure of components $\Gamma_1$ *and* $\Gamma_2$ simultaneously.

[see David and Moeschberger (1972)]. Now the failures from causes $C_{11}$, $C_{22}$, and $C_{12}$ are mutually exclusive events. Introducing hypothetical "times due to fail," $X_1$, $X_2$, $X_3$ from $C_{11}$, $C_{22}$, and $C_{12}$, respectively, and using the approach of Section 9.4, one can obtain the same result for $S_T(t)$, and calculate the crude probability functions in a straightforward way. We obtain

$$Q_{11}^*(t)=\frac{\lambda_{11}}{\lambda}(1-e^{-\lambda t}), \qquad Q_{22}^*(t)=\frac{\lambda_{22}}{\lambda}(1-e^{-\lambda t}),$$

$$\text{and} \qquad Q_{12}^*(t)=\frac{\lambda_{12}}{\lambda}(1-e^{-\lambda t}). \tag{15.25}$$

Clearly,

$$Q_1^*(t)=Q_{11}^*(t)+Q_{12}^*(t), \quad \text{and} \quad Q_2^*(t)=Q_{22}^*(t)+Q_{12}^*(t). \tag{15.26}$$

[For more detail, see David and Moeschberger (1972).]

### 15.5.2 Generalization of the "Fatal Shock" Model to *n* Components

Lee and Thompson (1974) generalized Marshall and Olkin's (1967) model assuming that a device is a series system of $n$ components and there are

$2^n - 1$ possible kinds of "shocks" corresponding to failure of each component alone; any pair of components, any triplet of components, and so on. Considering these as $(2^n - 1)$ mutually exclusive events, and applying the results of Section 9.4, one can immediately generalize the results discussed in Section 15.5.1 for $n > 2$.

Manton et al. (1976) adapted this model in studying the simultaneous occurrence of various diseases (or injuries) as they are listed on death certificates. The concept of "fatal shock" is replaced, in their paper, by the concept of a "disease pattern." It is assumed that there are $n$ possible faults and $(2^n - 1)$ possible "defects," which play roles of internal shocks. Each individual is supposed to be endowed from birth with one of these defects. Each defect is essentially represented by a set of diseases listed on the individual's death certificate, without taking into account their order of occurrence (i.e., without distinguishing whether a disease is an underlying, immediate, or contributory cause). A death certificate describes a "disease pattern" (defect) for a given person. Using the death certificate data for U.S. population 1967, the authors distinguish four major disease patterns as causes of death and try to predict how the expectancy of life would be changed if one (or more) of these patterns were to be "eliminated."

## 15.6 IRREVERSIBLE MARKOV PROCESSES IN ILLNESS-DEATH MODELING

### 15.6.1 Basic Concepts

So far we have considered models of chronic diseases and mortality which require failure of at least one component for initiation of a disease or for death. In those models, the order of occurrence of the failures was immaterial. Actually, we often observe progress of a disease through several successive *states* before death occurs.

In constructing models of mortality we may wish to take into account the whole process of a disease (which is the cause of death) including development, being cured, relapsing again, progressing, and finally leading to death.

Construction of models of this kind uses techniques appropriate to *stochastic processes*. We restrict ourselves here to stochastic processes in which the various recognizable states in the illness-death process are *discrete* (with changes occurring instantaneously). Further, we assume that the conditional probability, given that a system (an organism) is in state $(i)$, that it will next move to state $(j)$ does not depend upon which state(s) the system was in previous to entering state $(i)$. This is called the *Markov property*, and the stochastic process is called a *Markov process*.

Given that the system leaves state $(i)$ at time $t$, the probability $p_{ij}(t)$ that it moves to state $(j)$ is called a *transition probability*. Denoting by $\lambda_{ij}(t)$ the hazard rate for transition from state $(i)$ to state $(j)$ at time $t$ (often called *transition intensity*), we have

$$p_{ij}(t)=\frac{\lambda_{ij}(t)}{\sum_{k\neq i}\lambda_{ik}(t)}, \qquad (15.27)$$

[$k$ is summed over all states, *except* $(i)$]. (See Exercise 15.4.)

### 15.6.2 A Two Component Parallel System: Two Distinct States Before Failure

For convenience, we use the standard terminology (e.g., component, system), which is more appropriate in reliability theory. However, in biological situations, "system" can be replaced by "organism" or "set of organs, tissues, cells"; "component" can be replaced by a "specific organ, tissue cell, genetic locus," and so on.

Consider a system consisting of two components, $\Gamma_1$ and $\Gamma_2$. Let $T_1$ and $T_2$ be the lifetimes of $\Gamma_1$ and $\Gamma_2$, respectively, and assume that $T_1$ and $T_2$ are *independent* and *identically distributed.*

Suppose that the system only fails if *both* components fail. Suppose however, that when one component fails, the hazard rate for the other component changes (e.g., increases). A problem of this kind was considered by Gross et al. (1972). The two kidneys represent a two component system; death occurs when both kidneys have failed. It is assumed that $T_1, T_2$, each has an exponential distribution with parameter $\lambda_0$, say. When one kidney fails, the hazard rate for the other kidney changes to $\lambda_1$ $(\lambda_1>\lambda_0)$.

To derive the survival function of the system, we distinguish *two* states of the system, while it is functioning:

state $S_0$: both components are functioning

state $S_1$: only one component is functioning.

Let $F_i(t)$, $S_i(t)$, $f_i(t)$, and $\lambda_i(t)$ denote the CDF, SDF, PDF, and HF of lifetime of each (surviving) component in state $S_i$, $i=1,2$.

We have

$$\Pr\left\{\begin{array}{l}\text{system is in state } S_0\\ \text{at time } t\end{array}\right\} = [S_0(t)]^2 = P_0(t); \qquad (15.28)$$

$$\Pr\left\{\begin{array}{l}\text{system moves from state } S_0 \text{ to } S_1\\ \text{in time interval } \tau \text{ to } \tau + d\tau\end{array}\right\} = P_0(\tau)\lambda_0(\tau)\,d\tau; \qquad (15.28\text{a})$$

$$\Pr\left\{\begin{array}{l}\text{system is functioning at time } t\\ \text{and it is in state } S_1\end{array}\right\} = 2\int_0^t P_0(\tau)\lambda_0(\tau)S_1(t-\tau)\,d\tau$$

$$= 2\int_0^t \lambda_0(\tau)[S_0(\tau)]^2 S_1(t-\tau)\,d\tau = P_1(t). \qquad (15.28\text{b})$$

Let $T$ denote the lifetime of the system. Then the survival function of the system is

$$S_T(t) = P_0(t) + P_1(t). \qquad (15.29)$$

**Example 15.1** Let $S_i(t) = \exp(-\lambda_i t)$, $i = 1,2$. $(\lambda_0 < \lambda_1)$. We then have

$$P_0(t) = \exp(-2\lambda_0 t).$$

$$P_1(t) = 2\int_0^t \lambda_0 \exp(-2\lambda_0\tau)\exp[-\lambda_1(t-\tau)]\,d\tau$$

$$= 2\lambda_0 \exp(-\lambda_1 t)\int_0^t \exp[-(2\lambda_0 - \lambda_1)\tau]\,d\tau$$

$$= \frac{2\lambda_0}{\lambda_1 - 2\lambda_0}[\exp(-2\lambda_0 t) - \exp(-\lambda_1 t)],$$

provided $\lambda_1 \neq 2\lambda_0$.

Hence, the SDF is

$$S_T(t) = P_0(t) + P_1(t) = \frac{1}{\lambda_1 - 2\lambda_0}[\lambda_1 \exp(-2\lambda_0 t) - 2\lambda_0 \exp(-\lambda_1 t)], \qquad \lambda_1 \neq 2\lambda_0.$$

When $\lambda_1 = 2\lambda_0 = 2\lambda$, we have

$$P_1(t) = 2\lambda e^{-2\lambda t}\int_0^t d\tau = 2te^{-2\lambda t},$$

and

$$S_T(t) = e^{-2\lambda t}(1 + 2\lambda t).$$

The PDF is

$$f_T(t) = -\frac{dS_T(t)}{dt} = 2\lambda te^{-2\lambda t}(1+2\lambda t) - 2\lambda e^{-2\lambda t}$$

$$= 2\lambda te^{-2\lambda t}\cdot 2\lambda t = (2\lambda t)^2 te^{-2\lambda t}.$$

### 15.6.3 Extension to *r*-Component Parallel System with *r* States

It is left to the reader to show that if there are $r$ components, $\Gamma_1, \ldots, \Gamma_r$, with independent lifetimes, $T_1, \ldots, T_r$, each having hazard rate $\lambda_i(t)$ at time $t$ if the system is in state $S_i$—that is with exactly $(r-i)$ components functioning—we have

$$\left.\begin{aligned}
P_0(t) &= [S_0(t)]^r \\
P_1(t) &= r\int_0^t \lambda_0(\tau)P_0(\tau)[S_1(t-\tau)]^{r-1}\,d\tau \\
P_2(t) &= (r-1)\int_0^t \lambda_1(\tau)P_1(\tau)[S_2(t-\tau)]^{r-2}\,d\tau \\
&\cdots\cdots\cdots\cdots\cdots\cdots\cdots\cdots \\
&\cdots\cdots\cdots\cdots\cdots\cdots\cdots\cdots \\
P_{r-1}(t) &= 2\int_0^t \lambda_{r-1}(\tau)P_{r-1}(t-\tau)\,d\tau.
\end{aligned}\right\} \tag{15.30}$$

The survival function is

$$S_T(t) = \sum_{i=0}^{r-1} P_i(t). \tag{15.31}$$

### 15.6.4 A Model of Disease Progression

In clinical investigations of chronic diseases, various states in the progression of disease might be observed. For example, in studying mortality of patients with advanced cancer, one may have additional information on the size of the tumor. The model presented in this section was discussed by Lagakos (1976), though under more restrictive assumptions.

State $S_0$ represents the presence of cancer, State $S_1$ a growth of tumor (called by the author a *progression*), $S_2$ death (this is also the state in which the system remains forever, the so called *absorbing* state). Schematically, the model can be represented as in Fig. 15.1.

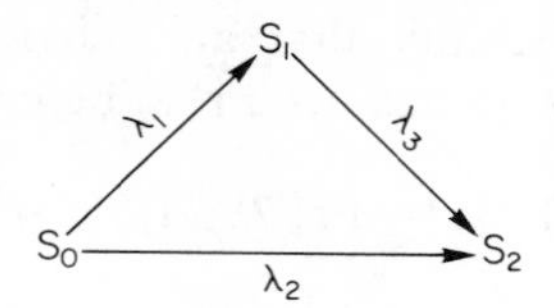


**Fig. 15.1**

Let $T_1$ denote "time due for progression" of the disease (further tumor growth) and $T_2$ "time due to die" without progression. The two events, "death" and "progression," may be considered at this stage as two *competing events*: death without progression is observed if death occurs first; or progression is observed if it occurs before death.

Let

$$S_{12}(t_1,t_2)=\Pr\{T_1>t_1,T_2>t_2\} \tag{15.32}$$

be the joint survival function of $T_1$, $T_2$, and let

$$h_i(t)=-\left.\frac{\partial \log S_{12}(t_1,t_2)}{\partial t_i}\right|_{t_1=t_2=t}, \qquad i=1,2, \tag{15.33}$$

be the crude hazard rates (see Section 9.4).

We wish to evaluate the probabilities of reaching each state. We introduce the following random variables

$$I=\begin{cases} 0 & \text{if } T_2\leqslant T_1 \text{ (death occurs first)}, \\ 1 & \text{if } T_1<T_2 \text{ (progression occurs first)}. \end{cases} \tag{15.34}$$

$$Y=\min(T_1,T_2) \tag{15.35}$$

or, equivalently,

$$Y=\begin{cases} T_2 & \text{if } T_2\leqslant T_1, \\ T_1 & \text{if } T_1<T_2. \end{cases} \tag{15.35a}$$

Let $T_3$ be the time elapsing to death after progression (i.e., after reaching state $S_1$), and let $T$ be time at death, that is,

$$T=\begin{cases} T_2 & \text{if } T_2\leqslant T_1, \\ T_1+T_3 & \text{if } T_1<T_2. \end{cases} \tag{15.36}$$

*At time t*, an individual can be in any of the states shown in Fig. 15.1.

1. *State* $S_0$. The probability that an individual is in state $S_0$ (i.e., neither death nor progression has occurred) before or at time $t$ is

$$S_Y(t) = \Pr\{Y > t\} = \Pr\{T_1 > t, T_2 > t\} = S_{12}(t, t) \tag{15.37}$$

[compare (9.4)].

2. *State* $S_1$. The probability that an individual has reached state $S_1$ before time $t$ and/or still is in state $S_1$, is

$$\Pr\{(T_1 \leqslant t) \cap (I = 1)\} = Q_1^*(t) = \int_0^t h_1(u) S_Y(u)\, du, \tag{15.38}$$

and

$$\Pr\{I = 1\} = \Pr\{T_1 < T_2\} = Q_1^*(\infty), \tag{15.38a}$$

so that

$$\Pr\{T_1 \leqslant t | T_1 < T_2\} = \frac{1}{Q_1^*(\infty)} \int_0^t h_1(u) S_Y(u)\, du, \tag{15.39}$$

and the conditional PDF of $T_1$ given $T_1 < T_2$ is

$$f_{T_1}^*(t | T_1 < T_2) = \frac{1}{Q_1^*(\infty)} h_1(t) S_Y(t). \tag{15.39a}$$

(See Section 9.4.3.)

3. *State* $S_2$ *without progression.* In method similar to that used for state $S_1$, we find the probability that an individual has reached state $S_2$ without progression before time $t$ is

$$Q_2^*(t) = \int_0^t h_2(u) S_Y(u)\, du, \tag{15.40}$$

and the conditional PDF of $T_2$ given $T_2 \leqslant T_1$ is

$$f_{T_2}^*(t | T_2 \leqslant T_1) = \frac{1}{Q_2^*(\infty)} h_2(t) S_Y(t)$$

$$= f_T(t | T_2 \leqslant T_1). \tag{15.40a}$$

4. *State* $S_2$ *and progression.* We are interested in

$$\Pr\{T \leqslant t | T_1 < T_2\} = \Pr\{T_1 + T_3 \leqslant t | T_1 < T_2\}.$$

Assume that $T_3$ is independent of $T_1$ (and of course, of $T_2$), and has the PDF $f_{T_3}(t_3)$. There is no great difficulty in generalizing the model by assuming that $T_3$ depends on $T_1$, that is, has the PDF $f_{T_3}(t_3|T_1=t_1)$. (See Exercise 15.7.)

The (conditional) joint PDF of $T_1$ and $T_3$, given $T_1<T_2$, is

$$f_{T_1T_3}(t_1,t_3|T_1<T_2)=f^*_{T_1}(t_1|T_1<T_2)\cdot f_{T_3}(t_3). \tag{15.41}$$

Thus the (conditional) PDF of $T$ given $T_1<T_2$ is

$$f_T(t|T_1<T_2)=\int_0^t f_{T_1T_3}(t_1,t-t_1|T_1<T_2)\,dt_1, \tag{15.42}$$

so that the probability of reaching state $S_2$ with progression occurring before time $t$, is

$$\Pr\{T\leqslant t|T_1<T_2\}\cdot\Pr\{T_1<T_2\}=\Pr\{I=1\}\int_0^t f_t(t|T_1<T_2)\,dt. \tag{15.43}$$

5. *Survival.* The PDF of time at death, $T$, defined in (15.36) is

$$f_T(t)=f_T(t|T_2\leqslant T_1)\cdot\Pr\{T_2\leqslant T_1\}+f_T(t|T_1<T_2)\cdot\Pr\{T_1<T_2\}, \tag{15.44}$$

and the overall survival function is

$$S_T(t)=\Pr\{T>t\}=\int_t^\infty f_T(u)\,du. \tag{15.45}$$

Note that $f_T(t)$ [and so $S_T(t)$] is a *mixture* of the two waiting time PDF's for reaching state $S_2$ in two ways.

The concept of disease progression can be elaborated by introducing more than one state of progression, leading to more complicated models.

**Example 15.2** Lagakos (1976) considered a model in which $T_1$, $T_2$, $T_3$ are independent exponential variates with parameters $\lambda_1$, $\lambda_2$, $\lambda_3$, respectively.

We have the joint SDF

$$S_{12}(t_1,t_2)=\exp[-(\lambda_1t_1+\lambda_2t_2)]$$

and the crude hazard rates

$$h_1(t)=\lambda_1, \qquad h_2(t)=\lambda_2.$$

Note that the hazard rates are proportional.

*State $S_0$.*

$$S_Y(t)=S_{12}(t,t)=\exp[-(\lambda_1+\lambda_2)t],$$

$$f_Y(t)=(\lambda_1+\lambda_2)\exp[-(\lambda_1+\lambda_2)t].$$

*State $S_1$.*

$$\Pr\{(Y\leqslant t)\cap(I=1)\}=\frac{\lambda_1}{\lambda_1+\lambda_2}\exp[-(\lambda_1+\lambda_2)t],$$

$$\Pr\{I=1\}=\Pr\{T_1<T_2)=\frac{\lambda_1}{\lambda_1+\lambda_2},$$

so that

$$f^*_{T_1}(t|T_1<T_2)=(\lambda_1+\lambda_2)\exp[-(\lambda+\lambda_2)t]$$

(the same as $f_Y(t)$—explain why?).

*State $S_2$ without progression.*

$$\Pr\{(Y\leqslant t)\cap(I=0)\}=\frac{\lambda_2}{\lambda_1+\lambda_2}\exp[-(\lambda_1+\lambda_2)t],$$

$$\Pr\{I=0\}=\Pr\{T_2\leqslant T_1\}=\frac{\lambda_2}{\lambda_1+\lambda_2},$$

so that

$$f^*_{T_2}(t|T_2\leqslant T_1)=af_T(t|T_2\leqslant T_1)=(\lambda_1+\lambda_2)\exp[-(\lambda_1+\lambda_2)t]$$

(the same as $f_Y(t)$—explain why ?).

*State $S_2$ and progression.* We have

$$f_{T_1T_3}(t_1,t_3|T_1<T_2)=(\lambda_1+\lambda_2)\exp[-(\lambda_1+\lambda_2)t_1]\times\lambda_3\exp(-\lambda_3t_3).$$

Hence

$$f_T(t|T_1<T_2)=\lambda_3(\lambda_1+\lambda_2)\int_0^t\exp[-\lambda_3t_3-(\lambda_1+\lambda_2)(t-t_3)\,dt_3$$

$$=\frac{\lambda_3(\lambda_1+\lambda_2)}{\lambda_1+\lambda_2-\lambda_3}[e^{-\lambda_3t}-e^{-(\lambda_1+\lambda_2)t}],$$

and so

$$\Pr\{T \leqslant t | T_1 < T_2\} = \int_0^t f_T(t|T_1 < T_2)\,dt$$

$$= 1 - \frac{\lambda_1+\lambda_2}{\lambda_1+\lambda_2-\lambda_3}\exp(-\lambda_3 t)$$

$$+ \frac{\lambda_3}{\lambda_1+\lambda_2-\lambda_3}\exp[-(\lambda_1+\lambda_2)t].$$

*Survival.*

$$f_T(t) = \frac{\lambda_2}{\lambda_1+\lambda_2}(\lambda_1+\lambda_2)\exp[-(\lambda_1+\lambda_2)t]$$

$$+ \frac{\lambda_2}{\lambda_1+\lambda_2}\cdot\frac{\lambda_3(\lambda_1+\lambda_2)}{\lambda_1+\lambda_2-\lambda_3}\{\exp(-\lambda_3 t) - \exp[-(\lambda_1+\lambda_3)t]\}$$

$$= \phi(\lambda_1+\lambda_2)\exp[-(\lambda_1+\lambda_2)t] + (1-\phi)\lambda_3\exp(-\lambda_3 t),$$

where

$$\phi = \frac{\lambda_2-\lambda_3}{\lambda_1+\lambda_2-\lambda_3},$$

so that the overall survival function is

$$S_T(t) = \phi\exp[-(\lambda_1+\lambda_2)t] + (1-\phi)\exp(-\lambda_3 t), \text{ provided } \lambda_1+\lambda_2 \neq \lambda_3.$$

Note that if $\lambda_2 < \lambda_3 < \lambda_1 + \lambda_2$, then $\phi$ is negative.

## 15.7 REVERSIBLE MODELS: THE FIX-NEYMAN MODEL

In all the Markov models so far discussed, movement between two stages has been in one direction only—that is, it was irreversible. In this section, we discuss a model in which movement is possible in either direction between two of the states.

A simple model of this kind was proposed by Fix and Neyman (1951). We will give some details of the derivation of survival functions, but those who find the mathematics too elaborate should concentrate on the general mode of analysis, and the results.

This model has four states:
$S_0$—under treatment for a disease (e.g., cancer);
$S_1$—died while under treatment;
$S_2$—apparently recovered from the disease, but still under observation;
$S_3$—lost from $S_2$, either by death or simply by no longer being available for observation.

The "reversible" part of the model is in movement between $S_0$ and $S_2$, reflecting the fact that a patient who has "apparently recovered" can have a relapse and once more come under treatment for the disease. Clearly, transfer to $S_1$ (died under treatment) is only possible from $S_0$; and transfer to $S_3$ (lost from $S_2$) is only possible from $S_2$. The model is set out schematically in Fig. 15.2.

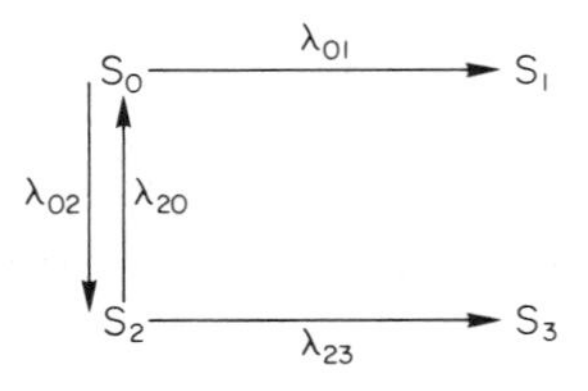


**Fig. 15.2**

In this scheme, with *constant* transition rates, $\lambda_{ij}$, from $S_i$ to $S_j$ ($\lambda_{03}=\lambda_{30}=\lambda_{10}=\lambda_{12}=\lambda_{21}=\lambda_{13}=\lambda_{32}=0$), the time spent in any one sojourn in $S_0$ is distributed exponentially with hazard rate $\lambda_{01}+\lambda_{02}=\lambda_0$; and in $S_2$ exponentially with hazard rate $\lambda_{20}+\lambda_{23}=\lambda_2$.

Further, whenever there is a departure from $S_0$

$$\Pr\{S_0 \rightarrow S_1 | \text{departure from } S_0\} = \frac{\lambda_{01}}{\lambda_0}, \tag{15.46a}$$

$$\Pr\{S_0 \rightarrow S_2 | \text{departure from } S_0\} = \frac{\lambda_{02}}{\lambda_0}. \tag{15.46b}$$

(See Exercise 15.4.) Similarly,

$$\Pr\{S_2 \rightarrow S_0 | \text{departure from } S_2\} = \frac{\lambda_{20}}{\lambda_2}, \tag{15.47a}$$

$$\Pr\{S_2 \rightarrow S_3 | \text{departure from } S_2\} = \frac{\lambda_{23}}{\lambda_2}. \tag{15.47b}$$

Consider now an individual in $S_0$ at time 0. The probability that there are at least $h$ returns to $S_0$ (after visiting $S_1$) before finally reaching $S_1$ or $S_3$ is

$$P_h = \left( \frac{\lambda_{02}}{\lambda_0} \frac{\lambda_{20}}{\lambda_2} \right)^h. \tag{15.48}$$

The event that the individual is in $S_0$ at time $t$ can be split up according to the number of times $(h=0,1,2,\ldots)$ the individual has returned to $S_0$ from $S_1$ since time zero. We introduce the notation:

$R_h(t) \equiv$ CDF of time of $h$th return to $S_0$, given that there *is* an $h$th return;

$D_h(t) \equiv$ CDF of time of $h$th departure from $S_0$ under the same condition.

(Clearly, if there is an $h$th return, there will be an $(h+1)-$th departure.) Then,

$$\begin{aligned} &\Pr\{\text{in } S_0 \text{ at time } t | \text{in } S_0 \text{ at time } 0\} \\ &\qquad = [1 - D_0(t)] + P_1[R_1(t) - D_1(t)] + \ldots \\ &\qquad = \sum_{h=0}^{\infty} P_h[R_h(t) - D_h(t)] = \sum_{h=0}^{\infty} \theta^h [R_h(t) - D_h(t)] \end{aligned} \tag{15.49a}$$

[with $R_0(t) = 1$].
Similarly,

$$\Pr\{\text{in } S_1 \text{ at time } t | \text{in } S_0 \text{ at time } 0\} = \frac{\lambda_{01}}{\lambda_0} \sum_{h=0}^{\infty} \theta^h D_h(t); \tag{15.49b}$$

$$\Pr\{\text{in } S_2 \text{ at time } t | \text{in } S_0 \text{ at time } 0\} = \frac{\lambda_{02}}{\lambda_1} \sum_{h=0}^{\infty} \theta^h [D_h(t) - R_{h+1}(t)]; \tag{15.49c}$$

and

$$\begin{aligned} &\Pr\{\text{in } S_3 \text{ at time } t | \text{in } S_0 \text{ at time } 0\} \\ &\qquad = \frac{\lambda_{23}}{\lambda_2} \cdot \frac{\lambda_{02}}{\lambda_0} \sum_{h=1}^{\infty} P_{h-1} R_h(t) = \frac{\lambda_{23}}{\lambda_{20}} \sum_{h=1}^{\infty} \theta^h R_h(t). \end{aligned} \tag{15.49d}$$

The reader should check that the four probabilities add to 1.

A set of formulas similar to (15.49a–d) for an individual in $S_2$ at time 0 is obtained by interchanging subscripts "0" and "2," and "1" and "3." The formulas also apply if times 0 and $t$ are replaced by $t'$ and $t''$ ($>t'$), replacing $t$ by $(t''-t')$.

To evaluate $R_h(t)$ and $D_h(t)$, we note that $R_h(t)$ is the CDF of the sum of $h$ independent exponential (with hazard rate $\lambda_0$) variables, and $h$ independent exponential (with hazard rate $\lambda_2$) variables. Also $D_h(t)$ is the CDF of the same sum *plus* a further independent exponential (with hazard rate $\lambda_0$) variable.

Since the sum ($Y$) of $h$ independent exponential (with hazard rate $\lambda$) variables has the gamma distribution with PDF

$$f_Y(y)=\frac{\lambda^h}{\Gamma(h)}y^{h-1}e^{-\lambda y}=g_h(y;\lambda),\quad \text{say,}\quad (y>0) \tag{15.50}$$

so $R_h(t)$ is the CDF of

$$Z_h=Y_{0h}+Y_{2h}, \tag{15.51}$$

where $Y_{0h}$ and $Y_{2h}$ are mutually independent and the PDF of $Y_{ih}$ is $g_h(y;\lambda_i)$ $(i=0,2)$.

Similarly, $D_h(t)$ is the CDF of

$$Z_h'=Y_{0,h+1}+Y_{2h}. \tag{15.52}$$

It can be shown that the PDF of

$$Z=Y_{0h_0}+Y_{2h_2} \tag{15.53}$$

is

$$f_{h_0,h_2}(z;\lambda_0,\lambda_2)=\begin{cases}\left(\dfrac{\lambda_0}{\lambda_2}\right)^{h_0}\displaystyle\sum_{j=0}^{\infty}c_{h_0,j}\left(\dfrac{\lambda_2-\lambda_0}{\lambda_2}\right)g_{h_0+h_2+j}(z;\lambda_2) & \text{if } \lambda_2>\lambda_0,\\ \left(\dfrac{\lambda_2}{\lambda_0}\right)^{h_2}\displaystyle\sum_{j=0}^{\infty}c_{h_2,j}\left(\dfrac{\lambda_0-\lambda_2}{\lambda_0}\right)g_{h_0+h_2+j}(z;\lambda_0) & \text{if } \lambda_0>\lambda_2,\end{cases} \tag{15.54}$$

where $c_{h,j}(w)$ is the $j$th term in the negative binomial expansion of $(1-w)^{-h}$.

Then,

$$R_h(t)=\int_0^t f_{h,h}(z;\lambda_0,\lambda_2)\,dz, \tag{15.55}$$

$$D_h(t)=\int_0^t f_{h+1,h}(z;\lambda_1,\lambda_2)\,dz. \tag{15.56}$$

We also note that

$$\int_0^t g_h(u;\lambda)\,du=1-\int_t^\infty g_h(u;\lambda)\,du=1-e^{-\lambda t}\sum_{j=0}^{h-1}\frac{(\lambda t)^j}{j!}. \tag{15.57}$$

(This leads to some simplification in the expression for $[R_h(t)-D_h(t)]$ when $\lambda_2<\lambda_0$.)

Chiang and Hsu (1976) give a detailed mathematical analysis of this problem, though not in the same terms as set out above.

From the formulas derived above, it is possible to construct the likelihood function corresponding to data by observing the numbers of individuals in an experience who belong to each of states $S_0$, $S_1$, $S_2$, $S_3$ at each of certain times of examination, $t_0=0$, $t_1$, $t_2$, $t_3,\ldots$. Maximum likelihood estimators might then be obtained.

It is natural to start with, $N_0$, say, persons under treatment (i.e., in state $S_0$) for the disease, and observe the numbers $N_{0j}$ in $S_j$ $(j=0,1,2,3)$ after a period of time $t$. These data alone, however, provide only three relevant statistics, since

$$N_{00}+N_{01}+N_{02}+N_{03}=N_0,$$

while there are four parameters, $\lambda_{02}$, $\lambda_{20}$, $\lambda_{01}$, $\lambda_{23}$, to be estimated. As Fix and Neyman (1951) pointed out, suitable additional information is provided by also observing a set of individuals initially in $S_2$. Such a set may not be available at the beginning of an investigation, but the $N_{02}$ individuals in $S_2$ at time $t$ can often be observed for a further period.

The schemes shown in Fig. 15.3*a* and *b* can be extended by addition of *progressive* (irreversible) states beyond $S_1$ and/or $S_3$, (see Fig. 15.3a), without notable increase in complexity of analysis. Subdivision of states $S_1$ and/or $S_3$ (see Fig. 15.3*b*) also does not lead to much increase in complexity. (The states $S_{11}$, $\ldots$, $S_{3b'}$ may represent different causes of death.) Schemes of the type shown in Fig. 15.3*b* were described by Fix and Neyman (1951) and included by Chiang and Hsu (1976) in their analysis.

Inclusion of further reversible movements between states, as in Fig. 15.3*c*, however, rapidly leads to considerable increase in complexity.

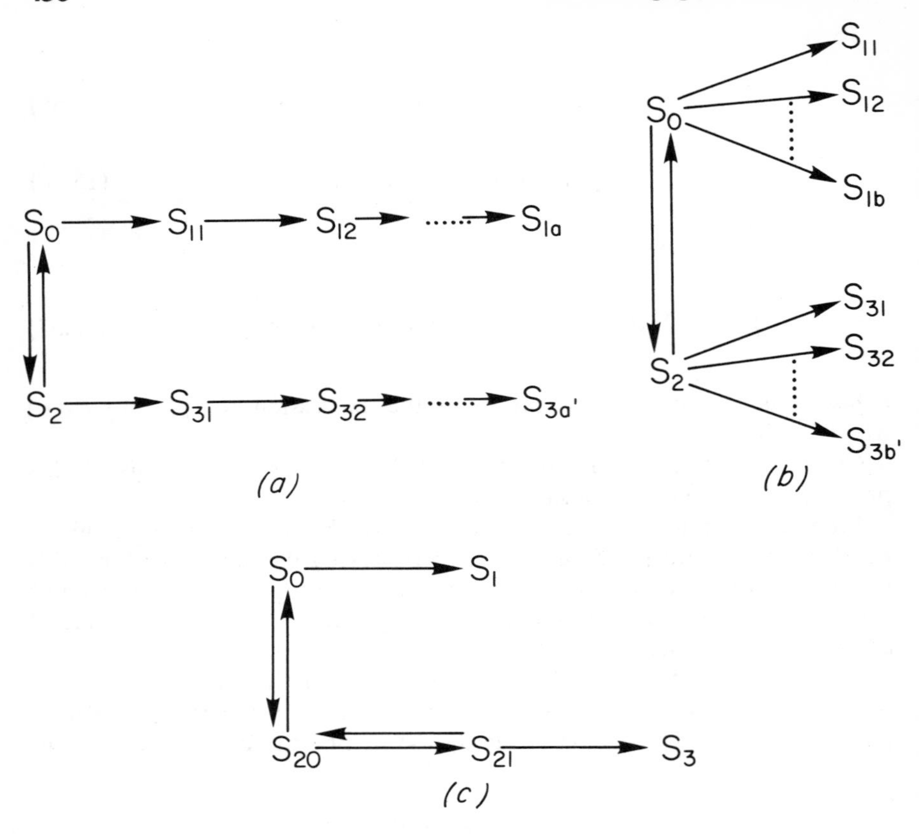


**Fig. 15.3** Modified Fix-Neyman schemes.

If it were possible to observe the lengths of sojourns in states $S_0$, $S_2$, it would be possible to estimate the corresponding $\lambda_0$, $\lambda_2$ directly. If this *were* possible, it would most likely be possible to observe the destination of each departure also, and from the proportions of departures going from $S_i$ to $S_j$, to estimate $\lambda_{ij}/\lambda_i$ ($i \neq j$, $i, j = 0, 2$). From the combined data, the four $\lambda$'s could be estimated very simply. To obtain these data, we would need records to be kept of each change of state for all individuals in $S_0$ or $S_2$.

## REFERENCES

Arley, N. (1961). Theoretical analysis of carcinogenesis, *Fourth Berkeley Symposium on Math. Statistics and Prob.*, Vol. 4, University of California Press, Berkeley, pp. 7–17.

Armitage, P. and Doll, R. (1954). The age distribution of cancer and a multi-stage theory of carcinogenesis, *British J. Cancer*, **8**, 1–12.

Armitage, P. and Doll, R. (1961). Stochastic models for carcinogenesis, *Fourth Berkeley Symposium on Math. Statist. and Prob.*, Vol. 4, University of California Press, Berkley, pp. 19–38.

Burch, P. R. J. (1966). Spontaneous auto-immunity: Equations for age specific prevalence and initiation rates, *J. Theoret. Biol.*, **12**, 397–409.

Burch, P. R. J. and Rowell, N. R. (1965). Systemic lupus erythematosus; Etiological aspects, *Amer. J. Med.*, **38**, 793–801.

Chiang, C. L. and Hsu, J. P. (1976). On multiple transition time in a simple illness-death process—A Fix-Neyman model, *Math. Biosci.*, **30**, 55–71.

Finch, C. E. and Hayflick, L. (Eds) (1977). *Handbook of the Biology of Aging*. Van Nostrand Reinhold, New York.

Fisher, R. A. and Tippett, L. H. C. (1928). Limiting forms of the frequency distributions of the largest or the smallest member of a sample, *Proc. Cambridge Phil. Soc.* **24**, 180–190.

Fix, E. and Neyman, J. (1951). A simple stochastic model for recovery, relapse, death, and loss of patients, *Hum. Biol.*, **23**, 205–241.

Gross, A. J., Clark, V. A. and Lis, V. (1972). Estimation of survival parameters where one of two organs must function for survival, *Biometrics*, **27**, 369–377.

Iversen, S. and Arley, N. (1950). On the mechansim of experimental carcinogenesis, *Acta. Path. Microbiol. Scand.* **31**: V, 27–45; VI, 164–171.

Lagakos, S. W. (1976). A stochastic model for censored-survival data in the presence of an auxiliary variable. *Biometrics*, **32**, 551–559.

Lee, L. and Thompson, W. A. (1974). Results on failure time and pattern for the series system, *Reliability and Biometry*, SIAM, Philadelphia, 291–302.

Manton, K. G., Tolley, H. D. and Poss, S. S. (1976). Life table techniques for multiple cause mortality, *Demography* **13**, 541–564.

Marshall, A. W. and Olkin, I. (1967). A multivariate exponential distribution, *J. Amer. Statist. Assoc.*, **62**, 30–44.

Moeschberger, M. L. and David, H. A. (1971). Life tests under competing risk causes of failure and the theory of competing risks, *Biometrics*, **27**, 909–933.

Nordling, C. O. (1953). A new theory on the cancer inducing mechanism, *Brit. J. Cancer*, **7**, 68–72.

Weibull, W. (1951). A statistical distribution function of wide applicability, *J. Appl. Mech.*, **18**, 293–297.

## EXERCISES

**15.1.** Suppose that a device consists of three components, $\Gamma_1$, $\Gamma_2$, $\Gamma_3$, and that failure of the device occurs as soon as any two components fail (not necessarily simultaneously).

(*a*) List the three *causes* of failure of the device in terms of failures of the components [e.g., $C_1 \equiv C_{(23)} \equiv (\Gamma_2, \Gamma_3)$, etc.] Let $T_j$ be the time to fail of component $\Gamma_j$, and assume that $T_j$'s are *mutually independent*, the $T_j$ having exponential distribution with the same parameter $\lambda$. Let

$$X_\alpha = X_{(ij)} = \max(T_i, T_j)$$

be the "time due to fail" from cause $C_\alpha = C_{(ij)}$, and let $X$ be the failure time of the device.

(*b*) Show that

$$S_X(t) = e^{-2\lambda t}(3 - 2e^{-\lambda t}), \qquad t > 0.$$

(*c*) Show that the conditional failure distribution, given the device fails from $C_\alpha$, is identical with $F_X(t) = 1 - S_X(t)$. Why would you expect this result, without actually carrying out the calculations? (Quote the appropriate theorem from Chapter 9.)

(*d*) Find the joint CDF of $X_1$, $X_2$, $X_3$ the marginal CDF's $F_\alpha(x_\alpha)$, and $F_{\alpha,\beta}(x_\alpha, x_\beta)$, $\alpha, \beta = 1, 2, 3$; $\alpha \neq \beta$.

(*e*) Find the joint SDF of $X_1$, $X_2$, $X_3$.

**15.2.** Consider the three component system as in Exercise 15.1, but assume that $T_1$, $T_2$, $T_3$ have a joint generalized Farlie-Gumbel-Morgenstern distribution [see Johnson and Kotz (1975), *Commun. Statist.*, **3**, 415] with PDF of the form

$$f_{T_1,T_2,T_3}(t_1,t_2,t_3) = \lambda^3\{1 + \theta[2\exp(-\lambda t_1) - 1][2\exp(-\lambda t_2) - 1]\}$$
$$\times \exp[-\lambda(t_1 + t_2 + t_3)]$$

where $0 \leqslant \theta \leqslant 1$.

(*a*) Show that the overall SDF, $S_X(t)$, is

$$S_X(t) = \{2 + (1 - 2\exp(-\lambda t))[1 + \theta(1 - \exp(-\lambda t))^2]\}\exp(-2\lambda t).$$

(*b*) Let $X_1$, $X_2$, $X_3$ be random variables as defined in Exercise 15.1.

Let $h_\alpha(t)$ be the crude hazard rate for cause $C_\alpha$ ($\alpha = 1, 2, 3$) associated with the joint distribution of $X_1$, $X_2$, $X_3$. Show that, for each $\alpha$,

$$h_\alpha(t)S_X(t) = \exp(-2\lambda t)[1 - \exp(-\lambda t)]$$
$$\times \{2 + \theta[1 - \exp(-\lambda t)][1 - 3\exp(-\lambda t)]\}.$$

**15.3.** Consider a two-component series system as described in Section 15.5.1, but assuming

$$\lambda_1(t)=R_1e^{at},\ \lambda_2(t)=R_2e^{at},\text{ and } \lambda_{12}=R_{12}e^{at},\ R_1, R_2, R_{12}, a>0, t\geqslant 0.$$

(*a*) Obtain an expression for the overall SDF, $S_T(t)$.
(*b*) Find the proportions of failures, $\pi_1$ and $\pi_2$ of components $\Gamma_1$ and $\Gamma_2$, respectively. Note that $\pi_1+\pi_2\geqslant 1$. Why?

**15.4.** Prove that, if the instantaneous rate of transition from state ($i$) to state ($k$) is $\lambda_{ik}(t)$ at time $t$ (the $\lambda_{ik}(t)$ being continuous functions of $t$), then given that a transition occurs at time $t$, the probability that it is a transition to state ($j$) is

$$\frac{\lambda_{ij}(t)}{\sum_{i\neq k}\lambda_{ik}(t)}.$$

(*Hint*: Use Bayes' theorem and the fact that any one transition is to only one other state. See also Exercise 13.5.)

**15.5.** Consider the model of disease progression discussed in Section 15.6.4. Suppose that a random sample of $n$ individuals is observed from time $t=0$. Further, suppose that *at time* $\tau$ we observe:

(*i*) $n_0$ alive who have not experienced progression.
(*ii*) $n_1$ alive who have had progression at times $t_{ij}$, ($j=1,2,\ldots,n_1$).
(*iii*) $n_2'$ who have died at times $t_j$ ($j=n_1+1,\ n_1+2,\ldots,\ n_1+n_2'$) without having experienced progression.
(*iv*) $n_2'$ who died at time $t_j$ having previously experienced progression at times $t_{ij}$ ($j=n_1+n_2'+1,\ldots,n_1+n_2'+n_2''$).

Of course, $n_0+n_1+n_2'+n_2''=n$.

(*a*) Construct the likelihood function (in general terms), assuming that $T_1$ and $T_2$ are not independent, but that $T_3$ does not depend on $T_1$ and $T_2$.
(*b*) What would be the likelihood function if the exact values of times of progression, $t_{1j}$, were not observed but it was known whether progression has occurred before death or time $\tau$, as the case may be ?
(*c*) What is the likelihood function if only times at death, $t_j$, were recorded?

**15.6.** Assume that $T_1$ and $T_2$ have a joint Gumbel Type $A$ distribution as defined in (9.49), with $S_{T_1}(t_1)=\exp(-\lambda_1 t_1)$ (exponential), and $S_{T_2}(t_2)=\exp[(R/a)(1-e^{at_2})]$ (Gompertz), ($\lambda_1$, $a$, $R$, and $t_1$, $t_2>0$).

Further, assume that $f_{T_3}(t_3)=\lambda_2\exp(-\lambda_3 t_3)$ $t_3>0$.

Obtain explicit formulas for the likelihood functions discussed in Exercise 15.5.

**15.7.** Obtain the likelihood functions discussed in Exercise 15.5, but assuming that $T_3$ depends on $T_1$, that is, we have $f_{T_3|T_1}(t_3|t_1)$.

**15.8.** The results obtained in Section 15.2 are still valid if all the hazard rates vary with time but in such a way that their *ratios* remain constant. Why is this so ?

**15.9.** Suppose that a disease has three stages such that the transition rates for movement from stage ($i$) to stage ($i+1$) are $\lambda_i$ ($i=0,1,2$). Stage (0) corresponds to healthy conditions (i.e., an individual is free of the disease). The force of mortality for an individual in stage ($i$) is $\mu_i$ ($i=0,1,2,3$).

Assuming that: (1) it is not possible to move from a stage with a higher number to one with a lower number, and (2) $\mu_0<\mu_1<\mu_2<\mu_3$, obtain expressions for the expectation of life in each of stages (0), (1), (2), (3).

**15.10.** Answer Exercise 15.9 under the conditions that there is a constant recovery rate, $\omega$, from stage (1) to stage (0), but a person who has recovered (one or more times) has a transition rate $\lambda_0+\delta$ ($\delta>0$) of once again moving to stage (1). (Recovery rate remains constant at $\omega$.)

**15.11.** Construct a model for the development of a disease $D$ on the following assumptions:

(*i*) Diagnosis of stage (1) of the disease is effectively instantaneous.

(*ii*) From stage (1) there might be: (a) apparent recovery to healthy stage (0); (b) progression to a more advanced stage (2); (c) progression directly to a chronic stage (3).

(*iii*) From stage (2) there will eventually be progression to stage (3).

(*iv*) From stage (3) there is eventual progression to death.

There are no other possibilities of change of status.

You should assume that the transition rates for each change of status subsequent to stage (1) are constant, but that the transition rate for return to stage (1) after apparent recovery depends on how many apparent recoveries there have been. As a practical approximation take $\mu_{01}$ for the rate subsequent to first, and $\mu_{01}^*$ the rate subsequent to each later apparent recovery.

Transition for a movement from stage (3) to death does not depend on whether the stage immediately preceding stage (3) was (1) or (2).

The model should ignore other causes of death—that is the model should be appropriate to "time due to die."

**15.12.** Suppose that in observations on disease $D$ of Exercise 15.11, you have available information on:

(*i*) Dates of first and subsequent diagnoses.
(*ii*) Date of initiation of stage (3).
(*iii*) Date of death.

This implies that only stage (2) is not directly observable.

(*a*) How do you test the hypothesis that stage (2) does not exist?
(*b*) Try to extend your work so that it can be applied when the date of initiation of stage (3) is known only approximately (e.g., between two successive examinations).

**15.13.** Suppose that the progression of a second disease, $D'$, is independent of that of $D$, but has the same structure as described in Exercise 15.11, [with $\mu$, $\mu^*$ replaced by $\mu'$, $(\mu^*)'$].

Given that $D$ is the first to be diagnosed, the conditional distribution of time between first diagnoses of the two diseases has a constant hazard rate $\lambda$. Obtain an expression for the probability that when $D$ has been diagnosed before $D'$, "time to die" from $D$ will precede that from $D'$.

**15.14.** Prove the results of Section 15.6.3, given by the set of equations (15.30).

**15.15.** Prove that if $X$ has a Gompertz distribution with parameters $R$, $a$, and $R$ has exponential PDF, $f_R(y)=(1/\theta)e^{-y/\theta}$ $(y>0,\ \theta>0)$, then the compound SDF is the truncated logistic,

$$S_X(x)=\frac{a\theta^{-1}}{a\theta^{-1}-1+e^{ax}} \qquad (x>0).$$

# Author Index


Abramowitz, M., 69, 75, 111, 120
Aitchison, J., 383
Anderson, T. A., 383
Anderson, T. W., 218, 220
Antle, C. E., 208, 220
Arduino, L. J., 354, 384
Arley, N., 416, 436, 437
Armitage, P., 262, 263, 385, 418, 437
Aroian, L. A., 206, 220
Ashford, J. R., 401, 437
Axtell, L. M., 118, 120

Bailar, J. C., 354, 383, 384
Bailey, R. C., 350, 371, 383
Bailey, W. G., 316, 319
Bain, L. J., 208, 209, 220
Barclay, G. W., 120
Barnard, G. A., 252, 262
Barnett, H. A. R., 214, 220
Barr, D. R., 217, 220
Basu, A. P., 274, 280, 285, 291
Batten, R. W., 38, 124, 125, 154, 175
Bayard, S., 367, 384
Beers, H. S., 120
Benjamin, B., 104, 120, 155, 175
Berkson, J., 171, 214, 220
Berman, S., 291
Bernoulli, D., 309
Bernstein, R. A., 407-411
Birnbaum, Z. W., 288, 291
Bliss, C. I., 204, 220
Breslow, N. E., 157, 175, 263, 361, 383, 385
Bryson, M. C., 147
Burch, P. R. J., 418, 419, 437
Byar, D. P., 217, 221, 354, 367, 372, 383, 384

Chiang, C. L., 14, 100, 104, 106, 120, 146, 167, 169, 171, 175, 288, 291, 316, 319, 327, 435, 437
Clark, V. A., 209, 220, 260, 262, 384, 424, 437
Cohen, A. C., 220
Colosi, A., 217
Conover, W. J., 217, 243, 262
Constantidinis, J., 131, 146
Corle, D. K., 372, 383
Cornfield, J., 383, 385
Cox, D. R., 204, 220, 263, 279, 291, 362, 366, 381, 383
Cramer, H., 214, 220
Crowley, J., 157, 158, 175
Cutler, S. J., 118, 120, 171, 175, 262

D'Alembert, J. le R., 309
Damiani, P., 320, 321
Darling, D. A., 218, 220
David, H. A., 288, 291, 340, 422, 437
Davidson, T., 217, 220
Dawson, M. M., 120, 126
Day, N. E., 383
De Vylder, F., 214, 220
Doll, R., 418, 437
Dorn, H., 38, 175
Drolette, M. E., 169, 170, 171
Duncan, D. B., 385
Duncan, L. J. P., 404, 411

Dyer, A. P., 376, 384

Ederer, F., 171, 175
Elandt-Johnson, R. C., 38, 170, 171, 175, 279, 281, 288, 291, 311, 312, 319, 381, 384, 407, 411
Elston, R. C., 131, 146, 201, 221
Elveback, L., 168, 169, 171, 175
Epstein, F. H., 384

Falconer, D. S., 404, 411
Farewell, V. T., 383, 385
Farr, W., 309, 319
Feigl, P., 355, 357, 384, 385
Finch, C. E., 437
Finlaison, J., 46
Finney, D. J., 204, 220
Fisher, R. A., 204, 220, 262, 418, 437
Fix, E., 431, 435, 437
Fluornoy, N., 383, 385
Freedman, L. S., 259, 262
Freeman, D. H., 411
Freeman, J. L., 411
Frost, W. H., 171, 175
Fürth, J., 135, 211, 220

Gaffey, W. R., 38
Gage, R. P., 171
Gail, M., 279, 291, 311, 319
Gehan, E. A., 146, 171, 175, 210, 220, 237, 262
Gershenson, H., 38, 154, 175
Ghosh, J. K., 274, 280, 285, 291
Giefer, E. E., 407-411
Gilbert, G. P., 237, 262
Glasser, M., 357, 384
Go, R. C. P., 131, 146
Gompertz, B., 61
Greenberg, R.. A., 367, 384
Greenwood, J. A., 140, 158, 173
Greville, T. N. E., 104, 105, 120, 316, 319
Grizzle, J. E., 211, 220
Gross, A. J., 208, 220, 260, 262, 384, 424, 457
Gumbel, E. J., 75, 283, 291

Haenszel, W., 262
Hakulinen, T., 279
Hankey, B. E., 375, 384
Harley, J. B., 384
Hartley, H. O., 235, 262
Haybittle, J. L., 259, 262
Haycocks, H. W., 104, 120, 155, 175, 316, 319
Hayflick, L., 437
Herman, R. J., 340
Hickman, J. C., 322
Hill, A. B., 38
Hill, I. D., 38
Hoel, D. G., 179, 192, 198, 201, 220, 336, 338, 339
Holford, T. R., 384
Hollander, M., 215, 221
Holt, J. D., 384
Homer, L. D., 350, 371, 383
Hooker, P. F., 104, 120, 314, 319
Howard, S. V., 263
Hsu, J. P., 435, 437
Hu, M., 158, 175
Huse, R., 354, 383

Iversen, S., 416, 437

Jenrich, R. I., 262
Johnson, N. L., 75, 170, 221, 282, 291, 384, 438
Johnson, W. D., 384
Jordan, C. W., 104, 120

Kalbfleisch, J., 383, 384, 385
Kannel, W., 385
Kaplan, E. B., 201, 221
Kaplan, E. L., 173, 175
Karn, M. N., 309, 319
Kay, R., 366, 384
Kerridge, D. F., 383, 402, 412
Keyfitz, N., 44, 104, 120, 123, 263
Kim, P. J., 262
Kimball, A. W., 135, 146, 211, 220, 221
King, G., 104, 121
Koch, G. G., 211, 220, 384, 411
Kodlin, D., 77, 221
Kotz, S., 75, 221, 282, 291, 438
Koziol, J. A., 221
Krall, J. M., 322, 384
Krane, S. A., 221

Kulldorf, G., 221
Kuzma, J. W., 170, 171, 175

Lagakos, S. W., 391, 429
Lawless, J. F., 384
Lazar, P., 180
Lee, L., 422, 437
Leppard, P., 118, 121, 385, 392, 412
Lis, V., 424, 437
Lomax, K. S., 224
Longley-Cook, L. H., 104, 120, 314, 316, 319

McCarthy, P. J., 367, 384
McCutcheon, J. J., 126
McPherson, K., 263
Mann, N. R., 260, 262
Mantel, N., 262, 263, 375, 384
Manton, K. G., 423, 437
Marshall, A. W., 420, 437
Meier, P., 173, 175, 262
Meinert, C. L., 250, 263
Mellinger, G. T., 354, 384
Merrell, M., 39, 104, 121
Miller, D. R., 279, 291
Miller, J. M., 39
Moeschberger, M. L., 288, 291, 339, 422, 437
Monson, R. R., 29, 38
Myers, M.H., 375, 384

Nádas, A., 274, 280, 291
Neill, A., 315, 319
Nelson, W. A., 175, 182, 196, 221
Neyman, J., 431, 435, 437
Nordling, C. O., 418, 437

Olkin, I., 420, 437

Patell, R. K. N., 340
Pearson, E. S., 235, 252, 262
Peterson, A. V., 383, 385
Peto, J., 252, 258, 262, 263
Peto, R., 252, 258, 262, 263
Pike, M. C., 259, 262
Pomper, I. M., 39
Poss, S. S., 423, 437
Prentice, R. L., 357, 358, 383, 384, 385, 386
Preston, S. H., 44, 121, 123, 263, 402, 412

Rahiala, M., 279, 291
Reed, J. L., 104, 121
Rényi, A., 221
Rimm, A., 407-411
Robison, D. E., 206, 220
Rowell, N. R., 419, 437

Sarfaty, G., 118, 121, 385, 392, 412
Schafer, R. D., 260, 262
Schoen, R., 44, 121, 123, 263, 402, 412
Schulman, L. E., 39
Schwartz, D., 180
Seal, H. L., 146, 223
Shryock, H., 38, 104, 121
Siddiqui, M. M., 147, 210, 220
Siegel, J. S., 38, 104, 121
Singhal, K., 384
Singpurwalla, N. D., 260, 262
Sirken, M., 104, 121
Smith, C., 404, 411
Smith, P. G., 263
Snee, R. D., 385
Snell, E. J., 366, 383
Sowden, R. R., 401, 437
Starmer, C. F., 211, 220
Stegun, I. A., 69, 75, 111, 120
Stephens, M. A., 216, 218, 219, 221
Storer, J. B., 176
Summe, J. P., 350, 371, 383
Susarla, V., 178

Tallis, G. M., 118, 121, 385, 392, 412
Thiele, T. N., 224
Thompson, W. A., 422, 437
Tippett, L. H. C., 418, 437
Todorov, A. B., 131, 146
Tolley, H. D., 384, 423, 437
Truett, J., 385
Tsao, C. K., 243, 263
Tsiatis, A., 279, 291
Tukey, J., 241, 263

Upton, A. C., 146, 211, 220

Uthoff, V. A., 384

Van Ryzin, J., 178
Vieira, J. J., 407-411

Walburg, H. E., 179
Walker, S. H., 385
Walter, S. D., 37, 38
Weibull, W., 418, 437
Werner, L. H., 407-411
Wilks, S. S., 75
Wold, H., 214, 220
Wolfe, D. A., 215, 221

Zelen, M., 355, 357, 384, 385
Zippin, C., 385

# Subject Index

A 1967-70 life table, 117, 120
$A^2$ statistic, *see* Anderson-Darling statistic
Abridged life table, 84, 93, 109-110
  construction, 104-111
Absolute rate, 12, 17, 51
Absorbing state, 426
Acceleration, 13
Accidents, 188, 301
Accuracy, 361
Acid phosphatase, 355
Actuarial assumption, 312
Actuarial estimator, 154, 157, 162, 165, 171
Additive model, 353, 355, 379
Additivity, of hazard rates, 273
Adequacy of model, *see* Goodness of fit
Age, 6, 115, 347, 350, 355, 371, 376
  adjusted rate, 23, 25
  at diagnosis, 133, 357
  last birthday (lbd), 21, 84
  at onset, 35, 392
  of onset distribution, 50, 392, 409
  specific death rate, 21, 103, 297
Aggregate, 17
  mortality, 117
Aging, 415
  models, 414
AG-negative, 356, 385
AG-positive, 356, 385
Analytical methods, 182
Anderson-Darling statistic, 218
Anniversary method, 152, 170
Approximation, 15, 70, 75, 99, 101-102, 110, 156, 167, 170, 183, 229, 314-317, 354, 366
  *see also* Interpolation
Artificial data, 263
Assumptions, 8, 29, 100-101, 162, 269, 270, 273, 288, 312, 319, 422
Asymptotic properties, 74
Auer rods, 356
Average population at risk, rate, 14

Balducci hypothesis, 124
Base SDF, 279, 290, 331
Bayes' theorem, 439
Bedridden, 355
Beginners, 152
Bias, 157
Binary response data, 204
Binomial distribution, 229
Binomial proportion, 131, 138, 170
Biometrics Seminar, 243
Birth rate, 17
Bivariate Gumbel Type A distribution, 283-285
Blood pressure, systolic, 376
Breast, cancer of, 118
Breathlessness, 401

Canada Life life table, 120
Cancer, 33
  of breast, 118
  of cervix, 327
  of digestive organs, 397
  of lung, 357, 387

of prostate, 354
Carcinogenesis models, 416-418
Cause of death, 269, 310
Cause specific death rate, 21, 297
Cell, 416
Censored data, 150, 207, 210, 216
Censored Kolmogorov-Smirnov test, 217
Censoring, 150, 153, 207, 240
  progressive, 151, 249
  random, 153, 157, 166
  time, 153
Census, 105
Census-type data, 420
Central death rate, 14, 52, 95, 99, 142, 154-156, 296
Central exposed to risk, 155
Cervix, cancer of, 327
Chi square, 75, 242, 367
  distribution, *see* Gamma distribution
  minimum, 182, 209, 213
  modified, 209, 213
  test, 219
    Pearson's conditional, 247-249, 250-251
Chicago Peoples Gas Light and Coke Company, 376
Cholesterol, 376
Chronic disease, 35, 310, 416, 418
Cigarette smoking, 376
Circulatory system, 301
Clinical trial, 6, 153, 240, 346
Clone, 417, 419
  forbidden, 417, 419
Coefficient, correlation, 10
  Lagrangian, 112
Cohort, 83, 128, 134, 323, 394
  data, 206, 332, 335
  hypothetical, 83
  life table, 83
Comparison of life tables, 225
  and mortality experiences, 225, 228
Comparison of mortality experiences, 225, 231
Competing causes, 163, 166, 270, 381, 420
Competing events, 427
Complete life table, 84, 111
Complete mortality data, 128, 196, 206, 209, 229, 336
Composite hypotheses, 368
Compound distribution, 67, 191, 201, 288-289, 379
Compounding distribution, 67
  *see also* Compound distribution
Concomitant variable, 3, 7, 115, 345, 367
  random, 379
  time dependent, 377
Conditional Chi square test, Pearson's, 247-249, 250-251
  distribution, 138, 210-211, 248
  expectation, 227
  expected deaths, 24
  likelihood, 329, 331-332
  probability, 52, 134, 227, 324, 406
  survival function (SDF), 421
  variance, 94, 140, 141, 158
Connecticut Tumor Registry, 118
Construction of likelihood function, 204-207, 349, 359, 363
Continuous process, variables, 346
Core SDF, 279, 281-282, 285, 290
Correlation coefficient, 10
Covariable, *see* Concomitant variable
Covariance, 10, 72, 138
Covariant, *see* Concomitant variable
Cox model, 361-362, 364
Cross-product ratio, 37
Cross-section data, 100, 139, 298, 395
Cross-section study, 36
Crude birth rate, 17
Crude death rate, 20
  central, 52
Crude hazard rate, 273, 420
  onset, 393
Crude residual, 366
Crude survival, function, 272, 331
  probability, 274-275, 289, 295
Cubic approximation, Greville's, 110
Cumulative distribution function (CDF), 50, 306
Cumulative hazard function (CHF), 51, 52, 174-175, 182, 196, 348, 366
Cumulative hazard paper, 366
Current prevalence, 32
Curtailed sampling, 247
Curve of deaths, 51, 94, 141, 172

Data, artificial, 363
binary, 204
censored, 150, 207, 210, 216
complete, 128, 196, 206, 209, 229, 336
failure, 7
grouped, 133, 154, 183, 206, 210-211, 219, 241, 329
incidence, 395
incomplete, 150, 200, 210, 236, 334
prevalence, 395-398
quantal response, 203
survival, 7
truncated, 150, 214, 216, 247, 326
Death rate, 18
age specific, 21, 103, 297
cause specific, 21, 297
central, 14, 52, 95, 99, 142, 154-156, 296
crude, 20
standardized, 22
Deaths, curve of, 51, 94, 141, 172
Decay, rate of, 13
Delta method, *see* Statistical differentials
Dementia, 131, 230, 233
Density function probability (PDF), 51, 143
*see also* Curve of deaths
Departure, *see* Withdrawal
Diabetes, 250, 263, 404-407
Diagnosis, age at, 133, 392
time since, 357
Differential mortality, 393, 399, 407, 410
Differentials, statistical, 69, 141, 142, 178
Direct standardization, 25
Discrete variables, 346
Disease, 35
Disease model, 426
pattern, 423
progression, 423, 426
Distribution, binomial, 229
chi square, 63
*see also* Gamma distribution
compound, 67, 191, 201, 288-289, 379
compounding, 67
*see also* Compound distribution
conditional, 138, 210-211
exponential, 60, 63, 67, 101, 164, 166, 168, 174, 282, 353, 429
extreme value (Type 1), 61, 183
Farlie-Gumbel-Morgenstern, 282
function, cumulative (CDF), 50, 306
hazard rate (HRF), 14, 51, 346, 352
survival (SDF), 50
future lifetime, 5
gamma, 62, 433
Gaussian, *see* Normal distribution
Gompertz, 61, 108, 113, 181, 183, 187-192, 199, 200, 210, 212, 218, 287, 339, 350, 416
hypergeometric, 252
improper, 286, 290, 331
logistic, 63, 184, 204
lognormal, 63, 389
Makeham-Gompertz, 61, 184
mixing, 67
mixture, *see* Compound distribution
multinomial, 138, 206, 324
normal, 204, 213, 379
onset-incidence, 393, 394
Poisson, 264, 418
posterior, 380
prior, 380
proper, 52, 271
randomization, 238
rectangular, 204
*see also* Uniform distribution
survival, 50
truncated, 53, 59, 60, 183, 217
uniform, 59, 66, 98, 100, 163, 166, 167, 168, 313
waiting time (onset), 394, 403, 407, 409
Weibull, 62, 181, 184, 188-196, 201, 208, 210, 417
Distribution-free, 129, 231
Dosage, 204
mortality technique, 203
Dosage variable, 379
*see also* Indicator variaable

Early decision, 240, 246
Eighth Classification, of causes of death, 301, 306

Elapsed time, 350
Elimination, of a cause of death, 271, 309
Empirical cumulative hazard function, 197
Empirical failure distribution, 129
Empirical survival function (SDF), 129, 130, 211, 215, 233, 332
Enders, 152
English life table No. 12, 127
English life tables, 294
Entries, staggered, 242
Entry-departure line, 26
Epidemics, 34
Equivalence, 277
  classes, 279
  theorem, 279
Equivalent models, 278
Estimate, 11
Estimation, 128, 145, 181
  of SDF, 158
Estimator, 128
  biased, 157
  first moment, 169, 171
  least square(s), 209
  maximum likelihood (MLE), 72, 73, 74, 163, 165, 168
  of $q_i$, 171
  unbiased, 157
Events, 5, 16
  competing, 427
  mutually exclusive, 273, 281
  repetitive, 16, 18
Expectancy, *see* Expectation of future lifetime
Expectation of future lifetime, 56, 92, 227, 229
Expected fraction, of last year(s) of life, 100, 106
Expected number of deaths, 23
  conditional, 24
Expected number of survivors, 83
Expected value, 55
  *see also* Expectation of future lifetime
Experience, 26, 104, 114, 202, 225, 228
Experimental-type data, 4, 128, 323
Explanatory variable, *see* Concomitant variable
Exponential distribution, 60, 63, 67, 101, 164, 166, 168, 174, 282, 353, 429
  standard, 282
Exponential growth, 14
Exponential interpolation, 69, 97
Exponential model, 353
Exponential piecewise, 206, 386
Exponential-type hazard function, 356
Exponential-Weibull model, 376
Exposed to risk, 18, 25-31, 105, 154
  central, 155
  effective initial, 156, 326, 406
Exposure to risk, 6
Extrapolation, 114
Extreme value (Type 1) distribution, 61, 183
Extreme value (Type 2) distribution, *see* Weibull distribution

Factor, risk, 35, 345
Factorization theorem, 278
Failure, 3
  distribution, 50, 58
  model, 64
Farlie-Bumbel-Morgenstern distributions, 282, 438
"Fatal shock" models, 420-423
Fault, inborn, 416
Fetal death, rate, 11
  ratio, 10
First moment estimator, *see* Moment estimator
First occurrence, *see* Onset
Fisher information matrix, 74
Fit, goodness of, 74, 214, 331, 365
  tests, 214
Fitting, 182, 214, 260, 416
  by eye, 190
  least squares, 182, 209
  maximum likelihood, 182
  minimum Chi square, 162
    modified, 209, 213
  moments, 169
  percentile points, 182
Fix-Neyman model, 431
Fixed intervals, 206
Fixed number of deaths, 135, 243
Follow-up, 152, 323, 326
  study, 151

Forbidden clone, 417, 419
Force of mortality, 51, 93, 142, 182, 226, 297, 314
 *see also* Hazard rate
Formula "G", 180
Fraction, 11, 155
 expected, of last year(s) of life, 100, 106
Frequency, observed, 36
 relative, 11
Future lifetime, 55
 distribution, 55
 expectation, 56, 92, 227, 229
 median, 56, 98, 227, 228
 percentiles, 57, 99
 variance, 56, 94

Gallbladder, 408-410
Gamma distribution, 62, 434
Gamma radiation, 135
Gaussian distribution, *see* Normal distribution
Gehan and Gilbert test, 237
Geneva life table, 146, 230
Geneva University Psychiatric Clinic, 131
Gilbert and Gehan test, 237
Gompertz, 61
 distribution, 61, 111, 113, 181, 183, 187-192, 199, 200, 210, 212, 218, 287, 339, 350, 416
 model with covariates, 358, 378
Good estimator, 172
Goodness of fit, 74, 214, 331, 365
 tests, 214
Gotha life table, 126
Graphical methods, 182, 193-198, 214, 226, 366
 test, 214
Gravity law, 13
Greenwood's formula, 140, 158, 173
Greville's cubic approximation, 110
Grouped data, 133, 154, 183, 206, 210-211, 219, 243, 329
Growth rate, 14, 16, 111
Growth time, 417
Gumbel Type A bivariate distribution, 283-285

Half-replicates, method of, 367
Hazard function, *see* Hazard rate function
 cumulative, 51, 52, 174, 182, 196
 paper, 187
 plot, 188
 rate, 51, 93, 142, 212, 415
  crude, 273, 420
Hazard function, log-linear, 357
 net, 273
 proportional, 280, 359, 422, 429
 underlying, 353
 vector, 272
  component, 272
Hazard rate function (HRF), 14, 51, 346, 352
Health function, 393
Heart transplant data, 158, 174, 177
Hemoglobin, 355
Heterogeneous population, 10, 288, 311
Histogram, 226
Hospital prevalence, 32
Hypergeometric distribution, 252
Hypothesis, Balducci, 124
 composite, 368
 null, 260
Hypothetical cohort, 83
 *see also* "Time due to die"

Identified minimum, 274, 277
Improper distribution, 286, 290, 331
Inborn fault, 416
Incidence, 31, 32
 onset distribution, 393, 395, 396
 rate, 33, 394
  average, 34
Incomplete data, 150, 200, 210, 236, 334
Independence, 241, 276, 279, 281, 292
Index, 10
 random, 274
Indicator variable, 346, 352, 354
Indirect standardization, 22, 23
Infinite likelihood, 208
Information matrix, Fisher, 74
Initial exposed to risk, 156, 326, 406
Instantaneous rate, 12, 14, 273
Institute of Actuaries, 45, 46
Insulin, 250

Intensity rate, *see* Hazard rate function
Interaction, 365, 372
  treatment-univariate, 365, 372
Interpolation, 67, 96, 114
  exponential, 69, 97
  Lagrange, 68
  linear, 69, 96, 296
  polynomial, 68
Irreversible Markov process, 423

Johns Hopkins Hospital, 40
Joint Mortality Investigation Committee, 120
Joint SDF, 329, 338, 427

Kaplan-Meier estimator, 173, 334
Keyfitz' method, 111
Kidney transplant data, 350, 371
Kolmogorov-Smirnov statistic, 215
  censored, 217, 243
  modified, 218
  truncated, 217
Kolmogorov-Smirnov test, 216-219
  two sample test, 231

lbd (last birthday), 21, 84
Lagrange interpolation formula, 68, 111
Lagrangian coefficients, 112
Last record, method of, 152
Least squres, 209, 212
  estimator, 209
  weighted, 210
Left hand truncation, 54
Leukemia, 356, 386
Lexis diagram, 27, 28
Life table, 83, 225, 228, 394
  A1967-70, 117, 120
  abridged, 84, 93, 104-111
  Canada Life, 126
  clinical, 134
  cohort, 83
  complete, 84
  England and Wales, 403
  English No. 12, 127
  Geneva, 146, 230
  Gotha, 126
  multiple decrement, *see* Multiple decrement life tables
Lifetime, 425
  distribution, *see* Survival distribution
  future, 55
  prevalence, 32, 395
Likelihood, 163, 165, 167, 172, 200, 201, 203, 204, 205, 222, 329, 335-336, 339, 349, 351, 352, 354, 359, 376
  conditional, 329, 331-332
  function, 73
    construction, 204-207, 359, 363
  maximum, 208
  partial, 362
  ratio, 260
    test, 72, 260, 368-374
Linear-exponential model, 354
  interpolation, 69, 96, 296
  motion, 13
  transformation, 58
Local maximum, 208
Location parameter, 58
Location-scale family, 58, 204, 207
Logistic distribution, 63, 184, 204
Logistic-linear model, 374, 376
Logit analysis, 204
Log-linear exponential, 359
Lognormal distribution, 63, 389
Logrank test, 258-259
Long-term studies, 25
Longitudinal studies, 242
Lost to observation, 151
Lung cancer, 357, 387
Lung tumor (mice), 179
*Lupus erythematosus,* 419

Makeham-Gompertz model, 62, 184
Malignant neoplasms, 301
Mantel-Haenszel test, 251-258
Marginal SDF, 271, 311, 331
Markov process, 424
  irreversible, 423
  reversible, 431
Mass-time units, 15
Matching, 346
Maximum likelihood equations, 208, 354
Maximum likelihood estimation, 72, 73, 74, 162-169, 171, 201, 204, 213, 325, 339, 350, 354, 360-361
  method, 208

MAXLIK (computer program), 201, 212
MDLT, *see* Multiple decrement life table
Median of future lifetime, 56, 98, 227, 228
Melanoma, 178
Melbourne Central Cancer Registry, 118
Metastases, 355
Mice data, 136, 177, 179, 193, 198, 201, 336
Midperiod (midyear) population, 19, 105, 106, 155, 195
Minimum Chi square, 182, 209, 213
  identified, 274
  modified Chi square, 209, 213
  nonidentified, 274
Mixing distribution, *see* Compounding distribution
Mixture distribution, *see* Compound distribution
Model, 301, 345-346, 351, 416
  additive, 352, 355, 416
  competing risk, 381
  exponential, 353
  exponential-Webull, 376
  "fatal shock," 420-423
  linear-exponential, 354
  logistic-linear, 374, 376
Model log-linear-exponential, 357
  of carcinogenesis, 416-418
  of disease, 426
  of failure, 64
  Makeham-Gompertz, 62, 184
  of mortality, 64, 301
  "mosaic," 418
  multi-hit, 417
  multiplicative, 356, 359, 382, 386
  multi-stage, 415
  one-hit, 416
  proportional hazard rate, 359
  quadratic-exponential, 355
  reversible, 431-436
  statistical, 73
Moment estimator, 169, 171
  generating function, 379
Mortality, 3
  aggregate, 117
  data, 7
    complete, *see* Data, complete
    differential, 399, 407, 410
    experience, 104
    force of, 51, 93, 142
    *see also* Experience; Hazard rate function
  grouped, *see* Grouped data
  incomplete, *see* Incomplete data
  model, 64, 301
  rate, *see* Death rate
  select, 115
  ultimate, 116
"Mosaic" model, 418
Multi-hit model, 417
Multinomial distribution, 138, 206, 324
Multiple deaths, 131, 332, 362
Multiple decrement life tables (MDLT), 294, 295-297, 300
  sample, 134
  select, 115, 119, 347, 400
  single decrement (SDLT), 312, 318
  standard, 107, 228
  U. S., 44, 86-91, 93, 109-110, 121, 187-191, 263, 294, 300, 307, 310, 314
  U. S. S. R., 263
Multiple failures, *see* Multiple deaths
Multiplicative model, 356, 359, 382, 386
Multistage model, 415
Multistage testing, 241, 254
Multivariate hazard rate vector, 272
Mutation, somatic, 417, 419
Mutually exclusive events, 273, 281

Negative binomial expansion, 417, 419
Negative selection, 115, 118, 350
Nelson's method, 196
New case, 35, 393, 403
New entrant, 152
Nominal significance level, 241
Nonidentifiability, 277, 280, 281
Nonidentified minimum, 274
Nonparametric methods, 324, 336
Normal distribution, 204, 231, 379, 386
Notation, 294

Null hypothesis, 248, 266

Obesity, 11
Observed prevalence, 32, 402
Occupational hazards, 25
Odds ratio, 37
One-hit model, 416
Onset, 35, 392
  age of, 35, 392
    distribution, 50, 393
  hazard rate, crude, 393
  incidence, 394
    distribution, 393, 394
  waiting time, 394
    distribution, 394, 403, 407, 409
Order, rank, 197
  statistics, 366
Overall survival function, 272, 331, 420, 429

Paper, cumulative hazard, 182-185
  hazard function, 187
  probability, 182-185
Parallel system, 65, 417, 424, 426
Parameter, 181, 206
  location, 58
  scale, 58
  shape, 58
Partial likelihood, 362
Pearson's conditional $X^2$-test, 250-251
Percentage, 11
Percentile of future lifetime, 57
Percentile points, method of, 182
Performance status, 250
Person-days, 161
Person-time units, 154
Person-years, 18, 24, 25, 155
Piece-wise exponential, 206, 386
Piece-wise survival function, 185, 186, 205
Placebo, 250
Planned termination, 151
  withdrawal, 151
Poisson distribution, 264, 418
  process, 420
Population, average, 19
  exposed to risk, 105
  midperiod, midyear, 19, 105, 106, 155, 395
  at risk, 19
  stable, 103, 149
  stationary, 103
Population target, 17
Posttive selection, 115
Posterior distribution, 380
Prevalence, 31, 402
  current, 32
  data, 395-398
  hospital, 32
  observed, 32, 402
  treated, 32
Primary variables, 3
Prior distribution, 380
  therapy, 357
'Private' probability of death, 271, 310, 334
Probability, 6
  conditional, 52, 53, 134, 227, 324
  integral transformation, 66
  paper, 182
  'private,' 271, 310
  'public,' 270, 306
  theory, 6
Probability Density Function (PDF), 51, 143
Probit analysis, 204
Product-limit estimator, *see* Kaplan-Meirs estimaor
Progression, of disease, 426-431
Progressive censoring, 151, 249
  failure censoring, 151
  life testing, 240
Proper distribution, 52, 271
Proportion, 6, 9, 11, 32
  binomial, 131, 138, 170
  and rate, 9
Proportional hazard rate, 280, 287, 298, 350, 359, 422, 429
Prospective study, 36
Prostate, cancer of, 354
"Public" probability of death, 270, 307
Purposive selection, 346

Quadratic-exponential model, 355
Quantal response, 203

Radiation, gamma, 135
Radix of life table, 83, 326
Random censoring, 157, 166
Random concomitant varialbes, 379
Random index, 274
Random sample, 361
Random variable, 50
  transformation of, 57, 66
Randomization, 243, 246
  distribution, 238
Rank, 131
  order, 197
Rankit analysis, 204
Rate, 9, 12
Rate, absolute, 12, 17
  age-adjusted, 25
  birth, 17
  central, 14
  crude, 17
  death, *see* Death rate
  growth, 14, 16
  incidence, 33, 394
  mortality, *see* Death rate
  and proportion, 9
  relative, 13, 17
  specific, 17
Ratio, 9, 10
  fetal, 10
  odds, 37
  sex, 10
Reaction velocity, 13
Recording unit, time, 7, 131, 362
Rectangular distribution, 204
  *see also* Uniform distribution
Regression, 211, 351-355, 368
Relative frequency, 11
Relative rate, 13, 17
Relative risk, 36
Reliability theory, 418
Repeated experiments, 241
Repetitive events, 16, 18, 32
Representativeness, 346
Residuals, 366
  crude, 366
Response, 204
  binary, 204
  quantal, 203
*Reticulum cell sarcoma,* 193, 336
Retrospective data, 403
Retrospective study, 36
Reversible models, 431-436
Right-hand truncation, 54
Risk, factor, 35, 345
  relative, 36
  set, 334, 361, 364
Robust estimator, 172

Sample, 128
  life table, 134
  selection, purposive, 346
    random, 361
Scale parameter, 58
SDF, 50, 84, 154, 158, 174, 182, 431
SDLT, *see* Single decrement life table
Select life table, 5, 115, 119, 346, 400
Select period, 115, 119
Selection, 114
  of concomitant variables, 367
  duration of, 115, 119
  negative, 115, 118, 350
  positive, 115
  purposive, 346
Sequential methods, 260
Series system, 64, 420
Serum cholesterol, 376
Sex, 6, 350, 371
  ratio, 10
Shape parameter, 58
Shock model, *see* Fatal shock model
Significance level, nomial, 241
Single decrement life table (SDLT), 312, 318
Slope, 193, 366
Sojourn time, 436
Somatic mutation, 417, 419
Speed function, 13
Stable population, 149
Staggered entries and withdrawals, 252
Standard deviation, 10
Standard error, 351, 354
Standardization, direct, 25
  indirect, 22, 23
Standardized mortality ratio (SMR), 22
Standard life table, 107, 228
Standard population, 22
Standard treatment, 357

Stanford Heart Transplantation Program, 158, 174
States, in disease progression, 423, 424, 428-429, 430-431
Stationary population, 149
Statistical differentials, 69, 141, 142, 178
Statistical model, 73
Step-down procedure, 368, 369
Step-function, 174
Step-up procedure, 368, 371
Stochastic process, 423
Stratification, 25-26
Survival, data, 7
  distribution, 50, 58
  function, 84, 154, 158, 174, 182, 431
    empirical, 130, 211
    joint, 329
    overall, 272, 331, 420, 429
  probability, 145
  time, 129, 133, 233
Survival Data Function (SDF), 50
Symptom, 345
System, 64
  parallel, 65, 417, 424, 426
  series, 64, 420
*Systemic Lupus Erythematosus,* 39
Systolic blood pressure, 376

*t*-distribution, 354
t-test, 351, 354, 376
Target population, 17
Termination, planned, 151
  time of, 150
  unplanned, 151
Test of hypothesis, 214, 228, 351
Therapy, prior, 357
*Third National Cancer Survey,* 33, 397, 412
*Thymic lymphona,* 193, 336
Tied observations, 8, 232, 235, 258
Time, elapsed, 350
  of follow-up, 152
  recording unit, 7, 131, 362
  since diagnosis, 357
  of termination, 150
"Time due for onset," 393
"Time due for withdrawal," 163
"Time due to die," 271, 393
Tolbutamide, 250
Tontine, 46
TOPS Club, Inc., 408
Transformation, 57, 279
  linear, 57, 58
  probability integral, 66
Transition probability (Transition intensity), 424
Transition rate, 432
Transplant data, heart, 158, 174, 177
  kidney, 350, 371
Treated prevalence, 32
Treatment-covariate interaction, 364-372
Trend, 227, 242, 303
Trinomial population, 149
Truncated data, 150, 214, 216, 247, 326
Truncated distribution, 53, 59, 60, 183, 217
Truncation, 54, 150, 214, 240
  from above "right-hand," 54
  from below "left-hand," 54
Tsao-Conover test, 217
Tumor, Adeno, 357
  large, 357
  lung, mice, 179
  small, 357
  squamous, 357
Tumor Registry, Connecticut, 118
Two-by-two table, 35, 247, 249
Type 1 extreme value distribution, 61, 183
Type 2 extreme value distribution, *see* Weibull distribution

Ultimate mortality, 116, 350
Unbiased estimator, 170
Underlying cause, of death, 270
Underlying hazard rate, 353, 356
Uniform distribution, 59, 66, 98, 100, 163, 166, 167, 168, 204, 313
U. S. S. R. life table, 263
Unit, person-time, 154
  time recording, 131
U. S. life table, 44, 86-91, 93, 110, 121, 187-191, 263, 294, 300, 307, 310, 314, 319
U. S. *Vital Statistics,* 21, 109-110, 121, 269, 298, 319

University Group Diabetes Program (UGDP), 250, 263
Unplanned termination, 151
Unplanned withdrawal, 151
Ureteral dilatation, 355

Variable, concomitant, 3, 7, 115, 345, 367
continuous, 346
discrete, 346
primary, 3
random, 50
Variance, 55, 140
conditional, 140, 141, 158, 165
of future lifetime, 56, 94
Variance-covariance structure, 74, 211
Veterans Administration Cooperative Research Group, 354, 383, 384
Vital statistics, 9, 21
*Vital Statistics,* U.S., 21, 109-110, 121, 269, 298, 307, 319

Waiting time, distribution, 311, 329
onset distribution, 394, 403, 407, 409
Weibull distribution, 62, 181, 184, 188-196, 201, 208, 210, 417
Weibull model with covariates, 358, 367, 376
Weight, 355
Weighted average, 22
Weighted least squares, 210
White blood count, 356
Wilcoxon test, 231, 234
modified, 236-240
Withdrawal, 151, 202, 326, 331, 334, 335-336
nonrandom, 151
planned, 151
random, 157
staggered, 242
times, 240
unplanned, 151

$X^2$, *see* Chi square

*Applied Probability and Statistics (Continued)*

DODGE and ROMIG • Sampling Inspection Tables, *Second Edition*

DRAPER and SMITH • Applied Regression Analysis

DUNN • Basic Statistics: A Primer for the Biomedical Sciences, *Second Edition*

DUNN and CLARK • Applied Statistics: Analysis of Variance and Regression

ELANDT-JOHNSON • Probability Models and Statistical Methods in Genetics

ELANDT-JOHNSON and JOHNSON • Survival Models and Data Analysis

FLEISS • Statistical Methods for Rates and Proportions

GALAMBOS • The Asymptotic Theory of Extreme Order Statistics

GIBBONS, OLKIN, and SOBEL • Selecting and Ordering Populations: A New Statistical Methodology

GNANADESIKAN • Methods for Statistical Data Analysis of Multivariate Observations

GOLDBERGER • Econometric Theory

GOLDSTEIN and DILLON • Discrete Discriminant Analysis

GROSS and CLARK • Survival Distributions: Reliability Applications in the Biomedical Sciences

GROSS and HARRIS • Fundamentals of Queueing Theory

GUPTA and PANCHAPAKESAN • Multiple Decision Procedures: Theory and Methodology of Selecting and Ranking Populations

GUTTMAN, WILKS, and HUNTER • Introductory Engineering Statistics, *Second Edition*

HAHN and SHAPIRO • Statistical Models in Engineering

HALD • Statistical Tables and Formulas

HALD • Statistical Theory with Engineering Applications

HARTIGAN • Clustering Algorithms

HILDEBRAND, LAING, and ROSENTHAL • Prediction Analysis of Cross Classifications

HOEL • Elementary Statistics, *Fourth Edition*

HOLLANDER and WOLFE • Nonparametric Statistical Methods

JAGERS • Branching Processes with Biological Applications

JESSEN • Statistical Survey Techniques

JOHNSON and KOTZ • Distributions in Statistics
- Discrete Distributions
- Continuous Univariate Distributions—1
- Continuous Univariate Distributions—2
- Continuous Multivariate Distributions

JOHNSON and KOTZ • Urn Models and Their Application: An Approach to Modern Discrete Probability Theory

JOHNSON and LEONE • Statistics and Experimental Design in Engineering and the Physical Sciences, Volumes I and II, *Second Edition*

JUDGE, GRIFFITHS, HILL and LEE • The Theory and Practice of Econometrics

KALBFLEISCH and PRENTICE • The Statistical Analysis of Failure Time Data

KEENEY and RAIFFA • Decisions with Multiple Objectives

LANCASTER • An Introduction to Medical Statistics

LEAMER • Specification Searches: Ad Hoc Inference with Nonexperimental Data

McNEIL • Interactive Data Analysis